Ecology of Marine Ports of the Black and Azov Sea Basin

A.K. Vinogradov · Yu.I. Bogatova
I.A. Synegub

Ecology of Marine Ports of the Black and Azov Sea Basin

Springer

Authors
A.K. Vinogradov
Department of Ecology of Marginal Communities
Institute of Marine Biology National Academy of Sciences of Ukraine
Odessa
Ukraine

I.A. Synegub
Department of Ecology of Marginal Communities
Institute of Marine Biology National Academy of Sciences of Ukraine
Odessa
Ukraine

Yu.I. Bogatova
Department of Water Quality
Institute of Marine Biology National Academy of Sciences of Ukraine
Odessa
Ukraine

Translated by Mariya V. Gelmboldt, Metz, France

ISBN 978-3-319-63060-1 ISBN 978-3-319-63062-5 (eBook)
DOI 10.1007/978-3-319-63062-5

Library of Congress Control Number: 2017946054

Original Russian edition published by Astroprint Odessa, Odessa, 2014

Printed on acid-free paper

This Springer imprint is published by Springer Nature
The registered company is Springer International Publishing AG
The registered company address is: Gewerbestrasse 11, 6330 Cham, Switzerland

Contents

Introduction

> Modern ecology is a science about ecosystems uncovering the laws of their composition, structure, functioning and evolution.
>
> V.D. Fedorov, T.G. Gilmanov

The present book proposes to the reader revised, expanded and improved material partly based on the previously published monograph "Ecosystems of marine ports aquatories of the Black–Azov Sea basin (introduction to marine ports ecology)" (Odessa, Astroprint, 2012). In the new sections of this edition, particular attention has been given to ecological peculiarities of organization and general ecological mechanisms of marine port (MP) aquatories ecosystems functioning.

Coastal and estuarine MPs are typically located within the sea shelf in "land–sea" and "land–river–sea" transitional zones. MPs constitute peculiar marginal ecosystems in which natural and anthropogenic components are variously combined. All MPs have unique individual features, but, at the same time, they are all created with the main objective of reducing the wave and wind influence on ships to an acceptable level. MPs of the classical type include three main components in their structure: (1) protected aquatories, (2) artificial hydrotechnical structures (HTS) located in the aquatories and protecting them; (3) access channels (AC) with depths matching those in aquatories.

In the Black–Azov Sea basin, the targeted exploitation of the shores and the construction of port cities started during the seventh–eighth centuries BC, during the period of the "great colonization" by the Greeks. After three to four centuries, several dozens of MPs were already operating on the shores of the Black and Azov Seas. The absolute majority of modern MPs are operating at the same locations as the ports existing in ancient times. Until the fourteenth–fifteenth centuries AD, the use of rowboats and sailing rowboats with draft up to 2–2.5 m has not resulted in an urgent need for the construction of harbours with deeper aquatories. During the sixteenth–seventeenth centuries, shifting to multi-decker sailing vessels with draft up to 5–6 m placed new requirements on MPs. During the period nineteenth–twentieth centuries, the appearance of steamships with iron and, later on, steel hulls provided an impetus for construction of berths with depths of 8–10 m in the Black and Azov Seas. Berths with depths of 15–20 m appeared in Novorossiysk, Odessa, Yuzhny and Constanza during the period from the end of the twentieth to the beginning of the twenty-first centuries. Invention of steam and diesel engines allowed to mechanize the necessary dredging activities. Over 2500 years, each of the ship dimensions (length, width and draft) has increased ten times, while draft and carrying capacity (deadweight) have increased threefold. At present, ships with lengths up to 300–320 m, drafts up to 20 m and displacements up to 220–230 thousand tonnes are considered the most economically viable for navigation in the Black Sea.

At the end of the eighteenth—beginning of the nineteenth century, merchant shipping started to develop on the Russian coasts of the Black Sea and merchant harbours were constructed along with military ones. The Treaty of Küçük Kaynarca (1774) established a "free and unrestricted navigation of Russian flagged merchant ships" sailing into the Black Sea via the Dardanelles and Bosphorus straits, thus making Russia a fully legitimate Black Sea state. The opening of the commercial ports in Kherson, Feodosiya and Sevastopol was announced in 1784 by Tsar's manifesto and the Russian Black Sea trade started in the 1790s. Odessa was proclaimed a free trade zone (porto-franco) in 1794. The Treaty of Adrianople (also called the Treaty of Edirne) with the Ottoman Empire (1829) granted Russia the freedom of commercial navigation in the Black Sea straits and the freedom of trade for Russian merchants within the Ottoman Empire. At the end of the twentieth century, after the collapse of the USSR, MPs located on the northern coast of the Black and Azov Seas and in Crimea, previously belonging to Russian Empire and the USSR, were acquired by Ukraine.

MPs were located in gulfs, bays, limans and estuaries already having some natural protection. Several ports were initially built for military purposes and later became merchant ports. During the period of their existence, all ports have undergone, and continue to undergo, periods of rise and fall with human activity ceasing for decades or centuries in some of them.

Dimensions, depths, peculiarities and general economical significance of each MP are primarily determined by its hinterland, or zone of economic attraction, and the influence of associated logistic. Ships, ports and cities in which they are located belong to systems centred on the flow of merchandise, cargo and freight. Shipbuilding has progressed along with navigation and port builders have been adapting ports aquatories and infrastructures to the newly originating requirements [9].

The analysis shows that there is a close direct and inverse connection between the sizes of a city and its port. Already in ancient time, MPs acted as "city-forming enterprises". They were providing not only connections, but also economic well-being of the maritime cities. Changing of conjuncture, resource deterioration, appearance of new itineraries of transportation could all lead to reduction of the role of a port. Increase of cargo traffic creates the need for reconstruction of old berths and construction of new ones as well as dredging in MP aquatories and their AC.

There are direct and indirect ecological links and interrelationships among ships (length, draft, number, etc.), ports (sizes, depths, access channels, number of artificial HS) and marine cities (population, communications and infrastructures). Changes in one of the above-mentioned components inevitably cause direct and indirect consequences in the others. Thus, the port city of Istria throve in ancient times and then totally declined due to sand accumulation in its navigating channel. In port cities, the growth of population and number of circulating vessels were increasing the volume of domestic effluents and faecal wastes directly discharged into the protected MP aquatories, which favoured processes of local eutrophication. The more numerous the populations of ancient port cities grew, the bigger their influence on terrestrial and aquatic ecosystems. The overgrowing of hydrobionts on ships' hulls during more than two thousand years favoured species exchange between the aquatories of various MPs.

Various manufactures (smelting and working of metal, production of pottery and glass, leather working, etc.) developed in wealthy port cities over the centuries. These first workshops did not have a pronounced negative influence on the environment until the eighteenth–nineteenth centuries. During the nineteenth–twentieth centuries, the big Black and Azov Seas' ports, which nowadays are located on the continental territory of Ukraine and Crimea, were connected by railway lines with industrial and agricultural regions. MPs such as Odessa, Ilyichevsk, Yuzhny, Nikolaev, Kherson, Kerch and Mariupol became big transport and industrial complexes, with evident negative impact on terrestrial and aquatic ecosystems including MPs aquatories and adjacent sea zones.

Ecosystems of the modern MPs of the Black–Azov Sea basin in general, and of Ukrainian MPs in particular, comprise three more or less autonomous subsystems: (1) the pelagial zone, (2) the periphytal zone, (3) the benthal zone [2]. In a given port, the pelagial connects all the subsystems into an integrated ecosystem. In spite of individual differences, MPs are constructed and functioning on the basis of some general principles. MPs ecosystems experience conditions of weakened hydrodynamics and water exchange. The artificial increase in MP depths opens the possibility for the vertical stratification of water masses and the formation of a stable pycnocline. The presence of high quantities of HS (hard substrata) in MP ecosystems stimulates biological production. Accumulation of organic matter (OM) takes place as a consequence of protection and lowered hydrodynamics of MP aquatories. In MPs, the coastal shallow waters and associated biocenoses are partially or totally destroyed and the conditions of the shallow shore are substituted by deep shore ones. On the bottom and the near-bottom layer of MPs, saprobiotic situations causing mass mortality occur quite often. At the same time, there is no fishing in MPs aquatories and their biota comprises hundreds of hydrobionts species [1, 2, 10], so that MPs ecosystems are a source of larval material for adjacent ecosystems. Given all these peculiarities, the authors consider that MPs aquatories should be regarded as complete, specific aquatic ecosystems.

Construction and operation of port moles and AC not only changes the bottom relief, but also directly and indirectly influences longshore drifts flow, processes of seashore abrasion and origin of new accumulative formations. Bottom drifts enter MPs aquatories mainly via passages in HTS, notably in the correspondence of ACs.

The problem of sediment accumulation in MPs aquatories is directly connected with that in AC.

The Romanian port of Constanza has the biggest aquatory area (26 km^2) among marine ports of the Black and Azov Seas located on seashore and artificially protected. The aquatory of the Bulgarian marine port Varna-West is formed by the interconnected lakes Varna and Beloslav and has a total area of more than 20 km^2. MP Sevastopol has an aquatory of the estuarine type (i.e. Sevastopol Bay) with an area of about 7 km^2. The aquatories of Ilyichevsk and Yuzhny MPs, located in Sukhoy and Grigorievsky Limans respectively, both have an area of about 6 km^2.

Because HTS in MPs are stationary, accumulation of drifts in aquatories and changes in water level have a significant impact on their normal functioning. Fluctuations of water level have a naturally determined character and are difficult to forecast. In MPs, offshore–onshore phenomena and seiches can be observed all year round. The speed of water level rise in the Black Sea is 1.83 ± 0.07 mm $year^{-1}$ or 18 cm for 100 years [6]. Ports in the Black and Azov Seas experience nearly invisible tides; contrary to many ports, located in different regions of the World Ocean, whose aquatories are regularly washed out by tides. Tidal amplitude is 7 cm in Constanza, 8–9 cm in Poti and in 5–6 cm Odessa. The tidal currents are scarcely observed [4].

In many regions of the Black Sea, anthropogenic changes affect not only MPs aquatories, but also the shores adjacent to them. From the 1960s to the 1990s, about 600 drifts-trapping groins (for a total length slightly exceeding 30 km) were constructed on the shore zone of the South Crimean Coast. In the zone of Odessa Gulf, shore protection works were carried out along more than 12 km from Cape Bolshoy Fontan to Cape Lanzheron [3, 7].

In the Azov Sea, construction of Glukhoy canal and Temryuk port caused accumulation of drifts carried by the Kuban River on the windward side of these structures. Consequently, the vast, so called Chaika sandbar has formed between Glukhoy canal and the Petrushin arm of Kuban River; with depths decreasing from 6 to 1.5 m at a distance of 1 km from the shore [11].

For many ports, the "Pilot chart of the Black Sea" [5] indicates the water depth at distances of 1 and 5 m from berths. Taking Odessa MP for example, depth changes with distance from the berths can range from 0.5 to 3.0 m; causing heterogeneous conditions at the bottom. A similar situation is observed in most ports of the Black and Azov Seas and can be connected with the constant deposition at the HTS basis of the matter formed in the fouling biocenoses of underwater surfaces. It could also be explained by the danger of damage to berths' underwater parts during dredging and by peculiarities of the water flows in the near-bottom layer. The round form of ships' submerged hulls is compatible with such differences because, when moored, the deepest draft is at the keel part of ships' bottom.

According to international classification, ports of I, II and III class are accepting ships with draft up to 20, 12 and 9 m, respectively [8]. The deepest water ports of Ukraine are Yuzhny, Odessa and Ilyichevsk MP can accept vessels with draft of more than 12 m and, at some berths of Yuzhny and Odessa MPs, up to 15–17 m.

In sea ports, the patterns of hydrobionts distribution, the flows of live and dead (suspended and dissolved) OM and the formation of aggregates of living material are mainly determined by physical, physical-chemical and chemical processes.

Up to now, the ecological role of marine ports located on the shores of tideless seas such as the Black and Azov Seas is poorly described in the scientific literature. However, according to tentative estimations, the total area of anthropogenically modified MPs aquatories in this basin comprises more than 50 km^2. Furthermore, the cumulative length of access channels (AC) to MPs is more than 350 km and the total extension of HTS in MPs exceeds 300 km; the area of their submerged parts being over 2 millions m^2. In MPs, moles, breakwaters, berths and floating docks represent hard substrata artificially introduced into marine coastal ecosystems and are classified as artificial reefs (AR) [2]. For fouling hydrobionts, wetted surfaces of ship's hulls also represent HS. The ruins of ancient ports and towns submerged by the sea a long time ago (Dioskoyrias, Kállatis, Sinop and other) also serve as HS for fouling organisms. Taking all this into account, the authors consider the elaboration of general recommendations to ameliorate the ecological situation in MPs aquatories as an imperative task.

Chapter 1
Main Characteristics of Marine Ports of the Northern Coast of the Black and Azov Seas and of Their Access Channels

In the Black–Azov Sea basin, natural sheltered areas (gulfs, bays, limans and estuaries) served as ports already B.C. In the seventeenth–eighteenth centuries, under the Russian Empire, the MPs were called “marine shelters”. Construction of commercial harbours on the Russian coast of the Black Sea started in the nineteenth century when the creation, in 1857, of the Russian Steam Navigation and Trading Company (ROPiT) gave a huge input to commercial navigation in the Black Sea.

A port (from the Latin “portus”—harbour, quay) is a seashore area with an adjacent aquatory, usually well protected from wind and wave action, and a territory equipped for ships’ anchorage, cargo storage and loading–discharging operations. In MPs, the natural protection of the aquatory is usually completed by an artificial one. During berths construction in MPs, the coastal line undergoes smoothing and simplification. Ports are provided with infrastructure for ships maintenance and repair. Terminals for handling of container and package cargoes, wheeled equipment and other specialized subdivisions for gaseous, liquid and bulk cargos (including toxics and explosives) are also part of MP zones. Berths, located at quays and moles, are the main HTS in MPs. Ports of the classical type are built in order to satisfy the principle of “bringing the ship alongside the berth”.

In MPs, shore zones protected by harbour walls and equipped by bitts, cargo-handling gear, railways, warehouses, etc., are called quays.

Moles (from the Latin “moles”—embankment) are HTS, whose external surfaces safeguard port aquatories from roughness. One end of the moles is connected to the shore. Berths are quite often located on the internal sides of protective moles. The intra-port moles are usually called piers and are equipped with berth on two or three sides. Protecting moles not connected to the shore are called breakwaters and have no berths. They are used to protect aquatories having a large width, as for example, in Odessa and Berdyansk MPs.

HTS are constructed by various methods, using different materials and construction units. For long time berths, moles and breakwaters were build mainly from stones and concrete blocks weighing from a couple of tonnes up to dozens and hundreds of tonnes. In hydro construction, artificial concrete blocks with a regular

A.K. Vinogradov et al., *Ecology of Marine Ports of the Black and Azov Sea Basin*, DOI 10.1007/978-3-319-63062-5_1

form (cubes, parallelepipeds, and pyramids) are called continuous pours. In some cases, the concrete blocks are placed in regular rows (regular laying), as in Reydovy (Raid) mole and the Stary (Old) breakwater in Odessa MP. In other cases, blocks are dumped or irregularly piled, as for example in the Novy (New) breakwater in Odessa MP. For wave-protection of the moles' external surfaces, monolith concrete walls are installed at zones adjacent to the shore such as in the case of the Container terminal at Karantinny mole in Odessa MP. Berths and moles are built using metal and concrete sheet-pile walls and different pile foundations from metal and concrete. Modern moles and breakwaters are continuous (gravitational) or pile constructions. To facilitate construction and to save raw materials, there where sediments availability permits, concrete masses are cast in the form of boxes, transported by floating to the zone of installation and sunk by filling with quarry stone or sand.

Where a random distribution is possible or necessary, concrete bodies with masses of 0.5–1.5 t are used. Tetrapods (quadrupted) and dolosse (two cylinders connected by crosspiece) are the most known, but hexalegs (cross-shaped solids composed from six elements) and others are also are used.

Protective constructions, built by regularly piling stones and blocks or by using sea concrete walls and sheet-pile walls, reduce the wave energy hitting against the external surfaces. In HTS built by dumping, the wave energy is dumped inside the construction by repeated reflection of the differently directed water jets. Irregularly dumped HTS create more diversified life conditions for hydrobionts.

Drifts accumulate at the external sea side of moles' basement and in pockets along the shore, while bank erosion occurs downstream of the port. This happens because the material of the longshore flow is temporarily retained by moles, breakwaters and AC until filling of the shore prominences formed by these structures. Periodical or constant dredging of ACs worsens the situation. Downstream from ports, erosion sometimes affects vast shore zones (kilometres, dozens of kilometres) as it happened after moles' construction in Sochi and Ochamchire MPs [20]. In 1915–1917, a continuous mole, about 150 m long and almost perpendicular to the shore, was constructed in Gagra. Towards 1923, as a consequence of sand accumulation on its western side, the shore line had shifted up to the mole's end while a beach line about 6 m wide was washed away over 0.8 km on its eastern side [17].

In 1978, the construction of a deepwater berth with a length of 200 m was completed in Yevpatoriya port. Within a couple of years, an artificial sandbank had formed on one side of the berth and beach erosion had started on the opposite side. The regular dredging activity in the AC to the MP became one of the causes of beaches erosion [10, 11].

The construction of jetties and AC to Ilyichevsk and Yuzhny MPs affected longshore drifts flow, intensifying the processes of beaches destruction at some shore zones and sand accumulation at others.

The aquatories of most MPs are usually accessible via one or more access channels (e.g. three for Odessa MP), which significantly limits water exchange with the adjacent zones and reduces currents' velocity in the water column of MPs. The highest hydrodynamical activity is usually preserved in the AC to the MP and adjacent zones. In the Ukrainian MPs located in gulfs shores, bays, limans and estuaries of small and "dying" rivers, vertical stratification according to density could quite often be observed with a series of negative consequences, especially in the near-bottom water layer. According to some data, the Coriolis effect is more marked in MPs than in aquatories with homogeneous water masses [24].

In MPs aquatories, the presence of protective structures, several HS, division of aquatories into separate harbours, artificial increase of depth and a series of other factors create a wide spectrum of conditions, different from those of the adjacent sea zones.

1.1 Main Characteristics of Aquatories of the Marine Ports of the Northern Coast of the Black and Azov Seas and Their Access Channels

Ecological peculiarities and ecological roles of the Ukrainian MPs aquatories are mainly determined by their position in the "sea–rivers" systems and by the salinity conditions for hydrobionts. Dissolved and suspended organic matter (OM) and biogenic matter (BM) produced in the aquatories of MPs located at rivers' mouths and in backwaters enter the lower reaches of the rivers and contribute to eutrophication in the ecosystems of riverine and coastal MPs.

Depending on their location, some MPs on the northern coast of the Black and Azov Seas belong to inner (riverine and estuarine) ports, with indirect influence on the ecosystems of coastal MPs. Examples of inner ports are Reni and Izmail MPs, as well as Vilkovo and Kiliya port-stations, on the Danube. Kherson MP, located on the Dnieper, Nikolaev MP, Oktyabrsk, Dnieper-Bug MPs on Yuzhny Bug can be classified as mouth MPs (Fig. 1.1).

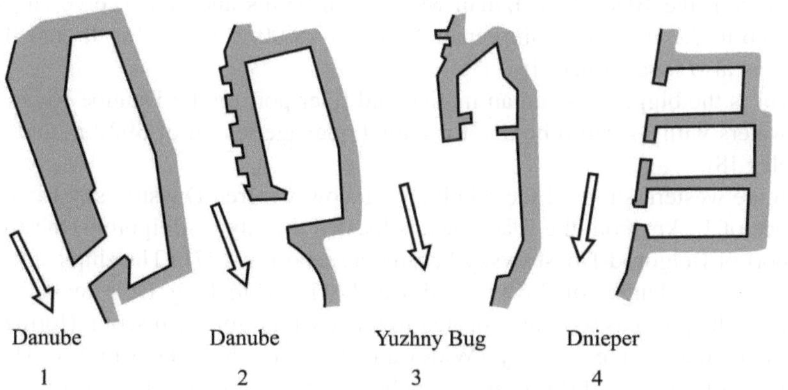

Fig. 1.1 Main scheme of the inner (riverbed) marine ports aquatories: *1* Reni MP, cargo zone № 3; *2* Izmail MP, cargo zone № 3; *3* Nikolaev MP aquatory; *4* Kherson MP, boat-yards №№ 1–3

Coastal MPs are located in gulfs and bays on the sea coast (Odessa, Skadovsk, Khorly, Yevpatoriya, Yalta, Feodosiya, Arshintzevo, Kerch, Krym, Genichesk, Berdyansk and Mariupol MPs) and on the coasts of big limans (Belgorod-Dnestrovsky MP in the Dnestrovsky Liman, Ochakov MP in the Dnieper-Bug Liman).

A number of narrow estuary type basins, formed by ancient or "dying" rivers, are used as MPs aquatories: Sukhoy Liman (Ilyichevsk MP, Fishing port), Grigorievsky Liman (Yuzhny MP, berths of "Transinvestservice" company), Uzkaya Bay (Chernomorsk MP), Sevastopol Bay (Sevastopol MP), Kamishovaya Bay (Fishing port), Balaklava Bay (Balaklava port station). MPs located in estuarine basins have the best natural protection. In such MPs, protective HTS are located at the mouth of the estuary and consists in two moles protruding towards each other from both shores and with a single passage between them. In the presence of bay-bars that separate MPs aquatories from the sea, access channels are constructed (e.g. in Yuzhny and Ilyichevsk MPs). Protective jetties, extending up to a certain distance into the sea to weaken sediment accumulation in the passage, are built parallel to access channels.

The Ukrainian MPs are mainly located on flat coasts with loose sediments. However, Crimean MPs Sevastopol, the fishing port in Kamishovaya Bay, Chernomorsk MP, Balaklava port-station, Yalta MP and Sudak port-station are located on rocky shores.

Ust–Dunaysk MP is located in the southern part of Zhebrianskaya Bay (Black Sea). It was built at the end of the 1970s mainly for handling of lighter aboard ships. It has a small aquatory and no berths. Vilkovo port-station, located in 20 km from the sea, is subordinated to Ust–Dunaysk port and has a berth with a length of 117 m and depths of 1.5–1.6 m depths. Kiliya port-station is also subordinated to Ust–Dunaysk MP and is located at 47 km from the Black Sea on Kiliya arm of the Danube. A berth with a length of 150 m and 2.8–4.5 m deep has been built in Kiliya. The shipbuilding-ship repair yard with berths 600 m long and a slipway dock with a length of 99 m is located in the town [8].

Izmail city is situated on the Kiliya branch of the Danube River at a distance of 80 km from the Black Sea. Izmail MP has 26 berths and can receive ships with drafts up to 7.2 m. Izmail Shiprepair Yard, with 600 m berths length and floating docks, is also located here [8].

Reni is the biggest Ukrainian marine and river port on the Danube River. It has backwaters with 37 cargo berths, for a total quayage length of 3927 m and a ferry complex [8].

On the western shore of the relatively shallow-watered Dnestrovsky Liman, at a distance of 18 km from the Black Sea, is located the city of Belgorod-Dnestrovsky. The port of Belgorod-Dnestrovsky became a seaport in 1971. The ships enter it via an AC with a length of 2.6 km and via the 17.1 km long Dniester–Limanskiy channel. The port has 9 berths, located in the Yuzhny Southern scoop (formed by a sand spit) and on the Shiroky (Wide) and Severny (Northern) moles. The total extension of berths is 1159 m with depths of 3.5–5.2 m near the berth walls. One passage leads into the MP aquatory [8, 18]. Bugaz port station, located at the entrance into Dnestrovsky Liman, is subordinated to Belgorod-Dnestrovsky MP and its berth has an extension of 127 m and depths up to 5.0–5.5 m.

The construction of Ilyichevsk MP started in 1957 on the western shore of Sukhoy Liman. An AC of 1.5 km channel was dug in the liman's bay-bar to connect this port with the sea. The depths near the berths were increased up to 10–14 m. The total number of berths is 28 and their cumulative length exceeds 6.0 km. The berths and floating docks of Ilyichevsk Shiprepair Yard are also located in Sukhoy Liman. Total berths' length is 2895 m, berths of the shipbuilding yard zone having a length of 210 m. The berths of ferry line and Ilyichevsk fishing port are located on the eastern shore of Sukhoy Liman. A series of small berths belongs to other enterprises. The total extension of berths and other HTS in Sukhoy Liman is about 10 km. One access leads into the port and the shipping channel with a maximum depth of 13 m passes along the liman axis [2, 8, 18].

Odessa MP is located along the western and south-western shores of Odessa Gulf and belongs to the MPs constructed on open marine shores. At present, in Odessa MP, there are 38 berths specialized in trans-shipment of various cargoes. Oil harbour and Container terminal berths occupy a special place. The berths of Odessa MP are located along the shore and on moles having various orientations with respect to the shore. Moles with piers divide the MP aquatory into 7 harbours, each one being connected with the waters of Odessa Gulf. The south-western part of the MP is protected from winds blowing from the sea and from the north by Karantinny (Quarantine) mole, Reydovy (Raid) mole and Stary (Old) breakwater. The western part of the MP is protected by Zavodskoy breakwater, Novy (New) breakwater and Neftyannoy (Oil) mole. Three passages and three ACs lead into the port aquatory. The depths of the different harbours vary from 3 up to 18 m. In 2013, the main berths and the protective constructions of Odessa MP extended over more than 9 and 5 km, respectively. Shiprepair yard "Ukraina", adjoining to Odessa MP, has HTS with depths up to 7.1 m and is equipped with three floating docks. A new Container zone with a berth length of about 850 m and a breakwater has been completed in the zone of Karantinny mole.

Among the main Ukrainian Black Sea ports, Yuzhny MP is the most deep-watered, rarely frozen, recent and fast developing. It is built on the shores and basin of Grigorievsky Liman (Maly Adzhalyk), 30 km north-east of Odessa city. As in the case of Sukhoy Liman, it is a naturally protected aquatory connected with the sea via an artificial AC. Freight-handling region № 1, on the western shore of Grigorievsky Liman, serves for handling of chemical cargoes. It consists of three specialized berths (total length 713 m) that mainly serve Private Stock Company (PSC) "Odessa Port Plant". One berth for sand trans-shipment, a terminal for handling liquid oils and Zernovoy (Grain) berth function on the same shore.

Freight-handling zone № 2, on the eastern coast of Grigorievsky Liman, includes two specialized terminals. The terminal for general and bulk cargoes trans-shipment has an extension of 915 m. The terminal for trans-shipment of chemical fertilizers, with a length of 566 m, is practically not used according to its original destination, but for grain and other bulk cargoes transfer.

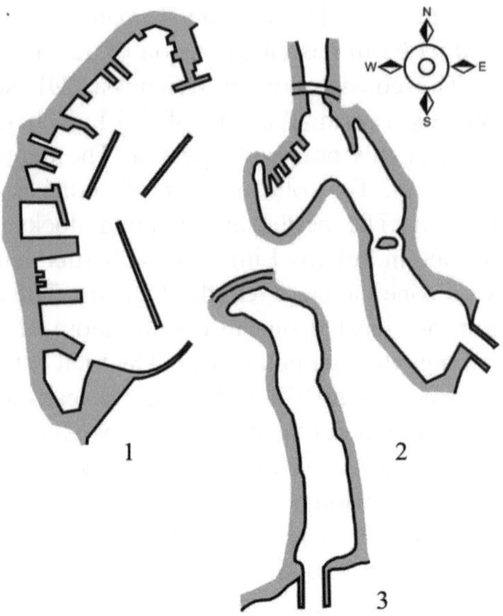

Fig. 1.2 Scheme of marine ports aquatories: *1* Odessa MP, *2* Sukhoy Liman–Ilyichevsk MP, *3* Grigorievsky Liman–Yuzhny MP

HTS of the oil terminal are located in the lower part of Yuzhny MP. In the upper part of the basin, “Transinvestservice” (TIS) company constructed a number of berths and one container terminal on the eastern shore. The depths near the berths and in the passage to them have been increased up to 15–17 m and further dredging up to 18–20 m and more is planned. About 18–19 million tonnes of various cargoes are trans-shipped each year via “TIS” berths and the total amount of cargoes transferred in Yuzhny MP exceeded 40 million tonnes in 2013.

In Yuzhny as well as in Ilyichevsk MP, the berths are located alongshore and, at present, their total length is about 7 km. According to an integrated development project, 47–48 berths with a total extension of 12 km should be constructed in Grigorievsky Liman and depths should be increased up to 20–21 m (Fig. 1.2).

Skadovsk MP is located on the northern shore of Dzharylhach Bay. The port presents a long basin with depths up to 6 m and its entrance is protected by two dykes. The port is equipped with 6 berths having a total length of 825 m [8, 18].

Port Khorly, subordinated to Skadovsk MP, is located on the southern part of the Gorky Kut peninsula and is protected by a mole from the west. Its three berths have a total length of 305 m with the depths near them ranging from 2.1 to 4.3 m [8, 18].

Big marine ports such as Nikolaev, Oktyabrsk, Dnieper-Bug, Kherson and Ochakov are located in the aquatory of Dnieper-Bug Liman and in the lower reaches of the rivers flowing into it. All of them have long access channels.

Ochakov port-station was founded in 1934 and received the statute of marine port in 1999. The port is located near the entrance into Dneprovsky Liman and on its northern shore. Ochakov harbour is formed by a double-knee mole protecting the aquatory from the south and by a breakwater. In Ochakov port, there are 5 berths with a total length of 726 m and with depths near them from 3.5 up to 5.5 m.

A berth of "bolverk" type with a length of 50 m was built in 1999 for the service of ferry vessels. Port Ochakov is under reconstruction and its further development is envisaged. A universal berth wall with a length of 140 m and a depth of 7.35 m was built in it and the deepening of AC and aquatory up to 13.5 m is foreseen [8, 18].

Nikolaev city is located at the confluence of Yuzhny Bug and Ingul Rivers. Nikolaev MP has 14 berths with depths up to 11 m, located along the left bank of Yuzhny Bug river oxbow at 3.5 km from the river mouth. The total berths length is 2362 m.

Besides the MP, Nikolaev city also has a river port located on Yuzhny Bug river with 900 m quayage and depths near the berths of 4.5–5.0 m [8].

Oktyabrsk marine commercial port is located 25 km downstream of Nikolaev city, on the left bank of Yuzhny Bug and near Krivaya Spit. This specialized port was created in 1965 and it consists of two harbours divided by a mole. It has 7 berths with total length of 1.9 km and with depths near the berths of up to 10 m [8].

The Dnieper-Bug marine commercial port was opened in 1978 on the left bank of the Bug Liman as a structural business unit of Nikolaev alumina refinery plant and was primarily designated for service of foreign bauxites. The port aquatory consists of one harbour with one passage. It is delimited by a protective dyke and coast-protecting constructions to the north and south. Initially, 6 berths with the depths from 5.5 m up to 11.5 m and total length of 810 m were functioning in the port. The berth line was later increased to 1540 m and the ports aquatory deepened up to 12 m [8] (Fig. 1.3).

Kherson MP is located on the right bank of the Dnieper estuary, 15 km upstream of its mouth, north-east of the Koshevaya river. The ten berths of the MP have a length of 1.6 km and are mainly located along the shore. On the left bank of the Dnieper there are three boat-yards, whose berths have depths between 5.5 and

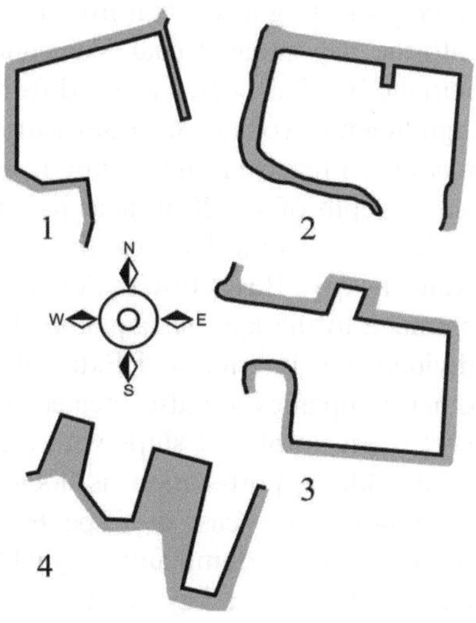

Fig. 1.3 Schemes of marine ports aquatories: *1* Belgorod-Dnestrovsky MP, *2* Dnieper-Bug MP, *3* Skadovsk MP, *4* Oktyabrsk MP

7.25 m. Kherson River port has a quayage of about 1 km. Kherson Shipyard and Shiprepair Yard, named after Kuybyshev, have their own berths.

Port-stations (with berths) subordinated to Kherson MP are located in the Dnieper river mouth and in the aquatory of Dnieper-Bug Liman. They include Golaya Pristan (on Dnieper arm Konka), Geroyskoe settlement (on the southern bank of Dnieper Liman, in front of mouth of river Yuzhny Bug), and settlements Stanislav, Rybalchye and Staraya Zburevka.

Yevpatoriya MP is located on the northern coast of Kalamit Bay in Yevpatoriya Bight. It has 6 berths for a total length of about 790 m. The port can receive car-ferry ships with draft of 5–7 m [8, 18]. Port-stations Steregushee, Mezhvodnoe, Nikolaevka as well as port Chernomorsk (or Chernomorskoe) and the cargo zone of Donuzlav Liman, located on the Crimean coast of Karkinit Gulf, are subordinated to Yevpatoriya MP. Steregushee and Mezhvodnoe port-stations are located in Baikalskaya Bay and Yarilgach Bay, respectively. Both port-stations are accessible for ships with 5–6 m draft. In Nikolaevka port-station, ships with draft up to 3 m can be moored.

Chernomorsk MP uses Uzkaya Bay as an aquatory. A mole was constructed on the western coast of the bay, at 600 m from its western entry cape, and a pier projects from the southern coast to a distance of 100 m from the mole extremity. The depth at the harbour entrance is 12 m and this port can receive ships with draft up to 4.5 m.

In 1961, a channel for ships' passage was dug in the bay-bar of Donuzlav Liman and a berth and an overpass were also constructed. Taking into account the presence of big depths (25 m), Donuzlav is considered a perspective place for MP construction with a possible annual freight turnover of up to 50 million tonnes. Donuzlav MP should be oriented eastwards and a line of big enterprises should be constructed on its shores.

Sevastopol MP is located in Sevastopol Bay and has a semiclosed aquatory of estuarine type elongated from east to west. In 1976–1977, the Severny (North) mole, protecting the aquatory from the western winds, and the Yuzhny (Southern) mole projecting towards it from the opposite shore were constructed at the entrance into the bay situated near Konstantinovsky Cape. As a consequence, the passage into the bay has been narrowed from about 1200 to 415 m and water exchange has significantly worsened. In Sevastopol Bay, the total extensions of berths and of all main HTS reach 11 and 16 km, respectively. The bay has an average depth of 12 m and a depth of 19–20 m near its entrance [8, 18].

In the second half of the twentieth century, a fishing port was built in Kamishovaya Bay, close to Sevastopol Bay. Eastern and Western moles protect the entrance to the harbour aquatory from storms. The aquatory of the fishing port is divided into Internal and External harbours by the Inner Eastern mole. Berths of other companies are also located in the fishing port along with floating piers. The port is accessible for ships with a draft up to 8 m [8, 18].

Balaklava port-station is located in a narrow and elongated bay situated 4.5 miles to the east of Cape Fiolent. Its aquatory has a natural protection. At present, it is predominantly a yacht harbour with depths decreasing from 34 m at the entrance of the bay down to 2–4 m at its apex.

Laspi Bay, located between the capes Aya and Laspi, is considered a good harbourage, especially at eastern winds. The depths in its central part are up to 9 m. A small mole serves as protective structure and a berth with depths of 1.7–2.2 m was constructed for small vessels on its eastern shore.

In Yalta city, the modern port was constructed in Yalta Bay at the end of the nineteenth century. The port is formed by the harbour with the protective moles and has a central basin and one passage with berths around the periphery. It is a so-called old port with a passenger vessels zone. The length of the berthing line is 955 m and the maximal depth is 9.75 m. Cargo-and-passenger Yalta port-station is located to the east of the passenger one in the zone of Massandra. It is protected by a groin and was put into operation in 1986. The berths length in the cargo-and-passenger zone is 850 m with a near-berth depth of 8.2 m. Several port-stations are subordinated to Yalta MP: Foros, Kastropol, Simeiz, Alupka, Miskhor, Lastochkino Gnezdo (Swallow's Nest), Zolotoy plyazh (Golden beach), Livadiya, Nikitsky botanical garden, Gurzuf, Frunzenskoe, Rabochy ugulok, Alushta, Malorepenskoe, Rybachie and Sudak. They have piers for small vessels. A basin with a berth having depths of 0.9–4.9 m close to it was built in Alupka town. In the western part of Sudak Bay, there are 4 piers designed for small passenger and fishery vessels with depths of 2.5–4.5 m near them [8, 18] (Figs. 1.4, 1.5).

Feodosia MP is constructed in the western part of Feodosia Gulf, on the southern coast of the Feodosia Bay apex. The port was built in 1892–1899 and it was reconstructed later on. The port harbour is delimited by Zashitny (Protective) and Shiroky (Wide) moles to the east and to the north, respectively. A quay stretches between them parallel to the shore. The port has 15 berths, but part of them is shallow-watered. The length of the main cargo berths (№№ 1, 2, 3, 14, 15) comprises 765 m with depths of 5–8 m near them. There are also floating piers and oil-piers with depths of about 5 m. Two single anchor leg mooring berths were built outside the port for trans-shipment of oil products. The shipbuilding association

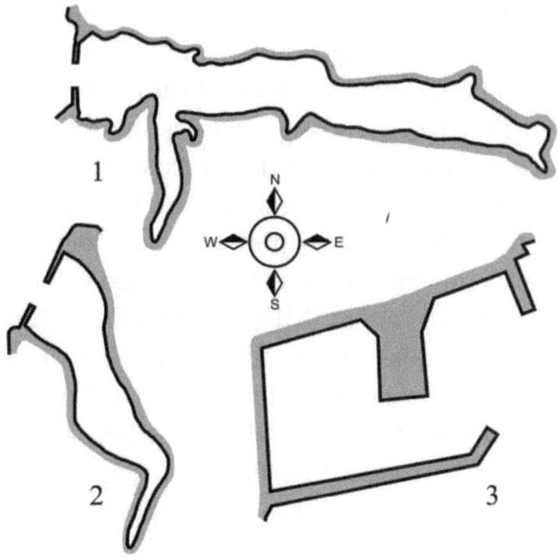

Fig. 1.4 Schemes of marine ports aquatory: *1* Sevastopol Bay, *2* Kamishovaya Bay, *3* Feodosiya MP

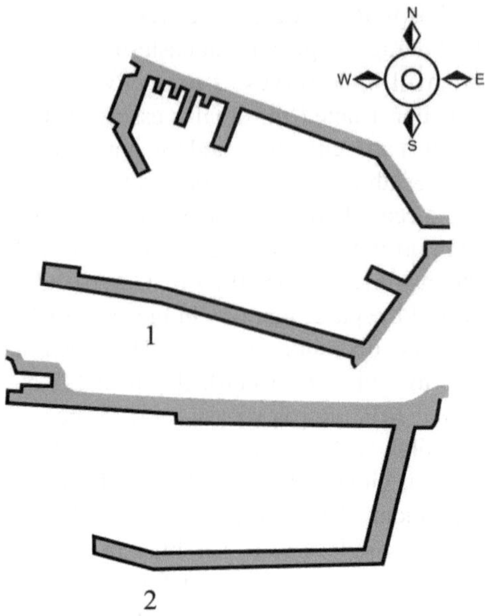

Fig. 1.5 Scheme of Yalta MP aquatory: *1* passenger zone, *2* cargo-passenger zone

"Morje" has its own berths and other HTS in Feodosia MP. Planerskoe (Koktebel), Krimskoe Primorie, Blizhnie Kamyishi and Dalnie Kamyishi port-stations are subordinated to Feodosia port. They have their own piers with depths from 1.5 up to 3.4 m [8, 18].

Kerch MP, located in the falciform, homonymous bay, has acquired its modern organization in the twentieth century. It consists of three parts divided by two moles: Shiroky (Wide) and Novy (New). The southern part of the port is protected from storms by the Genuezky (Genova) mole. There are 7 berths in the port, with a total length of 1224 m and a depth near them from 6.7 m up to 9.75 m. In Kerch are located a Fishing port, Kerch Shiprepair Yard (with a berth walls lengths of 1800 m), Shipyard "Zaliv" (Arshintzevo settlement), sea plant "Fregat" with its berths and other HTS. The specialized ports Krym and Arshintzevo (ex Kamish-Burun) are located on the Crimean coast of the Kerch Strait [8].

Port Krym mostly operates the train-ferry passage between Port Krym and Port Kavkaz. Its aquatory is protected by the South-Western and North-Eastern moles, with a passage between them. The length of berth line is 254 m and depths are 5.0–5.5 m. Arshintsevo port was built between Kamish-Burun and Zmeiny Capes for servicing the iron-ore plant named after Voikov [8].

The Ukrainian MPs of Genichesk, Berdyansk and Mariupol are located on the northern coast of the Azov Sea. Genichesk MP is subordinated to Skadovsk MP. It is located on the western shore of Utlyuksky Liman, in the Genichesk (or Tonkiy) Strait which connects the Azov Sea with the Syvash lagoons. The port is equipped with two modern berths with lengths of 164 and 200 m. The total length of the mooring line is 400 m and ships with the draft up to 3 m can be accepted here [8].

Berdyansk city and its MP are located in the north-eastern part of Berdyansk Bay (Azov Sea). The port comprises an Eastern and a Western basins and has a total of 9 berths with an overall extension of 1.7 km. The Eastern basin (berths №№ 1–6) is protected from wave action by a breakwater and by moles from both the west and the north. The port can accept ships with draft up to 7.9 m. Specialized berths should be constructed in the future for transfer of vegetable oils and mineral fertilizers along with a container terminal and overpasses for transferring oil products [8].

Mariupol city is built on the estuary of river Kalmius. Around 1889, three moles and two breakwaters with a total length of more than 3 km and a quay of 850 m length were constructed in Mariupol MP. Depths in the harbour were increased up to 4.25 m. The structure of the modern MP includes: a fishing port in Shmidt harbour (located in the Kalmius river mouth), the port of "Azovstal" Metallurgical Plant and the port of Azov Shiprepair Yard with three floating docks and berths of 760 m length. At present, Mariupol MP is the biggest and most equipped port of the Azov Sea. It has 16 berths with a total length of 3.2 km. The depths near them range from 5.8 to 7.95 m. Internal roadstead is bounded by the Eastern pier on the north, by the Northern, Southern and Eastern moles on the east and by the Western mole on the south. The internal roadstead consists of an outer port, Ugolnaya (Coal) and Khlebnaya (Grain) harbours and a Repair basin (scoop). The southern mole is not connected with the shore and, as a matter of fact, it is a breakwater. Two passages lead into the port: the first one is between the Northern and Southern moles, while the second is located between the southern extremities of the Southern and Western moles [8] (Fig. 1.6).

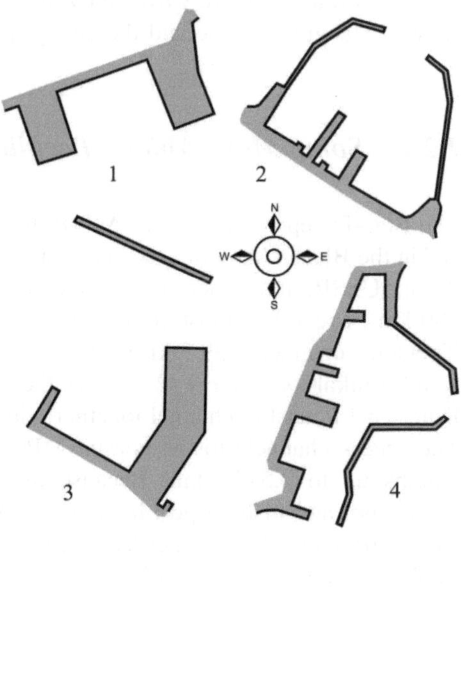

Fig. 1.6 Schemes of marine ports aquatories: *1* Kerch MP, *2* Krym MP, *3* Berdyansk MP, *4* Mariupol MP

More than 210 cargo berths with a total extension exceeding 36 km were built in the main sea commercial ports of the Black Sea northern coast and of the Azov Sea (as well as in their separate cargo regions). Their number constantly increases.

1.2 Main Abiotic and Biotic Peculiarities of the Access Channels to the Marine Ports of the Black–Azov Sea Basin

An AC is a man-made deepening of the sea bottom at the entrance into a port to allow the access of ships with big draft. ACs to MPs of the Black–Azov Sea basin are both an essential component of MPs transport hubs and an integral part of the ecosystems of their aquatories. Because of the shallowness of the Ukrainian shelf, the most extended ACs are located in the north-western part of the Black Sea (NWBS), with the Dniester and Dnieper-Bug Limans, the Kerch Strait and the Azov Sea. In general, ACs resemble wide-bottom channelled ravines or valleys.

An AC creates depths which are anomalously big and uncommon for the adjacent zones of the continental slope. Depending on wind directions, AC presence could lead to irregular wave propagation over the channels hollows and over the adjacent, shallow-water zones.

When reaching shallow waters, the waves' propagation is hampered by the sea bottom. As a result, waves become steeper, shorter and asymmetrical and their front slope breaks more easily. In ACs, the wave motion remains more fluid.

According to the laws of hydrodynamic, waves spread faster in deeper waters and could thus overtake waves moving outside the channel. The wave front could twist, its length increase and the energy per unit of front length decrease.

1.2.1 Some Main Abiotic Peculiarities

The Bug–Dnieper–Lymanskiy AC (BDLC) extends for 81.5 km and is the longest AC in the Black–Azov Sea basin although, when considering other southern seas of the ex-USSR, it is shorter than the Volga–Caspian AC whose length exceeds 100 km. The other most important ACs in the Black–Azov Sea basin are the Kherson channel (39.5 km), the AC of Mariupol port (27.8 km), the Kerch-Enikalsky channel (24.3 km), the AC of Berdyansk port (20.0 km) and the Dniester–Limanskiy channel together with the Dniester marine channel (19.7 km). The access channels to the coastal MPs of Odessa, Ilyichevsk and Yuzhny have lengths up to 2.5–3.0 km. Because of the quick increase in sea depth, Yalta, Sevastopol and Fishing port in Kamyishovaya Bay do not require access channels for the present. The channel leading to Nikolaev MP is more than 100 years old. Table 1.1 presents the morphometric characteristics of the Black–Azov Sea basin

Table 1.1 Morphometric characteristics of the access channels of the Black–Azov Sea basin

Channel name	Morphometric characteristics		
	Length, (km)	Width, (m)	Depth, (m)
Bug–Dnieper–Limanskiy channel or Nikolaev marine access channel	81.5	100	10.0
Kherson channel (along Dneprovsky Liman, Rvach River, arm Olkhovsky Dnieper and proper Dnieper)	39.5	100	8.0
Shipping channels of Varna ports	30.0	100	14.0
Access channel to Mariupol port	27.8	100	12.0
Kerch–Yenikalsky channel	24.3	120	10.0
Access channel of Berdyansk port	20.0	90	8.5
Dnestrovsky marine and Dnestrovsko–Limanskiy channels	19.7	60	4.5
Access channel of Taganrog port	19.0	90	5.0
Azov–Don marine channel	15.0	70	4.5
Access channel Sulina	7.5	100	9.5
Inlet of Berezansky Liman	6.8	100	2.5
Access channels of Odessa port (three)	6.5	100	16.0
Shipping channel in Grigorievsky Liman	6.5	170	16.0
Access channel of Khorly port	6.0	50	5.0
Access channel of Ochakov port	5.0	100	8.0
Shipping channel in Sukhoy Liman	5.0	170	13.0
Access channel of Yuzhny port	4.0	200	16.0
Shipping channel the Danube–Black Sea (arm Bystry)	3.2	170	8.0
Access channel of Constanza port	2.9	170	18.0
Donuzlav inlet	2.6	120	10.0
Access channel of Ilyichevsk port	2.2	180	14.0
Access channel of Skadovsk port	2.2	50	8.0
Access channel of Yeysk port	2.2	70	4.5
Access channel of Temryuk port	2.2	70	4.5
Access channel of Burgas port	2.0	170	15.0
Access channel of port Kerch	2.0	180	8.0
Access channel of Kerch fishing port	2.0	50	8.0
Access channel of port Arshintzevo	2.0	150	8.0
Access channel of Poti port	1.4	100	10.0
Channel Sevastopol Bay–mouth of river Chernaya	1.0	45	5.0
Access channel of Genichesk port	1.0	60	4.5
Access channel of Tuapse port	0.5	120	13.5
Access channel of Sochi port	0.5	100	9.0

ACs. In the Black and Azov Seas, the need for AC construction urgently emerged in the eighteenth–nineteenth centuries when ships' draft became equal to or higher than the depths of access ways to MPs near the berths. The deepening of AC is accompanied by an increase of their length.

Marine shipping channels (from the Latin "canalis"—pipe, gutter) are free-flowing, artificial water-ways constructed in the mainland and in bottom sediments of the water basins for secure and economic use of ships and MPs. Shipping channels can be distinguished into connecting, deepened (at shallow waters), by-pass and cut-off channels. Shipping channels represent water-ways (shipping ways) secure for navigation passages marked by buoys, spar buoys, navigational leading marks and other.

It is important to remember that during building of an AC, the marine bottom biocenoses always undergo changes that can be big or small in scale [16, 33].

Narrowing of river channel in estuaries (with the help of various protecting and guiding moles and dams) leads to an increase in the current speed and deposition of the river solid discharge at a greater distance than under natural conditions.

In the Danube delta, the jetties of Sulina channel in Romania and the longitudinal dyke of the Shipping way Danube–Black Sea in the Bystry arm have caused a large-scale reorganization in drifts distribution in the near-shore zone and in the delta, contributing to the genesis of accumulative formations.

In the middle of the 1950s, the construction of the AC to Ilyichevsk MP and the large-scale sand extraction from the underwater pits in the shore zone of Sukhoy Liman—rural settlement Chernomorka at a depth of 4–6 m led to the accumulation of sand drifts in the zone of artificial hollows over a length of 15 km. The bottom biocenoses were deteriorated or destroyed over an area of about 20 km^2. On the continental slope, the area covered by sandy drifts up to 0.3–0.4 m thick decreased by eightfold [30]. The changes took place not only on the sea bottom, but also on the shore. The frequency of landslides increased threefold and the speed of bottom abrasion by two to threefold up to depths of 6–7 m [30].

In Kuban delta, construction of Glukhoy channel moles and Temryuk port led to the formation of the so-called Chayka shallow-water on their windward side. Here, at a distance of 1 km from the shore, the depths have decreased from 6.0 to 1.5 m and the coastal line started to move forward with a speed of 5–6-m $year^{-1}$ [33].

In the 1970s, in the central part of Gelendzhik Bay, the sand intake for the city beach nourishment formed a pit in which organic matter (OM) started to accumulate and hydrogen sulphide appeared [12].

Starting from 2004, in the near-shore zone of Kiliya delta, Ukraine has constructed and developed the Danube–Black Sea Shipping way via the Bystry arm as a passage for big ships into the Danube, as an alternative to the Sulina arm (Romania). The marine channel is 3.2 km long, 170 m wide at the bottom and 8.0 m deep. On its northern side, the channel is protected from drifts by a rubble stone dam extending for 1.2 km from the contemporary shore. According to the project, a dam with the same characteristics should be constructed on the southern side of the channel.

Access channels to MPs of the Black–Azov Sea basin mainly appeared in the twentieth century and are relatively recent components of the ecosystems of MPs aquatories. Given a total extension of more than 350 km for all ACs in the basin, about 300 km belong to Ukrainian MPs. ACs to MPs are artificial negative forms of bottom configuration and therefore are constantly exposed to various natural factors directed to bottom equalization. For this reason, designing, constructing and maintaining the operating depth in an AC is a complex engineering task. Access channels, becoming a new element of the coastal aquatic ecosystems, exercise various influences on the ecosystems of MPs as well as on the adjacent aquatories and require special studies.

To some extent, all ACs are subject to sediment accumulation. In the forthcoming decades, the maximal depths required in ACs to the Ukrainian MPs are expected to be around 20–21 m. During designing of ports, it is extremely important to foresee the position of the port entrance, i.e. the place where the AC gets into the MP protected aquatory. Sediment accumulation in ACs and MP aquatories is maximal if the passage is located towards the longshore drifts flow. Dredging is used to contrast sediment accumulation in AC and can be realized by scoop dredge or by suction systems. Both methods destroy and partly withdraw both substratum and hydrobionts of bottom biocenoses [1].

The speeds of currents are also taken into account during the design of ACs. The speeds of currents are delimited by erosive and silting velocities, with the most intensive deposition of sediments occurring at minimal current speeds.

Each channel represents an artificial corridor dug inside the sediment so that its construction requires extraction and removal of sediment from the channel bed. Works at silty, sandy, clayey and rocky sediment are carried out by various methods and are accompanied by different ecological consequences. Negative influences can also be observed in the zones of dumping of dredged sediments.

The main method of AC protection from sediment accumulation is the incision of trenches (hollows) in the bottom, i.e. creation of the depths clearance. Permanent structures protecting ACs against drifts and roughness are also constructed in the form of continuous barriers. These barriers can be relatively short (straight jetties, up to a couple of hundreds metres long, such as in Ilyichevsk and Yuzhny MPs) or more extended dams (e.g. the 1.2 km dam protecting the Shipping way Danube–Black Sea via the Bystry arm).

On sandy bottoms, ditches (hollows), i.e. pockets having a semi-trapezoidal cross-section and designed to trap drifts, can be arranged near the AC exit from the port aquatory [20]. As a matter of fact, these are shallow-watered lateral parts of an AC. Ditches (hollows), which are dug in the marine bottom parallel to the AC, serve for intercepting part of the bottom sediments drifts. In zones with silty bottom, it is recommended to place such trenches (hollows) at a minimum distance of 100 m from the main channel [20]. It is important to notice that construction of trenches (hollows) has a negative effect on existing biocenoses, but, at the same time, increases the heterogeneity of bottom conditions.

The transverse section of a shipping channel (Fig. 1.7) usually has the shape of an inverted isosceles trapezium with a bottom bed and lateral slopes (or shores).

The channels slopes have different angles and different width according to the sediment type. As a rule, channels have a sediment bed, i.e. not rocky and not lined. In channels built in rock and native clays, the angle of the slopes is less than in channels dug in sandy or silty sediments [20] (Fig. 1.8).

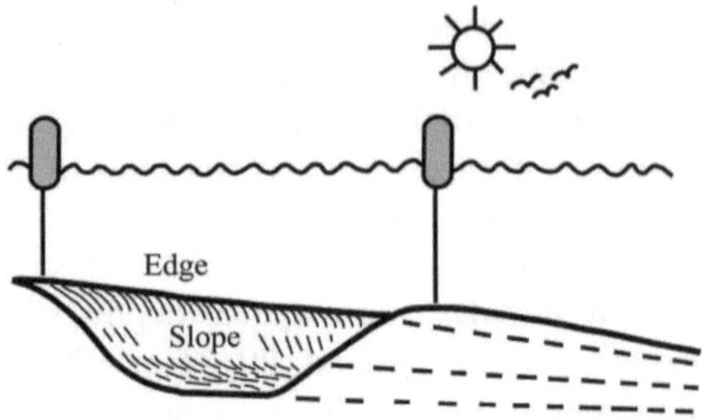

Fig. 1.7 General scheme of a channel hollow

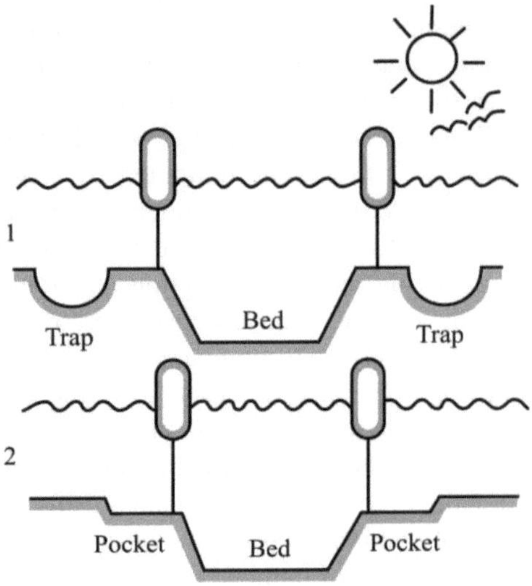

Fig. 1.8 Recommended methods for prevention of sediment accumulation in access channels (acc. to [20]): *1* with the help of hollows-traps, *2* with the help of hollows-pockets

In the Black–Azov Sea basin, the absolute majority of shipping channels is built on the bottom of the basins, and is dug in sediment. Hollows in bay-bars can be considered as connecting channels as in the case of the Black Sea–Sukhoy Liman, the Black Sea–Grigorievsky Liman and the Black Sea–Donuzlav Liman channels or the Danube–Black Sea Shipping way.

In the presence of an AC, the bottom relief undergoes a process of realignment. During this process, the edges of the channel slopes, dug in the soft and semi-liquid sediments, slip and channels gradually become wider in their upper part, but shallower. The insufficient consideration of the interference of ports and AC construction on the dynamic of natural systems can have various consequences including shores' destruction, excessive sediment accumulation in the channels and aquatories, washout of the sandy or silty-sandy sediments in the zones adjacent to channels up to exposure of parent rocks. In the 1970–80s, the sedimentation in the Prorva channel (Danube delta) sharply increased due to its deepening and plans about its use for navigation were abandoned.

Moles and breakwaters mainly divert drifts movement seaward after they have filled shore irregularities, while the ACs become traps for the longshore sediments drifts along their whole length and until levelling of the bottom configuration. AC interrupts the movement of part of the drifts. As a consequence, a certain drifts deficit originates at some shore zones and erosion processes are activated.

Being artificial negative forms of the bottom relief, access channels require regular maintenance of their operating depths and play an important role in functioning of the ecosystems of MPs aquatories and of the adjacent sea zones. Owing to AC, there is a constant interrelation between the ecosystems of MPs and those of the basins on whose bottom they have been built. Hydrobionts move, either passively or actively, with water masses across ACs. The processes linked to onshore–offshore wind phenomena are also occurring via ACs. The water masses in the channel hollow quite often do not take part in the process of wind-induced mixing of the water surface layer. Owing to the onshore–offshore wind phenomena, compensational currents form in ACs and the horizontal water transfer becomes more significant than it usually is in the surface layer.

Access channels at all depths are intersected by the longshore sand drift. It was determined that the increased sediment accumulation takes place only at longshore drifts displacement that comprises the zone up to the depth of more than 20 m. The drifts mainly move in the thin near-bottom layer. Experimental and visual studies showed that the movement of sand particles at the bottom quite often occurs discontinuously and, rarely, by rolling on the bottom surface. Five–ten percent of the total drifts flow loss is usually deposited in the channel hollow. The thickest drifts' layer is accumulated in the deepest part of hollow, but sedimentation takes place along its whole profile [23].

Sorting of sediment according to particles' size takes place in ACs during the displacement of the longshore drifts. This phenomenon plays an important role in the formation of hydrobionts settlements [26]. Accumulation of coarse-grained sand and broken shells in the bed and slopes of an AC increases roughness of the bottom surface and favour turbulization of bottom flows. The presence of microvortexes is essential for the life of the majority of benthic organisms [21, 22].

Conditions for the formation of a pycnocline, which is absent in the adjacent shallow waters, appear in ACs deeper than 8–10 m. Easily mobilized sediment comprising a high proportion (up to 25–30%) of organic matter accumulates in AC beds at a current speed of 5 cm s^{-1}. Aleuritic and pelitic particles are remobilized at a current speed up to 15–20 cm s^{-1} and sand grains with sizes of 0.1–1.0 mm at a speed of 20–30 cm s^{-1}.

Foraminifera and small-sized molluscs are transported with the longshore drift together with the sediment particles. Rolling on the bottom, they accumulate in some channel's zones where they form shells' accumulation as well as settlements (*Abra*, *Cerastoderma*, *Mya* and other).

The majority of the Ukrainian MPs are constructed in river mouths or near the mouths of big and small rivers. Aquatories of such ports are usually more influenced by drifts than Crimean ports that are not strongly affected by river run-off

(Sevastopol, Yalta, Chernomorsk, Balaklava MPs). ACs that cross seashores near river estuaries, or are located in them or in their vicinities, are drifted with intensity determined by river sediment load quantitative indexes [23]. The problem of sediments accumulation in navigation channels reduces sharply or is absent in those MPs whose navigation ways are located over submarine canyons and where the deep waters are found near the shore (e.g. Yalta MP).

The shallow-watered Azov Sea (up to 12–13 m) receives the inputs of 28 rivers with permanent flow and carrying solid discharge during the whole year [27]. That is why the longshore sediments drift is very high in this basin.

Ten ACs to MPs are located on the NWBS shelf. In general, they not only intensify the erosion (mainly of western coastal zones), but also retain part of the longshore drifts flow. In the case of Ilyichevsk MP, the drifts accumulation takes place westward of the channel. Drifts accumulation in the AC leading to Ilyichevsk port amounts to 35–40,000 m^3 per annum [4].

The intensification of navigation is accompanied by an intensification of sediments deposition in the channels. Erosion of the slopes and caving of the banks and slopes are observed. These phenomena are explained by the strong flows affecting the bottom and the slopes of the channel following ship traffic. To reduce the negative influence of navigation on channel profile, the ships' velocity in them is limited, or their movement is provided by shallow-draft tugs. In terms of exploitation, the optimal ACs are those without bends.

As it was already mentioned, a variable degree of sediment accumulation is typical to all shipping channels, including the channels bed, in the same way as part of the suspended matter deposits at the sea bottom. However, the longshore drifts flow plays the main role in sediments accumulation. The sediment movement occurs, first of all, under the influence of waves and currents. The wave action on sediments relatively quickly attenuates with the increase in water depth.

The depth of bottom outwash mainly depends on wave height and wave steepness [6]. The impact depth of wave energy on the sea bottom, i.e. the depth at which sediment particles resting on the bottom are set into motion, is different for each wave height. For the NWBS shelf it corresponds to 8–10 wave heights and passes, on average, along five metres isobath and along the twelve metre isobath in the case of storm waves. The strongest waves could stir up the bottom surface at depths up to 30 m [28]. All shipping channels of the Black–Azov Sea basin are located in the zone of relatively strong wave action on bottom sediments. The shipping channels have the same depth over their entire length, so that the force of waves' action on their bed and slopes changes moving towards the shore. Major storms induce significant changes in usual pattern of drifts flow and could involve considerable additional masses of traction material.

It was determined that, in the Black–Azov Sea basin, wind speeds of not less than 10 m s^{-1} must be considered when estimating sediment accumulation in AC located in flat coasts formed by loose sediments (silt, sand). In the case of steep shores, the wind speed to be taken into account is raised to 15 m s^{-1}. Special attention should be given to possible, uncommon wind directions during fierce storms [6].

The Black Sea is most turbulent in winter, when the frequency of waves with a height of 2 m and more is nearly anywhere 30%. The frequency of the same waves and more is 5–13% in summer and 15–17% in spring and autumn. In general, waves less than 2 m high are prevailing in the Black Sea. The frequency of waves less than 1 m high is 50–70% in summer and 27–45% during the rest of the year. The wave height can reach 4.5–5 m during especially severe storms [29].

Along the whole coast of the Black and Azov Seas, circulation currents are directed mainly counterclockwise in the majority of the gulfs and bays. The character of the coastal zone currents is significantly different for waves approaching the shore perpendicularly or sideways. When waves approach perpendicularly to the shore, the speed of the longshore currents is 20–30 cm s^{-1}. In the flows of a rip current, speeds exceed 2 m s^{-1} [23]. In semiclosed MPs aquatories, the currents speed in the surface layer usually ranges from 1 to 5–10 cm s^{-1} and is markedly less near the bottom. In MPs, sedimentation of even small-sized suspended particles is quicker because of the small current speeds. Only particles having sinking speeds of less than 1 m per day in calm water have could constantly remain in suspension [24].

For a wave orbital diameter of only 2.5 cm, a roughness with the period of 20 s prevents sedimentation of the finest silt particles and keeps them in suspension for a prolonged time [17].

The turbulent motion maintains solid (hard) particles suspended in water, thus contributing to their transfer. The water flow cannot carry particles by itself if the turbulence does not keep them suspended in the water column. When water stays in a still state, particles' sinking velocity is governed by Stokes' law. The bigger is the diameter of a particle, the quicker it sinks. The sinking velocity is proportional to the particle specific gravity and is inversely related to water viscosity.

Near the shore, the transfer of sand particles by roughness and currents is facilitated by the influence of the surface tension film. Foam, formed near the shores during the swash, causes flotation of the dried sand present near the shore and carries it at significant distances. The diameter of the sand grains carried during such process can exceed 3 mm, although grains with sizes of 0.5–1.0 mm predominate [17].

Bottom sediments consist of mineral and organic particles of various size and origin. Organic particles deposited on bottom sediments have a density of about 1 g cm^{-3}, while the density of mineral particles is two to threefold higher. For this reason, the wave size at which particles are resuspended into the water column and the speed with which they settle after roughness termination are different. It was estimated that, in bottom drifts, the main part of sand particles is displaced by traction or within the mobile sediment layer rather than in suspension. Big and heavier particles, not suspended by roughness, are carried by the near-bottom flow, are practically in continuous contact with the bottom and quasi roll over the channel bed. Because the flow speed drops in the hollow of channels, the heavier particles are retained in the zones with the steepest slopes. Part of the bottom drifts transported in suspension (mainly organic particles and most small-sized mineral particles) takes part in the process of advected-turbulent displacement and can easily escape the channels hollows. Thus, particles of various size and density are influenced by hydrodynamical conditions and bottom slope of a specific sea zone and AC during the process of longshore movement. The shipping channels, being new elements of the bottom relief, have a certain influence on sediments differentiation.

In the NWBS, the total content of suspended drifts in the near-bottom layer after a storm period declines sharply (12-fold) between the 6–7 m isobaths. The sedimentation of resuspended material after a storm is completed in two days, which leads to water clearing and increased transparency. An experiment using sediment traps showed that a sediment layer with a thickness of 27 cm deposited at the depths of 10–15 m during the entire autumn–winter–spring period. At the depth of 25–30 m, the process of wave-induced mixing of sandy material slows down and the sedimentation of silty material starts. The depths of silt particles sedimentation decreases at weaker roughness. The strong roughness affects and redeposits sediments at deeper depth. During a 4-grade storm (wave height up to 2 m), labelled sand on the outer part of the shore zone penetrated into sediment up to a depth of 90 cm, indicating that this layer was affected by waves' action [28]. All these processes have their influence on living conditions of hydrobionts dwelling on the slopes and on the bottom of shipping channels. Weak flows, transporting organic suspension, favour its accumulation in the deepest parts of the channels, which could worsen some hydrochemical indices.

The slopes of shipping channels, from the edge to the bottom, could be dozens and even hundreds metres wide and the drop in depth could exceed 10 m. On this area originate conditions which usually occur at greater distances from a shore. In fact, conditions of steep shores are artificially created on the slopes of ACs near ports.

In ACs, the deepest incut depth into the bottom is observed near the shore, where the sea depth is less. The widest slopes are formed in this zone, thus improving the stability of the channel profile over time. With increasing distance from the shore, the thickness of the water layer over the edges of slopes crests increases, while the width of slopes themselves and the depth of incut, decrease. The thickness of the water layer above the channel bed, i.e. channel depth, is maintained nearly equal along its entire length. Thus, big depths are brought directly towards the port via a channel. With the distance from the port, the AC hollow in the bottom relief becomes less pronounced until it completely disappears. It is evident that the major changes in drifts migration and in hydrobionts living conditions take place in the upper and middle part of the channels, where slopes are relatively high and the hollow is deep.

At big depths, where the near-bottom wave motion is much weaker, the hollows of the channels are shallower than in the coastal part of the channels. Slopes are also less pronounced than in the middle or coastal part of channels, where the hollows have a significant depth [6].

Various shipping channels can cause ecological consequences of different type and scale. In conditions of isolation from the sea, e.g. in Sukhoy and Grigorievsky Limans, the water salinity in the summer months can be significantly higher than in the sea. After the construction of connecting channels, a salinity gradient between the limans and the sea was established and hydrochemical conditions became more stable. The biota of Sukhoy and Grigorievsky Limans was enriched by dozens of species which were new for their ecosystems [7, 25, 31]. A steady double-layer stratification of water masses started to form in them. After the construction of the Black Sea–Donuzlav Liman channel, the vertical stratification of water masses also appeared in the liman. The hypersaline water (with a salinity of more than 90‰) was preserved in the near-bottom layer, while salinity approached that of the Black Sea in the surface layer.

After construction of the AC to Belgorod-Dnestrovsky MP, located in the strongly desalinated Dnestrovsky Liman, the salt marine water together with the marine hydrobionts started to penetrate farther into the liman aquatory. According to the observations carried out in 2010 in the Danube–Black Sea channel, at depths of 7–9 m, by the Odessa Branch of the Institute of Biology of Southern Seas of the National Academy of Sciences of Ukraine (IBSS, NAS Ukraine), salt waters together with various hydrobionts were penetrating up to 1 km inland via the Bystry arm. The phenomenon occurred despite one of the highest water volume run-off of the Danube during the whole history of observations. Here, the salinity in the channel at the depth of 6 m was 6.3‰. Construction of AC to MP via river mouths and their dredging usually favour the upstream movement of seawater in the bottom layer, sometimes to a point that even the waterworks intakes of settlements are threatened.

The salinity in Bug Liman is higher than that of Dneprovsky Liman. Salinization of Bug Liman waters mainly results from entrance of the seawater via the shipping channel up to the upper reaches of the liman in the form of a bottom counter current. Water salinization near the left and right banks is different and changes according to the course of the channel which passes closer to one or the other shore [3].

The connecting shipping channels and channels in the river mouths lead to reorganization in aquatic ecosystems and to changes in the qualitative composition and quantitative indices of their biological components.

As a rule, temperature, salinity, pH, oxygen content and biogenic matter do not show differences between the near-surface water layer above the shipping channels and the sea zones adjacent to channels. Significant differences in oceanological and hydrochemical indices originate in the near-bottom water layer, under the pycnocline which forms in the channels with depths of 8–18 m. In the upper and middle part of these channels, hydrogen sulphide can form at the bottom and, in some situations, the redoxocline can rise into the water column. Close to the ports accesses, the slopes of the deep channels can hamper horizontal and vertical water mixing [5, 32].

ACs are a constant specific biotope for some hydrobionts, while others find here temporary shelter. In numerous ACs, saprobiotic conditions form in the hottest months and hypoxia and anoxia are registered. Mass mortality of bottom fauna, with variable duration and extension, often occurs in ACs. Thus, accumulation of some decomposition products of organic matter (hydrogen sulphide, mercaptans, methane and others) is registered in the AC of Odessa MP during conditions of lowered hydrodynamics from July to September. At offshore phenomena with compensational currents, these toxic substances with a disagreeable odour enter MPs aquatory and, as a result of further upwelling, the atmosphere.

AC located in the Dnieper-Bug Liman are subject to the complex influence of the Dnieper, Yuzhny Bug, Ingul and Inguletz river run-off which carry, on average, about 168,000 tonnes of biogenic and 15,000 tonnes of contaminating substances per year [19]. As a result of the regulation of rivers' run-off, the previously observed annual "washout" of Yuzhny Bug and Dneprovsky Limans during the high water period does not occur anymore in years with average and low water flows [19]. Accumulation of biogenic and toxic substances takes place in ACs that are deeper than the adjacent bottom areas. It was determined that the speed of contaminants sorption on suspended particles depends on their size. The lower is the grain size of sediment drifts; the faster is contaminants' sorption. The concentration of oil products and some other strong toxicants increases in the pelitic sediment fraction [13] which, at weakened hydrodynamics, accumulates in ACs.

Already in the first part of the twentieth century, a strong impoverishment of oxygen concentration, and sometimes total anoxie, were registered in the near-bottom layers in the central part of the Dnieper Liman during July and August [3]. The situation has only worsened since.

In the past decades, hypoxia developed in the shipping channels of the Dnieper-Bug Liman and led to the formation of a stable zone contaminated by hydrogen sulphide where hydrobionts mortality occurs almost annually. Most often, the mass mortality phenomena takes place in August (27% of cases) and in July (19% of cases) and most rarely from October until March (0–8% of cases) [19].

Mass mortality is usually registered in the shipping channels leading to Nikolaev and Kherson MPs, but is quite rare in Kinburn Strait region where flow and hydrodynamic activity are high [19]. The highest biomasses of zoobenthos are registered here.

Except for ACs located in the lower reaches of rivers and for channels cut in land, flows speeds in shipping channels are relatively low and slightly differ from those in the waters adjacent to them.

In channels, flows in the surface and near-bottom water layers can be divergent and are opposite at offshore winds. Various phenomena originate in access channels and passages into MPs aquatories, including outsized waves, harbour oscillations and various water circulations connected with the morphometric peculiarities of an MP or arising from the meteorological and hydrological situations in the region of their location. Owing to currents, bottom sediments enter MPs aquatories via shipping channels and form sandbars near piers and breakwaters. The sand accumulative formations around the protective HTS increase biotope diversity in the ports. ACs to MPs contribute to sorting of sediment material in the longshore drifts flowing around protective constructions and in the drifts flows entering the aquatories in a tongue-like shape. At strong currents, big and heavy mineral fractions accumulate in the channels, while accumulation of the organic suspension takes place at weak flows.

On the slopes of ACs, conditions for particles sedimentation vary between the channel crest and bottom. Such differences lead to sediment differentiation in terms of size, density and OM content. The latter is less than 1% in sand, but it is 5% and more in silt [23]. Settlements of mass bottom hydrobionts in various parts of the channels slopes directly depend on the local patterns of sediments deposition. Such patterns reflect the hydrodynamical processes activity which also determines hydrochemical conditions. The patchy character of hydrobionts distribution is mainly connected with such processes.

An area located near the maximal depth of water turbidity is a favourable area for filter feeders [24].

During periods of lowered hydrodynamical activity, usually in summer months, accumulation of easily stirred up sediment takes place in ACs and its thick layer may become an obstacle for the heavy creeping hydrobionts such as crabs, hermit crabs and bivalves of the genus *Rapana*, *Nassa* and others.

In particular, schools of mass pelagic fish (sprat *Sprattus phalericus*, European anchovy *Engraulis encrasicolus,* haddock *Merlangius euxinus*) and dolphins enter port aquatories via shipping channels. Owing to this process, part of the OM produced in the MP aquatories is exported from their ecosystems.

In many shipping channels, the rules of shipping and navigation foresee that vessels' draft should be 20–50 cm less than the channels' depth. Various water flows originate during ship navigation along the channels and favour water mixing, disruption of water masses stratification and oxygen penetration in the water column. Water exchange between MPs aquatories and the adjacent sea waters occurs via shipping channels and passages. The majority of MPs of the Black–Azov Sea basin have only one passage where, as a rule, currents having opposite directions form in the near-surface and near-bottom layers. The intensity of water exchange mainly depends on the aquatory water volume and on the channel cross-sectional area, but is also affected by a wide range of factors, among which wind speed, direction and duration are the most important. At strong offshore phenomena, the flow of the colder and more saline marine water via the near-bottom layer of shipping channels fills up the MPs aquatories. In the course of this process, a variable quantity of OM, which has accumulated in the upper and middle part of the channels and is easily resuspended, enters the MP ecosystem. During these periods, channels function as specific "pumps", increasing the amount of BM. Upwelling occurs in MP aquatories, owing to the formation of compensational currents via the ACs during offshore phenomena. Water flows originating from the near-bottom layer bring not only BM into the water column and into the near-surface layer, but also hydrogen sulphide and other gases (methane, ammonia, mercaptans, etc.) which form during OM decomposition in anoxic and hypoxic conditions, especially in hot summer months.

The process of silting of various MPs aquatories extends over time (years, decades, centuries) to a variable extent. However, phenomena exist that lead to quick changes in depth (between 0.5 and 2.0 m) at various shores of the Black Sea. These are wind-induced processes and seiche oscillations of the water level. Catastrophic earthquakes and associated consequences also present a certain danger, but they are rare and not possible to forecast.

In the Azov Sea, the largest wind-induced fluctuations of water level [9], influencing the depth in ACs and MPs, are registered in autumn and winter and, less frequently, in spring; which is related to cyclonic activity.

In the region of Taganrog MP, during offshore phenomenon, water can withdraw from the shore line for up to 5.5–6.0 km. At onshore phenomenon, the port berths were sometimes submerged with the height of the water column reaching about 2.0 m above the mean sea level. In such situations, the territory adjacent to berths was also submerged. The most significant water fluctuations at offshore and onshore phenomena were registered in Taganrog Bay and in the western part of the Azov Sea. The maximum value of wind-induced fluctuations of water level is 6 m in region of Taganrog port and 4 m in the ports of Berdyansk, Kerch and Genichesk. The time span during which the water level stays above the dangerous level during the onshore phenomenon and below it at offshore phenomenon does not exceed 12 h in the majority of cases. Only in Genichesk, it sometimes lasts more than 2 days.

In Odessa MP region, the lowering of sea level at offshore winds could reach 1.75 m and the water rise at onshore winds 1.0 m. Changes in the sea level can happen quite quickly, reaching 20–25 cm $hour^{-1}$. The seiche fluctuations of the sea level in the aquatories of Odessa, Ilyichevsk and Yuzhny MPs rarely exceed 40 cm. During windstorms and big wave fetch, the sea level fluctuations can exceed 2.0–2.5 m and are bigger on the southern, eastern and western coasts of the Black Sea [5, 15, 18, 29].

In the Azov Sea, the lowering of sea level at offshore phenomena can reach 2–3 m and, given the shallowness at various shores, coastal areas of the bottom are totally emerged over some kilometres from the shore up to the edges of the channels slopes where water remains in hollows. Here, the permanent inhabitants of channels as well as vagile invertebrates and fish from the adjacent sea bottom find shelter for dozens of hours during such periods.

Unlike hollow-channels situated at the bottom of the basins, the edges of the slopes of slit-channels built on land emerge out of water and form biotopes for littoral organism's complexes. Artificial HS, such as jetties protecting the entrances into Sukhoy and Grigorievsky Limans, are used for reinforcing the slit-channels slopes. In specific zones of slit-channels, the slopes are reinforced by stones, concrete plates, sheet pilings and others. The presence of HS artificially placed in the shipping channels creates conditions for periphyton formation. Artificial HS are also present in hollow-channels in the form of navigational buoys. The overall biotope diversity is higher in slit-channels than in hollow-channels. In the Black–Azov Sea basin, the hollow-channels are cut mainly in the loose and soft sediments where, except for depth, there are suitable conditions for hydrobionts' life. The slit-channels and the hollow-channels are mainly dug through sandy, silty-sandy and silty sediments and occasionally cut through compact clays and hard-rocks (Fig. 1.9).

In the shipping channels, the prevalence of porous and soft sediments (sand and silt, differentiated according to size, density and composition) creates conditions for the development of organisms on the channels' slopes and bed. Consequently,

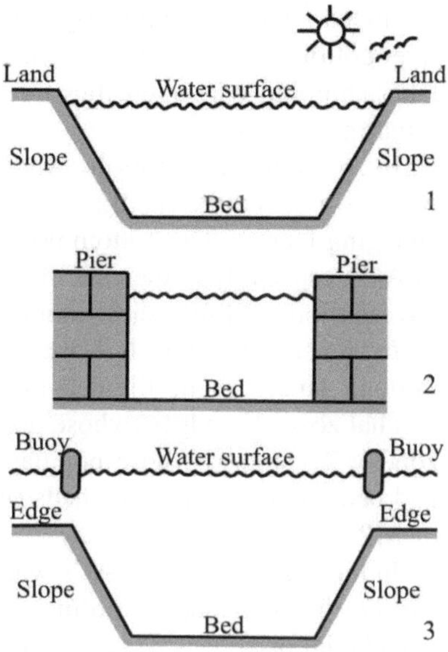

Fig. 1.9 Types (profile; side view) of shipping channels: *1* slit-channel, *2* passage-channel, *3* hollow-channel

epifauna (organisms inhabiting the sediment surface) dominates in the presence of coarse sand, while infauna (organisms inhabiting sediments) prevails in zones with silted sand or silt.

Organisms of attached and non-attached epifauna can develop in the channels with artificial HS zones. The zones of channels-passages delimited by two piers and leading into MPs aquatories possess a variety of specific properties. Their slopes are usually vertical or have a slight angle from the bottom to the surface. Here, conditions for the most intensive periphyton development are created because of the high hydrodynamic activity.

ACs represent dynamic artificial formations over their whole height and width, including slopes, and with which the natural basins ecosystems lead a continuous struggle to level-off the bottom configuration. At the same time, works for maintenance and deepening of channel depths and construction of new channels are carried out to provide conditions suitable for navigation. Channels' dredging inevitably leads to an increase in their length and, quite often, in their width.

In general, higher hydrodynamics is registered in the ACs of MPs than in their harbours and near the berths. The coarser fractions of the bottom sediments accumulated in ACs which include a quite wide spectrum of zones with various sedimentation rates.

1.2.2 Some Main Biotic Peculiarities

During analyses of AC biotic peculiarities, the main attention is given to ACs of Odessa MP and the Danube–Black Sea Shipping channel (Connection channel). Access channels of Odessa MP are a classical example of hollows in bottom sediments. The Shipping way Danube–Black Sea represents a hollow in sea and river bottom sediments on the one hand and a connecting channel on the other. This latter shipping way intensifies the reciprocal influence between the near-shore and estuarine ecosystems.

In AC hollows and in the water column above them, the pelagial biota species composition is quite changeable and subject to the influence of various and quickly changing factors. The bottom population better reflects the main biotic peculiarities of ACs, especially in the case of macrozoobenthos whose numerous representatives have prolonged life-cycles (months, years, decades).

In general, the absolute majority of ACs, dug in the basins' bottom or cut through land, can be divided as follows based on the character of biota in the benthal zone: 1—edges (whose population is identical to that of the adjacent bottom zones); 2—slopes (whose population differs to a variable extent from both edges' and beds' populations); 3—beds (whose population is usually rather similar to that of MPs deepwater aquatories).

In the pelagial zone of ACs with depths of more than 7–9 m, the following parts can be distinguished according to composition the phyto- and zooplankton: 1—epipelagial or above-pycnocline layer; 2—bathypelagic or below-pycnocline layer. In general, the epipelagial population of ACs is similar to that of the adjacent zones and the bathypelagial one is usually close to that of MPs.

Three ACs lead into Odessa MP aquatory. Their macrozoobenthos complex includes 18 taxa (worms—10, molluscs—5, crustaceans—3) at depths of 9.0–14.0 m. At reference stations (depth 9.0–14.0 m, average 11.2 m), 23 taxa were found (worms—7, molluscs—6, crustaceans—9, representatives of other groups—1) (Table 1.2).

In ACs and at reference stations, the bottom macrofauna is exclusively represented by euryhaline marine complexes. The taxa similarity coefficient (acc. to Jaccard—Alekhin) between them was only 36.7% and included 54.5% of the taxonomic groups of worms, 37.5% of molluscs and 20.0% of crustaceans.

The range of taxa number, density and biomass in AC and at reference stations is large (Table 1.3). The average number of taxa per station is nearly the same at both sites (9.3 vs. 8.2). However, the average density is nearly 2.5-fold higher in the AC (3476.7 ind. m^{-2}) than at reference stations (1384.2 ind. m^{-2}), while the corresponding values for the average biomass are 50.295 g m^{-2} and 117.919 g m^{-2}; i.e. 2.3-fold lower in the AC than at reference stations.

In the ACs of Odessa MP, the average density of macrozoobenthos was 3.8-fold and the average biomass was 4.4-fold higher than in the port aquatory (912.0 ind. m^{-2} vs. 11.520 g m^{-2}, respectively) [7].

Among the main taxonomic groups, worms were dominating in terms of quantity both in AC and at reference stations (89.5% and 86.2%, respectively). In terms of biomass, worms were also dominating (59.3%), while molluscs represented 78.5% of biomass at reference stations.

Detritophages were the predominant trophic group in AC and at reference stations both according to the number of taxa (9 in AC, 10 at reference stations) and quantity (88.8 and 91.8%, respectively).

Table 1.2 Comparative characteristics of the composition and quantitative indices (*N*—average density, ind. m^{-2}; *B*—average biomass, g m^{-2}) of macrozoobenthos in access channels of Odessa MP (2001) and at reference stations (1997–1999)

Taxon	Access channels		Reference stations	
	N	*B*	*N*	*B*
Coelenterata				
Obelia longissima (Pallas)	–	–	+	0.025
Vermes				
Turbellaria g. sp.	3.3	0.023	–	–
Harmothoe imbricata (L.)	30.0	0.130	6.7	0.077
Harmothoe reticulata Claparede	–	–	4.2	0.033
Neanthes succinea Leuckart	546.7	19.403	611.7	20.217
Nephtys hombergii Savigny	1.7	0.137	–	–
Nephtys cirrosa Ehlers	8.3	0.500	–	–
Spio filicornis (O.F. Muller)	16.7	0.013	–	–
Polydora cornuta Bosc	261.7	0.150	80.8	0.223
Heteromastus filiformis Claparede	2091.5	9.355	475.8	4.193
Melinna palmata Grube	1.7	0.003	7.5	0.063
Oligochaeta g. sp.	151.7	0.135	6.7	0.007
Mollusca				
Hydrobia acuta (Draparnaud)	170.0	0.427	24.2	0.055
Retusa truncatella (Locard)	–	–	0.8	0.015
Doridella obscura Verrill	1.7	0.005	–	–
Mytilaster lineatus (Gmelin)	–	–	8.3	0.533
Mytilus galloprovincialis Lamarck	–	–	13.3	72.457
Cerastoderma glaucum Poiret	168.3	10.153	16.7	7.855
Abra ovata (Philippi)	8.3	0.008	–	–
Mya arenaria L.	6.7	0.013	25.0	11.685
Crustacea				
Balanus improvisus Darwin	–	–	8.3	0.350
Crangon crangon (L.)	1.7	9.833	–	–
Iphinoe tenella G.O. Sars	–	–	13.3	0.028
Idotea baltica basteri Audouin	–	–	1.7	0.032
Gammarus insensibilis Stock	–	–	0.8	0.002
Melita palmata (Montagu)	–	–	3.3	0.007
Microdeutopus gryllotalpa A. Costa	6.7	0.007	61.7	0.055
Erichthonius difformis M.–Edwards	–	–	1.7	0.002
Corophium bonelli (M.–Edwards)	–	–	11.7	0.005
Total	3476.7	50.295	1384.2	117.919

In ACs, detritophages and sestonophages comprised 57.8% and 20.2% of the biomass, respectively; while the figures at reference stations were 78.7% for sestonophages and 21.1% for detritophages. The index of uniformity of the trophic

Table 1.3 Comparative characteristics of quantitative indices of macrozoobenthos at access channels of Odessa MP (2001) and at reference stations (1997–1999)

Index	Access channels	Reference stations
Number of samples	6	6
Sampling depth, m	9.0–14.0	9.0–14.0
Sampling depth, average, m	11.5	11.2
Taxa number, total	18	23
Including worms	10	7
• Molluscs	5	6
• Crustaceans	3	9
• Others	–	1
Taxa number per sample	8–12	5–11
Average taxa quantity per sample	9.3	8.2
Benthos quantity per sample, ind. m^{-2}	1230–6200	850–1940
Benthos average quantity, ind. m^{-2}	3476.7	1384.2
Including worms, ind. m^{-2} (%)	3113.3 (89.5)	1193.4 (86.2)
• Molluscs, ind. m^{-2} (%)	355.0 (10.2)	88.3 (6.4)
• Crustaceans, ind. m^{-2} (%)	8.4 (0.3)	102.5 (7.4)
• Other, ind. m^{-2} (%)	–	–
Biomass benthos per sample, g m^{-2}	14.500–89.440	33.940–342.230
Benthos average biomass, g m^{-2}	50.295	117.919
Including worms, g m^{-2} (%)	29.849 (59.3)	24.813 (21.1)
• Molluscs, g m^{-2} (%)	10.606 (21.1)	92.600 (78.5)
• Crustaceans, g m^{-2} (%)	9.840 (19.6)	0.481 (0.4)
• Other, g m^{-2} (%)	–	0.025 (0.02)
Index of uniformity of trophic structure	0.27	0.58
Infauna species, taxa number	13	10
• Quantity, ind. m^{-2} (%)	3283.1 (94.4)	1242.2 (89.7)
• Biomass, g m^{-2} (%)	49.733 (98.8)	44.328 (37.6)
Epifauna species, taxa number	5	13
• Quantity, ind. m^{-2} (%)	193.6 (5.6)	142.0 (10.3)
• Biomass, g m^{-2} (%)	0.562 (1.2)	73.591 (62.4)
Vagile hydrobionts, taxa number	18	19
• Quantity, ind. m^{-2} (%)	3476.7 (100.0)	1358.1 (98.1)
• Biomass, g m^{-2} (%)	50.295 (100.0)	44.596 (37.8)
Sessile hydrobionts, taxa number	–	4
• Quantity, ind. m^{-2} (%)	–	26 (1.9)
• Biomass, g m^{-2} (%)	–	73.323 (62.2)
Invasive alien species, taxa number	3	3
• Quantity, ind. m^{-2} (%)	270.0 (7.8)	114.2 (8.2)
• Biomass, g m^{-2} (%)	0.168 (0.3)	12.258 (10.4)

structure of the bottom macrofauna was twofold lower in AC (0.27) than at reference stations (0.58).

The number of taxa of in- and epifauna (10 and 13) was nearly the same at reference stations; while in AC the number of infauna taxa (13) was much higher than number of epifauna taxa (5). In terms of quantity, infauna representatives were dominating in AC (94.4%) and at reference stations (89.7%). However, with regard to biomass, organisms of infauna were dominating in AC (98.9%) and organisms of epifauna at reference stations (62.4%).

In AC, the bottom macrofauna was exclusively represented by vagile taxa. At reference stations, vagile hydrobionts also dominated in terms of taxa number (19 taxa) and quantity (98.1%), but sessile hydrobionts accounted for 62.2% of biomass, with the mussel *Mytilus galloprovincialis* comprising 72.457 g m^{-2} or 61.4% of the total benthos biomass.

Three invasive alien species (*Polydora cornuta*, *Doridella obscura* and *Mya arenaria*) were registered in AC. The polychaete *P. cornuta* was the most abundant with 261.7 ind. m^{-2}, compared with 80.8 ind. m^{-2} at reference stations. The average density of invasive species at reference stations (114.2 ind. m^{-2}) was 2.4-fold lower than in ACs (270.0 ind. m^{-2}), but their percentage share in the average benthos quantity was practically the same (8.2 and 7.8%) in both zones.

The average biomass of invasive species at reference stations (12.258 g m^{-2}) was 73-fold higher than in AC (0.168 g m^{-2}). Such differences are caused, first of all, by the high average biomass of the bivalve *M. arenaria* whose biomass comprised 11.685 g m^{-2} at reference stations and only 0.013 g m^{-2} in AC (Table 1.2).

Connecting shipping channels and channels in the river mouths lead to a reorganization in the aquatic ecosystems and to changes in the qualitative composition and quantitative indices of the biological components [20].

During 2004–2013, macrozoobenthos was collected during the same sampling campaigns in the shipping channel that connects the arm Bystry of the Danube with the Black Sea and in the adjacent sea zone. In the channel, the samples were taken at three stations: № 7, located directly at the border of river and sea, and № 7–1 and № 7–2 situated seaward of № 7 at a distance of 1 and 2 km, respectively.

On the channel bottom, at depths of 3.0–8.7 m (average 6.0 m), are deposited black silts which are sometimes semi-fluid and sometimes with an admixture of small-sized sand and detritus. These bottom sediments are absent at similar depths in the sea zone adjacent to the channel and depth was the main criterion for the selection of reference stations. At reference stations, at a depth of 3.8–9.0 m (average 7.6 m), the bottom sediments were more varied: sands, sometimes with admixture of silt and silts (sometimes with sand admixture).

The composition and quantitative indices of bottom macrofauna of the Shipping way (Connecting channel) Danube–Black Sea are under the influence of river run-off. The presence of fresh water was periodically registered at all three stations in the channel track, both at the surface and in the near-bottom water layer. Thus, the whole water column was fresh-watered at station № 7 (border of the Bystry arm and sea) in September 2005, August 2008 and August 2010; at station № 7–1 in September 2005 and in August 2010 and at station № 7–2 in September 2005.

During other surveys at the same stations in 2004–2013, the water was fresh or salty at the surface and salty in the near-bottom layer. The salinity gradient between the near-surface and the near-bottom layers was sometimes very steep; e.g. 0.3 and 15.29‰ in October 2011 at the station № 7; freshwater and 12.89‰ in October 2008 at the station № 7–1; 1.3 and 11.67‰ in June 2011 at the station № 7–2. Observations have shown that "tongues" of salty (sea) water can penetrate far into the Bystry arm at the channel bottom. During the period 2004–2013, the average salinity of the near-bottom layer was 5.69‰ at the station № 7, 4.69‰ at the station № 7–1 and 9.95‰ at the station № 7–2. The highest indices of the bottom macrofauna in the channel were registered during the periods of highest salinity of the near-bottom water layer.

A total of 29 taxa were found in the composition of the macrozoobenthos of the Shipping way: worms—9, molluscs—6, crustaceans—12 and other groups—2. At reference stations, 28 taxa were registered: worms—9, molluscs—10, and crustaceans—9 (Table 1.4).

Table 1.4 Comparative characteristics of composition and quantitative indices (*N*—average density, ind. m^{-2}; *B*—average biomass, g m^{-2}) of macrozoobenthos of the Danube–Black Sea Connecting channel and at reference stations (2004–2013)

Taxon	Connecting channel		Reference stations	
	N	*B*	*N*	*B*
Vermes				
Harmothoe imbricata (L.)	–	–	0.6	0.004
Neanthes succinea (Frey et Leuch)	161.6	4.343	255.9	7.017
Hediste diversicolor O.F. Muller	0.4	0.010	–	–
Nephtys hombergii Savigny	–	–	0.3	0.004
Spio filicornis (Muller)	80.0	0.128	23.9	0.053
Polydora cornuta Bosc	201.0	0.227	33.3	0.048
Prionospio cirrifera Wiren	1.4	0.001	0.6	0.001
Streblospio gynobranchiata Rice & Levin	1.4	0.001	–	–
Heteromastus filiformis (Claparede)	1.2	0.001	492.9	1.634
Capitella capitata (Fabricius)	47.6	0.072	3.5	0.005
Oligochaeta g. sp.	1.4	0.003	0.6	<0.001
Mollusca				
Mohrensternia lineolata (Michaud)	–	–	0.9	0.002
Hydrobia acuta (Draparnaud)	1.4	0.002	23.2	0.049
Bittium reticulatum (Costa)	–	–	0.3	0.002
Retusa truncatula (Lockard)	–	–	2.1	0.011
Anadara inaequivalvis (Bruguiere)	18.6	1.345	102.2	3.028
Mytilus galloprovincialis Lamarck	–	–	1.5	0.759
Cerastoderma glaucum Poiret	155.5	2.854	299.7	18.248
Spisula subtruncata (Costa)	–	–	3.5	0.440

(continued)

Table 1.4 (continued)

Taxon	Connecting channel		Reference stations	
	N	*B*	*N*	*B*
Mya arenaria L.	389.0	20.617	448.3	12.554
Lentidium mediterraneum (Costa)	194.5	1.212	1645.4	11.105
Sphaerium sp.	5.9	2.767	–	–
Crustacea				
Balanus improvisus Darwin	11.4	0.057	34.7	0.509
Crangon crangon (L.)	0.3	0.001	–	–
Macropipus holsatus Fabricius	–	–	0.6	0.053
Rhithropanopeus harrisi tridentata (Maitland)	0.7	0.018	–	–
Limnomysis benedeni Czerniavsky	–	–	0.3	0.002
Mesopodopsis slabberi (Van Beneden)	1.0	0.003	0.3	0.001
Pterocuma pectinata (Sowinskyi)	0.7	0.001	–	–
Iphinoe maeotica (Sowinskyi)	–	–	0.6	0.001
Jaera sarsi Valkanov	0.2	<0.001	–	–
Ampelisca diadema A. Costa	7.6	0.026	348.6	1.092
Perioculoides longimanus (Bate et Westwood)	–	–	0.3	<0.001
Marinogammarus olivii M.-Edwards	1.5	0.005	–	–
Cardiophilus baeri G.O. Sars	1.7	0.009	–	–
Pontogammarus maeoticus (Sowinskyi)	8.4	0.056	–	–
Pontogammarus sp.	2.1	0.022	–	–
Microdeutopus gryllotalpa A. Costa	–	–	0.3	0.001
Corophium volutator (Pallas)	51.4	0.127	0.3	0.001
Insecta, larvae				
Trichoptera g. sp.	1.0	0.006	–	–
Chironomus sp.	1.4	0.003	–	–
Total	1350.3	33.918	3724.6	56.623

The taxa similarity coefficient (according to Jaccard–Alekhin) between the Shipping way and the reference stations was only 39.0%, including 63.6% for taxonomic group of worms, 45.5% for molluscs and 23.5% for crustaceans. Three not abundant freshwater taxa (the bivalve *Sphaerium* sp., larvae of insects *Trichoptera* g. sp. and *Chironomus* sp.) all apparently carried into the sea by river run-off) and brackish water complexes (the crustaceans *Pterocuma pectinata, Pontogammarus maeoticus* and *Pontogammarus* sp.) were found in macrozoobenthos composition of the channel; the total quantity of their biomass comprising 19.5 ind. m^{-2} and 2.855 g m^{-2} biomass. Despite of the big riverine influence, the bulk of quantity (98.6%) and biomass (91.6%) of the bottom macrofauna of the channel was formed by representatives of euryhaline marine complex. At reference stations, macrozoobenthos was represented by euryhaline marine complex with the exception of not sparse specimen (0.3 ind. m^{-2}, 0.002 g m^{-2}) of the crustacean *Limnomysis benedeni.*

The variability of taxa number, quantity and biomass of the macrozoobenthos at one station in the Connecting channel and at the reference stations were high (Table 1.5), but the macrozoobenthos average indices were higher at the reference stations.

Thus, at reference stations, the average taxa number per sample (6.0) was 1.8-fold, the average quantity (3724.6 ind. m^{-2}) 2.8-fold and the average biomass (56.623 g m^{-2}) 1.7-fold higher than in the channel (3.4; 1350.3 ind. m^{-2} and 33.918 g m^{-2}, respectively).

In both zones, molluscs, among which the most numerous were bivalve *M. arenaria*, *Lentidium mediterraneum* and *Cerastoderma glaucum* (Table 1.6), were prevailing among the main taxonomic groups in terms of quantity (56.7% in the channel, 67.8% at reference stations) and biomass (84.9 and 81.6%, respectively). Despite significant number of crustacean taxa, in both zones, a significant number of crustacean taxa were present, but their overall share in the general indexes of benthos was insignificant but, compared with other taxonomic groups (Table 1.5).

With the exception of rare bivalves with shell length of more than 20 mm (*Anadara inaequivalvis* and *M. arenaria*), all invertebrates of the channel and adjacent bottom areas belong to forage component for fish [14]. The average biomass of the forage benthos was 1.6-fold higher at reference stations (52.837 g m^{-2}) than in the Connecting channel (33.332 g m^{-2}), but the same index expressed in percentage terms was slightly higher in the channel than at the reference stations (98.3 and 93.3%, respectively).

Considering the main trophic groups, detritophages were dominating in the Shipping way and at reference stations in terms of taxa number (17 and 13, respectively); seston feeders (57.4 and 68.1%) and detritophages (41.5 and 31.2%) in terms of quantity; and sestonophages (85.1 and 82.4%) in terms of biomass. The index of uniformity of trophic structure of bottom macrofauna was nearly the same in the channel (0.69) and at the reference stations (0.61).

Infaunal organisms dominated in the channel and at the reference stations in terms of taxa number (23 in channel, 20 at reference stations), quantity (98.8 and 98.3%) and biomass (99.7 and 97.7%) (Table 1.5). Organisms representatives of the vagile complex prevailed both in the channel and at reference stations with regard to taxa number (28 in channel, 26 at reference stations), quantity (99.2 and 99.0%, respectively) and biomass (99.8 and 97.7%, respectively) (Table 1.5).

A total of 6 invasive species from various taxonomic groups (*P. cornuta*, *S. gynobranchiata*, *A. inaequivalvis*, *M. arenaria*, *B. improvisus* and *Rh. harrisi tridentata*) was found at reference stations and in the Connecting channel, including all of them in the channel and four of them at the reference stations. The bivalve *M. arenaria* (389.0 ind. m^{-2} and 20.617 g m^{-2} in the channel and 448.3 ind. m^{-2} and 12.554 g m^{-2} at reference stations) was the most numerous in both cases. The total quantity (622.1 and 618.5 ind. m^{-2}) and biomass (22.265 and 16.138 g m^{-2}) of invasive species were relatively close in the two zones, but their percentage share in the total indices of benthos were 2.8- and 2.3-fold higher in channel than at reference stations (Table 1.5).

Table 1.5 Comparative characteristics of quantitative indices of macrozoobenthos of the Danube–Black Sea Connecting channel and at reference stations (2004–2013)

Index	Connecting channel	Reference stations
Number of samples	29	34
Sampling depth, m	3.0–8.7	3.8–9.0
Average depth of sampling, m	6.0	7.6
Taxa number, total	29	28
Including worms	9	9
• Molluscs	6	10
• Crustaceans	12	9
• Other	2	–
Taxa number per sample	0–13	0–15
Average taxa number per sample	3.4	6.0
Benthos quantity per sample, ind. m^{-2}	0–17,090	0–43,100
Average benthos quantity, ind. m^{-2}	1350.3	3724.6
Including worms, ind. m^{-2} (%)	496 (36.7)	812 (21.8)
• Molluscs, ind. m^{-2} (%)	765 (56.7)	2527 (67.8)
• Crustaceans, ind. m^{-2} (%)	87 (6.4)	386 (10.4)
• Other, ind. m^{-2} (%)	2 (0.2)	–
Benthos biomass per sample, g m^{-2}	0–637.470	0–251.790
Average benthos biomass, g m^{-2}	33.918	56.623
Including worms, g m^{-2} (%)	4.786 (14.1)	8.765 (15.5)
• Molluscs, g m^{-2} (%)	28.797 (84.9)	46.198 (81.6)
• Crustaceans, g m^{-2} (%)	0.325 (1.0)	1.660 (2.9)
• Other, g m^{-2} (%)	0.009 (–)	–
Average biomass of forage component (for fish), g m^{-2} (%)	33.332 (98.3%)	52.837 (93.3%)
Index of uniformity of trophic structure	0.69	0.61
Infauna species, taxa number	23	20
• Quantity, ind. m^{-2} (%)	1334 (98.8)	3663 (98.3)
• Biomass, g m^{-2} (%)	33.830 (99.7)	55.297 (97.7)
Epifauna species, taxa number	6	8
• Quantity, ind. m^{-2} (%)	16 (1.2)	62 (1.7)
• Biomass, g m^{-2} (%)	0.088 (0.3)	1.326 (2.3)
Vagile hydrobionts, taxa number	28	26
• Quantity, ind. m^{-2} (%)	1339 (99.2)	3689 (99.0)
• Biomass, g m^{-2} (%)	33.861 (99.8)	55.355 (97.8)
Sessile hydrobionts, taxa number	1	2
• Quantity, ind. m^{-2} (%)	11 (0.8)	36 (1.0)
• Biomass, g m^{-2} (%)	0.057 (0.2)	1.268 (2.2)
Invasive alien species, taxa number	6	4
• Quantity, ind. m^{-2} (%)	622.1 (46.1)	618.5 (16.6)
• Biomass, g m^{-2} (%)	22.265 (65.6)	16.138 (28.5)

Table 1.6 Comparative characteristics of composition and quantitative indices (*N*—average quantity, ind. m^{-2}; *B*—average biomass, g m^{-2}) of macrozoobenthos at the stations in the Danube–Black Sea Connecting channel (2004–2013)

Taxon	Station № 7		Station № 7–1		Station № 7–2	
	N	*B*	*N*	*B*	*N*	*B*
Vermes						
Neanthes succinea (Frey et Leuch)	151.4	5.048	241.4	6.487	109.6	1.235
Hediste diversicolor O.F. Muller	–	–	–	–	1.6	0.035
Spio filicornis (Muller)	–	–	72.9	0.087	226.3	0.386
Polydora cornuta Bosc	–	–	1.4	0.001	727.5	0.821
Prionospio cirrifera Wiren	2.9	0.003	–	–	–	–
Streblospio gynobranchiata Rice & Levin	2.1	0.002	1.4	0.001	–	–
Heteromastus filiformis (Claparede)	1.8	0.002	–	–	1.3	0.001
Capitella capitata (Fabricius)	–	–	–	–	172.5	0.260
Oligochaeta g. sp.	0.2	0.001	3.9	0.007	1.3	0.001
Mollusca						
Hydrobia acuta (Draparnaud)	–	–	–	–	5.0	0.008
Anadara inaequivalvis (Bruguiere)	38.6	2.786	–	–	–	–
Cerastoderma glaucum Poiret	0.7	0.126	2.9	1.486	560.0	8.828
Mya arenaria L.	12.1	1.364	1.4	0.200	1387.5	72.175
Lentidium mediterraneum (Costa)	–	–	728.6	4.700	67.5	0.283
Sphaerium sp.	2.9	0.021	17.1	11.143	1.3	0.243
Crustacea						
Balanus improvisus Darwin	–	–	8.6	0.157	33.8	0.069
Crangon crangon (L.)	–	–	–	–	1.3	0.003
Rhithropanopeus harrisi tridentata (Maitland)	–	–	1.4	0.069	1.3	0.005
Mesopodopsis slabberi (Van Beneden)	0.7	0.004	–	–	2.5	0.005
Pterocuma pectinata (Sowinskyi)	1.4	0.001	–	–	–	–
Jaera sarsi Valkanov	0.5	0.001	–	–	–	–
Ampelisca diadema A. Costa	3.6	0.011	4.3	0.009	17.5	0.068
Marinogammarus olivii M.-Edwards	3.1	0.011	–	–	–	–
Cardiophilus baeri G.O. Sars	–	–	–	–	6.3	0.033
Pontogammarus maeoticus (Sowinskyi)	3.8	0.023	22.9	0.157	3.8	0.025
Pontogammarus sp.	4.0	0.045	0.4	0.003	–	–
Corophium volutator (Pallas)	6.4	0.014	138.6	0.404	53.8	0.081
Insecta, larvae						
Trichoptera g. sp.	2.1	0.011	–	–	–	–
Chironomus sp.	–	–	–	–	5.0	0.013
Total	238.3	9.474	1247.2	24.911	3386.7	84.578

As shown by the results of 2004–2013, the average salinity of the near-bottom water layer increased from 5.69 to 9.95‰ moving seawards from the mouth of Bystry arm over a distance of 2 km (i.e. from station № 7 to № 7–2). The total taxa number showed little changes (18 and 21, respectively) and the average taxa number per station increased 2.5-fold, i.e. from 2.2 to 5.5 (Table 1.6). At the same time, the average density of organisms increased 14.2-fold (from 238.3 to 3386.7 ind. m^{-2}) and the average biomass 8.9-fold (from 9.474 to 84.578 g m^{-2}) (Table 1.7).

At all the three stations located in shipping way, the bulk of macrozoobenthos quantity (85.8–96.6%) and biomass (96.8–99.6%) was formed by worms and

Table 1.7 Comparative characteristics of the quantitative indices of macrozoobenthos at different areas of the Danube–Black Sea Connecting channel (2004–2013)

Index	Station		
	№ 7	№ 7–1	№ 7–2
Number of samples	14	7	8
Average sampling depth, m	7.0	5.1	5.1
Average salinity of near-bottom water layer, ‰	5.69	4.69	9.95
Taxa number, total	18	15	21
Including worms	5	5	7
• Molluscs	4	4	5
• Crustaceans	8	6	8
• Other	1	–	1
Average taxa number per sample	2.2	3.4	5.5
Average quantity of benthos, ind. m^{-2}	238.4	1247.1	3386.3
Including worms, ind. m^{-2} (%)	158.4 (66.4)	321.0 (25.7)	1240.1 (36.6)
• Molluscs, ind. m^{-2} (%)	54.3 (22.8)	750.0 (60.1)	2021.3 (60.0)
• Crustaceans, ind. m^{-2} (%)	23.5	176.2	120.3
• Other, ind. m^{-2} (%)	2.1	–	5.0
Average biomass of benthos, g m^{-2}	9.475	24.911	84.575
Including worms, g m^{-2} (%)	5.056 (53.4)	6.583 (26.4)	2.739 (3.2)
• Molluscs, g m^{-2} (%)	4.297 (45.4)	17.529 (70.4)	81.537 (96.4)
• Crustaceans, g m^{-2} (%)	0.109	0.799	0.289
• Other, g m^{-2} (%)	0.012	–	–
Average biomass of forage (for fish) component, g m^{-2} (%)	8.261 (87.2)	24.911 (100.0)	84.578 (100.0)
Number of main trophic groups	3	4	6
• Detritophages taxa	11	7	12
Index of uniformity of trophic structure	0.24	0.44	0.92
Invasive alien species, taxa number	3	5	4
• Quantity, ind. m^{-2} (%)	52.8 (22.1)	14.2 (1.1)	2150.1 (63.5)
• Biomass, g m^{-2} (%)	4.152 (43.8)	0.428 (1.7)	73.070 (86.4)

molluscs. However, the relative contribution of the two taxonomic groups to the total indices of macrozoobenthos was different. Thus, at the station № 7 (mouth of the Bystry arm–sea), worms dominated in terms of quantity (66.4%), although the biomass of worms (53.4%) and molluscs (45.4%) was rather similar. At stations № 7–1 and 7–2, situated at distances of 1 and 2 km from the river mouth, molluscs were dominating in terms of quantity (60.1–60.0%) and biomass (70.4–96.4%).

The polychaete *N. succinea* (64.6% of quantity and 53.3% of biomass) and the invasive bivalve species *M. arenaria* (41.0 and 85.3%, respectively) were the mass species at stations № 7 and № 7–2, respectively. At the intermediate station № 7–1, the bulk of quantity and biomass were formed by two different species of bivalves: the small-sized *L. mediterraneum* representing 58.4% of quantity and *Sphaerium sp.* constituting 44.7% of biomass. The fish forage benthos component comprised 8.261 g m^{-2} (or 87.2% of the total) at station № 7, i.e. at the mouth of Bystry arm, and 100% of the total at stations № 7–1 and 7–2.

Among main trophic groups, detritophages dominated at all stations in the channel with regard to the number of taxa. At station № 7, detritophages were dominant in terms of quantity and their biomass (45.4%) was rather close to that of the sestonophages (53.8%). At stations № 7–1 and 7–2, seston feeders dominated both in terms of quantity (60.8 and 60.5%, respectively) and biomass (71.0 and 96.5%). Compared with station № 7, the number of trophic groups of the bottom microfauna at the most seaward station № 7–2 increased twofold (from 3 up to 6) and the index of uniformity of trophic structure 3.8-fold (from 0.24 up to 0.92) (Table 1.7).

In the connecting channel, infaunal organisms dominated at all stations. Between 13 and 18 infaunal taxa were found at each station comprising 97.6–99.2% of quantity and 99.1–99.9% of biomass. Vagile organisms dominated the macrozoobenthos complex of the channel in terms of quantity (99.2%) and biomass (99.8%).

Sessile organisms were represented only by one invasive species; the barnacle *Balanus improvisus* (Table 1.6). However, this species was absent at station № 7 and, at stations № 7–1 and 7–2, its quantity (0.7 and 1.0%, respectively) and biomass (0.6 and 0.1%) were extremely low.

A total of 6 invasive alien species (3–5 species per station), comprising representatives of various taxonomic groups (*P. cornuta*, *S. gynobranchiata*, *A. inaequivalvis*, *M. arenaria*, *B. improvises* and *Rh.s harrisi tridentata*), was registered in the Connecting channel. The highest total indices of invasive alien species (2150.1 ind. m^{-2} and 73.070 g m^{-2}) were observed at station № 7–2; the most distant from the mouth. *M. arenaria* was the dominant species (41.0% quantity, 85.3% biomass).

It is possible to conclude that specific ecological conditions, distinct from those in the adjacent zones of the bottom, form in AC to MP and in the connecting channels. In channels, conditions of hypoxia, anoxia and more or less prolonged mass mortality of bottom fauna can periodically originate. Detritophagous vagile infauna dominates the macrozoobenthos composition in the channels leading to MPs.

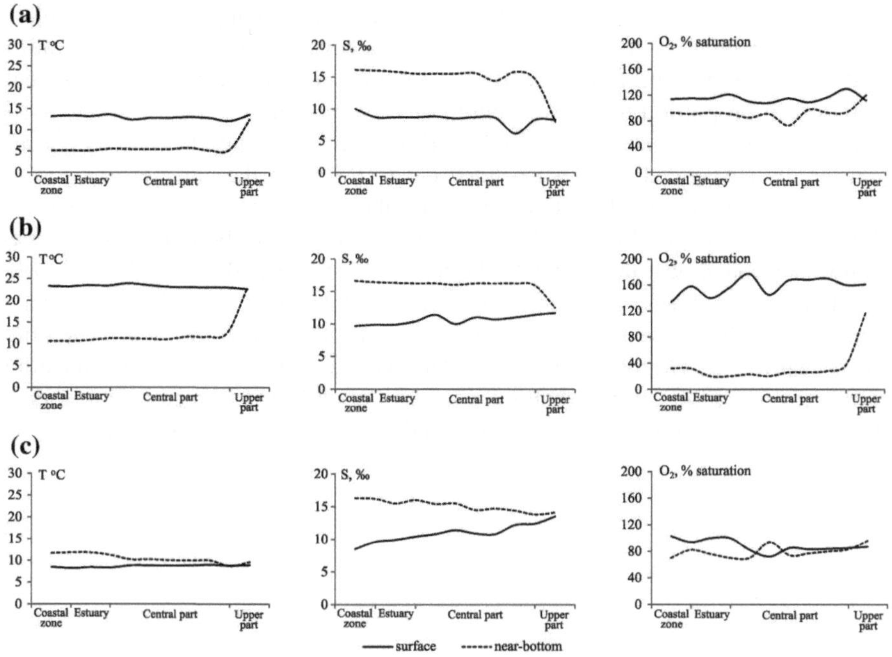

Fig. 1.10 Space distribution of temperature, salinity and water saturation by oxygen (%) at section coastal zone—estuary—central part—upper part of Grigorievsky Liman in April (**a**), August (**b**) and November (**c**) 2005

As a result of AC construction, detritivores substitute sestonophages as the predominant bottom population in most of the area occupied by the channels.

Many MPs are located in naturally narrow bays and basins of estuarine type in which three zones are distinguished according to depths: 1—the deepest water zone near the passage; 2—the middle zone with intermediate depths; 3—the apical zone with the lowest depths. The hollows of the shipping channels built through all the three zones of the basin have the maximal incut into the bottom in the apex part of the basin. In particular, unfavourable conditions for hydrobionts most often occur in this part of the shipping channel. Studies performed in 2005 in Grigorievsky Liman (Yuzhny MP aquatory) showed that temperature, salinity and oxygen conditions leading to mass mortality could originate in spring, summer and autumn (Fig. 1.10). Analogous situations are created in the port aquatories artificially protected by piers and breakwaters in which are built the shipping channels.

Particularly special conditions for hydrobionts life are formed in the shallow-water zones (0.1–1.0 m) adjacent to the crest of shipping channels in both the water column and at the bottom. These zones are subject to more or less regular influence of ships' waves caused by the ships entering and exiting the MPs aquatories.

In the shallow waters adjacent to shipping channels, shipping waves regularly stir up sediments from the depth up to 1 m. As a consequence, accumulation of the well-washed sand takes place and the amount of fine fractions in sediments increases in proportion to the weakening of wave action.

Ships' waves intensify the overall hydrodynamics in the shallow waters and transfer the suspended and resuspended dead OM from the channel towards the shore. Accumulation of dead OM occurs in places of weakened wave action and along the shore edge. As a result, the shallow-watered and bottom zones covered with washed sand near the channels alternate with stripes and zones of bottom near the shore covered with silty-sand sediment enriched with OM. The latter zones usually have the densest settlements of *M. arenaria*, *Abra ovata* and some polychaete worms. In Grigorievsky Liman, the average biomass of macrozoobenthos comprises 6–7 g m^{-2} in the shallow-water zones subjected to regular influence of ships' waves and about 250 g m^{-2} in the adjacent areas with similar depths.

Chapter 2
Pelagic Zone Subsystem of the Marine Ports Aquatories

In MPs aquatories ecosystems, the water mass is not only an environment in which the transport ways (waterways) for ships are built but also the biggest (by volume) and most dynamic biotope. The water column connects the main subsystems of ports ecosystems (pelagial, benthal and periphytal zones) and creates similar living conditions for the ecosystems' biotic components of MPs. The water volume in the Black Sea ports reaches dozens of million cubic metres and the thickness of the water layer can be 20 m and more.

The peculiarities of MPs ecosystems and their role in the entire marine ecosystem become evident when comparing them with the ecosystems of the adjacent sea zones in both present and past times.

Dredging and construction of hydrotechnical structures increase the depth, water volume and area of artificial hard substrata (HS) in MP, while the area of surface water and bottom sediments remains the same or even reduces. In some ports (Varna-West, Ilyichevsk and Yuzhny), construction of wide and deep permanent access channels, which connect the ports aquatories with the sea, has not only stabilized their hydrological and hydrochemical parameters, but also synchronized all main ecological processes with those of the adjacent sea zones. Major morphological peculiarities of the main Ukrainian Black Sea ports aquatories (Odessa, Ilyichevsk and Yuzhny), as well as Sevastopol Bay, are presented in Table 2.1. Figure 2.1 shows the schemes of Odessa, Ilyichevsk and Yuzhny MPs aquatories.

The ecosystems of all MPs of the Black–Azov Sea basin function in conditions of reduced water dynamics because of the natural or mixed (i.e. natural and artificial) protection of their aquatories. The distribution of organisms' complexes in all subsystems of MPs ecosystems is regulated, first of all, by light intensity, food quantity, hydrodynamics and oxygen content. Except for eutrophication, anthropogenic pollution of MPs aquatories usually has an episodic character and its own specificities in each MP. The food production is mainly determined by illumination. Oxygen content in water can be increased by: 1—adsorption from the atmosphere; 2—production by aquatic vegetation via photosynthesis; 3—input with rain.

A.K. Vinogradov et al., *Ecology of Marine Ports of the Black and Azov Sea Basin*, DOI 10.1007/978-3-319-63062-5_2

Table 2.1 Major morphological characteristics of the main Black Sea ports aquatories of continental Ukraine and Crimea

Morphometric characteristics	Port			
	Odessa	Ilyichevsk	Yuzhny	Sevastopol Bay
Aquatory peculiarities and connection with sea	Divided into harbours with independent ACs	Single, with single AC	Single, with single AC	Single, with single AC
Aquatory length, km	3.0	7.2	7.3	7.0
Aquatory width, km	0.8–1.2	0.2–1.3	0.8–1.2	0.8–1.2
Average depth, m	9.5	7.5	9.0	12.0
Maximal depth, m	16.5	14.0	19.0	20.0
Aquatory area, km^2	2.7	5.7	5.8	7.7
Water volume, mln. m^3	26.0	45.0	50.0	90.0
Length of berths and breakwaters, km	18.0	8.0	7.0	16.0
Sectional area of the passage on aquatory, m^2	11,300	2500	3000	11,000
Areas of underwater surfaces of hydrotechnical structures, m^2	170,000	80,000	90,000	190,000
Ratio of water volume to area of underwater surfaces of hydrotechnical structures, unitless	153	525	535	473

The presence of various equipments and constructions, ships moored at berths of MPs and HTS contributes to the formation of wave, wind and light shadows. This phenomenon directly or indirectly influences the oxygen enrichment of MPs waters via effects on: adsorption from the atmosphere, photosynthesis and equilibration of its concentration in the water column. In MPs, the amount of solar energy per one m^3 of water inevitably reduces during dredging.

Forces providing the natural hydrodynamics in MPs and forming currents (water motions) in the seas are usually divided into external and internal ones. The external forces are: 1—wind (wind-induced and drifts currents); 2—changes in the distribution of atmospheric pressure (baric currents); 3—changes in the relative position of Earth, Sun and Moon (tides); 4—influx of coastal and river waters and pile up of seawater, causing changes of the sea level at different shores and of the subsequent drainage flows.

In the Black–Azov Sea basin, the tidal phenomena are weak and do not usually exceed 10–15 cm [16, 108]. The internal forces causing flows are connected with the changes of physical state of water masses and, most of all, with changes of their temperature and salinity (gradient currents). In MPs aquatories, flows are created by moving ships (ship waves).

Flows of exclusively natural origin do not exist in seas. At detailed study, the strongest and most pronounced flows turn out to be complex ones [192].

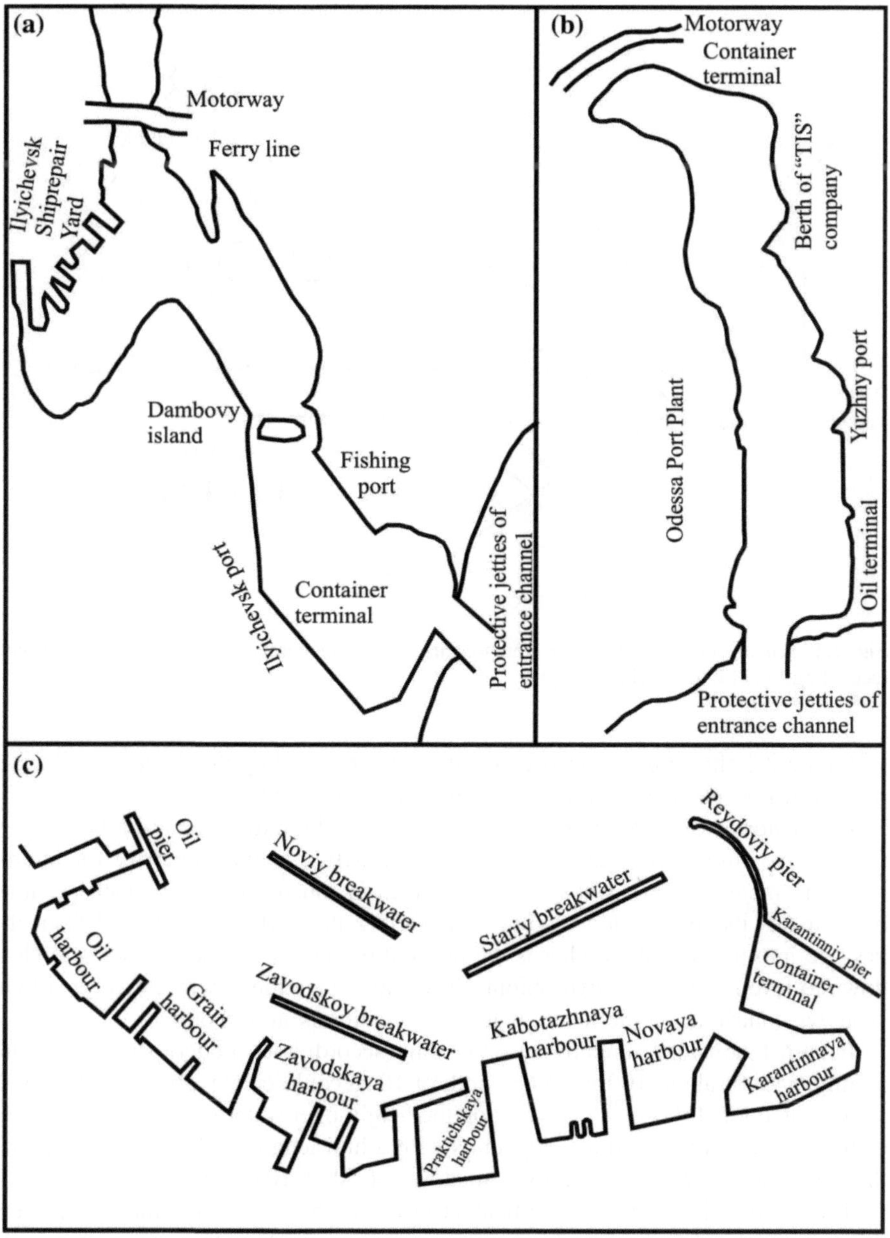

Fig. 2.1 Schematic plan of the main Ukrainian MPs aquatories: *A* Ilyichevsk, *B* Yuzhny, *C* Odessa

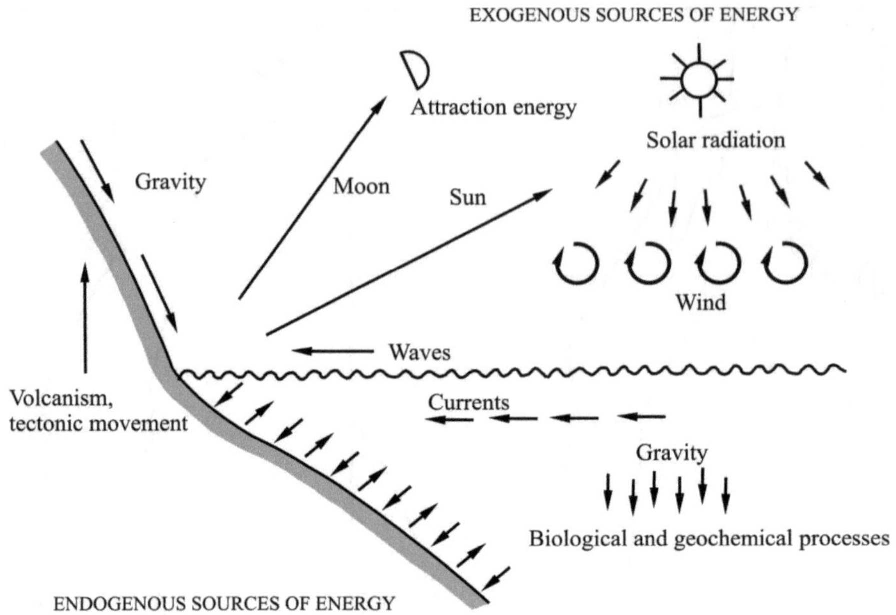

Fig. 2.2 Scheme of main energetic components of marine coastal shallow-water geosystems (according to [114, 133] with changes)

Figure 2.2 shows the main sources of energy controlling the dynamics of water masses in the sea coastal zones, including MPs aquatories.

In aquatories of MPs divided by moles and berths into separate harbours (e.g. Odessa, Novorossiysk, Tuapse and Constanza MPs), habitat diversity for hydrobionts is increased, but water hydrodynamics is much worsened. Each separate harbour has its special flow regime. The farther the harbour is situated from the passage into the MP, the higher is the possibility that its hydrodynamics will slow down to levels critical for hydrobionts. The above-mentioned wind, wave and light shadows contribute to the heterogeneity of conditions in harbours.

In an MP aquatory, a relatively small ship (according to modern standards) with a hull length of about 100 m and a width of 10 m could create a light shadow with an area of 1000–1500 m^2 depending on the height over water of the ship's hull and of the different Sun positions. Big ships with a hull length of 270–300 m entering the Black Sea ports create the light shadows of 5000–10,000 m^2.

The illumination of berths' vertical surfaces and of the water column near them depends on the duration of ships mooring at these berths. The more intensively berths are used, the less favourable are illumination and conditions for light-requiring micro- and macrophytes.

The light intensity and hydrodynamics are reduced on the inner side of HTS protecting MPs aquatories.

It was determined that, for a given light intensity, each macrophyte species grows better in a defined and rather narrow interval of water velocity. At lowering or increasing of water speed, biomass is immediately and significantly reduced. The same gradient of water movement exists along the vertical profile and changes not only the biomass, but also the size and age structure of population [2]. Thus, in MPs aquatories, the species composition and biomass of macrophytes, which depend on illumination and water velocity, can differ from those in the adjacent areas of the sea [87]. At the same time, some weakening of the hydrodynamics turns out to be favourable for many planktonic algae, whose development in the open sea and on open shores is limited by storms.

As it was already mentioned, the moored ships, various mechanisms and constructions at the berths and moles in ports form shadow zones that influence light intensity and duration of illumination, thus affecting the speed of photosynthesis and algal growth. In MPs, the light shadows can significantly influence these processes both in the water column (phytoplankton) and at the bottom and HTS lateral surfaces (microphytobenthos, macrophytes). The macrophytes biomass on the smooth-walled, vertical hydrotechnical constructions quite often does not exceed 10–50% of the biomass of the same species on natural HS [72].

In ecosystems, the artificial lowering of the dynamics of water masses always leads to mass development of some phytoplankton species, as it is clearly observed in reservoirs on big rivers. The lowered hydrodynamics of MPs aquatories is characterized by the absence of big waves and strong currents that damage plankton algae, thus frequently allowing them to reach quantities and biomass that cause water "blooms". The coefficient of relative transmission of solar radiation depends on a number of factors, but water turbidity plays the major role. The mass development of phytoplankton during dredging works reduces amount of energy entering the water column. All other conditions being equal, the division rate of planktonic algal cells is directly proportional to light intensity. The increase of water turbidity in the upper horizon and in the zone of the density gradient (pycnocline), along with the presence of light shadows in MPs aquatories, can modify the depths having the optimal illumination intensity for micro- and macrophytes.

In MPs divided into single harbours, many berths are situated on moles forming various angles with respect to the shore and are quite often surrounded by water on three sides. In Odessa Gulf, the Stary, Novy and Zavodskoy breakwaters protect Odessa port from roughness and are washed from all sides. In general, breakwaters, moles, berths and docks of Odessa MP, as well as the corresponding HTS of other MPs of the Black–Azov Sea basin, can be considered as AR systems.

Constanza, Odessa and Yuzhny MPs have average depths up to 10 m and more and maximal depths up to 18–22 m. In the sea, such depths usually appear at 1–2.5 km distance from the shore. The deep depths are brought directly into the "land—sea" contact zone in spite of increasing gradually as it is usual in the NWBS. The presence of these depths forms preconditions for the development of thermo- and halocline vertical stratification of water masses. At the same time, big depths directly at the shore prevent excessive heating of the whole water column, which is observed in the adjacent sea areas, during the summer months. MPs aquatories in the Black–Azov Sea basin artificially create conditions typical for the edge of the rocky deepwater shores, but with reduced wave action. The ships movements in MPs aquatories favour water mixing, which plays a positive role during the prolonged periods of calm in summer. However, the deeply immersed ships can also cause remobilization of the bottom sediments.

In MPs, a constant exchange of heat, gases, humidity and chemical substances (airions) occurs between the water mass and the atmosphere through the film of surface tension. The oil film (during oil products spills) and other films of different origin disturb these processes.

The water column is an environment connected with the food resources of MPs ecosystems: seston and dead organic matter (OM) in the form of suspended (SOM) and dissolved (DOM) organic matter.

In MPs aquatories, the most unfavourable conditions for inhabitants of the pelagial zone are found in the near-bottom layer of the deepwater zones and at zones of some harbours, dippers and inlets most distant from the passages and shipping channels. In many MPs aquatories, the hydrochemical and microbiological indexes of the water column are influenced by river run-off and discharges of rain, snowmelt waters, raw wastewaters and insufficiently treated sewage.

Except for the access channels to aquatories, the pelagial zone of MPs is confined on all sides starting from the film of surface tension to the surface of bottom sediments. In MPs, the different layers of the water column move with various speeds relatively to the HTS walls and periphytal. In the whole water column thickness, the water layer adjacent to lateral walls of moles and breakwaters, having a width of about 0.5–1.0 m, represents a contact, or ecotone, zone in which typical plankton organisms, larvae and young individuals of vagile invertebrates from the composition of periphyton and larval, newly hatched and juvenile fish occur. The distribution of various organisms in the MPs aquatories and in the water column is determined by hydrodynamics and by the organisms' ability to move horizontally and vertically as well as by the fear factor. The hydrobionts' communities of hyponeuston, plankton and nekton are distinguished in MPs pelagial. Phytoplankton composition includes phyto-, bacterio-, zoo- and ichtyoplankton.

According to various indexes, the MPs pelagial zone is vertically divided into the following horizons: 1—hyponeustal (0–30 cm)—the layer, immediately below the surface tension film; 2—epipelagial—the layer between the hyponeustal and the upper border of the pycnocline; 3—pycnocline layer; 4—bathypelagial—the layer between the pycnocline and the hydrogen sulphide zone; 5—the zone of hypoxia or anoxic zone—the near-bottom water layer, interested by formation of hydrogen sulphide. Settling, SOM concentrates along the isopycnic surfaces and at the bottom.

The pelagial zone connects all MP subsystems into ecosystems by unifying various general factors, abiotic and biotic connections and relationships.

During the analysis of abiotic and biotic peculiarities of biotopes and ecosystems of MPs of the Black–Azov Sea basin, attention is focused on the main Ukrainian

MPs aquatories: Odessa, Ilyichevsk and Yuzhny, located in the NWBS. Information about ecosystems belonging to the Russian Federation, such as Novorossiysk, Tuapse and Sochi MPs, located in the NEBS, was also used as well as information about Sevastopol Bay (located in Crimea), which serves as a base for military fleet and an aquatory for marine commercial port.

Odessa, Ilyichevsk and Yuzhny ports belong to the biggest MPs of the Black–Azov Sea basin and are located on the NWBS coast, which has an extension of about 50 km. This sea region is strongly influenced by the Dnieper, Yuzhny Bug, Dniester and Danube rivers. Odessa MP was constructed on the open seashore in Odessa Gulf and is protected from the sea by artificial HTS. After their artificial junction with the sea via constantly operating shipping channels, Sukhoy Liman (the aquatory of Ilyichevsk MP) and Grigorievsky (Maly Adzhalyk) Liman (the aquatory of Yuzhny MP) have been transformed into marine bays, with mainly natural protection from wind and wave impact.

According to hydrological and hydrochemical parameters, Odessa, Ilyichevsk and Yuzhny MPs belong to the Odessa region of the NWBS, which is delimited by the near-estuary zone of Grigorievsky Liman to the north and by the near-estuary zone of Sukhoy Liman to the south [20, 154]. Typical average values of hydrochemical parameters in this zone are as follows: salinity—16.3‰, oxygen content ucode Type="General_Pun">—6.75 mg dm^{-3}, oxygen saturation—69.5%, pH—8.21, phosphates—25.3 μg dm^{-3}, organic phosphorus—26.8 μg dm^{-3}, ammonia nitrogen—85.2 μg dm^{-3}, nitrites—2.9 μg dm^{-3}, nitrates—15.2 μg dm^{-3}, organic nitrogen—460 μg dm^{-3} and silicon—810 μg dm^{-3} [154]. The temperature regime is very similar along the entire coast from Grigorievsky to Sukhoy Liman. Here, more or less favourable conditions for development of phyto- and zooplankton in the water column are observed during all seasons. Productive–destructive processes decline during the winter period because of the low temperatures, while periodical phenomena of hypoxia and anoxia in the near-bottom layer and at the bottom occur in the summer months [16, 162, 196].

Shallow shores with soft and loose sediments predominate in Odessa region. Because of the relatively shallow water of the region, the aquatories of Odessa, Ilyichevsk and Yuzhny MPs have shipping ACs with lengths of 2.5–3.5 km and freezing of the sea is possible in winter.

Novorossiysk, Tuapse and Sochi MPs located in the NEBS are built on the coasts of Krasnodar Kraiy (a region of Russian Federation). Novorossiysk MP is the northernmost among them and Sochi MP is the southernmost. The ports of the NEBS are located in climatic conditions milder than those in the NWBS, but some climatic differences are also present in this part of the Black Sea. In Novorossiysk region, air temperatures below 10 °C can occur in winter, while negative air temperatures are very rare in the region of Sochi MP. During the coldest months (January and February), the superficial waters of Novorossiysk Bay can cool down to 0 °C. The sea hydrological regime in Novorossiysk and Sochi regions is close to that of the open sea. The waves' fetch could exceed thousands of kilometres and the roughness is quite often strong, typical for an open sea. The river influence has local character. Because of the high hydrodynamic activity, salinity is less variable and remains around 18.0–18.5‰. The longshore current is directed northwestwards and, in the area Novorossiysk—Sochi, the longshore drift flow has mainly the south-eastern direction [152, 161]. The NEBS ports are located in a region with deepwater shores and rocky sediments. In the aquatories of these MPs, wind influence and waves are weakened by artificial protective HTS.

Sevastopol Bay is located in the south-western part of the Crimean peninsula and, like Sukhoy and Grigorievsky Limans, is a water basin of estuarine type. In Sevastopol region, the shores are deepwatered with rocky bottoms.

All studied MPs are characterized by significant fluctuations of water level (1.0–2.0 m) caused by the strong offshore–onshore winds.

The ports of Novorossiysk, Tuapse and Sochi, as well as Sevastopol Bay, are located in regions with deepwater shores and, as a consequence, do not need long ACs. Channels with length of about 0.5 km are built in Tuapse and Sochi MPs. The shores here are rocky and edged by rocky bars, so that, unlike in the NWBS, there is no deficiency of HS of natural origin. Besides rocky sediments, clayey, silty, silty-sand, sandy, shell and silty-shell sediments are typical for the aquatories of Novorossiysk, Tuapse, Sochi MPs and Sevastopol Bay. In Novorossiysk MP, liquid silt without macrophytes is registered near the moles and berths at depths exceeding 8–10 m [83]. The black silt is characteristic for Odessa, Ilyichevsk and Yuzhny MPs, Sevastopol Bay, Novorossiysk and Tuapse MPs and, to a lesser extent, Sochi MP.

Only one passage lead into the aquatories of the above-mentioned MPs, except in the case of Odessa port that has three passages. The passages are delimited and protected from the wave action by moles of various constructions. In Sukhoy and Grigorievsky Limans, there are jetties protecting the passage. In Sevastopol Bay and Novorossiysk MP, relatively short moles project towards each other from both shores. The aquatories of Odessa, Tuapse and Sochi MPs are protected by extensive HTS.

Because of the discharge from major rivers, the NWBS is the less saline part of the Black Sea, where the mixing of freshwater and typical Black Sea waters varies according to hydrological situations in the various areas. Location and individual peculiarities of MPs aquatories bring their specificity into this process. However, in the warm period, the upper water layer is usually formed by the less salty and warmer waters. All concerned MPs are characterized by a more or less marked two-layered structure connected with the temperature and salinity differences between water masses. The freshwater and brackish water complexes of hydrobionts play various roles in pelagic and bottom communities of Yuzhny, Odessa and Ilyichevsk MPs. However, the distribution of the freshwater hydrobionts has a local character and representatives of marine Black Sea complex prevail during all seasons. The inflowing rivers have various influence on the aquatories of Sevastopol Bay, Novorossiysk, Tuapse, Ilyichevsk and Yuzhny MPs.

In MPs aquatories, anthropogenic activity artificially creates conditions which are typical for rocky deepwater shores, but with lowered wave action. The presence of depths of 8–20 m is a precondition for the formation of pycnocline and the characteristics of coastal shallow waters are partly or completely destroyed.

When designing new MPs and analyzing the environmental impact of existing ones, it is very important to collect information about the depth at which roughness stops its influence on bottom and ceases to stir up and displace not only sand, but also silty sediments. This depth is the so-called Cornaglia line or silt line [84, 104, 166]. Weakening of hydrodynamics in MPs aquatories leads to changes of Cornaglia line position, or to its disappearance.

Protection of MPs aquatories and OM accumulation in them frequently lower the water oxygen concentration to less than 2 mg dm^{-3}, which causes mass mortality phenomena [17, 40, 45].

Municipal sewage waters enter the aquatories of many MPs and, on average, carry faecal discharges of 75 g dry weight per person per day [17].

2.1 Main Abiotic Peculiarities

The MPs constructed on water bodies of estuarine type such as Varna and Beloslav Lakes, Sevastopol Bay, Sukhoy and Grigorievsky Limans have the largest aquatory areas in the Black–Azov Sea basin. The area of Constanza MP aquatory, located at sea coast, is 26 km^2. The following important biotopes can be distinguished in the aquatories of all MPs: 1—water column; 2—artificial HS; 3—coastal shallow waters sediments; 4—sediments of the deepwater zones (8–20 m). Ecosystems of MPs aquatories are formed from the subsystems: 1—pelagial; 2—periphytal; 3—benthal.

The large communities of hydrobionts (hyponeuston, plankton—epiplankton and batiplankton—and nekton) inhabit the pelagial subsystem from the surface to the bottom. The population of periphytal and benthal zones is represented by micro-, meio- and macrozoobenthos, and micro- and macrophytobenthos. Bacteria are present in all these three subsystems and larval stages of many benthic organisms can be found in the pelagial. Daily migrations of invertebrates take place from the periphytal and benthal into the pelagial zone.

The periphytal subsystem comprises a permanent component represented by HTS underwater surfaces and a temporary one to which belong the wetted surfaces of ships' hulls. In the benthal subsystem, significant differences are observed between the coastal shallow waters and bars and the deepwater areas with regard to sediments' feature and hydrochemical indices. Because hydrobionts that realize primary and secondary production, transformation and reduction of OM are present in each of the subsystems, matter can be cycled more or less autonomously in each subsystem.

Owing to gravity, dead OM and mineral particles of various origin settle from the pelagial subsystem into the benthal one. Plankton animals, which realize biosedimentation, also participate in the same process. Organisms inhabiting the periphytal zone satisfy their food requirements by filtering seston from the water column. They release metabolites, faeces and pseudofaeces into the water and transform the unused OM and ashes into sediments. In MPs, inhabitants of the pelagial, benthal and periphytal zones interact via numerous links.

The nekton representatives (fishes and dolphins) as well as the large bottom invertebrates that are capable to move from MPs ecosystems into the adjacent sea zones, unite them into larger scale systems and realize the transfer of matter and energy. These links are strengthened by shipping navigation. Aquatic and semi-aquatic birds realize similar functions.

In the ecosystems of MPs aquatories, abiotic and biotic processes are tightly connected and their division has a mainly operational character.

Hydrodynamics, i.e. the totality of horizontal and vertical movements of water masses, is an exceptionally important characteristic for the existence of biotic components in the whole ecosystems in MPs aquatories or in their separate parts.

As for any other fluid, two regimes are observed during the movement of seawater: laminar (or streamline flow) and turbulent. In laminar flow, the separate water layers move in parallel. On the contrary, the turbulent regime is more complex and comprises separate layers of laminar movement, transition regimes and totally turbulent regimes, i.e. swirling.

The transition from one regime to the other is important for organisms' movement in the flow, flow movement along attached or moving organisms and flow movement and interaction with substratum on which or in which organisms live.

As water flows around solid bodies and surfaces (or bodies move in water), a thin laminar film, often referred to as the boundary layer, forms on their surface. The thickness of this layer depends on roughness height and is usually equal to it. For organisms, the movement in laminar flow is energetically more advantageous than in turbulent flow. Organisms' movement in the water column is hampered by viscosity and facilitated by inertial forces that contribute to advancement. The shear force is always directed opposite to the direction of organisms and water flow.

Water mixing is mainly assured by turbulent motion, while laminar flow contributes to it only marginally. There are two types of turbulent mixing: frictional and convectional. The first one originates at the interface of two layers, which move at different speeds, when whirls caused by friction penetrate from one layer into the other. The intensity of friction mixing depends on the speed differences between separate layers and can occur in horizontal and vertical directions. The second one causes vertical flows and takes place as the result of changes in the specific water density connected with increase or decrease in temperature and salinity.

The transition from one water movement regime to another takes place at limiting, or critical, speeds specific for each layer.

The characteristics of viscous fluid flows can be described and compared using the Reynolds number (Re), which is dimensionless and defined as: $Re = pvl \cdot \mu^{-1}$, where p is the water density, v is the typical velocity of flow or organism, l is the specific linear dimension and μ is the water viscosity coefficient.

At low Reynolds numbers, pico-, nano- and microplankton organisms live in motion conditions which are close to laminar. The turbulent wind, commonly observed in seas, generates wind waves and flows on the sea surface, creates turbulence, forms a mixed (quasi-homogeneous) surface water layer and also heats or cools it off. In seas, the upper layer combines the paradoxical properties of highly developed turbulence and homogeneity from one side and complexly ordered movements from the other [101, 137].

In seawaters, one of the convection mechanisms is the formation of the so-called "salt fingers". It is caused by the contrasting influence of temperature and salinity on water density and by unequal speeds in turbulent heat conduction and salts' diffusion [36, 60].

Owing to various water movements, the subsystems of MPs ecosystems (pelagial, periphytal and benthal) are functioning as a single whole. At the same time, the pelagial subsystem of MPs is also under the constant influence of the benthal and periphytal subsystems.

2.1.1 Main Peculiarities of the Water Exchange

The processes of water exchange occur between water basins and inside a water basin following the replacement of one water mass by another between its separate zones. Water exchange controls the passive transfer of hydrobionts, the transportation of matter which is necessary for synthesis of primary OM, the mixing, dispersion and removal of metabolites, the transport of oxygen from the upper layers to the near-bottom ones and the diffusion of pollutants.

In MPs with narrow aquatories, deeply cut into the land and with a ratio of about 0.5 between the passage and the aquatory widths, two vertically separated layers with opposite water flows are usually formed in the inlet channel. In MPs with relatively wide passages, an annular longshore flow is most frequently observed. This flow enters MP via AC and exits it at different points. In the semiclosed aquatories of some harbours, proper circulations often originate because of the cumulative influence of wind and differences in temperature and salinity [80].

It is well known that the speed of water exchange in natural harbours and artificially protected aquatories of MPs is closely related with the ratio between the depth of the aquatory incut into land and the width of the passage or passages. Thus, for ratios between 1 and 5, the speeds of water exchange usually allow normal functioning of MPs ecosystems. At bigger ratios, the water exchange could be insufficient.

The fewer are the passages into a MP aquatory, the more independent is its water mass on its regime and vice versa. The proportion between the width and depth of AC and passages into MP is an extremely important characteristic of MPs ecosystems.

The ratio of incuts' depth into land to the passage width (i.e. the distance between natural delimiting capes or artificial protective moles) can be up to 10–20 in many MPs aquatories of the estuarine type such as Sukhoy and Grigorievsky Limans, Sevastopol, Kamishovaya, Uzkaya, Balaklava Bays and other. MP protecting structures located on the seashore (such as those found in Odessa, Yevpatoriya, Feodosia, Kerch, Berdyansk, Mariupol and other MPs) in fact model the shoreline of the natural aquatory into land. However, the ratio between the width of their aquatories to the passage width is, as a rule, significantly less than in MPs with aquatories of the estuarine type.

In the artificially protected aquatories of MPs, water exchange processes and generation of upwelling phenomena have regular character at offshore winds and manifest in the same way as in narrow, natural limans and bays deeply cutting into land [105].

The formation of thermo- and halocline (pycnocline) is a self-sustaining process, which interrupts only at significant changes of external conditions. Various studies show that, on average, 90–99% of the produced turbulent energy dissipates and only 1–10% is spent for destratification [101].

The thermocline stops developing and progressively disappears at weakening of the heat flow into the sea (winter period) and strengthening of wind to stormy values (autumn-winter period). Both factors change the energy budget of the mixed sea layer in favour of turbulence whose increase reduces the stratification energy accumulated in the thermocline (pycnocline).

In the most stratified upper part of the pycnocline, the vertical gradients of all parameters (temperature, salinity, flow average velocity, turbulent energy) are so steep that this layer is sometimes compared to the water boundary layer near hard surfaces [101].

In MPs, the processes of water exchange are willingly reduced. Thus, the construction of the Northern and Southern moles in the 1970s reduced the width of the passage into the Sevastopol Bay aquatory from about 1200 to 415 m and the water exchange intensity by 40–70%.

As in other ecosystems, chemical substances secreted by bacteria, protozoans, micro- and macrophytes, metazoans into MPs aquatories could favour the development of some species and depress other ones. When excreted into the water, these substances become factors of the external environment both for the organisms which secreted them and for other hydrobionts. Such chemical communicatory signals regulate intrapopulation, intraspecific and interspecific interrelations. Owing to the confined nature of MPs aquatories, dispersion of these substances is hampered and the biota representatives respond to them in a manner different from that of the adjacent ecosystems.

The internal water exchange depends on dynamic processes and includes horizontal and vertical exchange of water masses between parts of a water body. Water movement, or flow, is one of the main factors determining the behaviour of invertebrates and fish in MPs biotopes (pelagic, periphytal and benthal zones).

A laminar, boundary sublayer about 0.1 cm thick develops on the open surface of the natural basins. This layer originates because, close to the air–water interface, the speed of water mixing reduces from turbulent diffusion down to molecular diffusion. The laminar boundary sublayer is characterized by very high gradients of temperature and concentration. Usually, it coincides with, or is included within, the boundary cold film which is up to 1 cm thick. Both these layers are in turn included into the layer subject to surface microconvection, with a thickness of about 10 cm [1].

Even in the hydrodynamically active open waters of the North Atlantic, the content of labile dissolved organic matter (labile DOM) is nearly eightfold higher in the surface microlayer (about 0.1–1.0 cm thick) than in the underlying layers. This situation confirms that dissolved and suspended OM and living organisms accumulate at the "water—atmosphere" interface [96].

In the protected MPs aquatories, films formed by entrance of various substances of anthropogenic origin (especially oil hydrocarbons) into sea environment are quite often observed besides films of natural origin.

Formation of pollution surface films in MP's protected aquatories causes several serious consequences: 1—reduction of the surface tension; 2—reduction of evaporation; 3—intensification of foam formation; 4—changes in the characteristics of capillary waves; 5—reduction of water surface roughness; 6—reduction of wind tension; 7—disturbance of thermal balance; 8—increase of matter ejection into atmosphere; 9—impact on wave processes and drift flows.

Benthic invertebrates inhabiting in quickly flowing water layers, as well as pelagic ones, are sensitive to changes in water masses dynamics. For them, the flow speed is an ecological factor of primary importance, because it is connected with renewal of the oxygen reserve in the near-bottom layer. The higher the flow speed, the quicker and bigger the inflow of oxygen to bottom-dwelling organisms [107]. For hydrobionts, the intensity of water mixing in water basins sometimes has even more importance than a high level of water oxygen saturation. Cases of invertebrates' mass mortality during the flow's interruption are known in basins with sufficient oxygen saturation. Death occurred as a consequence of poisoning by the organisms' own metabolites [107].

Flow's speed at the sediment–water interface and character of bottom sediments in MPs aquatories play a key role in the distribution of bottom invertebrates in biotopes. Their influence could be direct and indirect. Organisms' quantity and biomass per unit of water volume, bottom area or artificial HTS area depend on flow velocity. It is considered that the highest productivity of invertebrates occurs at water flow velocities between 0.3 and 0.9 m s^{-1} [107]. Notably, the flow speeds in MPs aquatories are one- to twofold lower [16, 196]. The division of the MPs into separate harbours (such as in Odessa, Novorossiysk and Tuapse MPs) further weakens the internal water exchange.

The surface flows in the Black–Azov Sea basin are mainly induced by the wind pattern. They have the character of anticyclonic circulations and move in the Black and Azov Seas as a peripheral annular flow, which is directed counterclockwise with speeds up to 40 cm s^{-1}. Near the shore, the zone of coastal circulations extends up to 5 miles far from the coast itself. In this zone, the shoreline and shelf bottom configuration determine the direction and velocity of flows. The average flow speeds of the coastal circulations, ranging between 20 and 25 cm s^{-1}, have various impacts on MPs aquatories and on the shelf zones adjacent to them [161].

Wind also contributes to waves' formation on the water surface. Ripples, or ripple waves, appear at wind speeds up to 0.7 m s^{-1} and are 3–4 mm high and 40–50 mm long. Gravity waves start forming at stronger winds. The size of wind waves depends on wind speed, duration and fetch, but factors related to water bodies morphometry, coasts configuration, depth and bottom character are also important.

In the Azov Sea, superficial flows originate in the estuaries of Don, Kuban and other smaller rivers. In the Black Sea, the discharge current originating in estuaries of the big rivers (Danube, Dnieper-Bug and Dniester Limans) and in Kerch Strait contributes to the formation of superficial flows.

In the Black Sea, the following systems of flows are distinguished: Anatolian—in the region from the Bosphorus Strait to the traverse of Kolkhida lowland; Caucasian—from the traverse of Kolkhida lowland to Kerch Strait; Crimean—from Kerch Strait to Tendrovskaya spit; Rumelian—from Tendrovskaya spit to Bosphorus Strait. From Cape Kerempe (Turkey) one of the streams of the Anatolian current, going eastward from the Bosphorus, deviates northward towards Crimea. This branch was used by ancient seafarers.

In the coastal waters, wind-induced waves are the main source of energy ensuring water exchange. In MPs, the external sides of moles and breakwaters dampen the waves' strength and, inside MPs aquatories, wind impact on surface water masses is further reduced by different structures such as constructions, storehouses, moored ships, machinery and others. In MPs protected aquatories, the flows velocities are usually below 5 cm s^{-1} and rarely exceed 10 cm s^{-1} [154, 196].

In the NWBS, strong winds are observed all year round, but their frequency and duration depend on season; the strongest winds being registered from November until March. On average, 44–57 cases per year of intensive storm winds (more than 10 m s^{-1}) can be registered. Strong winds last, on average, 18–21 h, but storm duration could reach 5–6 days in some cases and up to 2 days in summer months [188]. Intense, sustained winds mix water masses and weaken the pycnocline. At prolonged calm weather, the pycnocline becomes more marked. Mixing of water masses is closely related with roughness. In the NWBS, roughness with wave height of 0.5 m have the maximal frequency (68% of the time), while storm waves with heights of 1.25 m occur for about 10% of the time [188]. In the region of Yuzhny MP, the storm winds are more frequent in the period October–March, with a frequency of 4–7 days per month.

Comparison of the abiotic components of MPs aquatories ecosystems showed that they have individual peculiarities as to the regime and velocities of flows, but also numerous common features connected with the specific character of their functioning and purpose. It was already said that the main common feature of MPs is the weakening of wind-induced waves. The wave height substantially depends on the strength and duration of wind action and its fetch [77, 191]. For example, possible wave fetch at coastal zones adjacent to Odessa MP could exceed 600 km and its striking energy could be about 5–6 t m^{-2}. Waves collapsing on the western and eastern coast of the Black Sea have the maximum fetch. Wave action is felt to the maximum extent in the region of Sochi MP on the eastern coast and in Varna-East MP on the western coast.

The aquatory of Odessa MP is protected from big waves' action by a system of moles and breakwaters. In the naturally protected aquatories of Ilyichevsk and Yuzhny MPs, the possible wave fetch is only some kilometres, i.e. two orders of magnitude lower than that of Odessa MP. Inside the aquatories of some harbours of Odessa MP, the possible wave fetch is equal to hundreds of metres and rarely exceeds 1 km. Wave action manifests to its full extent mainly on external surfaces of HTS protecting MPs.

The protective function of moles and breakwaters in MPs is of great importance for navigation and port activity. When correctly positioned, these structures damp up to 90% of the wave energy [153]. As a consequence, aquatories of MPs with 3–3.5-fold artificially increased depths and water volume (e.g. Odessa MP) receive about 10% of the wave energy which enters the shallow zones of the neighbouring shores. The naturally protected aquatories of Ilyichevsk and Yuzhny MPs slightly differ from Odessa MP with regard to this characteristic.

Because the influence of wind and roughness is repeatedly diminished in MPs, oxygen water saturation is not as rapid as in the adjacent sea zones. In Odessa MP, flows speed measured in various harbours at various horizons exceeded 8 cm s^{-1} only in about 10% of the cases [19].

The structure of flow fields in Grigorievsky Liman has a typical two-layered structure in all seasons and the liman water dynamics is characterized by temporary currents. Maximal flow speed does not exceed 27 cm s^{-1} at the surface and 14 cm s^{-1} at the bottom [148].

All harbours of Odessa MP are more or less freely connected with the waters of Odessa Gulf. The total sectional area of all passages into Odessa MP aquatory comprises about 11,300 m^2. Ilyichevsk and Yuzhny MPs are connected with the sea via one channel with cross-sectional areas of 2500 and 3000 m^2, respectively. The cross-sectional area of the passage into Sevastopol Bay is about 8000 m^2.

Offshore–onshore winds are frequently observed in Odessa Gulf and give intensive impulse to water renewal in Odessa MP [16]. The same phenomenon contributes to water exchange in Sukhoy and Grigorievsky Limans and in aquatories of other MPs of the Black–Azov Sea basin. Even at relatively weak winds of 4–7 m s^{-1}, the water exchange could take place during dozens of hours via the deepwater channels of Grigorievsky and Sukhoy Limans. The frequency of onshore–offshore phenomena in the region of Grigorievsky Liman is, on average, five times per season [196] and a similar situation is observed in the zone of Sukhoy Liman. Water exchange between MPs of Odessa region and the adjacent sea zones involves, first of all, the surface water masses (3–5 m) of their aquatories.

During offshore–onshore winds, the highest flows velocities in the surface horizon of MPs aquatories are observed in the access channels. These winds cause also fluctuations in the water level. Maximal fluctuations of the sea level in Odessa MP are mainly connected with nonperiodic changes determined by offshore and onshore winds. Offshore wind phenomena causing a decrease in the water level are more often observed because of the prevalence of north-western winds. These changes do not usually exceed 30–40 cm, but can be very rapid (20–25 cm h^{-1}) because of relatively shallow depths in this region. In Odessa MP, the highest decline of sea level caused by offshore wind was 175 cm, while the highest water setup at onshore winds was 100 cm [16].

In Yuzhny MP, the level fluctuations are connected not only with the offshore–onshore phenomena, but also with the run-off of the Dnieper and Yuzhny Bug rivers. During the warm period, fluctuations of water level do not exceed 0.5 m, while they reach 1 m in the cold period. The maximum variation of water level in the Black–Azov Sea basin is registered near the north-eastern coast of the Azov Sea, where fluctuations can reach ±3 m. In Taganrog Bay, onshore winds prop up river waters during the period of the Don River floods and a significant increase of the water level takes place. At offshore–onshore phenomena, the water level could vary by 1.5–2.0 m at different shores of the Black Sea.

In Odessa MP, seiches (oscillations of the water level) ranging from 2 to 45 cm are registered along with offshore–onshore phenomena. The long-term average seasonal differences in sea level for Odessa MP are 19 cm [16]. The seiches in the aquatories of Sukhoy and Grigorievsky Limans rarely exceed 30–40 cm and have smoother and more lasting character than in Odessa MP because of their morphological peculiarities and protection from winds. In MPs aquatories, the changes of water level have their impact, first of all, on the composition of fouling biocenosis in the near-surface water layer. Internal waves forming in the thermocline layer favour horizontal mixing.

Turbulent mixing of water masses, temperature, solar radiation and sedimentation play a determining role in maintaining the water quality and biological productivity in aquatic ecosystems [145]. The role of these factors in MPs is not less important, but here they are subjected to certain changes. The seawater surface cools constantly. Its first centimetre layer absorbs only 20% of the incoming radiation and, owing to evaporation and long-wave length radiation, gives away 100% of heat absorbed by the boundary layer. In the uppermost layer with the thickness of a couple of millimetres, the water temperature is lower than that of the underlying water by fractions of a degree and sometimes by degrees. This temperature gradient at the water surface creates density instability and microconvection which, in calm water conditions, usually extends down to the depth of 25–30 cm [137]. In MPs aquatories, conditions of calm water develop more often than in adjacent sea zones.

In MPs aquatories, and especially in the access channels to some of them, phenomena of rips and oscillations are observed. They are oscillating water motions which involve a significant part of the water column.

As mentioned earlier, during the summer period, the mixing of surface and deepwater masses in MPs takes place owing to thermal microconvection, which involves the surface layer down to 30 cm [1]. However, the sea warms up not only from above, because of the atmosphere and solar radiation, but also because of the thermal heat flow near the bottom. The deep-sea thermal heat of Earth originates convection with a pulsing character in the near-bottom water layer. A thermal film warms up near the bottom, develops and then tears away from it. In conditions typical of the majority of big MPs of the Black Sea, its thickness is about 1.8 cm and its lifetime about 53 min. Every year the water volume interested by convection mechanisms at the bottom approximately corresponds to a layer of 100 m [1]. Even in the Black Sea ports with maximal depths of 18–20 m, the whole water column could therefore, theoretically, pass through the mechanism of near-bottom convection

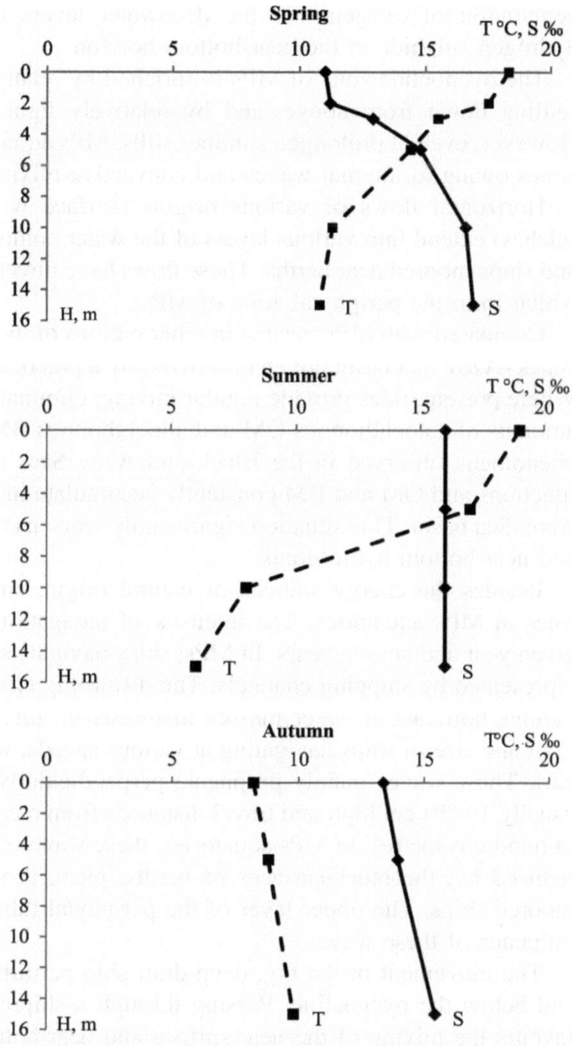

Fig. 2.3 Examples of pycnocline formation (*T* temperature, °C; *S* salinity, ‰) in Yuzhny MP (Grigorievsky Liman) during various seasons

in 2 months. On the flat area of the bottom, at a depth of 10–20 m in the Black Sea, the water velocity varies from 1 to 18 cm s^{-1} in the boundary layer with the thickness of 20 cm adjacent to the bottom [1]. In MPs aquatories, such range of average velocities is registered above the pycnocline layer during most of the year [16, 154, 196].

In Odessa, Ilyichevsk and Yuzhny MPs, vertical water homogeneity is usually registered in mid-winter, after the strong storms inducing wind mixing from the water surface to the depth of 20–25 m in all the NWBS.

In MPs, the vertical water exchange is hampered during most of the year owing to aquatories protection, artificial increase of depths and formation of the pycnocline, which breaks in winter months (Fig. 2.3). The pycnocline obstructs penetration of oxygen into the deepwater layers and favours accumulation of hydrogen sulphide in the near-bottom horizon.

The pycnocline zone of MPs is enriched by relatively heavy sediment particles, settling down from above, and by relatively light ones rising up from below. However, even at prolonged summer stills, MPs aquatories are not entirely stagnant zones owing to internal waves and convective mixing [154].

Horizontal flows of various origins (surface wind flows, internal flows and seiches) extend into various layers of the water column, but are dampened by HTS and ships moored near berths. These flows have direct influence on internal surfaces which form the periphytal zone of MPs.

Compared with MPs located in other regions of the World Ocean, the MPs of the Black–Azov Sea basin are characterized by a practically complete absence of tides. Where present, tides provide regular mixing, elimination and dispersal of excessive amounts of autochthonous OM and allochthonous BM. The offshore–onshore wind phenomena observed in the Black and Azov Seas do not sufficiently fulfil these functions and OM and BM constantly accumulate in MPs aquatories of the Black–Azov Sea basin. This situation significantly worsens the living conditions of bottom and near-bottom hydrobionts.

Besides the energy sources of natural origin, ships' navigation plays various roles in MPs aquatories. The intensity of navigation varies between periods of a given year and among years. In MPs, ships navigate along defined fairways, usually represented by shipping channels. The disturbing effect of the ships' motion on the various horizons of water masses manifests in different ways. Irrespective of the draft and size of ships navigating at various speeds, waves are initiated at their bow part. These waves mainly propagate perpendicularly to the shipping channel, are usually 10–30 cm high and travel distances from a couple of dozens up to a couple of hundreds metres. In MPs aquatories, these waves come into contact with, and are reduced by, the inner surfaces of berths, piers, protecting HTS and hulls of the moored ships. The upper layer of the periphytal (about 20–50 cm) falls under the influence of these waves.

The movement of the big, deep-draft ship perturbs water masses above, within and below the pycnocline. Passing through a shipping channel, a deep-draft ship favours the mixing of the near-surface and near-bottom layers along with oxygen penetration into the near-bottom layer. Part of the energy flows initiated by ships is assimilated and damped by the slopes of shipping channels. Ships' propellers also create various turbulences and water flows which stir up detritus from the slopes and bed of navigating channels.

2.1.2 Water Motion

The famous hydrobiologist Zernov [79] underlined that life intensity significantly decreases in aquatic ecosystems where water exchange is absent or weak.

Water motion comprises flows and roughness initiated by various phenomena. Energy is spent on the formation of flows which then become a source of energy themselves. In rectilinear flows and horizontal and vertical circulations, the transfer of water masses occurs together with that of small organisms inhabiting them. During wave propagation in the open sea, water particles move on orbits and eventually return back to their initial position. Near the shores, the movement of water particles is disturbed and they are displaced without returning to the initial position.

In the upper part of the shelf, where MPs aquatories are located, wind currents and waves (whose power in MPs is artificially weakened) are the main source of energy ensuring water exchange.

Waves caused by local winds are called wind waves. Wind blowing over the sea intensifies evaporation, carries substances with water droplets and creates surface waves and flows which are usually perceptible to the observer. However, in MPs, water motion also occurs in the water column and is determined by numerous factors which could reciprocally mask their effects.

All flows are divided into laminar and turbulent. When water layers move without mixing we deal with laminar, organized motion. On the other hand, the more widespread turbulent flow is characterized by rapid, irregular motion of water particles along complex trajectories. Heat exchange and mixing are more intense in turbulent flow. The transition from laminar to turbulent flow is quicker than the reverse one. In the surface (mixed) sea layer, turbulence is created by wind-induced shear stress. In aquatories of the Ukrainian MPs, located in the tideless Black and Azov Seas, the following flows are distinguished: 1—superficial (drift), mainly initiated by wind action, river run-off, seiches and ships' waves; 2—deepwater (or inner), having an impact on water the column and resulting from the influence of various natural and anthropogenic factors; 3—near-bottom, mainly compensatory and usually originating at offshore winds, sharp changes of atmospheric pressure, deepwater emission of wastewaters, submarine exit of fresh waters from land and other peculiar processes.

These flows are, in turn, divided into horizontal, vertical and mixed. All types of vertical and horizontal water motions directly or indirectly affect the distribution of biogenic matter, live plants and animals in the water column as well as habitat selection by hydrobionts in the benthal and periphytal zones. Furthermore, water movement redistributes, concentrates and disperses BM and food items necessary for hydrobionts as well as their metabolites and other toxic substances.

As in natural bays and harbours, a fully developed counterclockwise superficial water circulation, with various scales of horizontal and vertical circulations, periodically originates in MPs aquatories during the year (Figs. 2.4 and 2.5).

In MPs aquatories of the estuarine type, wave action can form water run-offs at leeward coasts and water accumulation at windward coasts (Figs. 2.6 and 2.7). The resulting difference in water levels causes compensational flows inside MPs. During winds blowing from the shore, a part of the superficial water mass moves offshore from the MPs aquatory, thus originating a near-bottom compensational flow which propagates into the aquatory along the AC (Figs. 2.8 and 2.9).

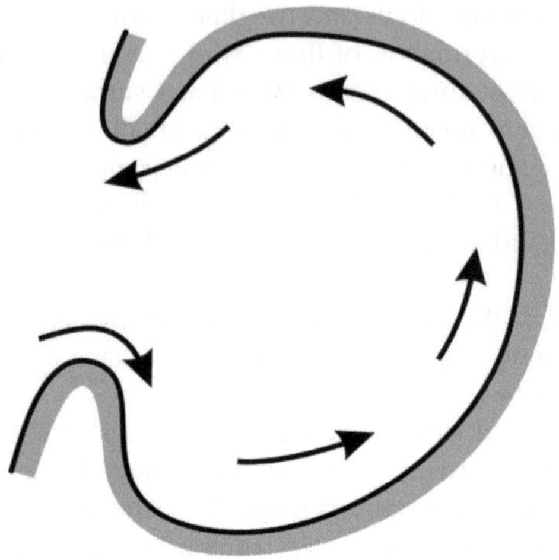

Fig. 2.4 Scheme of the circular flow in a natural bay, located in the northern hemisphere, with a wide passage, in which the incoming and outgoing flows move along right and left shores of the passage

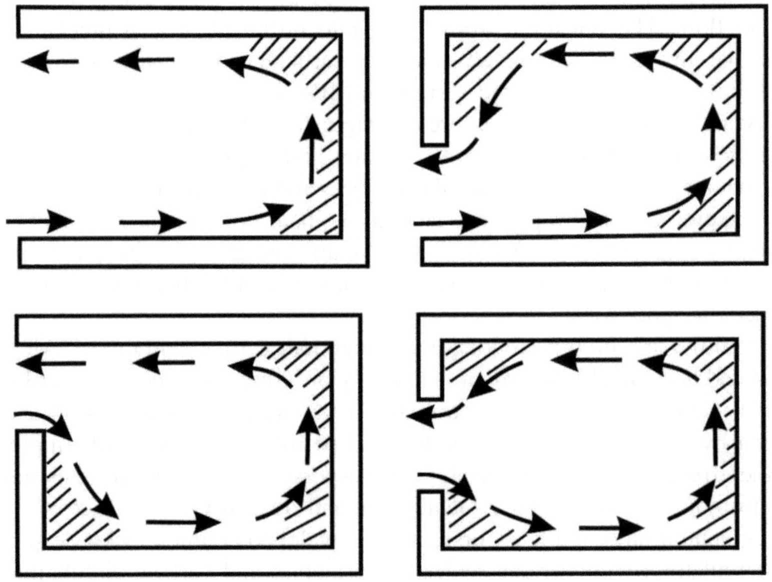

Fig. 2.5 General scheme of water flows and stagnation zones in port harbours having various configurations in the northern hemisphere

Rip currents can occur in MPs aquatories with preserved zones of natural, coastal shallow waters. Rip currents are water flows moving away from the shore through temporal channels which cross zone where waves are destroyed. Rip

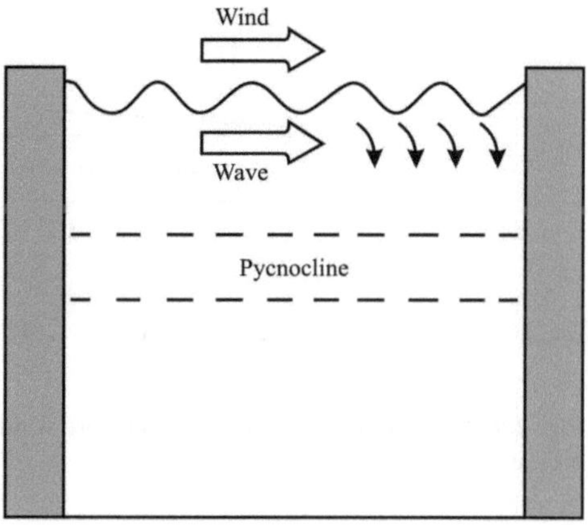

Fig. 2.6 Scheme of waves' impact induced by onshore winds on superficial water masses in marine ports

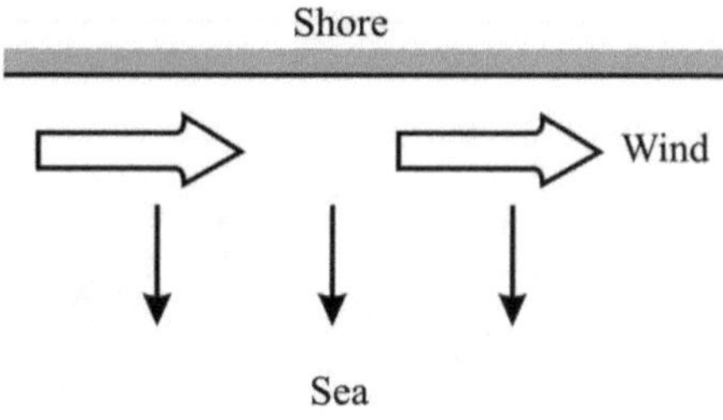

Fig. 2.7 Offshore phenomena in the upper layer of the sea coastal zone

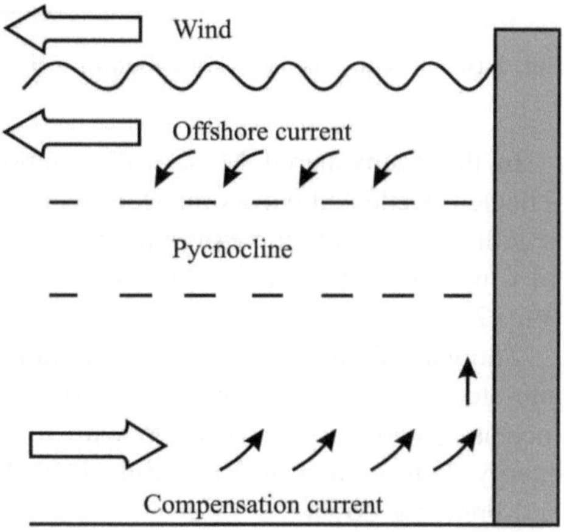

Fig. 2.8 Scheme of the main water flows in marine ports at offshore winds

currents reach high velocities and are formed as a result of water accumulation on the shore. They usually last a couple of minutes, can reach a speed of over 100 cm s^{-1} and are observed up to distances of a couple of dozens metres from the shore (Fig. 2.10).

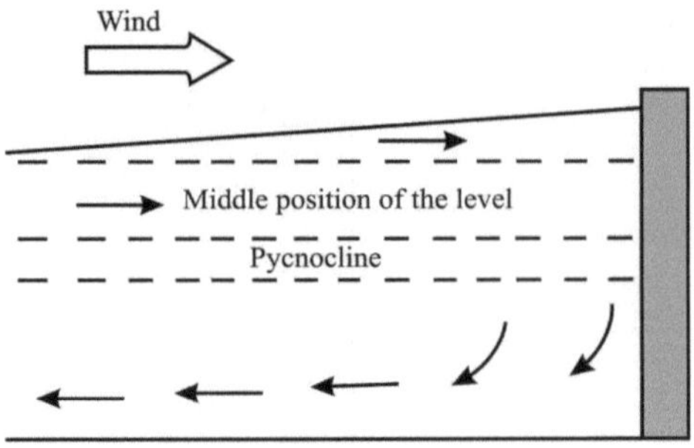

Fig. 2.9 Scheme of water run-on at onshore winds creating circulation and water flows in marine port and access channel

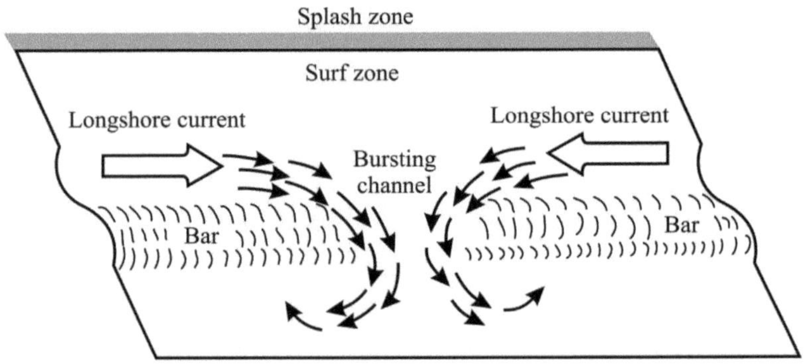

Fig. 2.10 General scheme of formation of a rip channel in the coastal bar zone

In the deepwatered MPs, waves approach vertical HTS, strike them and are reflected. Reflected waves interfere with the oncoming ones, forming a sequence of standing waves. On the external protection of Sochi MP, the wave force can reach 14 t m^{-2}, while the maximal registered wave force was 6 t m^{-2} in Tuapse port [78] (Fig. 2.11).

The waves' energy grows faster than the corresponding increases in their height and steepness. For example, a twofold increase in wave height leads to fourfold increase in its energy. A wave with a height of 3 m has two- to fivefold more energy than a wave of 2 m height [60]. Waves give up energy when overturning and breaking down.

The smooth ripples, which are slowly rocking ships or running on the shore, are what remain of the waves after they leave the wind influence zone. Even slow changes in the water level of MPs aquatories are waves. When entering natural or artificial narrow passages, such as those into the aquatories of some MPs, waves can significantly increase their height and intensify energy.

Besides the superficial wind waves and ripples, inner waves, seiches and ship waves are also registered in MPs aquatories.

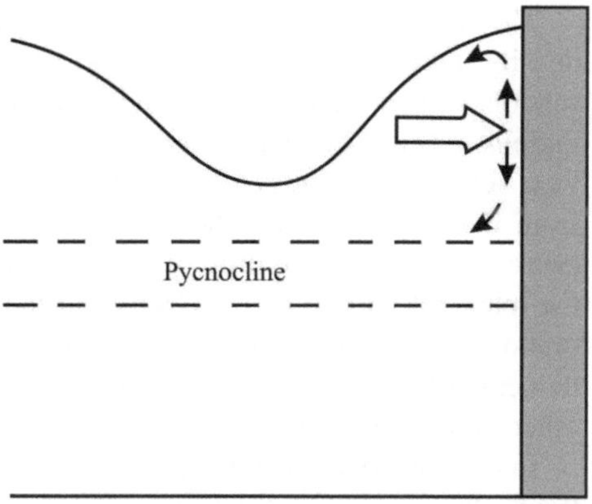

Fig. 2.11 Scheme of waves' energy consumption in marine ports aquatories at collision with hydrotechnical structures

Wind amplifies roughness when blowing in the same direction as wave propagation and weakens it when blowing in the opposite direction. Sea roughness and waves' height and steepness are proportional to wind duration, strength and acceleration distance, i.e. the distance over which the wind and roughness directions coincide. Smaller waves and ripples are superposed on the principal ones. Specifically small waves and ripples are mainly a transmission mechanism of kinetic moment (momentum) from wind to sea. In MPs aquatories, winds, waves and currents are diverted and weakened by natural and artificial obstacles that they meet on their way.

The frequency of wave run up is characterized by the wave period, i.e. the propagation time of two consecutive peaks (or trough) through the same point. Ripples, i.e. waves about 1 cm high emerging under wind influence on a previously smooth surface, have a period of about 1 s. The period of surf waves, having height up to a couple of metres is 4–10 s, while internal waves and seiches have periods from dozens of seconds to a couple of minutes and from a couple of minutes to a dozens of hours, respectively.

The speed of the wind-induced flow on the water surface corresponds to about 3% of the wind speed. Thus at a wind speed of 10 knots (18.52 km h^{-1}), the speed of surface flow is about 0.3 knots (55.5 m h^{-1} or about 1 m min^{-1}) [36].

Dynamic, wind-driven convection distributes energy from the top-downwards. In such situations, mixing is not much determined by the wind itself, but rather by the size and steepness of the waves that form and break, i.e. by gravitation.

Very slow mixing processes connected with the heat motion of molecules (molecular convection) prevail in water masses during calm weather. Flows displacing water masses with hydrobionts from one point to another, accomplish molecular convection, which plays a leading role in functioning and long-term life of aquatic ecosystems.

In MPs aquatories with relatively shallow depths (up to 20 m), the water movement has a three-dimensional character controlled by the energy entering with vortex-type formations (mainly from the upper water layers), the transformation of flows by bottom roughness and the HTS surfaces.

The molecular convection occurs during friction (dynamic) from differences in density, created by evaporation and owing to biological activity of hydrobionts, schools of fish, aquatic birds and mammals. Convective mixing occurs when, for various reasons, the upper water layers become denser than those located below, or vice versa. In both cases, vertical vortexes and currents are created and penetrate from one layer into another, thus mixing the water masses. At density mixing caused by temperature drop of surface waters, the vertical gradient of salinity plays the main role in the process, while in the case of density mixing caused by evaporation, the main role belongs to the vertical gradient of temperature. Mass settlements of filter-feeding invertebrates, schools of fish, dolphins, aquatic birds and other organisms realize biological convection.

In contrast to surface-driven phenomena, mixing connected with compensational flows affects a certain thickness of water over the bottom. The height of this mixed layer is determined, first of all, by flows speed above the bottom and by the vertical density gradients of the near-bottom waters. The higher near-bottom speed gradients correspond to the lower vertical density gradients. In such situations, the vortexes forming near the bottom and mixing the near-bottom layer can rise higher into the water column. Roughness and bottom irregularities contribute to the formation of such turbulent vortexes.

The shallower are MPs aquatories and the stronger is the wind, the easier is the total vertical mixing of waters.

In MPs aquatories bordered by high shores and protected by HTS, the wind force can be significant, but only weak currents and roughness can develop. In such situations, waves have no acceleration although the water level rises up in some zones, and lowers in others, because of wind-induced water accumulation.

At strong winds, a near-bottom flow of variable strength usually forms in MPs aquatories and ACs. At offshore winds, irrespective of direction, part of the surface water mass may press against the right or left bank and form a surface flow (directed into the adjacent sea zone) near the passage. At onshore winds two types of impact could be observed on MPs aquatories: first, part of the surface water masses are shifted to the zones most distant from the passage; secondly, water masses are transferred from adjacent sea regions into aquatories via AC. This transfer of water increases the pressure on the deepwater layers of MPs and originates a backflow carrying part of the near-bottom water mass from MP into the sea. When the wind drops, the opposite situation is observed with the superficial water mass flowing into the sea and a compensation current sucking bottom waters into the MP near-bottom layer. Density stratification, typical of the Ukrainian MPs, increases the transport capacity of compensation currents inside aquatories through ACs. This is why freshly settled OM, accumulating in the near-bottom layer of ACs, is easily stirred up and carried directly into MPs through the passages.

In the Northern hemisphere, the moving bodies, including air and water flows, are diverted to the right under the influence of Earth rotation. This effect was studied by French physicist Gaspard-Gustave de Coriolis and named after him. Coriolis acceleration is perpendicular to the velocity vector and as has values of about 0.1 cm s^{-1} for winds and 0.005 cm s^{-1} for major oceanic currents. In the case of marine currents and flows originating in MPs aquatories, the Coriolis acceleration is one- to threefold lower than in oceans. However, it is a constant force that participates in preserving natural biocenoses on the right shore of MPs and influences periphyton communities on artificial HTS.

Owing to the Coriolis force, water in MPs located in the northern hemisphere does not drift directly downwind, but is shifted at 30–35° rightwards to wind direction. Wind, blowing parallel to the open shore, either pile up water towards it, or drives it away from it. In the northern hemisphere, the anticyclonic circulations are directed clockwise and cyclonic ones anticlockwise. Vertical gradients of speed are observed in any flow; as a consequence of which the Coriolis force causes some turbulence and transverse circulations in homogeneous surface water masses. If wind direction is perpendicular to the open shore, the pile up and drive off phenomena are not observed.

Zones of flows convergence and divergence exist in water basins. In convergence zones, water sinks down following the convergence of surface flows. The sinking water enriches the deeper layers in oxygen (downwelling). In divergence zones, water rises up from deeper layers towards the surface and enriches the surface layer in BM (upwelling).

In standing waves or seiches forming on sea surface, the water particles do not move on circular orbits as it happens in wind surface waves. The impact of standing waves in the surface layer extends to a depth equal to wave height, i.e. seiches usually affect a water layer from a couple of centimetres up to 20–50 cm. Seiches belong to the long "closed" waves. They originate at undulation of water in closed basin, according to vivid expression of Shuleykin [193], "similarly to water poured into bowl to which was communicated a shove". While water is raising up at one shore of the basin, it lowers down at the other one. The so-called nodal line, at which the water particles stay motionless, passes somewhere in the middle (Fig. 2.12).

Zubov [80] brought an example of a sharp reduction of water level in Sevastopol Bay where the sea level lowered by 58 cm for 50 min during the storm of 25th August 1911 and then raised up to its initial level.

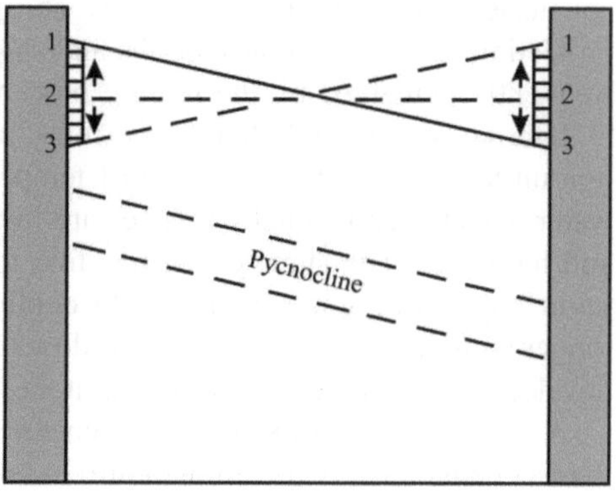

Fig. 2.12 Scheme of changes of sea level due to seiches influence on water masses of marine port (*1*, *3* seawater level at seiches oscillations; *2* water level position at calm weather)

In the Black and Azov Seas, seiches frequency ranges from 6–7 up to 22–23 h. As it was already mentioned, the wave period corresponds to the propagation time of two consecutive peaks (or troughs) through the same point. Seiches period depends on aquatory length and depth. Inside MPs aquatories, seiche waves usually have periods up to a couple of minutes.

Seiches have various causes: strong blasts of wind, sudden disappearance of wind action, sudden changes of atmospheric pressure, seismic activity, explosion, etc. They are usually observed in relatively closed gulfs, bays, straits and port harbours where the water masses can fluctuate as a whole. In some situations, seiches can be caused by the passage of a big ship.

After termination of force whose action started the seiche, the water level tends to occupy an intermediate position, i.e. an equilibrium position. In practice, under the influence of initial forces, water masses pass through the equilibrium position and causes progressively decaying fluctuations in the form of standing waves. Predominantly low amplitude vertical fluctuations of the water level are observed near the MPs shores and their protection structures.

Seiches cause an oscillatory motion of the whole water mass so that the surface of the entire aquatory acquires an inclination towards one or the other side. In MPs aquatories, each change of surface inclination is accompanied by a change in the inclination of isopycnic surfaces. Even if the position of the pycnocline reacts to seiches, the mixing of different layers of the pelagial zone is weak and most marked in the near-surface layer (about 0.5 m) interacting with the natural shore or with HTS lateral surfaces. Seiches have various influences on the near-surface belt of macrophytes and on barnacles, molluscs and hydroids settlements.

It is know that only surface waves, whose amplitude diminishes with depth, form in homogeneous water masses. In MPs, the seawater is usually a heterogeneous environment where internal waves, not visible at the surface, form between water layers of different densities.

The increase of atmospheric pressure lowers the water level, while its reduction increases the water level. Such fluctuations in atmospheric pressure are transmitted to water masses and induce water movement.

Outside MPs, the inner waves gradually weaken by spending energy in interactions with the water layers located above and below them. In protected MPs aquatories, inner waves are usually abruptly dampened by interaction with the protecting structures, moles, berths and ships. The striking energy of these waves is frequently perceived by the fouling biocenoses.

Internal waves and density currents in MPs are determined, first of all, by non-uniform vertical distribution of temperature and water salinity, both affecting water density. Thus, internal waves are frequently confined to the zones of thermo- and halocline. The summer thermocline forms in MPs when the turbulence minimum appears at some intermediate depths and vertical turbulent diffusion is no longer sufficient to transmit the heat flow to the underlying water layers without the development of a significant temperature gradient.

A change of 1% in salinity influences the seawater density about 5 times more than a change of 1 °C in temperature. In relatively small aquatories of the Ukrainian

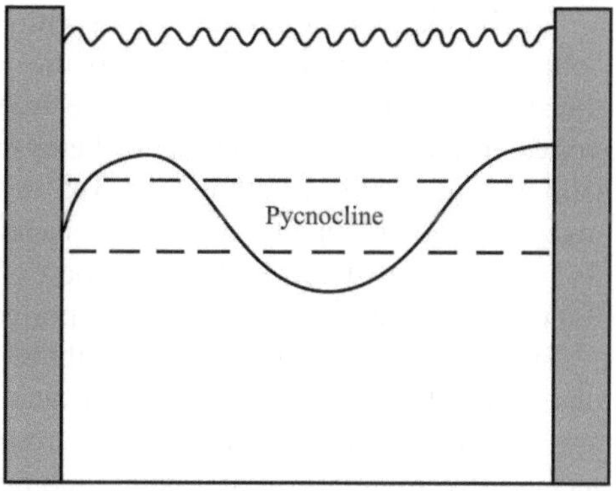

Fig. 2.13 Scheme of internal waves impact on water masses in marine ports

MPs influenced by river water discharges, differences in salinity between the surface and near-bottom layers usually influence water density more than differences in temperature. In Yuzhny, Odessa and Ilyichevsk MPs, the salinity of the surface layer (uppermost 3–5 m) could be 10‰ lower than that at the bottom and temperature differences between surface and bottom waters can exceed 10 °C. Under such conditions, the presence of a pycnocline and vertical stratification are steady phenomena.

In the thickness of stratified water masses, the energy of internal or subsurface waves propagates much similarly to that of the surface waves. Internal waves spread through isopycnic surfaces, which form as a result of density differences (thermocline, halocline). Internal waves, propagating from the pycnocline layer, initiate mixing that develops upwards and downwards and gradually fades away.

In the absence of motion, the boundary surfaces and the sea surface are horizontal. Internal waves originate at water masses borders and only in the presence of a vertical density gradient (Fig. 2.13). The density gradient is controlled not only by temperature and salinity, but also by significant inputs of river and drainage waters of various origin (e.g. industrial discharge, rain waters discharge) which are much less dense than seawater.

As shown by the example of Peter the Great Gulf (Sea of Japan), internal waves originate at the upper limit of the thermocline and reach heights ranging from 20 cm during calm weather up to 1 m during storms [118]. Usually, the heights of inner waves range from several dozens of centimetres to 2–3 m and the speed of their propagation from 5.0 to 35 cm s^{-1} at the depths up to 20 m [100].

In MPs, inner waves are higher than surface waves because the energy required to rise up the water layer in the air is much bigger than that necessary to rise up a water layer in an environment of similar density. The height of internal waves quickly diminishes when moving away (upward and downward) from the pycnocline. The position (depth) of the pycnocline, where internal waves form, changes depending on the height of the internal waves themselves.

Cases when the height of inner waves reached 20–30 m are known for ocean waters. Inner waves with an amplitude of about 10 m usually pass through the pycnocline with a period of about 20 min [36, 100].

The breakdown of internal waves with lengths of a couple of metres, periods of a couple of minutes and phase velocities of a couple of centimetres per second transmits the characteristics of the surface layer through the thermocline zone via vortex formation. Observations in Peter the Great Gulf showed that the characteristic size of visible formations is 3–5 m for internal waves and 5–20 cm of vortexes. Moreover, intense vertical descending motions with typical dimensions of 2–3 m were registered in the lower part of thermocline [118]. This phenomenon of "salt fingers" has been previously reported [36].

In protected aquatories such as those of MP, standing internal waves can be observed. The standing waves do not actually transport energy themselves, i.e. they are immobile with respect to the environment. In the water column, water movement is divergent above and below the pycnocline. Inner waves cause fluctuations of pycnocline. Although the amplitudes of inner waves can exceed those of surface waves, inner waves are usually slower. Formation of the standing inner waves is conditioned by size, configuration and depths of aquatories. All these waves are especially marked in aquatories where flows abruptly change their direction and speed. Such phenomena are observed in access channels and passages in MPs divided into separate basins and harbours. In MPs, inner waves originate in water layers with different temperature and salinity. They have heights from a couple of dozens centimetres up to a couple of metres and stretch horizontally from dozens of metres up to a couple of kilometres.

A ship slowly moving in a port aquatory generates internal waves at the border between the surface layer (upper 3–5 m) of strongly desalinated water and the denser deep-sea water. Such inner waves sharply reduce the ship's speed and create phenomena known under the term of "dead water".

Inner waves influence all components and processes that are affected by changes in temperature, salinity, water density; e.g. hydrobionts distribution, chemical and physical water properties, flows, bottom topography, etc. [100].

The famous oceanologist Zubov [81] emphasized that, regardless of their period, internal waves are particularly important for the periodical "renovation" of deep-waters and bottom water layer of the semiclosed aquatories, thus favouring the development of life in these zones.

In marine ports aquatories, navigation of different types of vessels takes place more or less regularly either autonomously or with the help of tug boats. Ship navigation originates visible surface ships' waves that consist of slanting (or diverging) and transverse waves. The angle between a ship's bow and the slanting waves formed by it depends on the ratio between the ship's speed and the depth and is, on average, equal to 18–20°. The front of transversal waves is perpendicular to the direction of ships' movement [191]. Figure 2.14 shows the scheme of distribution of ships waves in MPs aquatory.

In the stratified water masses characteristic of Ukrainian MP aquatories, ships form not only surface waves but also the so-called internal waves, which are not visible to the naked eye.

When turning around in MPs aquatory, a deep-draft ship causes a turbulent water movement along all the submerged lateral part of the ship.

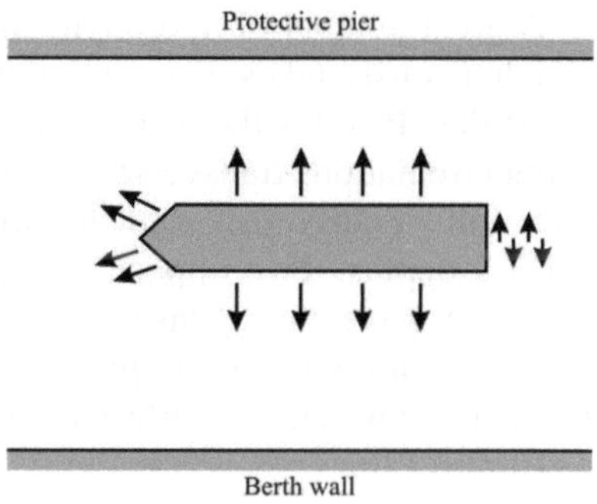

Fig. 2.14 General scheme of flows created by a ship in motion

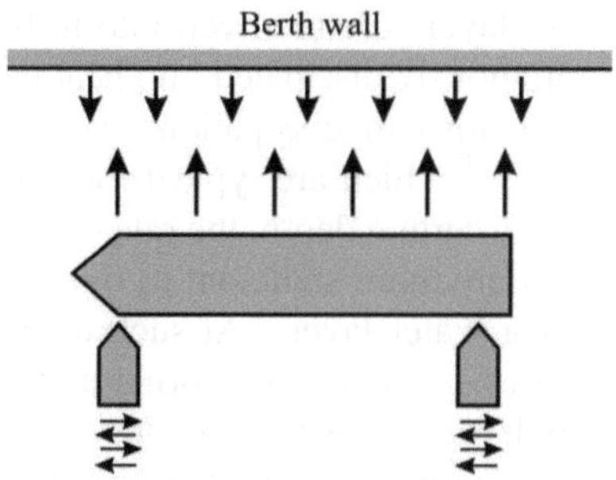

Fig. 2.15 General scheme of flows created at mooring to harbour wall by berthing tug

In particular, a ship pushes a water layer equal to the length of its hull and up to a depth corresponding to its draft. A similar situation takes place during a ship's mooring by tugs to the wharf wall. Pushing a ship with a length of 250–300 m and a draft 10–14 m to the wharf wall, the berthing tug destroys the pycnocline at a distance of a dozen metres from the wall and over a length equal to the underwater part of the ship. Both surface and internal waves form during berthing and, in aquatories with small dimensions compared with the ships' draft, seiches (Fig. 2.15).

The rotation of a ship's propellers forms a strongly turbulent stern wake whose influence on the water mass depends upon many circumstances, such as number and diameter of propellers, the speed of their rotation, the ship's draft and other. The stern wake could initiate surface and internal waves. The stern wake gradually dampens its energy in all directions by forming vortexes of various dimensions. In the stern wake, a dilution up to 1000-fold of the polluted waters discharged from the ship is achieved under the influence of ship's propeller [45]. The stern wake, saturated by oxygen, plunge up to a certain depth. Because the ships in MPs aquatories move along well-defined fairways, the water movements initiated by them propagate towards HTS along defined directions and angles. The angles at which waves approach HTS following navigation in harbours are different during ships' calling in and leaving from the port and such differences could reach 60–70°. In ports, some HTS are regularly subject to the influence of ships' waves, while others experience it only episodically.

A great number of hydrodynamic processes of different spatial-temporal scales simultaneously takes place in natural basins in general and in MPs aquatories in particular. Besides the usual drifts flows, wind creates a series of different scale spiral circulations transverse to its direction and with horizontal axis parallel to its direction. Transverse vertical circulations penetrate the water column up to various depths. Two opposite processes connected with accumulation and dissipation of wave energy take place in the water column. Upon breaking down, the wind-driven roughness originates small-scale movements of vorticity in the water. In open seawaters, the life span of energy containing vortexes could comprise 1–10 min [26].

In seawater, the vertical and horizontal components of turbulence significantly differ on scale and intensity. The largest energy exchange takes place when all water layers are involved into turbulent transition. The energy of turbulent flow is continuously dissipated into heat by the water molecular viscosity. In the ocean, the fluctuations of dissipation speed become insignificant at flow speeds from 1 to 6 cm s^{-1} which are typical for various water movements in MPs.

At a certain depth, the rate of turbulent energy formation reduces so much, that it is not anymore sufficient to overcome the resistance caused by density differences among water layers. At such depth, at the border of the mixed layer, a transition zone (the pycnocline) forms and a more or less sharp density gradient with low turbulence intensity develops.

The upper mixed layer and the pycnocline, two very important components of MPs aquatories ecosystems, form and evolve under the effect of several mechanisms, including vertical mixing (as small-scale turbulence), thermal (gravitational) microconvection connected with Benard cells, Ekman heterogeneity, collapse of surface and internal waves and Langmuir circulation [147].

With regard to their scale, the main processes of interaction between the sea surface and the atmosphere in MPs could be divided into micro-, small- and local scale. They diffuse to depths of no more than 10–20 m and, horizontally, up to 10–100 m and last no more than a couple of minutes [137]. Small- and middle- (or meso-) scale water movements are usually connected with MPs aquatories. Their spatial sizes range from dozens of metres to a couple of kilometres and their temporal scales vary from hours to days [137].

As already mentioned, the smallest waves, originating under wind influence on previously smooth surface with the height of about 1 cm, are called ripples and their period lasts less than a second. Scales of metres and periods of minutes are typical for the internal waves and seiches.

As it was already said, the sea surface constantly cools down (its topmost 1 cm layer adsorbs only 20% of incoming solar radiation and gives away, due to evaporation and long-wave length radiation, 100% of the energy adsorbed by the border layer with the thickness from a couple of metres up to a few dozen metres). The temperature of the sea surface is lower than that of underlying waters by fractions of a degree and, sometimes, by degrees. This particular feature creates density instability at the water surface and convection which usually propagates up to the depth of 25–50 cm in conditions of calm water. Thus, thermal, or gravitational, microconvection is the basis of water motion in the surface layer.

The phenomenon of microconvection originates because the surface tension exists not only between air and water, but also between waters masses of different temperature, i.e. between the cold surface film (0.1–1.0 cm) and the underlying water layer. About 1 cm^3 of colder and denser water sinks from the surface up to the depth 25–50 cm in the form of water thread jets with a diameter of about 1 cm and a speed of 0.2 cm s^{-1}. The masses of heavier water rotate during sinking [1].

Vertical jets arising because of temperature differences create microconvection and gradually acquire the form of regular hexagons in whose centres the water, in the form of concentrated jets, is directed towards the zones of increasing molecular viscosity; i.e. towards lower temperatures. The opposite, compensational water movement in the upward direction takes place along the periphery of the hexagons. Overall, a closed circulation originates in the form of cells known as Benard cells.

Directly in the near-surface layer of MPs, the exchange of matter and energy between the atmosphere and water masses takes place thanks to displacement of microscopic particles of environment, leading to transportation of mass, heat and other physical quantities. Transmission of energy in the form of heat (convective heat exchange) takes place in a non-uniformly warmed liquid medium and is caused by temperature and density gradients.

When the overlying water particles cool down because of the process of evaporation and become heavier than underlying ones, they sink down and lighter ones rise up in their place. This is a gravitational convection phenomenon that often takes place in the form of the already mentioned hexahedral Benard cells. The liquid sinks down along the edges of such cells and rises up in its centre. If a horizontal wind flow is superposed onto cells, cells line up along it and form convergence streaks, which resemble Langmuir circulation discussed below, but are limited to a thinner near-surface water layer [147].

In cases of lasting and stable wind, the so-called Ekman wind-driven currents can develop in a much thicker part of the upper water layer. Under some conditions, the instability of Ekman wind-driven currents also manifests in the form of flows transverse to wind circulations, also resembling Langmuir circulation. The typical time scale of Ekman instability is a couple of hours [147]. However, this time is sufficient for mixing of the near-surface layer over a thickness of dozens of metres.

MPs aquatories have substantially smaller horizontal and vertical dimensions than the seas where they are located. They are placed on the shelf in coastal zones with relatively small depths. In MPs aquatories, the processes of energy supply to the water column are slightly different than in the open sea, owing to deterioration of the macroscopic component of the water motion spectrum. In MPs, movements with a scale up to 10–100 m, connected with the transfer of energy during roughness provoked by winds, acquire a major role.

Water movements in MPs interact with the atmosphere, the bottom, lateral surfaces of hydrotechnical structures, ships' hulls and water layers with different density. Such interactions inevitably generate zones of turbulent flows and form vortexes of different scales. Displacement effects, determined by vertical and horizontal gradients of flows, the destruction of surface and internal waves, the character of vertical stratification, water density and temperature play an important role in controlling the genesis, size and intensity of the vertical vortexes. In aquatic ecosystems, turbulent flows in the form of small-scale vortexes are more ecologically relevant than laminar flows. Microvortexes with sizes from a couple of millimetres up to a few dozens centimetres have the greatest importance for some organisms. Vortexes with sizes from a couple of metres up to dozens of metres sustain the existence of populations and hydrobionts communities.

In MPs, the strongest "concentration of life forms" occurs in zones characterized by frequently repeated small-scale turbulence such as: 1—under the film of surface tension, 2—coastal shallow waters, 3—the upper part of lateral surfaces of hydrotechnical structures, 4—areas of the bottom with increased roughness, 5—the pycnocline zone.

Microvortexes are connected with feeding of hydrobionts, dispersion of metabolites, reproduction, fertilization and development. They are necessary for the inhabitants of all MPs biotopes to various degrees. The highest biomasses of attached and sediment-dwelling sestonophages and hydrobionts predators are usually concentrated in the zones that are most hydrodynamically active thanks to the presence of microvortexes [68, 139, 145].

In MPs, the water turbulent motion plays an important role in exchange processes (transformation) of momentum, heat, suspended particles, dissolved nutrients and different toxic substances. The marine surface is a transit zone where the exchange of matter and energy takes place between the water mass and the atmosphere owing mainly to processes of molar diffusion, i.e. microvortexes, rather than molecular diffusion.

In the open sea, the strong, wind-induced waves can generate vorticity in a water layer up to 10–20 m thick, i.e. up to the maximum depths of MPs. However, in MPs aquatories, the wind-driven waves usually induce vorticity in a water layer up to 1–3 m thick and the "life span" of energy containing vortexes does not exceed 1–3 min [61, 110, 137].

The turbulent water flow is characterized by complex curvilinear trajectories and vortexes continuously changing direction and velocity in each point of the flow. It causes intensive mixing of water masses and therefore exchange of characteristics. Turbulent exchange takes place in all directions, but vertical displacements are especially important because they favour penetration of oxygen into the MPs water column.

Turbulent vortexes of various sizes form when water flows around isolated obstacles such as HTS, ships or pile foundations. Shadow zones of various sizes, i.e. zones in which water movement becomes less noticeable or even disappears, form behind the obstacle depending on its size and form.

Turbulization of water column by moving ship's propellers favours vertical mixing [178].

It is considered that productivity of invertebrates is highest at flow velocities of 30–90 cm s^{-1} [145]. Interaction with the bottom starts at flows velocity of about

7.2 cm s^{-1} [153]. In MPs, turbulization of the flows in hydrobionts habitats depends on the roughness of the surfaces with which they are interacting.

During colonization of HTS surfaces by hydrobionts, bacteria and microalgae form a slimy layer that decreases HTS roughness and weakens turbulence. The appearance of macrophytes and attached invertebrates in fouling increases the roughness of the whole settlement. Owing to bioturbation and accumulation of faecal pellets, the surface of bottom sediments becomes rough and creates turbulence of flows. The increase in surfaces roughness ameliorates the energy supply of both biocenoses and individual organisms.

In the coastal stripe with large-sized sediment components (e.g. stones, pebbles, broken shells and coarse sand) turbulence could be excessive and depress the development of many hydrobionts. In zones with optimal turbulence and corresponding granulometric composition (sand, silted sand), the biomass of bottom population reaches dozens to hundreds of grammes per 1 m^2. In MPs, the mass development of periphyton takes place on hard substrata located in the zone of optimal vorticity. Organisms of benthos and periphyton of a defined size not only contribute to the generation of turbulent flows, but also create microvortexes of a certain size during their feeding and respiration [209, 228, 234].

Observations in Ilyichevsk, Yuzhny, Odessa and other MPs aquatories showed that, during windy weather, convergence streaks of foam, air bubbles, suspended matter and garbage, fragments of macrophytes, phyto- and zooplankton usually form parallel to the direction of the wind. It turned out that this phenomenon is connected with unusual water movements known as Langmuir vortexes, Langmuir circulation, Langmuir convection and Langmuir effect. Paired convectional cells transverse to wind circulation, also bearing the name of Langmuir cells, form owing to convection processes usually initiated by the wind.

American physicist and Nobel Prize winner Irving Langmuir postulated that the horizontal distribution of live and dead OM on the water surface and in the water column is mainly determined by physical causes. He confirmed his hypothesis that convectional processes are frequently lying at the basis of these phenomena by field surveys [212].

It was found that an ordered convective-drift circulation forms when wind-drift currents having common velocity gradients develop in the convection layer. Such circulation is an important mechanism of heat, kinetic moment and matter transfer, having significant influence on regime formation of all upper layer of water masses [61]. It is considered that Langmuir circulation is the result of a complex interaction between the wind flow and surface gravitational wind waves. The wind blowing over the water surface generates transverse circulations when frequencies "resonating" with water masses of given sizes are present in the wind turbulent spectrum [125, 126].

Langmuir vortexes represent a typical form of highly organized and at the same time randomly variable movement. In particular, it is assumed that Langmuir vortexes (convection) are the prevailing form of wind-driven flows.

Except in still weather conditions, several metres of the upper water column are constantly moving in the form of a vortex-type flow, which consists of small-scale

Langmuir vortexes with a distance of 3–6 m between convergence lines. At moderate to strong winds (5–15 m s^{-1}), Langmuir vortexes with different scales are observed [101]. In aquatories with sizes similar to those of MPs, Langmuir cells are typically 5–10 m wide, i.e. their size is limited by the depth of the pycnocline [118].

At moderate wind conditions, the coexistence of small-scale vortexes and wind spiral of Ekman type was observed. Langmuir circulations are located at the surface and Ekman circulation is located under it. At large-scale Langmuir vortexes, an Ekman spiral does not form [101]. The transfer of kinetic momentum from wind through the smaller vortexes to the bigger ones creates an effective connection between these vortexes. This type of link explains the quick response of the mixed layer to wind impact. A linear dependence between the wind speed and the water sinking rate was established. Various authors showed that practically any wave motion causes rectilinear translation of water in the direction of wave propagation, although this phenomenon is practically insignificant for flat waves [101].

It was discovered that, after some inertial period, the zones of convergence acquire anticyclonic rotation upon occurrence of still weather following a period of moderate winds [101].

Friction force, applied to a horizontal surface, leads to the formation of Ekman flows (Ekman spirals). This effect is a consequence of the Coriolis force action which shifts moving objects rightwards in the northern hemisphere and leftwards in the southern hemisphere. As a consequence, when wind blows over the ocean surface in the northern hemisphere, surface flows diverge rightwards from wind direction. Because of water viscosity, the diverging surface water layer puts into motion an underlying layer, which also diverge rightwards from the above one and so on. The flow weakens gradually with the increase of crook. In the presence of Langmuir vortexes, the conditions necessary for the origin of Ekman spirals are shifted to the lower border of Langmuir flow, which assumes the same role as the wind with respect to underlying waters and can thus form flows of Ekman type in these waters [101].

It was determined that the pattern of transverse circulations and convergence streaks (created by wind and marked by various suspension, bubbles and foam) is tightly connected with the geometry of a water basin. Only vertical circulation of a certain size can form within a specific water basin or in protected aquatories [101, 118, 147].

Irving Langmuir [125, 126] studied values of horizontal and vertical speeds in spiral circulation. At a wind speed of 6 m s^{-1}, vertical descending speeds of about 1.6 cm s^{-1} were observed under the convergence streaks. At the same time, the horizontal speed of water motion at the surface was about 15 cm s^{-1}. It was determined that spiral circulations caused by the wind penetrate as far as the depth of the thermocline layer. Distances between the well-marked streaks of convergence are usually similar to the penetration depth of transverse circulation. Based on numerous data, Langmuir circulations are an important source of turbulent motions in the surface water layer where horizontal and vertical transfer of elementary microvolumes of water simultaneously takes place [118, 125, 126, 147]. These water movements are extremely important for the organisms of the epipelagial zone [228, 234].

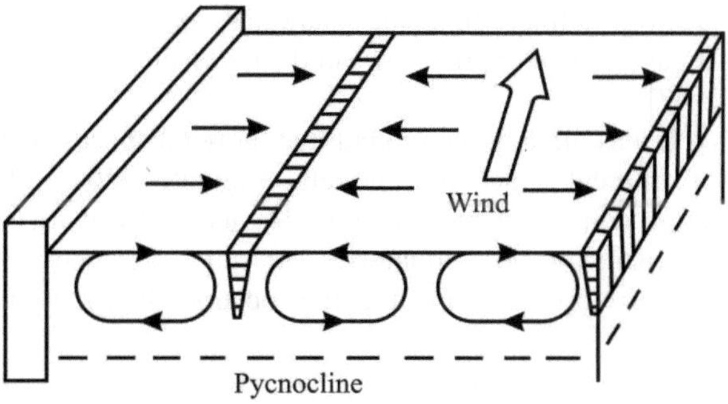

Fig. 2.16 Scheme of formation of Langmuir convection cells pair caused by wind influence (*hatched areas* identify the streaks of suspended matter accumulation)

Usually the wind-driven streaks marking flow circulations already become visible when wind speed exceeds 3–5 m s^{-1} and generates the associated wind flows. The distance between neighbouring streaks ranges from dozens of centimetres up to dozens of metres and more. Thus, Langmuir circulations are vortexes perpendicular to the wind direction and developing in the upper layers of a basin (above the pycnocline). Near the surface, water particles move in the direction of wind streaks where they sink down, creating the so-called downwelling. On the contrary, particles rise up between the windrows thus creating upwelling. At the bottom of Langmuir vortexes, the direction of water circulation reverses with respect to the surface. Two adjoining wind streaks (convergence bars) delimit a two-dimensional circulation cell comprising two oppositely directed vortexes (Fig. 2.16).

On the water surface, each pair of counterrotating Langmuir convection cells forms a stripe of suspended matter in the zone of convergence of surface flows.

In addition, it was found that dissolved OM also concentrates along the convergence line in the thin surface film at the "water–air" interface [36].

Following the pattern of Langmuir convection, flow circulation has opposite directions at depth and at the water surface where regular accumulation of suspended matter can also occur. In convergence zones, the concentration of phyto- and zooplankton can exceed that in the adjacent zones by twofold [36, 209].

In the correspondence of wind streaks, the vertical descending speeds reach about 10 cm s^{-1} and slightly exceed the ascending speeds. Within a circulation cell, vortexes are asymmetrical and thus each cell also has a longitudinal velocity component directed alongside the wind direction, besides the transverse one. The longitudinal component is higher in wind streaks than between them. At sudden changes in wind direction, streaks rearrange along the new direction within a couple of minutes. Typical time scale for this type of circulation spans a couple of minutes, while the corresponding spatial scale is measured in metres as it was already mentioned.

Langmuir circulations originate near the surface and then propagate into the depth. Clouds of air bubbles, which form at wave breaking, concentrate along wind streaks and are carried by Langmuir circulation within the limits of the upper mixed layer. Despite the significant variability in cells' sizes, their dominant dimensions are well-defined. In small aquatories, for example lakes, the distances between two streaks can be up to 35 m, but distances of 13.5 and 12.0 m are most often observed [147].

It was already mentioned that, at wind speeds of 5–15 m s^{-1}, a whole hierarchy of different-scaled motion cells originates [200]. The width of the average circulation cells is in the range 35–90 and 5–10 m for small-sized ones that are characteristic for MPs aquatories. At cells' destruction, their energy is transmitted to smaller vortexes. This situation occurs when Langmuir cells interact with HTS of MPs.

Thus, in water basins, there are the short-lived zones of flows convergence whose persistence is measured in hours or days. Convergence zones are also characterized by the highest concentrations of neuston organisms, which disperse as soon as the forces that gathered them together stop acting. Such temporary zones of convergence and water sinking are convergence streaks, or storm streaks. They are elongated in parallel lines along the wind direction. The distance between the streaks is proportional to depth and diminishes in shallow water. In convergence streaks, the neuston accumulations represent narrow (less than 1 m), long streaks extending downwind [71].

Planktonic organisms, whose buoyancy is practically neutral, are easily carried down and it is one of the main causes of macrozooplankton abundance in the down flow of convergence zones.

Short-lived and small-scale accumulations of neuston organisms appear under the influence of vortex-type flows formed in the shadow zones of different HTS and ships.

Contrary to turbulent wind mixing, formation of Langmuir convection cells increases the heterogeneity in the distribution of dead suspended OM and dissolved OM in surface water masses on the one hand and controls it for a more or less prolonged time on the other. Functioning of Langmuir convection cells favours the formation of "accumulations" of dead and alive organic matter in mixed layers having relatively homogenous temperature and salinity.

In natural water basins, various hydrodynamical processes with different spatiotemporal scales usually take place simultaneously and superpose one upon another. Nonetheless, it can be surmised that Langmuir circulations and the surface thermal (gravitational) microconvection are the physical driving forces leading to the increased concentrations of phyto- and zooplankton organisms in the hyponeuston and epiplankton layers. The concentration of fish pelagic eggs in convergence zones obviously provides the best feeding conditions to their larvae and fries after hatching.

In MPs aquatories, Langmuir circulations not only form accumulation of suspended OM at the surface, but also create the preconditions for accumulation of SOM and DOM near HTS lateral surfaces.

For small-sized planktonic organisms, the movement of water particles in vortexes with a diameter of about 5–10 m and more practically has a rectilinear and laminar character. Microvortexes with a size of 1–10 cm have a scale similar to the hydrobionts inhabiting the water column, the bottom and the HTS surfaces of MPs.

These microvortexes do not damage the inhabitants of the water column and wash out of their settlements without harming individual organisms.

Numerous studies showed that gas bubbles originated by sea roughness, photosynthesis, decomposition and other phenomena adsorb organic substances and transport them to the surface when rising up in the water column. After this phenomenon, changes of organic matter dispersity take place in the organic coatings of these bubbles. In practice, truly dissolved or colloidal organic substances form particles or aggregates whose composition and size allow their consumption by microheterotrophic organisms [71, 74].

Gas bubbles cover about 3–4% of sea surface at any moment. The number of bubbles increases in the presence of vertical transverse circulations of various scales. The ratio among substances contained in bubbles escaping into the atmosphere considerably differs from the ratio of the same substances in seawater because fractionation of some substances takes place during bubble formation. In exploding bubbles sprays, ammonium concentration is hundreds of times higher than in typical water samples and phosphates can be enriched from 3 to 600 times. OM concentration in bubbles exceeds that in the seawater by up to 2000-fold. A drop forming following the explosion of a bubble of one millimetre is projected upwards with a speed of about 10 m s^{-1}, flies up to 10–15 cm and quite often evaporates in the air, transforming into a salt particle. These particles can then be carried away by the wind. During emersion of inert gases bubbles in the water column, the "dissolved inorganic" phosphate transforms into particulate "organic" phosphate. Such phosphate-containing particles serve as food for some planktonic organisms [110].

In the water column, part of the dead bodies of planktonic hydrobionts, especially crustaceans, acquires positive buoyancy and rises up during decomposition. The above-mentioned flotation process which continuously takes place in the sea also contributes to this phenomenon. The presence of this "anti-rain" of dead bodies causes the constant presence of alive and dead organisms in the near-surface layers. Depending on the different water circulations, the ratio between dead and alive organisms in the upper mixed layer can experience significant fluctuations in convergence and divergence streaks.

As a result of the dead bodies "anti-rain", which directly affects the hyponeuston layer, a significant part of dead hydrobionts and fragments of their bodies concentrates in the film of surface tension and in foam. The rests of phyto- and zooplanktonic organisms are also found in foam [71].

Zooplankton density under convergence streaks could exceed the density in the adjacent zones by 100-fold [36]. In convergence streaks, concentration of pelagic eggs of fish, passively retained in the near-surface water column, provides favourable foraging conditions to larvae after embryos hatching and exhaustion of yolk stocks.

In the mixed layer, the relatively low speeds of water particles in vertical circulations permit to big-sized diatoms and unicellular flagellates to actively move in water. Flagellates usually swim upwards towards the light, while diatoms (which are mass species in MPs aquatories) strive to go into the deeper layer using the vertical circulations. Flagellates concentrate near the water surface at weak turbulence and low descending water movement.

Significant accumulation of dead and alive planktonic organisms in the surface layer, as well as rains, hamper the development of roughness and vertical circulations at the surface of MPs aquatories.

Convergence streaks on the surfaces of aquatories are formed from floating particles of different origin such as fragments of macrophytes, phyto- and zooplankton, and dead insects. These streaks constitute peculiar biotopes for some invertebrates, larvae, newly hatched young fries and juvenile fish. The oldest hyponeustal stages of invertebrate larvae and fish fries are common among accumulations of flotsam which, as a rule, are typical for convergence zones [71].

In MPs aquatories, various horizontal and vertical water movements, including Langmuir circulations, favour SOM, DOM, phyto- and zooplankton accumulation in the near-wall layer of HTS and in the harbours most distant from ACs.

Zaitsev [71] described a situation in which the water surface under the foam and in the spaces between its flakes was swarming with hyponeuston. Fries of the leaping mullet and dragonet, larvae of combtooth blennies, sand sole, anchovy, zoeas of decapods and megalopes of crabs, isopods and pontellidae, as well as numerous smaller-sized hydrobionts not perceptible by unaided eye, were swimming among algal fragments, wood chips, land-living insects, seeds and various garbage. Larvae and early fries of garpike raising to the surface were also connected with this biotope.

In MPs, wind and surface flows play an important role in the neuston distribution. Neuston reacts sensitively to the constant and temporary water motions of different scales [71] observed in MPs harbours.

In MPs, interconnection of various phenomena and processes in the pelagial, periphytal and benthal zones and, most importantly, the cause–effect relationship among them are mainly determined by the action of microvortex-type water motions, i.e. turbulence. In MPs, the processes supplying energy to the water column, the HTS near-wall water layer and the near-bottom water layer are slightly different than those in the open zones of the shelf, because the macroscopic spectral components of water movements (wind waves and flows, Langmuir circulations, seiches, inner waves, ships' waves and other) are reduced.

2.1.3 Salinity

Water salinity is a major ecological factor which, in MPs aquatories, is determined first of all by the salinity of incoming water masses and their interactions.

In the open regions of the Black Sea, the salinity of surface waters is 17.5–18.5‰. However, salinity of the surface horizon reduces significantly in the NWBS, in the Kerch pre-strait zone and near the mouths of big rivers, especially during the flood period. In various regions of the Azov Sea, the water salinity of the surface layer experiences strong fluctuations, but its average values are about 12.0–13.0‰.

In Sevastopol Bay, salinity usually stays around 17–18‰ at the surface and it is slightly higher near the bottom. However, salinity can decrease to lower values in the estuary of river Chornaya which flows into the bay [3]. In a significant part of Novorossiysk port, adjoining the Tsemes River estuary, water salinity changes from 8 to 16‰ during the year [51].

Before the digging of a 5 m deep channel in 1909, Varna Lake was a freshwater ecosystem. In August 1911, the superficial salinity reached 11.11‰ and hydrogen sulphide appeared near the bottom. In 1976, Varna Lake was connected to the sea by a second channel with a depth of 10 m. In 1976–1985, the salinity had increased throughout the water column to an average of 14.81‰ (range 11.13–16.31‰) in the surface layer and to 15.86‰ (range 13.91–17.50‰) in the near-bottom layer. The general tendency of increase in salinity was accompanied by the decrease of river run-off which, during summer months, was less than evaporation [144]. In 1990, a salinity of 17.08–17.23‰ was registered in Varna Lake, while values of 16.54–16.94‰ occurred in the connected Beloslav Lake [194]. The yearly hydrochemical regime in Varna Lake is not as distinctly marked as in the sea due to its water exchange peculiarities which are mainly influenced by offshore phenomena. The increase of seawater entry into the lake ameliorates its regime [144].

In Odessa MP, the average monthly salinity minimum of about 11.5‰ occurs in April–May, while the corresponding average monthly salinity maximum is usually registered in August–September with values in the range 15.6–15.7‰. However, salinity fluctuations from 3.4 to 18.4‰ were registered in the period 1999–2001. Such values indicate that water in the Odessa port aquatory is a mix of freshwaters from the Dnieper-Bug Liman and the deepwaters of the Black Sea. Persistent east quarter winds favour the entrance of desalinated waters into the Odessa Gulf and further westwards. During the period of spring floods, the seasonal halocline in the Odessa region of the NWBS is located at a depth of 3–5 m. The halocline further deepens at the beginning of summer before gradually weakening in the second part of summer and disappearing in late autumn. In winter, associated with homothermy, a mild halocline gradually forms at a depth of 7–9 m because of the periodical entry of water from the Dnieper-Bug Liman region [154].

The minimum salinity values (6.0–7.0‰) in the surface horizon of Grigorievsky Liman are registered in spring. In the same way as in Odessa MP, this situation is connected with the entrance of waters from the Dnieper-Bug Liman during the high water period. A two-layer structure is established in the liman. In the upper layer with a thickness of 3–5 m, salinity is in the range 7–8‰ and in the near-bottom layer it is 15–17‰. In Grigorievsky Liman, a weak variability of salinity near the bottom is observed all year round [196]. The desalinating influence of Dnieper-Bug Liman waters in Odessa MP frequently manifests much more strongly than in Grigorievsky Liman, albeit the latter is located noticeably closer to the source of fresh water. Such situation occurs because Grigorievsky Liman is connected with the sea via only one channel with a cross section of about 3000 m^2, while Odessa MP aquatory and Odessa Gulf are connected to the sea by three passages with a total cross-sectional area of about 11,300 m^2.

In the central part of Sukhoy Liman, the amplitude of salinity fluctuations is 5.28–18.06‰. In this basin, differences in salinity between the surface and the near-bottom layers reach 10–12‰ and salinity in the near-bottom waters is higher than at the surface throughout the year.

In MPs of the NEBS, similarly to Sevastopol Bay, significant differences in salinity between the near-surface and the near-bottom horizons are not observed.

2.1.4 *Temperature and Ice Formation*

The temperature regime of the Black Sea is typical for a temperate zone basin. The marine area is characterized by stronger winds than the surrounding land. The prevailing winds blow along the north-western, western and south-western directions, which is connected with the atmospheric circulation at mid-latitudes in the Northern Hemisphere. During the summer months, winds blow from the sea to the land (onshore) during the day and from the land to the sea (offshore) at night. In winter, the surface waters of the north-western and north-eastern regions of the Black Sea cool down to zero and an ice cover can form in the shallow-water shelf zones, bays and limans where ports are located. The coldest area is the north-western part of the Black Sea. Similar water temperatures, or sometimes even lower ones, are registered in the Azov Sea. The Black Sea annual average water temperature varies from 11.0 to 11.5 °C near Odessa to 15.5–16.5 °C in the Sochi region. In the winter period, waters of the northern part of the sea are subject to cooling to zero, while those in the southern part preserve temperatures around 8.0–9.5 °C. In winter, Bulgarian and Romanian coasts are under the influence of much warmer Mediterranean cyclones. The average air temperatures in winter are in the 0–0.6 °C range for the coastal region from Sulina to Mangalia and vary from 0 to 5 °C along the coasts from Mangalia to the Bosporus strait.

During winter, the deepwater regions of the Black Sea have surface water temperature of 5–9 °C. In summer, the sea surface warms up to 28–30 °C near the shores and to 21–23 °C in the deepwater regions. According to monthly mean data, the annual variations in temperature of the surface water layer are about 17–20 °C in most of the Black Sea coastal regions. In Varna Bay, the maximal range of temperature fluctuations reaches 30 °C [208]. Similarly, in Odessa, Ilyichevsk, Yuzhny, Sevastopol, Novorossiysk MP and others, the water temperature fluctuations are in the range of 28–30 °C, while they are significantly lower (about 15–20 °C) on the southern coasts.

In the NWBS and NEBS, the thermocline starts to form during the period of spring warm up (March–April). In many zones of the Black Sea coast, spring arrives lately because of the slower heating of seawaters in comparison with land. In the same way, autumn is warmer and sunnier in the coastal regions of the Black Sea than in those farther from the coast.

In MPs, the warming up of water to 26 °C is critical for many hydrobionts. However, vagile hydrobionts have the possibility to plunge down into the deeper water layers to take advantage of the quick decrease in temperature with depth. In all ports with depths above 8–10 m, a two-layer structure of water masses can form in the near-bottom layer. In Sevastopol Bay, the differences in temperature between the surface and near-bottom layers sometimes reach 12–15 °C [3]. In the aquatories of Odessa, Ilyichevsk and Yuzhny MPs, the temperature gradient can sometimes exceed 10–12 °C [154]. In MPs, the formation of the thermocline has various consequences.

The light shadows observed in MPs worsen the warm up of the upper layer and sometimes cause local temperature differences of up to 2–3 °C.

On the northern coast of the Black and Azov Seas, air temperatures up to 30 °C are registered. On the southern coast of the Black Sea, the air temperature in January is everywhere positive with values of 5–6 °C above zero. In the northern regions, the water cools down to 1–3 °C below zero. At such negative temperatures, the water surface in the NWBS, in Dniester and Dnieper-Bug Limans and in the Azov Sea is sometimes completely covered with ice having a thickness of up to 50–60 cm.

The freeze-up period negatively influences the living population of water bodies, including MPs aquatories. The ice cover isolates water from air and interrupts the diffusion of atmospheric oxygen into the water. Furthermore, snowfall on the ice surface decreases the penetration of solar radiation into the water column, sharply slowers down oxygen production by photosynthesis and thus decreases oxygen concentration in water. In this context, oxygen consumption by bacteria, aquatic plants and animals for maintenance of their vital activity can contribute to the development of hypoxia and anoxia.

Division of MPs aquatories into separate harbours as, for example, in Odessa MP accelerates the formation of ice covers. To maintain ship-pilotage, ice in the shipping channels and in MPs aquatories of the Black and Azov Seas is periodically broken down with the help of icebreakers and powerful tugs. These measures also ameliorate the oxygen regime in the water column and at the bottom. In places where such operations are not performed, winter mass mortality of bottom fauna is not rare.

The aquatory of Novorossiysk port occasionally freezes in some years, but ice covers do not form in other ports situated southward of it. Mariupol port (Azov Sea) is open all year round, but the average duration of navigation with ice- channelling during harsh winters comprises 86 days from December until March.

In Odessa MP, ice formation does not occur every year and most often is limited to partial freezing of the harbours. The earliest complete freezing was observed at the end of second decade of December and the latest one in the first decade of March. The latest date of final cleaning from ice in Odessa MP was observed in the first days of April [16]. During severe winters, ice in MPs aquatories could mechanically damage hydrobionts-foulers in the near-surface horizon.

The internal variation of temperature in the surface layer of Grigorievsky Liman shows the maximum value in August (peak of the summer hydrological season) and the minimum one in February (peak of the winter hydrological season). This pattern coincides with the average long-term annual variation of temperature in the adjacent sea region. Maximal temperature near the surface is 26–28 °C, but can reach 30 °C in shallow waters [196]. This feature is also characteristic for the aquatories of other MPs. In Grigorievsky Liman, ice most often appears in December and complete freezing is observed in the second part of January. The cast ice thickness is 40–45 cm during moderate winters and 55–60 cm during severe ones. The aquatory cleaning from ice usually occurs at the end of February, although ice can persist until the end of March in severe winters. Ice formation in the aquatory of Sukhoy Liman follows the same pattern [154].

In Odessa MP, ice formation can sometimes be tightly connected with slack ice carried from the Dnieper-Bug Liman. The presence of three passages into Odessa port favours the entrance of slack ice which, driven by wind and flows into the aquatory, can form a continuous ice cover at low temperatures. The entrance of slack ice from the Dnieper-Bug Liman into Grigorievsky and Sukhoy Limans was not observed.

In aquatories of the NWBS, the slack ice is carried from the Dniester Liman and from the near Danube region. It does not usually diffuse farther south than Kaliakra Cape, even if it may be found up to Bosporus strait in very cold winters.

Near the Crimean shores, ice can be formed up to Tarkhankut Cape and slack ice can reach Yevpatoriya port. The slack ice carried out from the Kerch strait may float up to Anapa to the east and up to Feodosiya to the west.

Seawater ice contains less salt than seawater before freezing. Salinity of the lower ice layer, submerged into the water, is higher than that of the upper layer, even for the freshwater ice carried into the sea because its lower part is soaked with seawater. Ice melting leads to reduction of water salinity in the surface layer of MPs aquatories.

2.1.5 pH

In the Black Sea surface waters, the pH value ranges from 8.1 to 8.5 units. The pH value depends first of all on the quantity of carbon dioxide dissolved into the water and on the content of hydrogen sulphide. At relatively high hydrogen sulphide contents, such as those observed in the near-bottom water layer of MPs during periods of phytoplankton die-off after mass vegetation, the pH value can decrease to 7.6. In pore waters, the pH value sometimes reduces to 6.0. With the increase in temperature of the water surface layer and intensification of photosynthesis, the pH value increases because the water carbon dioxide content decreases.

In the eastern part of Varna Lake, the pH value varies from 7.20 to 9.85 (average 8.10) [144].

In Sevastopol Bay waters, pH ranges from 8.23 to 8.53 in the surface horizon and from 7.91 to 8.49 in the near-bottom layer. Average pH values of 8.20–8.30 are typical for bay waters in the winter–spring period, which could be also connected with river run-off influence [3].

In 1998–2001, the pH values in the surface layer of Odessa MP aquatory have changed from 7.80 to 8.90 during different seasons [16].

The range of pH fluctuations in Grigorievsky Liman waters is quite considerable: 7.50–9.20. The pH values in the surface horizon of liman waters reach maximal values of 8.80–9.20 in zones with active photosynthesis, during the warm period at mass development of phytoplankton and water "blooms". In the near-bottom horizon, the pH value reduces to minimum values for marine environment (7.50–7.80) during OM destruction and hypoxia. In the cold period, pH values are relatively stable at 8.30–8.40 from the surface to the bottom [196]. In Sukhoy Liman, the pH values are in the range 8.18–9.19 [154].

In MPs, water usually preserves the slightly alkaline milieu near the lateral surfaces of HTS under the vegetal fouling in the near-wall. Fouling by animals, especially by molluscs, is accompanied by the formation of acidified waters in the near-wall zone because of the release of carbon dioxide.

2.1.6 Vertical Water Exchange and Vertical Stratification of the Water Column

Vertical and horizontal gradients play an important role in coastal marine ecosystems, including MPs aquatories that experience strong freshwater inputs from rivers, land run-off, and industrial and municipal drainage. Because flows and general hydrodynamics in MPs aquatories are weakened by artificial protections, the importance of vertical water exchange increases. Owing to gravity and other factors, the near-surface and near-bottom water layers, adjacent to the film of surface tension and directly overlying bottom sediments, play a particular role in functioning of MPs aquatories ecosystems with regard to the distribution of live and dead OM. Isopycnic surfaces in the water column are also involved in the same processes.

Significant vertical gradients exist in the coastal part of the Black and Azov Seas, and in MPs aquatories with depths up to 20 m, for many important parameters such as temperature, salinity, redox potential, concentrations of gases (e.g. oxygen and hydrogen sulphide) of importance for hydrobionts, and biogenic substances such as phosphates, ammonium and others. The position of gradients in the MP water column vertically reflects the interaction of hydrodynamic factors, first of all the vertical and horizontal motions of water masses. In MPs, anthropogenic activity, mainly connected with ships movement (draft, velocity, number, frequency, etc.), has a mitigating effects on gradients. The smaller are vertical density gradients, the easier mixing extends to higher depths and vice versa.

In MPs, vertical stratification (i.e. division of the water column in temporally stable pelagial layers) is determined by a series of factors. In spite of the relatively small depths, a thermocline and halocline can form in MPs aquatories of the Black–Azov basin because of the significant differences in temperature and salinity between the surface and the near-bottom water layers. A water density jump, called pycnocline, is observed in correspondence of such gradients. Along with other consequences, the pycnocline hampers oxygen penetration into the near-bottom layer, thus favouring the accumulation of hydrogen sulphide near the bottom. In MPs, detritus accumulation occurs in the water column at the isopycnic surfaces. The lowered hydrodynamics and the presence of natural and artificial aquatories, along with deep ACs, speed up the pycnocline formation compared with the adjacent sea zones. In MPs, the pycnocline formation is usually determined by temperature differences and has a seasonal character. However, in some aquatories, e.g. the Donuzlav Liman, the halocline could be preserved throughout decades.

In the Black–Azov Sea basin, many MPs represent, properly speaking, temporary or permanent meromictic aquatories. These aquatories have a two-level stratification of the water column with the water of the lower layer being colder and saltier (and therefore denser) than the water of the upper layer. In such aquatories, the circulation usually takes place only in the less dense surface water layer.

In marine waters, density-related water stratification is stable and resistant to vertical mixing. Under stability is intended a vertical density gradient and correction for adiabatic temperature change. In surface layer and in the coastal waters adiabatic correction in comparison with vertical gradients of density is very small and it can be neglected [81]. The value of conditional density is a number derived from temperature and salinity and it is usually lower in the surface than in the near-bottom layer.

In the NWBS open waters, vertical density gradients could exceed 6 units of conditional density per metre of water thickness and even winds with speeds of 15–17 m s^{-1} cannot destroy the existing pycnocline [75]. In Grigorievsky and Sukhoy Limans, as well as in Odessa Gulf, the density gradient could reach 8–10 units. In the coastal desalinated regions of the Black Sea, the average water density is about 1.0115 g cm^{-3} with variations between 1.0100 and 1.0105 g cm^{-3} near the rivers mouths and in the surface layer of MPs during the summer period. As a rule, the water density is notably higher in the AC and in the near-bottom layer of an MP than near the surface. Owing to such differences, the near-bottom layer is characterized by a maximal vertical stability and gravitational mixing of heavy and light water masses becomes impossible. Densities of water layers in MPs are primarily determined by temperature and salinity, wind-related factors, input of river and surface run-off and water discharge from land.

During periods of strong offshore–onshore phenomena, both strong floods and artificial weakening of river run-off entering MPs aquatories could contribute to the penetration of saltier deepwaters into the near-bottom layer as a compensation flow. In such situations, the pycnocline position in the water column raises and its thickness decreases. As a consequence, vertical gradients of density and stability of water masses increase.

As it was already mentioned, the mixing and homogenisation of MPs water masses result from temporally and spatially variable turbulent motions. Weakening of such movements leads to vertical stratification of the water column. In MPs, the primary sources of micro- and mesoscale turbulent energy, which play the most important ecological role by influencing vertical water exchange and stratification of water masses, are mainly located at the sea surface. Generation of turbulent energy originates from dispersion of ordered average velocity of flows [1] fading with increasing of water depth. In the aquatories of MPs of the Black–Azov Sea basin having various levels of protection, the temperature of the surface waters is usually 1–3 °C higher than in the adjacent zones during spring, summer and autumn. In summer, the surface waters temperature decreases to some extent because of evaporation and of the increase in salinity caused by wind mixing. As a consequence, the density of surface waters could be even higher than that of the underlying layers. In spring and summer months, which are hotter and less windy than the average, a more marked development of the thermocline is observed, resulting in maximal stability of the MPs surface and near-bottom water layers.

During windless days in summer months, the nearshore and MP waters can warm up to 25–26 °C and, occasionally, to 28–30 °C in the surface layer. In AC and at MP bottom, water temperature can be around 10–13 °C [16, 154, 196]. Concrete and metallic construction, mechanisms and ships heat up during the day and give away heat to air and water masses during the night; which can influence the local distribution of temperature in MPs aquatories.

Water basins heat up owing to inputs of solar energy to the water surface and into the water column. Heat exchange with the near-bottom water layers depends on the temperature of the water surface. As a result of differential heating and cooling of water, the water column stratifies at various depths of the water basins. Stratification is firstly caused by physical characteristics and then by chemical and biological ones. Changes in water temperature along aquatory and depth are diurnal, seasonal, annual, and long term. They depend, first of all, on inflow regime and absorption of solar energy. The warmed-up superficial water layer is mixed with the deeper layers owing to various hydrodynamical processes [79, 85, 89, 145].

The temperature regime of water basins includes periods of daily and spring–summer warm up and night and autumn–winter cooling. These temperature fluctuations trigger the dynamic mixing of water masses. The artificial shading of various zones of MPs aquatories causes water temperature gradients. In MPs, the presence of light and wind shadows contributes to uneven water heating, formation of microflows and convective mixing. Convective water motion is connected with energy transmission in the form of heat.

As already mentioned above, such motion is observed in the near-surface water layer (0–30 cm) owing to interaction between the atmospheric air and solar heat and also in the near-bottom layer (0–20 cm) owing to geothermal processes.

MPs aquatories located in the NWBS are largely influenced by large rivers such as the Danube, Dnieper, Yuzhny Bug and Dniester. In these cases, the role of salinity in the pycnocline formation is extremely important. In spring, marine water from the regions near river mouths is desalinated by river waters and spreads over the surface of the saltier, colder and denser seawater, gradually mixing with it. Spreading of such waters takes place rather quickly. Under favourable wind conditions, the desalinated waters from the Dnieper-Bug Liman, reach the Odessa Gulf and enter the aquatory of Odessa MP within 2–3 days. A desalinated layer of about 3–5 m can be stable for most of the year and only autumn–winter storms will cause its mixing with the lower, saltier water layer.

In the Black Sea and in the NWBS, the pycnocline position changes during the year. Thus in the NWBS and in MPs, the pycnocline lowers to the bottom and disappears in winter due to the reduction of temperature in the upper water layer, the decrease in river run-off and the intensive mixing. In the aquatories of Odessa, Yuzhny and Ilyichevsk MPs, the differences in temperature and salinity between surface and near-bottom waters could exceed 10 °C and 10‰ during the summer months, but become very small in winter. The water column becomes more homogeneous. In the NWBS, the seasonal pycnocline extends over a layer of 5–15 m and regularly forms in MPs aquatories. In Sukhoy and Grigorievsky Limans, the pycnocline can form earlier, and disappear later, than in the adjacent zones of sea because of several factors. A seasonal pycnocline is registered in Sevastopol and Novorossiysk Bays and in other MPs with depths above 8 m and which receive inputs of river water to a different extent.

The oxygen concentrations in seawater usually range from 1–2 to 9–10 mg dm^{-3}.

The processes leading to seawater enrichment in oxygen are distinguished in O_2 diffusion from the atmosphere, the so-called "old" oxygen which enters the water column due to mixing of the near-surface water layer, and O_2 production during photosynthesis in the near-surface water layer, the so-called "young" oxygen. Processes lowering oxygen content of the whole seawater column include O_2 evasion to the atmosphere, its consumption for oxidation (biological and biochemical processes), organisms respiration and oxidation of various contaminating substances of natural and anthropogenic origin.

In semiclosed MPs aquatories, the formation of a sharp pycnocline hampers oxygen diffusion into deepwaters and favours its almost total or total depletion in the near-bottom layer. The absence of oxygen leads to diffusion of hydrogen sulphide from the bottom sediments into the water column.

The absence or presence of a certain degree of hypoxia in the lower water horizons is an indicator of the intensity of vertical water exchange in MPs. In recent decades, in the NWBS, hypoxia starts to develop in May and persists until September in the near-bottom layer. Hypoxia extends from the bottom up to depths of 12–15 m, for areas with total depths of 17–23 m, and up to depths of 5–6 m for areas with total depths of 10 m. In the middle of this period, the thickness of the near-bottom water layer contaminated by hydrogen sulphide varies from 2 to 6 m and H_2S concentration reaches 0.95 ml dm^{-3} [75]. In general, the thickness of the water column with conditions favourable for hydrobionts reduces to a couple of metres of the epipelagial zone.

In Odessa region, a strong pycnocline forms regularly in spring–summer because of the warming up of surface waters, the influence of river run-off (in spring) and the weakening of wind action (in summer). This pycnocline blocks mass and gas exchange between the surface and deep layers. Under such conditions, the wind-induced coastal upwelling of waters, which develops at offshore winds, acquires an important ecological role [82, 172].

In the NWBS, hypoxic zones coincide with zones of suppressed vertical exchange in 73% of cases. In other cases, the differences can be ascribed to inertia of the physico-dynamical and biochemical processes [75].

In Sevastopol Bay, the difference in water density between surface and bottom waters was reported not to exceed 0.4 units of conditional density in spring during the period of pycnocline disruption. At the beginning of summer, the whole water column was weakly stratified and well mixed. In July, the difference in density between the surface and the near-bottom layers had raised to 2.4 units of conditional density and the possibilities for mixing worsened [123].

Balaklava Bay serves as an interesting and illustrative example of water masses behaviour in relatively deep aquatories with limited water exchange. Its length (incut depth into land) is about 1400 m with a width of about 300 m in the middle part and less than 200 m near the entrance. The bay is 34 m deep at the entrance and 2–4 m in its apex part. The salinity in its apical zone is 0.2–0.5‰ lower than that in the major part of the aquatory due to the influence of river Balaklavka run-off and surface land run-off. The Georgievsky source (a submarine discharge of fresh waters) is located at the south-western coast of Balaklava Bay, while a municipal collector of sewage waters, which locally causes desalination, is located on its south-eastern coast near Balaklava Cape [105, 106].

In warm months, the temperature profile shows a marked vertical stratification in most of the bay's aquatory. In the apical part, with depths of 4–5 m, the temperature of the water column remains uniform. On the other hand, in the rest of the bay's aquatory, the typical situation consists in the presence of an upper quasi-homogeneous layer (about 10 m thick and with an average water temperature of 19.8–20.0 °C) overlying a pycnocline showing a 0.5 °C m^{-1} gradient between the depths of 10 and 15 m. During the cold half-year period, the vertical stratification of the temperature profile is smoothed by convective flows. Vertical stratification of the temperature profile appears in April–May, remains stable until autumn, and disappears in October–November [105].

During offshore phenomena with associated upwelling, the nutrients' concentration in Balaklava Bay can decrease, by one to twofold during 0.5–17 days, to the concentrations typical for the Black Sea surface water thanks to turbulent diffusion. This happens even in extremely rare situations of heavy nutrient pollution (floods, heavy showers, dredging). The time to homogenize, via mechanisms of turbulent diffusion and processes of biological degradation, the concentration of nutrients in water masses interacting during upwelling processes varies from 0.2 to 11 days in situations of extreme pollution and from 0.2 to 2 days—for reference situations. In Balaklava Bay the whole upwelling cycle is completed within 1–10 days [106]. These observations provide information on the time scale of upwelling phenomena and confirm that they represent an important mechanism for reducing biogenic matter concentrations in MPs aquatories with depths of 8–10 m and more via water exchange, mixing and diffusion.

The upwelling phenomenon is also observed in the open sea, in particular, in the region of cape Kaliakra [62]. On the 28th of August 1986, the water temperature at the surface was 26 °C in spite of the strong south-western wind. After three days, the water temperature was only 7–8 °C. At the same time, all metal objects and dredges immersed in water turned black, which testified the presence of hydrogen sulphide. Oxygen was indeed lacking in the whole water column.

The pycnocline hampers the diffusion of oxygen into the near-bottom layer, weakening vertical mixing during warm months. In ports of the NWBS, NEBS and in Sevastopol Bay, the months of July, August and September are the most critical with respect to this aspect. The period of temperature stratification, during which the vertical circulation of the waters does not take place, is called stagnation period. During stagnation, an oxygen dichotomy phenomenon appears and oxygen content in the near-bottom layer is lower than above the pycnocline. Destruction of the pycnocline facilitates water exchange in the near-bottom layer and favours the increase of dissolved oxygen concentration into it.

During the spring–summer–autumn period, the high position (at 10–15 m depth) of the thermocline in the Black Sea ensures the increased stability of water masses in the upper part of the eutrophic zone and sustains development of diatom populations in well illuminated conditions [161]. Accounting for the presence of lower depths, this regularity also manifests in MPs aquatories.

In MPs, the processes connected with excessive eutrophication are exacerbated by protective structures, lowered horizontal water exchange, and formation of the pycnocline. This is why the worsening of vertical water exchange and prolonged stratification of the water column causes extremely negative, long-lasting consequences in their ecosystems.

Deepening of MPs aquatories to 8–20 m by dredging creates conditions for prolonged stratification of the water masses. In MPs, the presence of the deepwatered ACs favours the entrance into their aquatories of "tongues" of saltier seawaters, with thickness of a couple of metres, during the offshore phenomenon. In some situations, this phenomenon can contribute to the onset of upwelling and also intensify vertical stratification. When the low water layer is denser and the vertical mixing is hampered, less oxygen enters into the layer below the pycnocline and conditions become favourable to the formation of hydrogen sulphide, first within the sediments and then in the water column. The intensity of vertical water exchange is one of the major factors regulating hydrochemical and biological processes in the near-bottom water layer and in the benthal zone, particularly in MPs aquatories. The intensity of vertical gradients of water masses in the NWBS is one order of magnitude higher than in the open sea [154].

In the 2000s, the oxygen minimum in the NWBS was registered in the water layer having a thickness of 2–3 m and located directly below the pycnocline [174]. Hypoxic and anoxic layers can extend through the whole water column below the pycnocline. Such situations were observed in Mariupol port [85], in Varna Lake [142, 143] and in other MPs. In 1998–2001, regular phenomena of prehypoxia and hypoxia were registered in the near-bottom layer of Odessa MP from June to August [127].

2.1.7 Colour, Transparency and Illumination of the Water Column

The presence of ships of various sizes and in variable number inside MPs aquatories decreases the total area of the water surface impacted by sunlight and through which matter and energy can be exchanged. At the same time, the bottom area remains constant.

In MPs, the colour of water depends on season, depth, sediment character, quantity and characteristics of the suspension, transparency and illumination as well as on the port geographic position. The prevailing water colour usually has greenish-yellowish tints in ports of the NWBS and the Azov Sea, while blueish- and blueish-green tints are more common in ports of the eastern and southern parts of the Black Sea and at Crimean shores.

The NWBS coastal waters, particularly those of MPs aquatories, generally have greenish, yellow and yellow-green colours throughout most of the year; which is usually an index of high phytoplankton production. In MPs aquatories, the inflowing rivers, the discharges of waste and drainage waters, and the surface run-off carry differently coloured sediment particles which have various influences on the colour of water and on some hydrochemical and hydrobiological parameters. Many MPs are surrounded by cities and industrial constructions. An artificial increase of MPs aquatories protection increases the influence of the superficial flow from the drained areas into their ecosystems. Such influence reaches its maximum during floods, heavy showers and snow melting. During such events, the amount of sediments entering port aquatories is higher than during other periods. The Bolshoy Dalnik river waters flow into the aquatory of Ilyichevsk port, the Maliy Adzhalik river into the aquatory of Yuzhny port, the river Chernaya into the aquatory of Sevastopol Bay, the Tsemes river into Novorossiysk port aquatory, and the Tuapse river into the eponymous port. The suspended and dissolved substances of different origin entering MPs aquatories with the river run-off change the water colour and its transparency.

The ability of water to transmit sunlight depends on the characteristics of the water itself and on its transparency. The water in MPs aquatories can have various colours depending on season and meteorological conditions. The colour of water directly depends on the colour of substances dissolved into it as well as on suspended mineral and organic particles and microorganisms (ultranano- and nanoplankton) inhabiting the water column. In MPs, the mass development of small-sized planktonic hydrobionts, usually planktonic algae and protozoans, is quite often observed and causes water "blooms" phenomena. During planktonic blooms, water sometimes acquires a brown or reddish colour along with its usual green and yellow shades.

In MPs, the stirring of bottom sediments caused by strong storms, dredging, hydrotechnical construction and movement of deep-draft vessels causes changes in water colour and transparency that can last up to 2–3 days. In such cases, water acquires grey, yellowish or brown shades. During erosion of shores in the NWBS, the water acquires a reddish colour because of the presence of red clays and usually clears up over two days after the storm has ceased. Pale blue and grey clays give a corresponding colour to water.

In Sevastopol Bay, the water colour changes from blue-green (or green) at the bay entrance to yellow in its apical part [3]. In the river Tsemes estuary, located in Novorossiysk port aquatory, and in the adjacent north-western part of the port, the water has a constant lemon-yellow colour [51]. Water in Odessa, Ilyichevsk and Yuzhny MPs aquatories generally has a yellow, green or dimmed-yellow colour, although blue tints appear in winter [16, 154, 196].

Seawater often turns toward yellowish colours in the presence of high concentrations of the so-called "yellow substance" which is a mixture of dissolved OM of various origins; mainly dead cells of phytoplankton organisms [63, 161, 181]. In MPs, the water transparency depends on the concentration of suspended particles of various size and origin and on the presence of the aforementioned "yellow substance".

In the Black Sea coastal zone and in MPs aquatories, upwelling is frequently accompanied by a water "albication" phenomenon. It occurs when, in the presence of surface run-off phenomena, the cold deepwaters (which during anoxia become slightly acid and contain hydrogen sulphide) react with bicarbonate, forming carbon dioxide and water. During the subsequent upwelling episode, the deepwaters bring a high quantity of dissolved carbon dioxide and calcium ions towards the surface where the water heats up. Upon heating, carbon dioxide evaporates into the atmosphere, thus increasing the water pH and forming numerous fine crystals of aragonite ($CaCO_3$). These crystals gradually sink to the near-bottom horizon, where they dissolve again in the slightly acid environment of the near-bottom water, or settle onto the sediment. The seawater with the suspended aragonite fine crystals is opalescent.

In MPs aquatories, the water transparency has a seasonal character. The highest transparency is registered during periods of slow phytoplankton development, no release of reproductive products of molluscs-foulers and absence of strong storms. The relative water transparency in the open part of the Black Sea varies from 10 to 18 m in winter and spring, from 15 to 25 m during summer and from 15 to 18 m in autumn. In the coastal zone, water transparency for the same seasons ranges from 6 to 10 m, from 10 to 15 m and from 5 to 10 m, respectively. In the NWBS, near the estuaries of big rivers such as Rioni, Kizilirmak, Yesilirmak and Sakarya, transparency usually varies between 2 and 4 m during the whole year [161, 189]. It is considered that the highest recorded transparencies in the Black and Azov Seas were about 25 and 6 m, respectively [73].

In waters with reduced hydrodynamics such as those in MPs, mineral suspended particles usually sink to the bottom quicker than the other components, so that the bulk of suspension comprises living vagile forms and relatively recently died cells.

In MPs, the vertical distribution of waters with different transparency is connected with the position of the pycnocline where plankton, detritus and other organic and inorganic suspensions accumulate. In the near-bottom layer under the pycnocline, the water transparency can sometimes exceed that near the surface by twofold.

In various ports of the Black–Azov Sea basin, the water transparency varies widely; from 0.5 to 7–8 m. The water transparency in the Azov Sea and in the NWBS ports is usually lower than in Crimean ports, the NEBS and near the coasts of Turkey.

In Sevastopol Bay, the water transparency changes in the different zones of the bay, but it is usually less than that in the open sea. The gradual reduction of transparency towards the upper part of Sevastopol Bay is a general feature during summer and autumn. Thus, transparency is 7.5–6.5 m near the entrance into the bay, 5.5–4.0 m in its central part—and about 1.5 m in the apical part. In the zone adjacent to the bay entrance, the average concentration of total suspended matter in surface waters is 0.57 mg dm^{-3} (range 0.02–1.8 mg dm^{-3}). Inside Sevastopol Bay, the average concentration of total suspension in the surface layer is 1.14 mg dm^{-3}, with a range of 0.3–5.0 mg dm^{-3} [44]. In a significant part of Novorossiysk port aquatory, the water transparency often does not exceed 0.5 m [51]. In Odessa, Yuzhny and Ilyichevsk MPs, the water transparency varies between 2 and 5 m, sometimes dropping to less than 1 m [16, 154, 196].

As already mentioned, the variability of seawater transparency is substantially connected with the quantitative distribution of pico- and nanoplankton along with 'net phytoplankton' and composition of its dominant species. There is a direct and indirect link between transparency, illumination of the water column and phytoplankton development, so that a certain autoregulation of primary production is achieved in the photic zone [119, 161]. This autoregulation mechanism, along with others, is also present in MPs aquatories ecosystems. In MPs, clarification of the water column occurs during intensive grazing of phytoplankton by zooplankton which, in turn, becomes food for mass fish species (anchovy, sprat, horse mackerel, whiting, etc.).

The sunlight represents one of the energy forms that enter MP aquatories. Besides the natural sources, artificial light sources (e.g. coastal and ships' illumination) are present in MPs. Most of the light energy absorbed by seawater transforms into thermal energy, increasing water temperature. A relatively small share of the incident light is used by plants for the transformation of inorganic carbon (CO_2) into organic carbon during photosynthesis; which maintains life in the marine environment. Because the photosynthetic process is initiated mainly by solar radiation, the amount of light energy available to pico-, nano-, micro- and macrophytes determines the quantity of OM primary production.

The amount of light energy entering aquatic ecosystems is determined, first of all, by illumination, i.e. the light quantity per unit of surface, and transparency, i.e. the ability of water to transmit radiation, including visible light. Light is required for the normal photic activity of autotrophs as well as for the normal development of reproductive products, reproduction, development and growth of heterotrophic organisms. In MPs, illumination decreases quickly with depth and the intensity of oxygen production by photosynthesis weakens accordingly.

In MPs, the seawater is a polydisperse system of chaotically oriented, heterogeneous organic (living and dead) and inorganic particles; all of them with the ability to disperse and absorb light. Accumulations of particles, frequently observed in the pycnocline zone, can cause dispersion of light upwards, thus increasing light exposure in the epipelagial zone. Phytoplanktonic organisms, actively migrating according to diurnal rhythms, order this process to a various degree. Part of the light energy directed from the water column upwards is subject to various degrees of refraction and does not go out from water, but it is again reflected downwards from the water surface although with weakened intensity.

A flow of radiant energy from the sun and sky reaches the sea surface. Sunlight comprises the main part of total radiation. Light enters the water surface in a form of direct and diffused radiation. A ray of light falling on the water surface is subject to diffraction, polarization and spectral separation, besides reflection and refraction. Furthermore, while passing through the water column, light is absorbed and reflected from different particles suspended in water, so that a different amount of light energy reaches the different water horizons. In MPs water column, the light spreads not only in the top-down direction, but also bottom-up and horizontally, also getting to the lateral surfaces of HTS in a weakened state. Thus, only a part of the light entering water becomes a source of energy for hydrobionts.

In the water column of MPs, the sunlight penetrates in the same way as in the adjacent regions of the Black and Azov Seas, i.e. half of the radiant energy is absorbed in the upper half metre of the water layer. In MPs the process can be affected by light shadows formed by mechanisms, various constructions, warehouses located on piers and berths as well as protecting HTS, berths and ships. Thus, the concentration of various suspensions and live planktonic organisms is often higher in MPs aquatories than in the open sea.

The major part of the energy falling onto the sea surface is transformed into heat. In MPs, part of the heat is spent on heating the big mass of HTS and ships. At night, this heat partially returns to the water masses and to the atmosphere.

The light rays are refracted at the interface between two different milieus. In seawater, this phenomenon occurs near the surface. Falling onto the water surface, a ray of light is partially reflected into the atmosphere and partially penetrates into the water. According to the laws of optics, the incident and reflected rays form identical angles with respect to the normal to the reflection surface, while rays that enter into the water are bent towards the normal by a certain angle. The angle between the direction of the refracted ray in water and the normal to the water surface is called the angle of refraction. This angle is always lower than the incidence angle that is the angle between the ray direction in air and the perpendicular to the water surface at the point of the ray penetration into water. A refraction angle never exceeds 48.5° which causes the formation of shadow zones (Fig. 2.17).

Thus, upon crossing the water surface, a light ray refracts and enters into the water at an angle of no more than 49°. At such an angle between light and the vertical, the border between light and shade already appears. Passing from a water layer with one salinity value into another, the rays of light are again refracted. The refraction index is different for rays with different wavelength and has the lowest

Fig. 2.17 Scheme of light transmission in the central part of MP aquatory and in the zones adjacent to HTS (*hatched area*: shadow zones of periphytal)

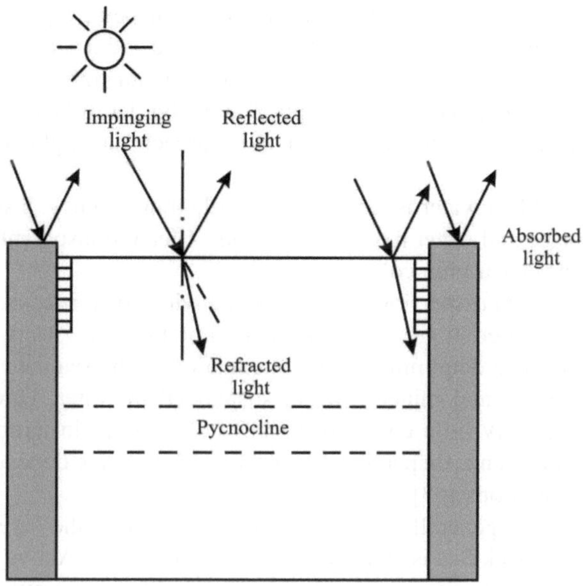

value for red light. Rays of this part of the spectrum, which is necessary for photosynthesis, are already absorbed in the upper water layer (about 5 m thick), yellow ones around the depth of 10 m, and green ones around 15–20 m. In general, the salinity of seawater does not markedly reduce light intensity. On the other hand, sea roughness changes illumination because of the formation of a water layer with a high quantity of air bubbles on the sea surface and can cause losses up to 15% [85]. In MPs aquatories, roughness is artificially weakened and illumination is slightly ameliorated compared with the open sea. The roughness causes fluctuations of the rays' propagation directions in water column because rays are refracted by a disturbed water surface. The roughness influence on the interaction of light with the sea surface becomes noticeable at a sun height below 20°. At low sun height, the reflection coefficient drops sharply because of roughness and more light enters into the water column.

Near all the infrared and the major part of the ultraviolet radiation of solar emission are already absorbed in the first metre of the water column. Since the long-wave part of spectrum (red, orange, yellow colour) carries the highest amount of energy, its absorption causes a more marked warming up of the upper metre of the water column. At lower depths, the role of light and chemical energy increases and the role of heat decreases sharply [36, 60, 63, 85].

In the relatively thin, but turbid layer of sea coastal waters inside MPs, complete (100%) absorption of long (infrared) as well as short (ultraviolet) wavelength is observed because of the presence of various suspensions. The minimal absorption occurs in the blue sub-spectrum near 0.47 μm. In the open ocean, a water column with a height of 10 m absorbs from 33 to 80% of the blue light, while the corresponding figure for the coastal waters in MPs is 94–100% [36].

The rays of light often refract and disperse repeatedly while diffusing from one water layer into another. Suspended live and dead particles and also the fine bubbles of gases constantly present in water favour light dispersion. Since accumulation of suspended particles takes place at the border of layers with different densities, layers differing in their degree of transparency and light dispersion appear in the water column.

When water is turbid, i.e. of little transparency, a strong dispersion of light takes place and water has a green colour. Clean, transparent water dissipates light weakly and has a blue colour.

Light dispersion in the marine environment is caused by the cumulative action of two different processes: dispersion by the water itself, which is a constant and main factor in determining light attenuation in the sea, and dispersion by the particles of organic and mineral origin suspended in water. Direct diffusion by water varies relatively little and only because of changes in temperature and pressure. On the other hand, dispersion by suspended particles depends on variations in their concentrations [63].

It is generally agreed that during summer the layer of optimum illumination in the Black Sea is located at the depth of 5–7 m. Above this depth, the photosynthesis of planktonic algae could be inhibited by excessive illumination and below it by lack of sufficient light [161]. In MPs, the optimum layer is close to surface because of poor illumination due to a variety of causes.

As it was previously mentioned, a significant part of the solar energy penetrating into water is already absorbed in the upper water layers. The amount of light energy absorbed by a water layer with a thickness of 1 m is about 92% for a water transparency of 0.7 m and a sun altitude of 12°, up to 76% at water transparency of 1.2 m and sun altitude of 39° and up to 46% at a water transparency of 1.6 m and sun altitude of 58°. At water transparencies of 2 and 7.5 m, the photosynthesis is most intensive at the depths of 1.5 and 6.3 m, respectively [145]. In MPs, because of the peculiarities of their ecosystems, depths from 0.2–0.5 to 1.5–3.0 m are optimal for photosynthesis.

Exposure to solar irradiation from above is a function of the sun elevation. Scalar irradiance in MP depends on sun altitude, light wavelength, water transparency and depth, presence of light shadows, orientation of the HTS rising out of the water and presence of ships standing at berth. In the winter period, the presence of an ice cover, possibly covered by snow, weakens the light penetration into the MP water column. The amount of solar energy entering the ecosystems of MPs aquatories depends on daylight duration and irradiance rate. The span of time during which aquatories are exposed to solar radiation depends on shores' height and harbours' orientation, especially at low sun elevations in the morning and evening hours. These conditions are different in each port.

The lower the sun elevation above the horizon, the more light rays are reflected by the water surface and do not enter the water column. At a sun height of 10°, less than two thirds of the light energy enters into the water. At such height, the distance that a ray must cross to reach a certain depth is 1.5-fold the value of that depth, compared with values of 1.32-fold for a sun height of 30° and 1.15-fold at a sun height 50° [85]. All other conditions being equal, the longer is the path of a ray, the more it is weakened.

The irradiance of berths, moles and breakwaters, which usually have vertical surfaces, represents the so-called horizontal irradiance constituent and change within limits of 75–85% of total irradiance [63]. Light shadows in MPs aquatories cause even more reduction of HTS lateral surfaces irradiance. This circumstance permits to consider that, in MP aquatories, inclined HTS surfaces, oriented southward, south-east and south-west, and to a lesser degree, eastward and westward are a priori more favourable for macrophytes foulers. Heliophobic macrophytes can be found at shallower depth in MPs than outside them.

2.1.8 Oxygen and Hydrogen Sulphide

The film of surface tension is the border between air and water. Water surface tension can be explained by forces of inter-attraction among the molecules in surface layer molecules, owing to which water experiences additional pressure.

An intensive exchange of gases usually occurs from the atmosphere to the marine environment (top down) and exchanges of water and salt take place from the aquatic environment to the atmosphere (bottom-up). Dilution in seawater of gases (e.g. oxygen, carbon dioxide, nitrogen, ammonia and hydrogen sulphide) that affect hydrobionts is determined by pressure, temperature and salinity.

Intensification of gas exchange occurs at breaking of wind waves and formation of water mist and sprays, as well as at appearance of ripples on the surface of seawater. The latter causes periodical tension and compression of the surface film and changes in the thickness of the viscous boundary water layers and air [137].

In the open sea and near natural shores in MPs, a multitude of "water–air" boundary surfaces is formed during strong storms, collectively creating a cloud of sprays and water mist with thickness up to a couple of metres. In such situations, gas, heat and moisture exchange are particularly active. In MPs aquatories, such effective mechanism of exchange is generally strongly weakened and is limited to the HTS zone.

In MPs, the appearance of various films, due to oil products and other substances, on the water surface influences gas exchange as does the accumulation of various small particles (suspension from live and dead planktonic organisms) that disturb the film of surface tension.

In MPs aquatories with reduced hydrodynamics, diffusion and submersion of surface waters (downwelling) provides partial entrance of oxygen into the near-bottom water layer. In the aquatories of many MPs, the superficial horizon is desalinated so that salinity could be 10‰, or more, higher in the near-bottom layer. In such cases, it is important to consider that oxygen in desalinated water is dissolving quicker than in saltier ones.

Oxygen and hydrogen sulphide play an extremely important role for hydrobionts inhabiting MPs aquatories. In these ecosystems, as well as in open sea regions, oxygen enters the seawater column via diffusion from the atmosphere, mixing caused by wind and wave action, rain water and as a result of the photosynthetic activity of micro- and macrophytes. In coastal regions of the Black Sea, where MPs are located, the value of daily oxygen production is 1.5–2-fold higher than in the deepwater regions [161]. In MPs aquatories, reduced hydrodynamics along with reduced wind and wave action hamper the diffusion of oxygen from the atmosphere. However, the same peculiarities favour mass development of some planktonic algae. In ports, conditions for photosynthetic activity of micro- and macrophytes in the near-surface layer remain quite favourable despite lowered water transparency and weakened illumination.

Atmospheric air is a main source of oxygen to water. Plants excrete a significant amount of oxygen during the photosynthetic process, during which water oxidation and reduction of carbon dioxide occur. Oxygen is essential to the absolute majority of aquatic hydrobionts from various systematic groups and ecological complexes. Furthermore, not only the presence of oxygen, but also its actual concentrations are important for hydrobionts.

Solubility of atmospheric oxygen in water depends on temperature, salinity and atmospheric pressure. Oxygen exchange between the aquatic environment and the atmosphere has a dynamic character and consists of two processes: invasion (entrance of oxygen from air into water) and evasion (passage of oxygen into the atmosphere following its oversaturation in the surface layer because of the plants' photosynthetic activity). These processes speed up at turbulent exchange between water masses and wind influence on the water surface.

The maximum oxygen water saturation cannot exceed 100% due to diffusion alone. However, oxygen saturation can exceed 100% and reach values of 150–200% and more as a result of algal photosynthesis. The brackish waters of the Azov and Black Seas can reach 250–300% oxygen oversaturation. This phenomenon appears when an especially high number of live phytoplankton concentrates in the surface layer at calm weather. Intensification of water mixing leads to the disappearance of oversaturation and even to partial under saturation [85].

Values of oxygen water saturation below 100% indicate unfavourable conditions for its diffusion, reduction of the intensity of its formation due to photosynthesis and significant consumption for oxidation and biological decomposition of OM. All these conditions can be found in the ecosystems of many MPs aquatories. In aquatic ecosystems, oxygen is indispensable for the decomposition of DOM, died plants and animals (detritus). During the normal decomposition process, the complex organic compounds are transformed into simple ones (carbon dioxide, water, nitrogen) and are cycled back into the sea matter turnover. In aquatic ecosystems, decomposition of OM mainly takes place owing to bacteria and fungi.

Bacterial degradation of OM depends on dissolved oxygen concentration. It proceeds normally at oxygen concentrations of 8 mg dm^{-3} and higher. The speed of decomposition reduces by 10% at a concentration of 6 mgO_2 dm^{-3} and by 25% at 4 mgO_2 dm^{-3}. At 2 mgO_2 dm^{-3}, the decomposition speed is only 40% of the corresponding value 8 mgO_2 dm^{-3} [145]. In the MPs near-bottom layer, oxygen content often drops to less than 1 mg dm^{-3}, especially in summer months. In ecosystems of MPs aquatories, the production of primary and, often, secondary OM is higher than in the adjacent regions of the sea and the elimination of excess OM is hampered. The combination of such conditions hampers the aerobic bacterial destruction of OM and favours the formation of zones of hypoxia and anoxia close to the MPs bottom.

Near the steep shores of the Black Sea (as it is the case in MPs), the onshore–offshore wind phenomena can cause the rise of coldwaters with low oxygen content from under the pycnocline up to the water surface. The rise up of waters from the depth of 20 m occurs during a couple of hours [85]. These phenomena are observed in many ports of the Black–Azov Sea basin. During offshore winds, waters from the depth of 10–20 m are pulled up to the shore and into MP aquatories via ACs. Rising bottom waters are always saltier than the surface ones and, in summer, also colder. In the warm period, water temperature near the shore can decrease by 10–12 °C within a day under the influence of offshore winds. During such days, water with an odour of hydrogen sulphide can approach the shores in summer and autumn [74]. This phenomenon is even more distinctly observed in MPs aquatories.

Two processes are the source of hydrogen sulphide in MPs: decomposition of sulphur-containing organic substances and reduction of sulphates (sulphurous compounds) by bacteria. Either process can dominate depending on specific environmental conditions.

Production of hydrogen sulphide during the decomposition of proteins also leads to the formation of sulphur-containing compounds such as mercaptans. In the absence of oxygen, the oxidation of hydrogen sulphide is done by sulphur bacteria and thiobacteria.

In Sevastopol Bay, the measured content of dissolved oxygen is in the range 4.94–8.28 ml dm^{-3} (78.0–143.7% saturation) in the surface layer (0–1 m) and 3.81–7.11 ml dm^{-3} (66.2–113.7% saturation) in the near-bottom layer. High levels of surface water saturation by oxygen were registered from May until September. The lowest content of, and degree of water saturation by, oxygen in the near-bottom layer were registered during the same months, which testifies about excessive amounts of suspended and dissolved OM along with insufficient water dynamics. Oversaturation of water by oxygen in May–June indicates an increase in the intensity of production processes and trophicity of bay waters [3, 88].

Various factors create preconditions for an oxygen deficit in the near-bottom waters of Sevastopol Bay during the summer period: reduction of wind mixing intensity, decrease in the horizontal water dynamics, relatively high water temperature, discharge of waste waters into the surface layer, unfavourable chemical and biological factors (increased BM content in surface waters, oxygen consumption for the oxidation of excess OM) [3].

In the middle part of Varna Lake, in 1976–1985, the oxygen concentrations varied from 4.36 to 13.31 mg dm^{-3} (average 7.30 mg dm^{-3}) near the surface and from 0 to 9.81 mg dm^{-3} (average 4.37 mg dm^{-3}) at the depth of 16 m. In the same period, the corresponding values for water oxygen saturation varied from 60.6 up to 212.5% (average 109.2%) at the surface and from 0 to 122.8% (average 59.4%) near the bottom [126]. In Varna and Beloslav Lakes, the oxygen content in the near-bottom layer is frequently reduced practically to zero during warm months [126, 189, 194].

A reduction of the oxygen concentrations to 1.0–2.0 mg dm^{-3} was registered in the near-bottom layer of Sukhoy Liman in 1970–1980. In 2001–2003, the maximal value of oxygen in the surface layer of Sukhoy Liman was 15.7 mg dm^{-3} (189.1% saturation). In some periods, oxygen content decreased to 0.80–1.10 mg dm^{-3} (10.0–15.0% of saturation) in the near-bottom layer [154].

In different periods of various years, for example in August 1995 and at the beginning of October 2007, a series of offshore winds reduced the oxygen water saturation to values not exceeding 30% in the near-bottom horizon of practically all the Grigorievsky Liman aquatory, creating a critical situation for many benthic animals [196]. During the period of intensive photosynthesis, the content of dissolved oxygen reached 15–16 mg dm^{-3} and up to 190% of saturation in the surface layer of the liman (Fig. 2.18). Oxygen concentration can change very rapidly during the same day. Thus, in August 2005, the measured oxygen concentrations decreased by 2 mg dm^{-3} 2 h after sunset. The oxygen content further decreased almost linearly with a speed of 0.4 mg dm^{-3} per hour overnight. Minimum water saturation was registered at 6 a.m., but had already increased by 12% 30 min after the sunrise [196] (Fig. 2.19).

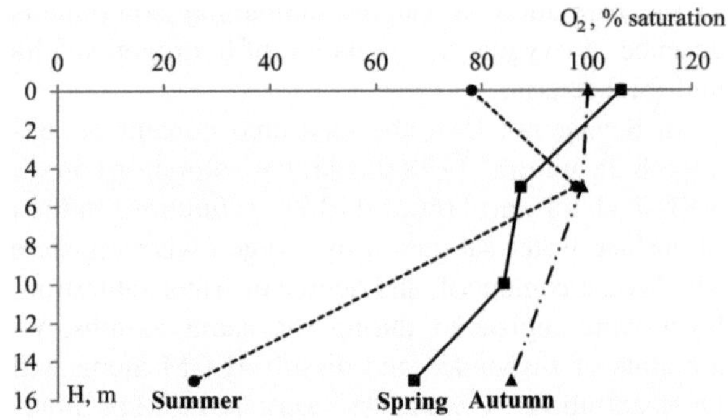

Fig. 2.18 Water oxygen saturation in Yuzhny MP (Grigorievsky Liman) in various seasons and at different depths

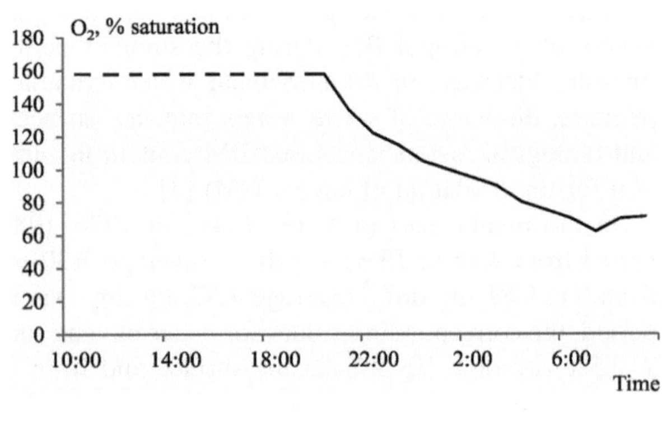

Fig. 2.19 Diurnal dynamics of saturation of the upper water layer with oxygen in the frontal zone of Yuzhny MP (Grigorievsky Liman)

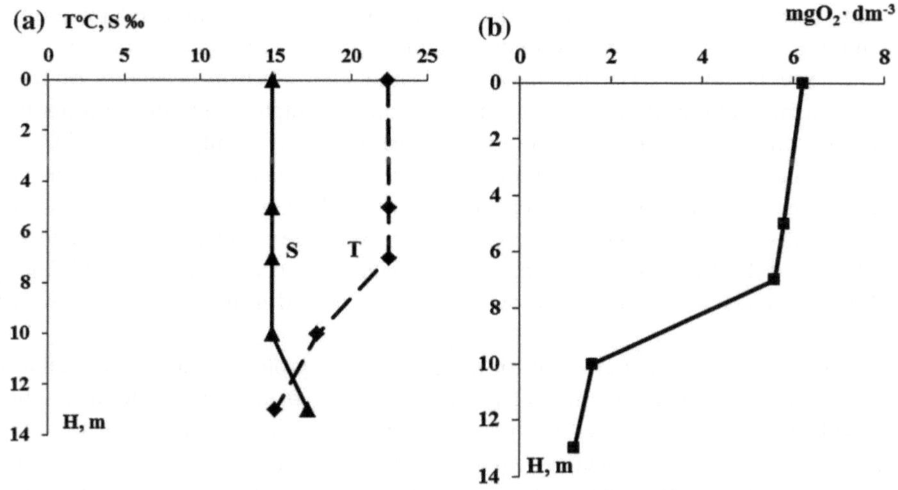

Fig. 2.20 Pycnocline formation (**a**) and the near-bottom hypoxia (**b**) in the Voennaya harbour of Odessa MP in August 2001

In the near-wall water layer, oxygen content increases during the day and decreases at night owing to the photosynthetic activity of micro- and macrophytes present in the composition of periphyton on MPs HTS.

The temporary presence of ships near berths in MPs can create wind shadows that reduce the possibility of oxygen penetration into the water column. Light shadows formed by ships also weaken the illumination of the water column and HTS lateral surfaces, thus influencing oxygen production by micro- and macrophytes to different degrees.

In many MPs (Odessa, Ilyichevsk, Yuzhny, Mariupol and others), the presence of hydrogen sulphide in the water column is observed after oxygen has disappeared in the near-bottom water layer. In the near-bottom layer of Odessa MP, dissolved oxygen content decreases to 0.5–1.0 mg dm^{-3} in certain periods, which leads to hypoxia and to the presence of hydrogen sulphide at concentrations of 0.5–1.0 mg dm^{-3} [16] (Fig. 2.20).

In eutrophic aquatories such as those of many MPs in the Black–Azov Sea basin, physical and chemical characteristics of the environment change significantly. BM and OM content increases, while the level of water saturation by oxygen decreases and anaerobic zones appear in the near-bottom water layers. At the same time, turbidity increases and water transparency decreases. The extent and rapidity with which eutrophication develops are determined by the inputs of biogenic elements as well as by other parameters including illumination and intensity of vertical and horizontal water exchange, water basin depth, water volume and degree of oxygen saturation. In MPs, depths are artificially increased and the water exchange weakened. In deep basins with intensive water exchange, eutrophication is slower than in shallow basins with low hydrodynamics.

In water basins, the anaerobic decomposition of OM accumulated in bottom sediments is a source of hydrogen sulphide. In the absence of oxygen, the sulphate-reducing bacteria use sulphates as the final electron acceptors of their metabolism to obtain energy and discharge hydrogen sulphide into the surrounding environment. Sulphate reduction consists in the reduction of sulphate ions (SO_4^{2-}) to hydrogen sulphide (H_2S) without oxygen participation. These processes are characteristic for practically all basins with high biological productivity where bottom silty sediments accumulate and water exchange is reduced. In aquatories with a good degree of oxygenation, sulphate reduction occurs only in the thickness of silty deposits and hydrogen sulphide is not registered in the near-bottom water layer.

Sulphur is an essential element for the synthesis of proteins in animal organisms and during the formation of sulphur-containing OM in plants. Thiobacteria are able to oxidize hydrogen sulphide to sulphur while releasing energy that can be used for metabolic processes; as it occurs with the use of solar energy during photosynthesis. Hydrogen sulphide belongs to the easily oxidizable compounds and it does not accumulate in waters where oxygen can easily diffuse.

Sulphate-reducing bacteria oxidize OM in the absence of free oxygen. They use the oxygen included in sulphate compounds for OM oxidation. These bacteria transform the sulphate ions (SO_4^{2-}) into sulphide ions (S^{2-}) and oxidize OM to carbon dioxide (CO_2) and water. When reacting with protons (H^+), sulphide ions form hydrogen sulphide. Slowly moving up in the oxygen-containing part of the water column, hydrogen sulphide oxidizes to sulphate. BM is released during OM decomposition by sulphate-reducing bacteria.

The process of hydrogen sulphide formation is most intense in the topmost 1–2 cm of silts, where the most active and numerous populations of sulphate-reducing bacteria also occur. Deeper into the silt bottom layer, their quantity and activity quickly drop [160]. The silt thickness 5 cm deeper under surface is often sterile [160, 161]. At a depth of 10 cm within a silty bottom layer, sulphate-reducing bacteria are usually absent.

At hydrogen sulphide concentrations of about 0.3 mg dm^{-3} in water, there is practically no oxygen left. In 2000, the hydrogen sulphide concentration in the NWBS was 1.7– 2.0 mg dm^{-3} not only in MPs zones, but also in the near-bottom layer of other areas [174].

Hydrogen sulphide generation from sulphates starts when the oxygen level in the near-bottom boundary water layer becomes less than 0.14 mg dm^{-3}; which occurs at low circulation, stagnation and abundant input of OM; i.e. typical conditions in MPs. The redoxcline moves from the sediments surface into the water column, leaving below itself an almost lifeless space containing hydrogen sulphide and ammonia [1]. In MPs, the upper level reached by the redoxcline in the zone of various berths can be inferred by the depth at which fouling hydrobionts are no longer present. The upper border of hydrogen sulphide dynamics determines the living vital water volume for oxybionts. Besides the long-term, annual and seasonal fluctuations, diurnal fluctuations of hydrogen sulphide level also occur under the influence of internal waves [143].

Academician Knipovich [85] described a situation that was already registered in Mariupol MP near the coal pier in the 1930s. The intensive formation of hydrogen sulphide was taking place in the silt which covered the bottom and silt was having a strong odour of this gas. In summer months, the oxygen content of waters having a total depth of 7.75–9.0 m was already reduced at a depth of 5.0 m and totally absent near the bottom. Sometimes, already at that depth, water was having a hydrogen sulphide odour. It was determined that a real hydrogen sulphide layer of local origin with a thickness up to a couple of metres can form in the waters of MPs aquatories.

Over the years, hydrogen sulphide in bottom waters of Varna Lake usually appears in summer and disappears at the end of autumn, although it sometimes persists through winter. In 1976–1985, the average near-bottom concentration of hydrogen sulphide was 1.38 mg dm^{-3} with an absolute maximum of 4.25 mg dm^{-3}. The total depth of Varna Lake reaches 18–19 m, but hydrogen sulphide was already registered at a depth of 10 m. In Varna Lake, the offshore rise up of waters containing hydrogen sulphide towards the surface is also observed at the lee shore and is often accompanied by fish mass mortality [144].

In Beloslav Lake, the lowest oxygen content was registered in summer close to the berths of Varna-West MP. As a result of summer stratification, oxygen is exhausted near the bottom and hydrogen sulphide appears. Hypoxia and anoxia lead to summer mass mortality of bottom organisms [194, 195, 225].

The river waters flowing into Varna and Beloslav Lakes favour vertical stratification by density. This stratification hampers vertical water exchange, especially in summer, when the less salty surface water has the highest temperature. Stagnation in the near-bottom layer leads to complete use of oxygen. Hydrogen sulphide is present in the near-bottom layer of lakes in summer and during warm winters and even all year round at low wind activity. In the deep part of Varna Lake, the bottom is covered with black silt containing hydrogen sulphide [189]. In Varna Lake, a biocenoses of hydrogen sulphide silt, where life is possible if hydrogen sulphide is absent, can exist at depths of more than 10 m during winter and early spring.

For most of the bottom black silts of Odessa, Ilyichevsk and Yuzhny MPs, oxidizing and reducing conditions alternate each other seasonally or over the years, which hampers the development of long-lived hydrobionts. In some seasons, the oxygen content in the near-bottom layer of Odessa MP reduces to 0.5–1.0 mg dm^{-3} and hydrogen sulphide concentrations increases to 0.5–1.0 mg dm^{-3} [16].

As a consequence of the redoxcline rise from the bottom, the thickness of MP water column suitable for the normal existence of hydrobionts diminishes, i.e. the water volume for oxybiont organisms practically becomes lower than the total volume.

When the diffusion depth of vertical flows and water exchange in MPs are limited by temperature and salinity stratification, oxygen under the pycnocline is consumed without the possibility to replenish its stock. As a consequence, oxygen concentration drops to values insufficient for hydrobionts respiration. Anaerobic bacteria develop below the pycnocline after total oxygen consumption.

2.1.9 Biogenic Matter

Living organisms, generally use about 40 elements for protoplasm synthesis; with carbon, nitrogen, oxygen, hydrogen, phosphorus, silicon, sulphur and iron being the most important ones. Substances containing these elements participate most actively in the vital activity of aquatic organisms. They are called biogenic matter (BM), from the Greek *bios*—life—and *genos*—birth. The remaining elements are contained in seawater and are required in lower amounts with calcium, potassium, magnesium and sodium being the most important. The named elements alternately pass from inorganic matter into living one and then into dead OM, participating in biogeochemical cycles of various complexity [111, 210, 231].

In general, the BM turnover in MPs aquatories ecosystems takes place in the same way as in the coastal marine ecosystems adjacent to them. However, there are some peculiarities that are specific to MPs. With regard to biogeochemical cycles, the atmosphere serves as the main reservoir in the turnover of gases (carbon, nitrogen, oxygen, water vapour). Elements included in the composition of sediments (phosphorus, sulphur, iron, silicon, etc.) are carried to the sea via surface and river run-off in both dissolved and particulate form. In the latter case, elements can accumulate in bottom sediments and return back into the biotic turnover following various processes.

The hydrobionts require carbon, hydrogen, oxygen, nitrogen, phosphorus, silicon and iron in relatively large amounts. The living organisms modify the content of phosphorus, nitrogen and silicon in seawater to the maximum extent. The turnover and distribution of these elements in aquatic ecosystems are controlled, first of all, by photosynthesis and OM decomposition [37, 95, 146, 202].

In seawater, BM is mainly present in a form of phosphates ions (PO_4^{3-}), nitrates (NO_3^-) and silicic acid (SiO_3^{2-}). In particular, nitrogen and phosphorus ensure the normal functioning of marine ecosystems, but at the same time they trigger various negative processes and consequences at extremely low and at extremely high concentrations. In BM compounds composition, the biogenic elements participate in the synthesis of OM via primary production, i.e. in photosynthesis and chemosynthesis. Biogenic matter (BM) comprises the mineral compounds of nitrogen (NH_4^+, NO_2^-, NO_3^-), phosphorus ($H_2PO_4^-$, HPO_4^{2-}, PO_4^{3-}) and silicon ($HSiO_3^-$, SiO_3^{2-}), as well as the organic compounds of nitrogen (N_{org}) and phosphorus (P_{org}), forming during the processes of vital activity of hydrobionts and during destruction of dead OM [146, 199].

In water, nitrogen also occurs as the molecular form N_2, whose importance is not high. As in the case of most terrestrial plants, marine plants are usually unable to directly assimilate molecular nitrogen, diffusing into water from the atmosphere and released during OM decomposition. This chemical form of nitrogen can be directly used only by some bacteria. Photosynthesis requires 1 part of phosphorus and 8 parts of nitrogen. During formation and decomposition of living organism's tissues, nitrogen and phosphorus are incorporated or released in a ratio close to 16:1. These elements are present in clean seawater at approximately the same ratio. According to the classical scheme, nitrogen and phosphorus are absorbed from the sea surface layer by phytoplankton and macrophytes. All other organisms use the primary production of phytoplankton and macrophytes, including nitrogen and phosphorus incorporated in their composition after uptake from water, via food chains of various complexities. The cycle ends when hydrobionts die and BM is released from OM during the process of decomposition by microorganisms and fungi. In the classical scheme, the role of micro- and mesoplankton in BM turnover is underestimated [159, 198, 213, 222, 223].

In MPs, the development of biological activity in the surface water layer is accompanied by uptake of BM and decrease in its concentrations. At a later stage, the formation of dead OM leads to BM enrichment in the near-bottom water layer and bottom sediments. At the same time, biological utilization of oxygen near the bottom for the decomposition of dead OM lowers its concentration. This is the reason why the continuity of biological processes in MPs aquatories directly depends on oxygen supply to the whole water column, i.e. on vertical water exchange.

The dead OM (accumulated in bottom sediments or present in the water column in suspended (SOM) or dissolved (DOM) form) is a potential source of mineral compounds of nitrogen and phosphorus. Destruction of the dead OM occurs both in the presence and absence of oxygen. The scheme of OM decomposition is presented below [46].

(1)	Aerobic oxidation $(CH_2O)_x(NH_3)_y(H_3PO_4)_z + (x+2y)O_2 \rightarrow xCO_2 + yHNO_3 + zH_3PO_4 + (x+y)H_2O$
	Nitrification $NH_3 + 2O_2 \rightarrow HNO_3 + H_2O$
(2)	Denitrification $(CH_2O)_x(NH_3)_y(H_3PO_4)_z + 0.8HNO_3 \rightarrow xCO_2 + yNH_3 + 0.4xN_2 + zH_3PO_4 + 4H_2O$

Aerobic mineralization of dead OM occurs at oxygen concentrations above 0.3 mg dm^{-3} and produces nitrates and orthophosphates. In anaerobic conditions, OM decomposition leads to the formation of ammonia nitrogen and nitrites [46].

In the last decades, a conception was elaborated based on fact that, in marine ecosystems, a large amount of BM circulates in the water column over prolonged periods (i.e. without settling onto bottom sediments) owing to nutrition and vital activity of pico-, nano-, micro- and mesoplankton [10, 11, 29, 55, 161, 198, 210].

The quantity of nitrogen and phosphorus compounds entering into MPs aquatories depends on many conditions, including the degree of development of plant cultivation, cattle and poultry stockbreeding and population density in the coastal zone. On average, each inhabitant introduces 2–4 g of phosphorus per day into aquatic ecosystems [45]. It is commonly believed that 1 g of phosphorus can provide an algal production equivalent to 50 g of organic carbon. Bacterial decomposition of dead OM requires 150 g of oxygen per 50 g of carbon [45].

In MPs aquatories, several zones offer abundant food for various uni- and multicellular hydrobionts, namely: the coastal zone of preserved natural shores, the HTS near-wall water layer, the near-surface water layer adjacent to the film of surface tension, the isopycnic surfaces in the zone of density jump, the surface of bottom sediments and the associated nepheloid layer.

Hydrogen sulphide (H_2S), which forms in bottom sediments under conditions of oxygen deficit, belongs to the strong reducing agents during OM decomposition. While reducing nitrites and nitrates, hydrogen sulphide causes the liberation of free nitrogen, thus explaining why some degree oversaturation by nitrogen always occurs in the reducing zone of MPs [232].

Ammonification of MPs aquatories, i.e. the accumulation of ammonium ions (NH_4^+), takes place when an aquatic environment enters a stable reducing state due to oxygen deficit [40, 68, 139, 157]. The nitrite ion (NO_2^-) acts as an intermediate compound in the chain of reactions transforming NH_4^+ into NO_3^-. The nitrite ion is a product of the reduction of nitrate (NO_3^-) or of the oxidation of nitrogen-containing compounds in lower redox states; i.e. NH_4^+, NH_2OH, N_2O, NO. Nitrite ion is very toxic [12].

Offshore wind phenomena can initiate the development of hypoxia in the near-bottom layer of the coastal zone, particularly in Odessa region [82, 171].

In summer, the development of hypoxia in the near-bottom layer favours the intensification of ammonium and phosphates fluxes from bottom sediments, slows down nitrification and causes a decrease in nitrates levels via denitrification processes. As a result, maximal concentrations of phosphates and ammonium are registered in the near-bottom water layer during the summer months [82].

In Odessa region, during the spring-summer period, concentrations of nitrogen and phosphorus organic compounds are exceeding those of their mineral compounds by, on average, two to threefold for phosphorus and three to sixfold for nitrogen. In autumn, the ratio between organic and mineral compounds of phosphorus reduces to one or less, while it remains the same for nitrogen. The maximal concentrations of organic nitrogen (730 μg of N dm^{-3}) and phosphorus (27 μg of P dm^{-3}) in the coastal waters are observed in autumn. Over a year, the average ratio between the concentrations of mineral compounds of nitrogen and phosphorus in water (N_{min}: P_{min}) is 10:1, while it increases up to 30:1 for their organic compounds (N_{org}: P_{org}) [13, 82, 171].

In coastal waters in general and in MPs aquatories in particular, the process of nitrogen regeneration from OM to mineral forms available for aquatic vegetation is slower than the process of phosphorus regeneration.

The regime of BM in MPs aquatories depends on the inputs of these compounds from natural and anthropogenic sources (external and local) and, in particular, from bottom sediments. Water temperature, illumination and hydrodynamics play an important role in determining the intensity of vital activity of aquatic organisms and the related processes of formation, transformation and destruction of OM. In eutrophic ecosystems, to which MPs aquatories belong, seasonal differences in BM content disappear because they offer the necessary conditions for an almost year-round mass production of primary OM by phytoplankton [154, 196].

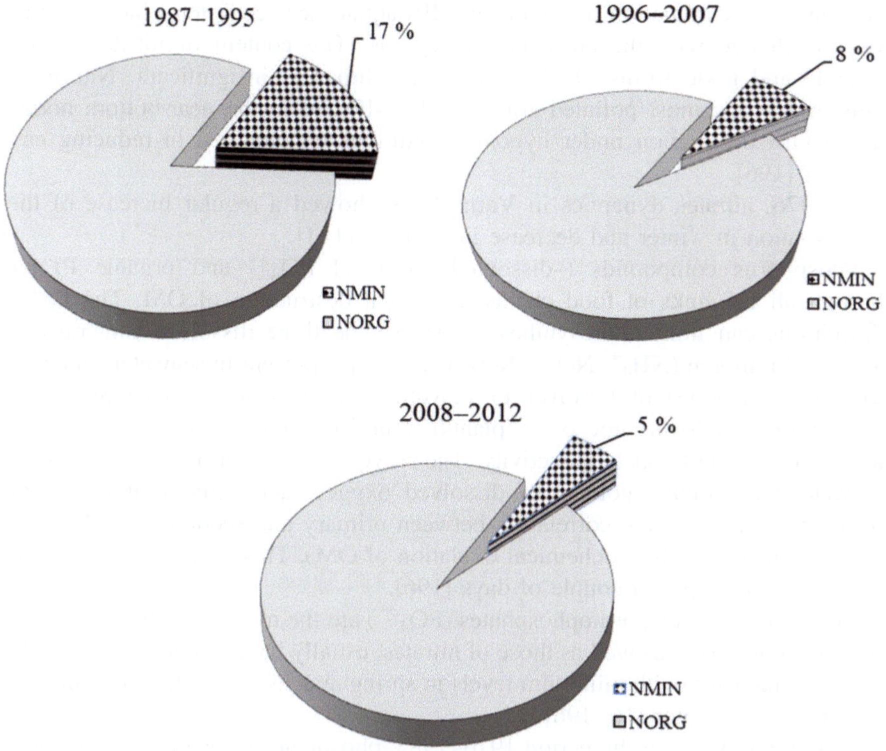

Fig. 2.21 Increase in the content of organic compounds of nitrogen (N_{org}) along with decrease in mineral forms (N_{min}) during eutrophication of Yuzhny MP aquatory (Grigorievsky Liman)

The major forms of nitrogen determining the biological productivity of aquatic ecosystems (i.e. the vital activity of micro- and macrophytes) are the inorganic ionic forms of nitrogen: ammonium (NH_4^+), nitrates (NO_3^-) and nitrites (NO_2^-). The total amount of nitrogen compounds is an indicator of the eutrophication level of an aquatic ecosystem. In MPs, organic nitrogen (N_{org}) is the dominant form and can account for up to 90% of the total as in the case of Grigorievsky Liman [196] (Fig. 2.21). The ratio between the mineral and organic compounds of nitrogen characterizes the relative rates of OM production and destruction processes. In Grigorievsky Liman, the average content of N_{org} had increased nearly sixfold towards the middle of the 2000s compared with the 1990s. Maximal values were registered during the warm period in the whole water column. The decrease of temperature slows down OM destruction and leads to its accumulation in the near-bottom horizon and in sediments [196].

The concentration of nitrates (NO_3^-), i.e. the oxidized mineral form of nitrogen which is more actively used by micro- and macrophytes during the formation of new OM, has slightly increased in Grigorievsky Liman waters during the last decades. This phenomenon can be connected with various processes, but it confirms the constant accumulation of nitrates in MPs aquatories ecosystems having limited water exchange with the adjacent sea regions. The content of nitrites (NO_2^-), unstable and toxic forms of nitrogen, in the liman is insignificant. Nitrites are registered in the most polluted zones of the MP and in the near-bottom horizon during OM destruction under hypoxia conditions encountered in reducing environment [196].

In 1976, nitrates dynamics in Varna Lake showed a regular increase of their concentration in winter and decrease in summer [144].

Phosphorus compounds (–dissolved, –mineral PO_4^{3-} and organic P) pass through all the links of food chains until total destruction of OM. The lack of phosphorus can limit photosynthesis, because the three dissolved mineral compounds of nitrogen (NH_4^+, NO_2^-, NO_3^-) are always present in seawater. In aquatic ecosystems, a constant turnover or recycling of phosphorus takes place from mineral forms into organic ones (plants), transformation along food chains and destruction owing to bacterial activity. The recycling speed of phosphorus depends on water temperature, content of dissolved oxygen, and activity of micro- and macroorganisms (i.e. the correlation between primary and secondary OM production; the latter causing biochemical oxidation of OM). This speed can vary from a couple of hours up to a couple of days [196].

In MPs ecosystems, orthophosphates (PO_4^{3-}) are the main phosphorus form and their concentrations, as well as those of nitrates, usually have a more or less marked seasonal character, with minimum levels in spring and summer and maximum ones in autumn and winter [16, 196].

In Varna Lake, over the period 1976–1985, phosphates showed a regular and, in the case of MPs, typical reduction in spring and summer, i.e. during the period of intense photosynthesis [144].

In the Black Sea, the surface layer contains an average of 3–4 mg dm^{-3} of dry OM [161]. In coastal waters, and especially in MPs aquatories, the content of dry OM could be one to twofold higher [154]. When OM decomposition occurs via aerobic oxidation, the decrease in oxygen concentration in seawater is directly connected with the increase of BM inputs into the water column [215, 226].

In the anionic form of silicic acid (SiO_3^{2-}), silicon is an important element in marine ecosystems, because it is necessary for the vital activity of diatoms and peridinian algae as well as of some zooplanktonic organisms. Diatoms and dinoflagellates use silicon for building their shells and exoskeletons. The river, rain and snowmelt waters, as well as the products of the decomposition of dead algae, are the main sources of silicon to marine waters. In some cases, diatoms and peridinian algae in marine ecosystems incorporate all the silicon available from the surface waters, thus slowing down their further development. Other microalgal species, which have no need for silicon, can then develop. This is an example of a self-regulation mechanism in aquatic ecosystems. Theoretically, in MPs surface waters, silicon could limit mass development of some microalgal groups such as diatoms and dinoflagellates. However, river waters, which are one source of silicon renewal, flow into many MPs, thus making the occurrence of silicon deficit unlikely. Furthermore, silicon accumulates in the sediments of MPs aquatories ecosystem and can return back into the water column from this reservoir.

A certain seasonal variability of silicon content is observed in the waters of MPs aquatories. This variability is closely linked to the development of diatom microalgae which are its main consumers. In Grigorievsky Liman aquatory, the ranges of silicon concentrations are larger than in the adjacent sea region because production–destruction processes are more active in the liman. According to the results of long-term observations, the silicon content in Grigorievsky Liman waters is higher than in the adjacent sea zone and ranges from 2.3 to 4.8 mg dm^{-3} during the year [196].

In 1998–2001, an increasing trend of silicon content was found in the waters of Odessa MP [127]. In Sevastopol Bay, the silicon content varies within 30–1500 μg dm^{-3} (average 382 μg dm^{-3}) in the superficial waters and 150–900 μg dm^{-3} (average 450 μg dm^{-3}) in the near-bottom layer. These values are significantly higher than the typical concentrations in the 0–20 m layer of open sea zones where silicon content is in the range 0–390 μg dm^{-3} [3].

In 1977–1985, the silicon content in the western part of Varna Lake varied from 0.5 to 6.12 mg dm^{-3} (average 2.08 mg dm^{-3}). In Varna Lake, the silicon concentration decreases in spring and summer and increases in autumn and winter in most cases [144].

In coastal aquatic ecosystems in general, and in the MPs aquatories ecosystems in particular, bottom sediments reflect the processes that take place in the water column. The solutes contained in the bottom sediments interstitial waters also characterize the state of bottom sediments and are an indispensable component in the BM balance of ports ecosystems. Studies in Odessa MP and Grigorievsky Liman showed that OM of allochthonous and autochthonous origin accumulates in the bottom sediments and, therefore, in interstitial waters.

Benthic organisms are particularly important in delivering OM to bottom sediments during their vital activity. They also transfer, use and convert OM, assist the aeration of the sediments upper layer, and participate in decomposition of OM and its diffusion into the near-bottom water layer. The OM destruction in bottom sediments of MPs takes place both in the presence and absence of oxygen. In Grigorievsky Liman, the main amount of labile BM (93–98%) in pore solutions is located in the upper ten centimetres layer of bottom sediments. This situation creates significant gradients of concentrations at the sediment–water interface and controls migration of BM from the bottom sediments into the near-bottom water layer [58, 196].

In bottom sediments, as well as in water, the aerobic mineralization of dead OM leads to the formation of nitrates (nitrification) and orthophosphates. Under hypoxia or anoxia conditions, OM decomposition takes place under reducing conditions with formation of ammonia nitrogen (ammonification) and nitrites. The speeds of OM destruction and BM turnover (recycling) in bottom sediments, as well as in water column, depend on temperature, oxygen content and vital activity of micro- and macroorganisms.

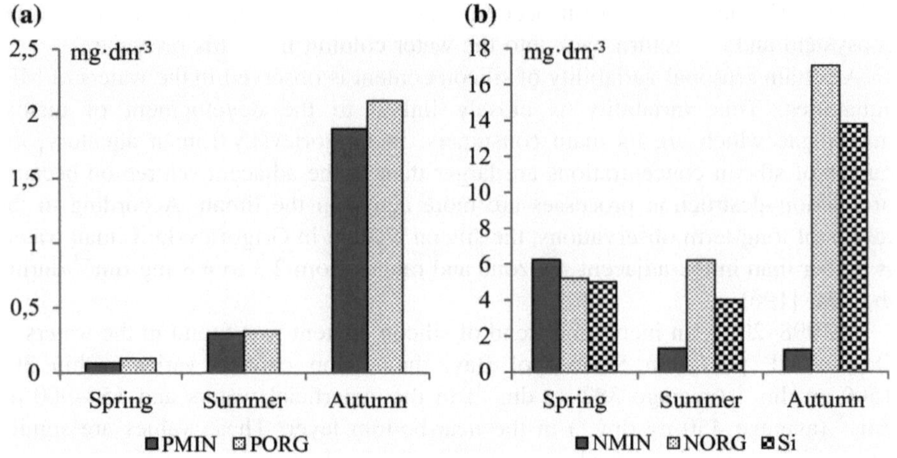

Fig. 2.22 Seasonal variability of nitrogen, phosphorus and silicon compounds in pore waters of the bottom sediments of Yuzhny MP (Grigorievsky Liman) in 2003

Long-term studies in Grigorievsky Liman determined that the BM concentrations are one order of magnitude (or more) higher in sediment interstitial waters than in the water column [196]. In MPs bottom sediments, the seasonal variations in the concentrations of nitrogen and phosphorus compounds are characterized by an increase from spring to autumn, especially for organic forms, which is connected with accumulation and destruction of OM (Fig. 2.22).

In MPs, concentrations of mineral and organic compounds are dozen times higher in the pore waters of bottom sediments than in the water column, so that pore waters serve as a source of additional inputs of BM into pelagial. BM release from pore waters can intensify the eutrophication of aquatories.

An increased content of BM is the main peculiarity of the north-western part of Novorossiysk port, into which flows the river Tsemes. In this aquatory, the amount of phosphates, nitrates and nitrites reach 44, 900 and 40–70 µg dm^{-3}, respectively. The highest concentrations of BM coincide with the period of rains and floods, in February–March, which favour mass development of phytoplankton. Oxygen content varies in the range 5.7–6.9 mg dm^{-3}, oxidability between 0.9 and 2.4 mgO dm^{-3} and BOD_5 between 1.37 and 2.20 mgO_2 dm^{-3} [50, 182].

BM content in the aquatories of Novorossiysk, Tuapse and Sochi MPs considerably exceeds that of the adjacent regions of the Black Sea. Chemical oxygen demand in the aquatories of Novorossiysk, Tuapse and Sochi MPs is usually two- to threefold higher than in the adjacent regions. The content of phosphates in water is two to fourfold and that of nitrates 4–10-fold higher than in the open water areas [52].

In studies carried out in Odessa MP in 1998–2001, an increasing trend was observed for phosphates, total phosphorus, ammonium, nitrates, and total nitrogen, while dissolved oxygen content and pH showed a decreasing trend. Each year, prehypoxia and hypoxia phenomena were recorded in the near-bottom layer during the summer period (July–August) [127].

In Sukhoy Liman, which is the aquatory of Ilyichevsk MP, a constant presence of BM was recorded in water, thus allowing a high level of OM production. Dominance of oxidized forms of nitrogen, i.e. nitrates (NO_3^-), over reduced ones, i.e. ammonium (NH_4^+), was observed only in early spring. In summer, the content of OM reduced, but the content of reduced nitrogen forms sharply increased in connection with increased intensity of the decomposition processes (increase of BOD_5) [129].

In Grigorievsky Liman, which is the aquatory of Yuzhny MP, typical natural cycles of phytoplankton development (spring and autumn maxima) are only weakly marked. This situation reflects the lack of nitrogen and phosphorus limitation for phytoplankton development, because excessive amounts of mineral N and P are present in the liman waters all year round [148].

The variation range of BM in Grigorievsky Liman is markedly higher than in the adjacent region of the Black Sea, despite similar values of hydrological parameters (but hydrodynamics is lowered in the liman). Production–destruction processes are active in the liman and they are more marked than in the adjacent regions of the sea. In Grigorievsky Liman, the value of BOD_5—an index of the potential bacterial oxidation of DOM over five days—changes from less than 1.0 to 6.0–7.0 $mgO_2\ dm^{-3}$ in various seasons depending on water temperature, initial concentration of dissolved oxygen and OM content. It is commonly accepted that the admissible limit values of BOD_{total} for marine waters should not exceed 3.0 $mgO_2\ dm^{-3}$. In 2004–2006, the COD in Grigorievsky Liman waters ranged from 0.5 to 6.0 $mgO_2\ dm^{-3}$ and was, on average, 3.0 $mgO_2\ dm^{-3}$. In 2007, it was 2.0–6.4 $mgO_2\ dm^{-3}$ for an average of 3.8 $mgO_2\ dm^{-3}$ [196]. These indices testify a high content of OM in the liman waters. In spring and summer, the content of labile OM in the surface layer of Grigorievsky Liman is twice than in the near-bottom, while OM accumulation in the near-bottom horizon is mainly observed in autumn [196].

The average phosphate content in Grigorievsky Liman is rather stable. The permanent summer enrichment of the surface horizon by phosphates and their accumulation in the near-bottom layer are connected with intensification of OM destruction and with phosphates desorption from bottom sediments. The latter phenomenon is characteristic for reducing conditions occurring during hypoxia and anoxia in the near-bottom horizon. In autumn, the content of organic phosphorus (P_{org}) decreases and becomes vertically homogeneous following the reduction of the vital activity of many aquatic organisms, OM mineralization and active water mixing [196].

In Varna Lake, the average BOD_5 was 5.02 $mgO_2\ dm^{-3}$ in the period 1976–1985. During the same period, in the central part of Varna Lake, the COD changed from 0.41 to 8.14 $mgO_2\ dm^{-3}$ (average 3.0 $mgO_2\ dm^{-3}$) in the upper layer and from 1.25 to 7.40 $mgO_2\ dm^{-3}$ (average 2.36 $mgO_2\ dm^{-3}$) near the bottom [144]. In Varna-West aquatory, the COD near the berths had an average value of 2.80 $mgO_2\ dm^{-3}$ in the autumn period of 1990–1998 [194, 195].

In Sevastopol Bay, the waters of river Chernaya, receiving wastewaters and sewages from various discharges, are an important source of BM. During the period 1960–1990, significant changes in BM concentrations and in the structure and quantitative indices of phytoplankton took place in the bay, witnessing the evolution of its ecosystems from an eutrophic phase to a hypertrophic one [3]. In Sevastopol Bay, the level of BM concentrations is 10–100-fold higher than in relatively clean coastal waters and in the photosynthetic layer of the open sea [123]. In the bay, the peak of BM accumulation is recorded in winter and is followed by a significant reduction of BM at the beginning of spring [59]. In summer months, the phosphates concentrations in the surface waters are higher than in the near-bottom, while the opposite situation occurs in late autumn. The water enrichment in phosphorus is connected with the intensification of OM destruction processes leading to the formation of mineral phosphorus [59].

In MPs, part of the DOM can be directly removed from the water column together with biogenic elements and assimilated by plants, as well as by other hydrobionts from different systematic groups, without undergoing total degradation [1, 2, 40, 156].

In MPs, water temperature is an important factor controlling the formation and destruction of OM primary production and the concentrations of BM in various horizons of the water column. Indeed, each algal species reaches its maximal photosynthetic rate within a defined temperature range [145, 180, 181]. Fluctuations of temperature outside this optimal range in semiclosed MP ecosystems lead to a reduction of photosynthetic production. Thus, BM concentration is indirectly connected with temperature. Mass development of planktonic algae leads to an increase of their metabolites and DOM in water and, on the other hand, to the possibility of DOM consumption by algae themselves.

In the coastal marine ecosystems, the foam forming in the surface water layer contains increased DOM and BM concentrations [71, 190]. The carry out of foam on flat shores provides support to organisms inhabiting the sea psammocontour on the one hand and is a mechanism for eliminating excessive amounts of dead OM and BM from the water column on the other hand. This mechanism is reduced to a minimum in the conditions typical of many MPs.

MPs aquatories become sedimentation traps owing to the artificial reduction of hydrodynamics, weakening of water exchange and presence of natural or artificial protection. At the same time, they trap dissolved BM, entering from the adjacent sea zones, which is used for cell construction by phytoplankton. At phytoplankton death, the remaining suspended OM is only partly removed outside the boundaries of MPs aquatories and its main mass settles to the MP bottom because of gravity and following the activity of various filter-feeding organisms (biological sedimentation). Sediments rich in OM are a food source not only for bacteria and fungi, but also for many protozoans and multicellular animals. While moving along the food chain, sediments turn into faecal pellets of different sizes which, in turn, become a food supply for other hydrobionts. Bacteria and fungi conclude the final destruction of OM. Because of the relatively small depths and weakened hydrodynamics, organic suspended matter settles to MPs bottom relatively quickly (hours, dozens of hours) as a component of detritus. MPs bottom sediments are a reservoir of BM.

2.2 Main Biotic Peculiarities

The MPs aquatories ecosystems are formed by natural and artificial components and differ in many respects from those of the adjacent sea zones. Along with the reduction of hydrodynamics, conditions for the increase of various carbon containing compounds and decrease in dissolved oxygen content are artificially created in MPs. In MPs, situations typical of ancient and modern ecosystems with saprobiotic (or similar) conditions often originate owing to anthropogenic impact.

It is well known that CO_2, O_2 and OM play an important role for the maintenance of modern life in seawater, while a harsh solar radiation is destructive for all living organisms on the planet. However, on the first stages of hydrosphere and atmosphere formation, solar radiation favoured the formation of complex organic (carbon containing) compounds, foregoing the origin of life, by abiotic reactions. At the same time, harsh solar radiation also contributed to the formation of free oxygen in the atmosphere. Free oxygen, although slowly, was formed chemically by water photodissociation under the influence of light. In the primitive atmosphere, a thin layer of O_3 already formed at relatively low oxygen concentration (about 0.001% of its modern level) and has since played a role of protective screen from harsh solar radiation until our time. In the same period, carbon dioxide (CO_2) was present in the reducing atmosphere and hydrosphere. Because of the processes described above, preconditions for the origin and development of life and for the synthesis of OM by heterotrophic living organisms appeared first in the aquatic environment and later on land.

Carbon is the main component of chemical organic compounds. Carbon has not only played a key role in the origin of life on Earth, but also in its sustention. Carbon content characterizes the state of aquatic ecosystems. In coastal seawaters, organic carbon enters in the composition of DOM, detritus and live OM approximately in the ratio 90:9:1 [32].

The major characteristic of carbon is its ability to form an immense amount of compounds, comprising a practically unlimited number of atoms. Carbon enters into the OM of all plants and animals in the form of various compounds and its average content in living matter is about 18%. In water, modern algae and aquatic plants obtain carbon by assimilation of CO_2 that diffuse into water from air or is released in water as a result of hydrobionts respiration or bacterial activity. Animals fulfil their carbon needs via feeding on live or dead OM. The total decomposition of OM ends with the formation of CO_2 and H_2O, i.e. carbon returns into the environment surrounding hydrobionts in a form available for autotrophs. Life in the aquatic environment in general, and in MPs aquatories in particular, can be viewed as a continuous transformation of carbon compounds.

The carbon dioxide present in the atmosphere plays a key role in the carbon cycle in the hydrosphere. Carbon dioxide not only dissolves in water like other gases, but also reacts with water forming carbonic acid and its dissociation products which, on the whole, constitute the so-called carbonate system of seawaters [27, 28, 34, 138]. The carbonate system has a dynamic character and anthropogenic perturbations of this system significantly disturb the life conditions of hydrobionts.

In seas, primitive aquatic ecosystems already originated with the appearance of the first life forms. These primitive ecosystems functioned as systems of reciprocally connected and regulating populations performing various functions (primary production, reduction, OM mineralization), organisms and components of environment, including carbon, oxygen, nitrogen, phosphorus and other elements.

The process of primary OM formation by photosynthesis did not appear immediately because it is a quite complex one. Many scientists consider that the primary marine ecosystems consisted of anaerobic heterotrophic organisms, mainly receiving energy from abiotically generated OM (i.e. the analogue of modern allochthonous OM at that time) which, gradually, was completely included into the biotic turnover [56, 124, 139]. With the origin and distribution of photosynthetic organisms, a source of oxygen became available for aerobic hydrobionts.

With the gradual increase of oxygen amount thanks to the activity of plants (autotrophs), various aerobic organisms (from protozoans up to mammals) appeared and spread into the seas. However, at the same time, some ancient anaerobic life forms have survived until modern times in certain niches. In general, the efficiency of aerobic metabolism is 30–40% higher than that of anaerobic one [139]. Disintegration of one molecule of glucose via fermentative processes (anaerobes) yields 2 calories compared with the 38 calories obtained during aerobic respiration [32]. This fact played, and still plays, a significant role in competition among various life forms and in their evolution into separate taxonomic groups of hydrobionts. In marine ecosystems, it is important that the functions of big biotic groups accomplishing matter turnover remain at equilibrium [32, 227].

During the evolution of the World Ocean and marine ecosystems, anaerobes were preserved, and continue to be preserved, in specific ecosystems reservoirs to ensure the continuity and circularity of biotic processes in the case of onset of reducing (anaerobic) conditions. For these reasons, energy (external sources, food) and matter (organic as well as mineral) are required by all hydrobionts, aerobes as well as anaerobes. In MPs aquatories ecosystems, unicellular (protozoans), as well as multicellular organisms, play various roles. The protozoans' cell is a whole organism and independently fulfils its vital functions: nervous, secretory, excretory and muscular. In contrast, multicellular organisms have specialized cells organized into tissues, organs and systems. According to classical concepts of life evolution, multicellular organisms originated in seas on the basis of various unicellular structures. Amplification of the organization created new possibilities in the competition for food and other resources.

The presence of regulation mechanisms of metabolic processes intensity permits to free-living and attached protozoans to switch into anabiosis at unfavourable environmental conditions. By accumulating in stagnant waters, the metabolites of protozoans act as a control mechanisms, hampering or suspending their reproduction when their population increases excessively [145]. Food availability is another important limiting factor regulating protozoans' quantity. Protozoans usually have a wide food spectrum which permits them to easily switch from one food item to another, thus avoiding trophic competition [218, 224].

In modern aquatic ecosystems, there is a great variety of hydrobionts differing on size, reproductive and locomotive strategies, feeding methods and food preferences. However, the mechanisms for digesting food and extracting energy from it are much less diversified.

Autotrophic and chemotrophic organisms extract from solution ions and other compounds they need via the process of mineral nutrition and formation of the primary OM. Food digestion by animals also requires the ingested food to stay in a liquid condition.

Nutritional strategies of heterotrophic organisms reflect their progressive adaptation to an increasingly more complete use of the possible food sources. It is believed that evolution of nutritional strategies went from a primary osmotic feeding towards a holozoic one. At the same time, marine invertebrates belonging to 11 taxa were shown to consume DOM (aminoacids) with 10 of them showing osmotrophy. Osmotrophy partly satisfies their food requirements, but nevertheless the main feeding strategy of heterotrophic hydrobionts is the holozoic one and food digestion takes place inside the organism. Some exceptions exist and food can be also digested outside the body, as in the case of *Rhizostoma pulmo*, commonly known as the barrel jellyfish [89].

For heterotrophs with holozoic feeding, food processing consists in mechanical fragmentation followed by chemical dissolution. The first stages of fermentation and absorption take place in an acidic environment, which was the common situation in the primary ocean.

Because dead and live OM consist of a limited number of main chemical compounds (proteins, fats and carbohydrates), the corresponding enzymes realizing the actual chemical reactions for their decomposition were also necessary. Such enzymes are necessary for all nutritional strategies (endocellular, extracellular, cavitary, membranous) and at all levels of living OM organization.

Aminoacids and small peptides, fats (glycerine and fatty acids) and monosaccharides, which are subject to absorption, are the principal final breakdown products of proteins, fats and complex carbohydrates. Cells, tissues and organs specific compounds are synthesized in organisms starting from these fundamental building blocks.

During extracellular nutrition, which is widespread among invertebrates, the enzymes synthesized in cells are excreted into the environment where they hydrolyze nutrient substances. In multicellular organisms, the extracellular digestion takes place in specialized body cavities and intestines.

The majority of highly organized animals have membranous or parietal digestion realized by enzymes, localized on membranes of the intestine cells. These systems accomplish the initial and final stages of food fermentation and the initial phases of its absorption.

Some symbiotic groups of bacteria and protozoans participate in the digestion process of both invertebrate and vertebrate animals. Their role consists in the secretion of digestive hydrolytic enzymes and a series of other substances. The symbiont's enzymes assist molecule cleavage by enzymes secreted by the cells of the host organisms itself during the digestion of food in digestive tracts. Food preferences (i.e. specificity of ingested objects or prey items) determine the set of digestive enzymes. The mechanisms of disintegration and digestion already found in the first stages of life evolution in seas are presently preserved by modern hydrobionts in various forms.

As food advances along the intestinal tract and nutrients are adsorbed, the gut contents are compacted into faecal pellets and thread-like faeces of various sizes and forms that are excreted outside. They serve as substratum for bacteria and food supply for other hydrobionts in the water column or on bottom and lateral surfaces of artificial HS in MPs.

Studies performed over the last decades testify about the extremely important and previously undervalued role of picoplankton, which is not caught by planktonic nets, in the trophodynamic processes of marine ecosystems [29, 55, 92, 159]. In the photic layer, picoplanktonic organisms (0.4–1.0 μm) can reach densities of 10^4–10^5 cells (or more) per mL of water. They account for up to 60% of all phytoplankton primary production, including that of the net phytoplankton. Picoplanktonic organisms perform photosynthesis at even lower levels of illumination than net phytoplankton. In MPs aquatories, this capability confers them a certain advantage. Picoplanktonic organisms, along with bacteria, represent the smallest food objects available for feeding to microheterotrophes and are thus part of the first links in the trophic chains of MPs ecosystems.

Studies of algal functioning showed that, whatever mineral carbon forms are used, pure autotrophy is quite rare among algae. Usually, the photic assimilation of carbonates (photosynthesis) is somehow accompanied by the use of organic metabolites dissolved in water. It was also determined that heterotrophy is accompanied by dark fixation of CO_2. Among marine animals, there are practically no “clear-cut” types of nutrition, with each species having a more or less mixed food spectrum. OM dissolved in the water serves not only as a food for animals, but some dissolved metabolites are specific regulators of growth and morphogenesis. The most primitive life forms are able to use the carbon dissolved in water as carbonates for their metabolism which uses an external source of energy. In seas, animals are connected with all the carbon forms of ecosystems, but the role of each carbon form in biological production is extremely different. In the marine environment, animals actively participate in OM destruction (carbon compounds), along with bacteria and fungi [21].

The DOM share in the ration of various plankton and benthic invertebrates varies from 1–2 to 90%. DOM plays the greatest role in the feeding of organisms inhabiting the interstitial waters and the near-substratum water layers of fouling communities. Quite often, CO_2, DOM, detritus and biomass of other species are simultaneously used in the formation of the biomass of various hydrobionts. The ratio of the various forms of carbon used by organisms differs on each trophic level. On lower levels, dissolved forms of carbon constitute the basis of the ration, while slime and particulate carbon suspended in water are the main ration for the higher trophic levels. In ecosystems, both depletion and renewal of the “stock” of each form of carbon are linked to all other forms and their transformation into each other.

The trophic value of dead organisms' remainants reduces according to their degree of decomposition. At the same time, chemical transformation of organic metabolites takes place. "Old" detritus and DOM are usually distinguished from "young" ones formed by recently dead and scarcely decomposed organisms and metabolites, which have not yet undergone significant chemical changes [21, 165, 181].

In aquatic ecosystems, about 10 groups of bacteria can be distinguished on the basis of their main functions. The following are particularly important: nitrogen-fixing, denitrifying, hydrogen oxidizing, methane-oxidizing, methanogenic, thion bacteria and iron bacteria [68, 69, 139, 145]. The various bacteria use photosynthesis or oxidize hydrogen, methane, ammonium, nitrites, sulphur compounds and iron to obtain energy. Apparently, these are one of the most ancient mechanisms of energy generation, used in the processes of chemosynthesis and photosynthesis of OM and also destruction of OM of abiogenic, as well as biogenic origin. In the World Ocean, bacteria and fungi have a main role in the mineralization of dead and, sometimes, alive OM. In fact, microorganisms control the biogeochemical cycles of nitrogen, phosphorus, sulphur, iron and other important biogenic elements. In MPs aquatories, various groups of bacteria and fungi are present in higher numbers than in the open sea waters, both in the water column and on the sediments' and HTS surfaces [32, 40, 140].

Situations, connected with mass mortality of the bottom fauna and many organisms living in the water column periodically arise in the ecosystems of many MPs aquatories. This phenomenon leads to mass mortality of aquatic organisms and is caused, first of all, by a reduction of oxygen content in water to less than 3–4 mg dm^{-3} because of worsened aquatory aeration, anthropogenic pollution by OM or accumulation of decaying vegetation [57, 141]. Mass mortality is indicative of a breach in the balance between the amount of live and dead OM and oxygen content in the ecosystems of MPs aquatories; i.e. a breach in the balance of important factors in the origin and evolution of aquatic ecosystems [6, 72].

In fact, in ecosystems of MPs aquatories, mass mortality is a mechanism of autoregulation that adjusts the quantity and quality of live OM to critical abiotic conditions represented by hypoxia, anoxia and appearance of hydrogen sulphide. At a later stage, following an increase in oxygen concentration, the functional structure of MPs aquatories ecosystems recovers owing to preservation of viable biotic components in the form of resting stages in different microniches and to colonization by different hydrobionts coming from neighbouring ecosystems.

Studies over the last decades clearly demonstrate that many important trophic-dynamic processes in aquatic ecosystems are carried out and regulated by microorganisms, including viruses, unicellular prokaryotes and eukaryotes and small multicellular organisms. Along with bacteria, fungi, which are heterotrophic eukaryotic organisms, have an enormous role in the destruction of OM in the water column and on the surfaces of sediments and HTS. Some fungi are unicellular, reproduce by spores and have a branched mycelium. Fungi without mycelium are observed in water. The maximum number of fungal species is registered in aquatories under strong anthropogenic impact [134]. For example, 56 species of higher fungi were found in the aquatory of Grigorievsky Liman [196].

The wide distribution and extremely rich enzymatic abilities of fungi (including the ability to saprotrophism and parasitism) determine their huge and not yet fully understood role in marine ecosystems [15]. In ecosystems with increased productivity, limited water exchange and lowered hydrodynamics such as MPs, the role of fungi in the ecosystem's functioning is much higher than in open sea regions. Fungi are the most important transformers of dead OM.

Polychimism is a peculiar characteristic of fungi. It consists in the ability to synthetize a high number of organic substances of different classes (sugars, organic acids, alcohols, aminoacids and others) and also biologically active substances such as vitamins, antibiotics and toxins. Some fungi use petroleum hydrocarbons and oil products, frequently detected in MPs aquatories, as the only source of carbon [15]. This ability has an extremely significant importance in the ecosystems of various MPs polluted by oil products. Representatives of marine fungi from the genus *Penicillium*, *Verticilium*, *Trichoderma* and *Cladosporium* are able to grow on all oils and oil products and representatives of the genus *Aspergillus* and *Mucor* on most of them [15].

Marine fungi are able to decompose very resistant substances such as cellulose and chitin.

By destroying oil hydrocarbons, cellulose and chitin, otherwise resistant in the marine environment, fungi transfer the organic carbon of these compounds into the protein of their body which are available to bacteria and various heterotrophs. Fungi thus return carbon incorporated in refractory compounds into the general carbon circulation of marine ecosystems [15].

In marine ecosystems, fungi belong to various trophic levels, first of all to decomposers of dead OM of vegetal and animal origin. They can themselves serve as a food chain link when different animals belonging to plankton, nekton and benthos consume their cells, spores and mycelium. In many cases, fungi act as parasites and affect different marine plants and animals, thus controlling their quantity and biomass and causing further reorganizations in other trophic levels.

In terms of size, fragments of mycelium and spores of fungi are comparable with nanoplanktonic algae and some protozoans, thus being a valuable food supply. Mycelium and spores of fungi from the genus *Penicillium* contain 36.31–37.81% of protein, 0.46–0.85% of sugar, 4.18% of fat and 10.80–12.05% of mineral matter [15].

Fungi play a noticeable role in the formation of the slimy biological film on fouling of HTS. Marine fungi, belonging to various groups, settle and develop on bacterial–detritus and bacterial–algal films.

Artenchuk [15] gave a general overview on the role of fungi in the Black Sea trophodynamic processes during the spring-summer period. The mass hatching of larvae belonging to important species of MPs benthos and periphyton starts in March (e.g. *Mytilus galloprovincialis*, *Mytilaster lineatus*) or April (e.g. *Balanus improvisus* and *B. eburneus*). Their free-swimming stage lasts 3–4 weeks. In April, a great number of invertebrates' larvae consume fungi in water, thus reducing their amount by fourfold, which likely causes the spring minimum of fungi in aquatic environments. In May, the larvae settle down and start to consume fungi on bottom sediments and from the macrophytes' surface together with other components of their ration. This situation may explain the corresponding maximum of fungi in water and their minimum in bottom sediments and on macrophytes in the same period. A second peak in the reproduction of *M. galloprovincialis*, *M. lineatus* and *Balanus* is usually observed in June–July. In this period, the quantity of larvae is usually much higher than in spring and the reduction of fungal quantities in water (by almost fourfold compared with May) is likely connected with the associated summer grazing.

It is necessary to stress that MPs aquatories offer extremely favourable conditions for the life and reproduction of *Mytilus*, *Mytilaster* and *Balanus* as well as for fungal development. The mechanisms observed by Artemchuk [15] are characteristic for these ecosystems.

In MPs, marine fungi actively participate in the processes of ecto-metabolism, especially in benthal and periphytal communities.

During mass mortality and sedimentation of dead phytoplankton cells, compact fungi settlements with an area from a few to a couple of dozens square metres sometimes form on the sediment surface. Such zones are characterized by whitish, yellowish and pinkish colours.

A high diversity of ecomorphs and life forms is characteristic for the inhabitants of the water column (including MPs) and linked, first of all, to two features determining the general organisms' architecture: the character and type of movement and the feeding strategy [139]. All constant or temporary plankton representatives have various adaptations for prolonging their residence time in the pelagial. In MPs, interaction between plankton and benthos takes place in the near-bottom water layer, while interactions between plankton and periphyton occur in the near-wall or the near-substratum water layer of the periphytal zone. Peculiar features concerning the movement and feeding strategies of the various representatives of pelagic populations in MPs are discussed in the respective sections. In plankton composition, several groups are distinguished depending on the organisms's size.

As already mentioned, picoplanktonic organisms (0.4–1.0 μm) are able to carry out photosynthesis at much lower light intensities than larger representatives of the phytoplankton [141]. This ability gives them an advantage to produce OM in the water layer under the pycnocline and in the shadowed zones of MPs aquatories.

Ultra-nanoplankton (including picoplankton) is represented by organisms smaller than 2 μm (viruses, bacteria, algae). Nanoplankton, microplankton and mesoplankton are formed by organisms having sizes between 2 and 20 μm (algae and flagellates), 20–200 μm (protozoa, rotifers and algae) and 200–1000 μm (protozoa, algae, larvae of benthic organisms, crustaceans and others), respectively. Macroplankton is formed by organisms with size from 1 cm to a couple of centimetres, while organisms larger than 10 cm belong to megaplankton. In MPs aquatories, some comb-jellies and jelly fishes can be ascribed to megaplankton.

Given the small sizes and relatively small velocity of displacement of planktonic organisms, the water flow regime around their bodies has a laminar character in the majority of cases. This circumstance has an impact on all their locomotive apparatus and feeding strategies.

Plankton organisms have relatively short life cycles and, once dead, form SOM particles, dissolved OM and slime blobs. Faeces and different metabolites of planktonic organisms play an important role in trophodynamic processes. Their survival in MPs ecosystems is favoured by the presence of resting stages, frequently able to preserve their viability for a couple of months and up to couple of years.

The body density of various nano-, micro- and mesoplankton organisms (given the composition of their capsules, valves and shells) is practically equal to or slightly higher than the density of seawater in which they live. After death and destruction of cellular membranes and external solid body parts, the protoplasm is released into the environment where it forms fine drops of slime less dense than the formerly living whole organisms. These drops, clods and slime clots occupy an intermediate position between DOM and SOM, concentrate in the near-surface water layer and become a food supply [181].

Because food is present in different states in the water column, plankton organisms have developed different methods to use dead OM. Some of them osmotically use dissolved OM, others swallow OM in the form of slime, and others use phagotrophic feeding, based on filtration and selective sizing of separate hard food particles and whole organisms.

Plankton communities are spatially and temporally dynamic. Such dynamics manifest in quick changes of their taxonomic composition, abundance and biomass. Many planktonic organisms perform horizontal and vertical nocturnal migrations. The significant expenses of energy necessary for vertical migrations are compensated by the fact that the upper water layer is richer in food with a high caloric content [139]. In the surface water layer, migrating planktonic organisms are less subject to grazing risk at night than in daylight hours.

Planktonic communities have specific spatiotemporal structures. Under the influence of various factors, plankton forms certain aggregations and condensations, which can also be observed on a small scale [135, 136, 139].

In aquatic ecosystems, groups of mass parasitic organisms are present and participate in controlling quantity and biomass of many plankton, nekton and benthos forms by leading to their more or less their simultaneous death and substitution with other forms.

In aquatic ecosystems, phages and viruses regulate the processes of production and dying of primary OM, determining the end of microalgae water "blooms" [69]. Viruses are present in all aquatic ecosystems and there are no hydrobionts, starting from smallest picoplanktonic forms up to fish, dolphins and birds, which are not subject to contamination by them. The viruses' size varies from 0.20 to 200–300 nm. Viruses are usually many times smaller than the majority of bacteria and their quantity in water and at the sediment surface can exceed that of bacteria by many folds. Viruses also regulate bacterioplankton development [94].

Molecular particles similar to viruses and infecting cells of living organisms are also known under the name of viroids. Viroids distinctly differ from viruses with regard to size, are depleted of the protein coat and consist only of circular RNA. Viroids are significantly smaller than the smallest viral particles. Even much smaller protein prions particles do not contain molecules of genetic substance. Subviral particles, the so-called virus satellites, consisting of nucleic acid molecules were also found. All listed particles reproduce themselves inside the cell of a living host and can cause infectious diseases leading to the death of host organisms. It is commonly believed that viruses are carriers of interspecific genetic information, genetically connecting all live beings into a single global ecosystem—the Earth biosphere—in spite of their diversity. Viruses are important participants in the evolutionary process on the planet [42].

In marine ecosystems, the richest communities in terms of viruses' genotypes are found close to upwelling zones, while those with the highest density occur in surface waters with hundreds of millions of virus particles per litre of water. Such conditions are typical in MPs. Viruses can significantly influence the state and quantity of bacteria and phytoplankton communities. All cellular forms of life are sensitive to viral contamination and any marine organism is a host of at least one type of virus. All the World Ocean could contain about 10^{30} viral particles. Viruses are frequently a starter for bacterial and fungal infections [42, 216, 230].

In last years, numerous viruses' genotypes were discovered. They affect different species of commercial and non-commercial organisms belonging to different systematic groups. In particular, bivalves, which are widespread in MPs, are not only infected by viruses, but also act as their reservoir transferring them further along trophic chain or diffusing them in the ecosystem via faeces and pseudofaeces [42].

Among bacteria, there are photosynthetic ones using bacteriochlorophyll and bacteriorodopsin pigments. Some bacteria are able to fix atmospheric nitrogen. Some of them can consume not only dead OM, but also live bacteria. Many bacteria are pathogenic for hydrobionts.

In MPs, parasitic flagellates, protozoans, worms and line of other groups of animals and fungi actively participate in the regulation of the abundances of many hydrobionts.

The composition of microalgae, protozoans and some fine-sized multicellular organisms includes autotrophs, heterotrophs and mixotrophs which are able to use both ways of feeding and energy production. Many macrophytes and multicellular invertebrates also utilize dissolved OM. Some protozoans feed and produce energy with the help of microalgal–symbionts, which is called algae-trophism. In MPs, such organisms exhibit certain universality in trophodynamic processes.

In both natural and anthropogenically modified ecosystems such as MPs aquatories, the transfer of energy has the form of a pyramid, as a result of the energy loss during each transition from one trophic level to the next one according to the laws of thermodynamics. The productivity of coastal waters, in which are located MPs aquatories of the Black–Azov Sea basin, usually varies between 0.5 and 5.0 g of carbon per 1 m^2 per day, although significant deviations on both sides can sometimes be observed. In ecosystems of MPs aquatories, the primary production is controlled by limiting factors such as presence, concentrations and ratios of biogenic elements, the intensity of solar radiation, hydrodynamics and the presence and distribution of various viruses. In MPs aquatories, the quantity and distribution of separate biotic components depend on primary productivity, inputs of allochthonous matter, predation, growth and migrations beyond the ecosystem borders.

Spatial-temporal changes and heterogeneity are characteristic for the ecosystems of MPs aquatories. The causes of such heterogeneity can be natural factors and processes as well as anthropogenic activity, which complicates the study, modelling and control of these ecosystems [173].

Each theory concerning the functioning of natural ecosystems and understanding of natural phenomena is inevitably based on some simplifications and therefore it is, in some way, approximated [103]. This caveat fully applies to the conception of biotic and ecological structures, organization and functioning of MPs aquatories ecosystems in general and to the corresponding subsystems in particular.

In MPs, the water column population is subject to diurnal, seasonal and inter-annual changes. It is connected with temperature, salinity, content of oxygen and biogenic matter, pH, hydrodynamics and vertical stratification to the fullest extent.

Organisms belonging to hyponeuston, plankton and nekton communities inhabit the MPs water column. The planktonic community is represented by bacterio-, phyto-, zoo- and ichthyoplankton. In MPs the planktonic complexes, as the similar complexes in the ecosystems of the Black Sea adjacent to them, have a relatively simple large-scale spatial structure. The relatively high homogeneity of their biota is connected, first of all, with mixing of water masses. Nevertheless, the vertical stratification of water masses, favoured by artificially created conditions, has a certain influence on the small-scale distribution of plankton in MPs along with many other abiotic and biotic factors.

In MPs, the thermocline is usually located closer to surface and more sharply marked in summer. In the aquatories of ports receiving inputs of river waters such as Sukhoy and Grigorievsky Limans and Novorossiysk and Sevastopol Bays, a horizontal heterogeneity in the distribution of planktonic communities is often present in connection with local processes.

In most cases, the hydrological regime is the main geographic factor in the formation of marine ecosystems structure [18, 22, 60]. In the MPs pelagial zone, as well as in the Black Sea water column, a two-layer distribution of organisms' complexes is evident below the hyponeuston micro-horizon (0–30 cm). In the Black Sea coastal waters, the first epiplanktonic complex inhabits the upper 5–20 m layer above the thermocline. This layer is well illuminated, relatively warm and has high rates of phytoplankton production during most of the year. It is inhabited by organisms which, as a rule, do not migrate and never leave this layer. The second complex is the bathyplanktonic one which inhabits the depths of more than 5–20 m below the thermocline and comprises mainly migratory organisms [131, 132].

At temperature gradients of 3–4 °C m^{-1}, the thermocline layer already becomes an almost insurmountable barrier for the majority of epiplanktonic and bathy-planktonic zooplankton organisms. At temperature gradients lower than 0.6 °C m^{-1} in the thermocline layer, the border between the two complexes is lost and is registered in the thermocline zone which becomes an ecotone [161]. Primary and secondary production of OM above the thermocline occurs at higher rates than under the thermocline [53].

In MPs, a "liquid bottom" phenomenon can originate during density-related vertical stratification of the water masses. The phenomenon consists in the accumulation of dead SOM particles and small-sized planktonic organisms at the border between the surface and deepwater masses [161]. It is favoured by the reduced MPs hydrodynamics.

During the hottest months, the redoxcline plays an important role in the hydrobionts' distribution in the deepwater zones of MPs aquatories. Because of the confined nature of aquatories, the raise of the redoxcline from the bottom into the water column reduces the aquatories volume suitable for the normal life of oxyphile species.

Planktonic organisms distribute in MPs pelagial zone in tight connection with SOM and DOM. Many pelagic planktonic animals feed on SOM, DOM and protoplasmic slime, formed during phytoplankton decomposition, i.e. detritus and aggregated DOM. However, in MPs aquatories, benthic animals inhabiting benthal and periphytal zones are the main consumers of the detritus and faecal pellets of planktonic organisms [76]. The faecal pellets are also colonized and used by nanoplanktonic organisms [233].

It was already mentioned that the majority of organic products of hydrobionts metabolism are present in the dissolved form in seawater along with seston and detritus, especially in coastal regions. These products include various carbohydrates, organic acids, proteinaceous compounds, amino acids, vitamins, hormones and other substances [33].

In seawater, the most substantial contribution to the total DOM content comes from macrophytes, first of all from the species dominating their biomass. The total quantity of OM excreted by a macrophytes' community during its vital activity reaches 2.6 kg per annum per 1 m^2 of substratum surface. The secretion speed of organic metabolites among actively growing bacteria could be 2.5–3-fold higher than that of bacteria with inhibited growth [33], and thus manifests certain seasonality.

Many algae inhabiting MPs aquatories have pelagic stages during their life cycle. Even if macrophytes are attached to HS, their thalli develop in the water column, either in the near-bottom layer or in the near-surface one if they form settlements on piers, berths and breakwaters. In MPs, the macrophytes belt growing on lateral surfaces of HTS usually forms at the depths between 0.2–0.5 and 1.0–1.5 m.

In the Black Sea coastal waters, the highest quantity of fungi spores and mycelia is observed in the near-surface water layer. In the water surface foam, the quantity of fungi is often two times higher than in the subsurface horizon. This phenomenon could be connected with the accumulation of SOM and DOM under the film of surface tension owing to hydrodynamics peculiarities [154]. The spores and fungi mycelia present in the hyponeuston constitute the miconeuston.

In MPs aquatories ecosystems, fungi play an important role by actively participating in OM destruction during conditions of high production and high concentrations of SOM and DOM. In MPs aquatories, the inhabitants of benthal and periphytal zones from various taxonomic groups anyhow participate in the formation and utilization of DOM.

In seawater, the distinction between SOM and DOM is operationally defined, but DOM plays an extremely important role in trophic chains owing to its lability and biological activity. According to some estimations, the level of DOM in seawater could exceed by more than 100-fold the amount of OM contained in all living marine organisms as a whole [48]. Average DOM concentration is 3.0 mg dm^{-3} in the NWBS open waters and 10–25 mg dm^{-3} in the NWBS coastal zone [181].

In MPs, DOM concentration increases at the sediment–water interface, in zones adjacent to the underwater surfaces of HTS and in the hyponeuston layer of the water mass. In protected MPs aquatories, owing to hydrodynamical processes, DOM is a cumulative, more or less averaged metabolite. An important role in DOM formation belongs to bacteria and diatoms, as well as to various zoo- and phyto-organisms [48].

Although the majority of planktonic organisms do not have a strong ability to move actively, many of them can perform significant migrations in the water column, especially vertically. Such displacements usually take place due to the work of locomotor organs or via changes in specific body gravity. The absolute majority of sea hydrobionts is heavier than water. However, many planktonic species from various systematic groups are able to create neutral and positive buoyancy with the help of gaseous inclusions or prolong their presence in the water column owing to various body appendices and active movement. The hydrobionts' feeding is also connected with displacement. In MPs aquatories, planktonic and benthic organisms (and their larvae) with morphofunctional adaptations for the creation of water flows (flagella, cilia and sifonal channel through which water is being pumped) have a competitive advantage [79, 89, 145].

Planktonic organisms are able to submerge at increased roughness, thus protecting themselves from mechanical damage. Gravity facilitates immersion, but planktonic organisms must then spend energy to overcome it when rising up again. However, upwards vertical migrations can be favourable because of better living conditions.

Figure 2.23 schematically shows the main pelagial biotopes in MPs.

For planktonic organisms, micro- and small-scale water motions play an important role, while large-scale ones could have catastrophic consequences. The frequency of interaction between planktonic predators and preys notably differs in "non-turbulent" and turbulent environments, being significantly higher in the latter. The microscale turbulence also influences the production rate of eggs and metabolism intensity, in particular among copepods. Heterogeneity and scale of turbulent processes in the water column (from the surface until the near-bottom layer) influence organisms' behaviour during vertical diurnal migrations as well as their distribution [135, 136].

Numerous studies showed that spatial and temporal distribution of planktonic organisms is not random, but depends on the combination of hydrodynamic conditions, specific physical and chemical factors and absence or presence of other organisms with which plankton interacts directly or indirectly. For planktonic animals, a patchy and layered distribution in the water column is most often typical and is also connected with isopycnic layers [36, 135, 136].

The formation of specifically organized accumulations of plankton led Wiebe [235] to conclude that their formation is tightly connected with physical processes,

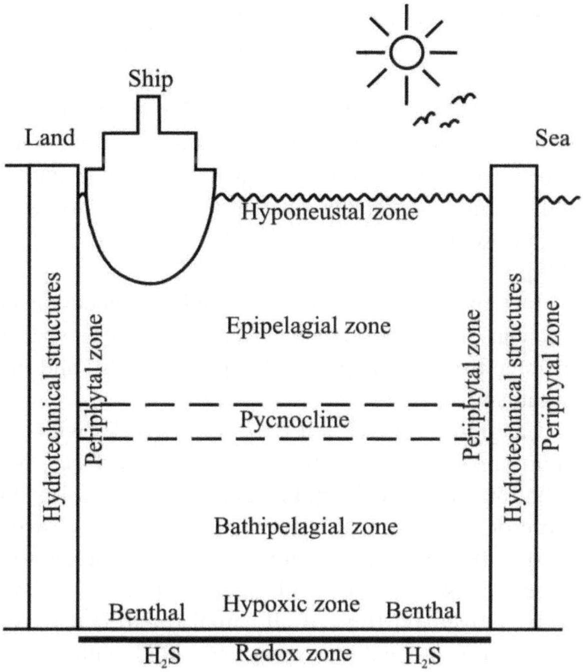

Fig. 2.23 General scheme of a marine port ecosystem of the classical type

in particular Langmuir circulation, occurring between the surface horizon until and the pycnocline.

It was already mentioned that wind, friction and the Coriolis force are the main factors causing movement and water rotation, i.e. the formation of vertical and horizontal circulations on various scales, including those of MPs. The smaller is the area of a MP harbour aquatory and the shallower its depth, the bigger is the role of frictional forces. The speeds of water circulations most distinctly reduce at shores, near the internal surfaces of MPs HTS and also near the bottom, while they are preserved over longer time spans in the central parts of aquatories. This pattern influences the distribution of both planktonic and benthic organisms.

In MPs aquatories, the dynamics of various frontal formations and circulations, including Langmuir vortexes, generates parasitic water circulations with different scale in the epipelagial zone. These parasitic circulations collect live and dead organisms from phyto-, zoo- and ichthyoneuston, and phyto-, zoo- and ichthyoplankton. Mono- and polyspecies plankton accumulations of various densities are registered in MPs harbours.

Langmuir [212] determined that a significant part of dissolved OM produced by vegetal cells concentrates in the thin near-surface film at the air–water interface. Later on, it was shown that Langmuir vortices concentrate OM present in the subsurface film along convergence lines on the water surface [229].

During the process of lateral compression in convergence zones, OM forms small organic conglomerates which, owing to surface microconvections and

Langmuir vortices, sink to various depths and accumulate at isopycnic surfaces. The aggregates formed from the dissolved OM, along with the bacteria adsorbed on them, serve as food for zooplankton [36].

Zooplanktonic organisms of various sizes are influenced by Langmuir circulation and their spatial distribution is eventually determined by the ratio between the speeds of water circulations and the speeds and direction of the movement of individuals. Organisms concentrate in zones of waters' divergence and raising at high flow speeds and in zones of convergence and sinking at low flow speeds [209, 228].

In Langmuir cells, the water sinking speeds usually range from 4 to 8 cm s^{-1}, but can sometimes reach 27 cm s^{-1}. These speeds already notably exceed the average movement speeds of copepods [136] and of most of the other zooplankton organisms.

Langmuir vortices not only form streaks with increased concentrations of OM and planktonic organisms on the water surfaces, but also determine plankton distribution in the MPs epipelagial zone and form plankton accumulations, with a more or less spherical shape, resembling cloud structures.

In MPs pelagial, accumulation of zooplankton in forms resembling cloud structures is connected with physical processes. The most important processes are water movement (first of all, small (metres) and average (dozens of metres) scale circulations at small or average Reynolds numbers of 1–10) and the regular active behaviour of zooplanktonic organisms. Mono- and multispecies aggregations form, first of all, during the feeding, reproduction and development processes at low Reynolds numbers.

Besides Langmuir circulation, gravitational internal waves [136] and ships' waves can influence the distribution of zooplankton in MPs over defined spatiotemporal scales. Plankton concentration in slick zones formed, for example, by internal waves could be very significant. In particular, concentration of fish larvae in the slicks can exceed the concentration outside them by 13–36-fold [221].

In MPs pelagial zone, all zooplankton aggregations have their own lifetime. They form over a period of time, preserve their main characteristics for a defined period and then disappear under the influence of various abiotic and biotic factors. Formation and destruction of variously sized accumulations of zooplankton in the water column is one of the main mechanisms for the normal functioning of pelagial subsystem in MPs. The constant process of redistribution of the alive and dead OM and energy in this subsystem takes place thanks to the cyclic nature of such accumulation. For zooplankton organisms, the presence in accumulations is an absolute requirement during their whole life cycle or at some of its stages [32, 198, 211, 220].

According to some hypotheses, Langmuir circulation could have had a significant role in the sustention and evolution of marine life already in the Precambrian time. At that time, the atmosphere was reducing, while seas were warm, shallow-watered, salty, and inhabited by cyanobacteria. In the Precambrian seas, oxygen was completely consumed by biotic components of ecosystems in the epipelagial zone and was entirely spent on the oxidation of minerals near the bottom, i.e. reducing conditions occurred both near the surface and at the bottom. The oxygen produced by cyanobacteria, even in small amounts, could concentrate, and not completely exhausted, only in the pycnocline layer because of the low speed of diffusion via isopycnic surfaces. Thus, the pycnocline was an ecological niche in which organisms requiring oxygen could survive and develop. Owing to Langmuir circulation, the pycnocline zone was also enriched in dead OM from the near-surface water layer. The first representatives of many hydrobionts types could have appeared in the pycnocline and later on quickly diffuse in the oxidizing conditions of the Cambrian period [36].

Thus, the pycnocline has played an important positive role at a certain stage of life evolution in the World Ocean. At present, in conditions of excessive eutrophication, the pycnocline quite often hampers aeration of the near-bottom water layer and is one of the factors of fauna mass mortality and simplification of biota composition in MPs aquatories ecosystems.

In the modern Black Sea MPs, downwelling of surface waters causes only an insignificant input of oxygen into the water column below the pycnocline.

2.2.1 Phytoneuston and Phytoplankton

In the Black Sea, according to the classical scheme, the mass development of diatoms is observed in spring (February–March), while dinophytes become dominant in summer (June–September). A new bloom of diatoms occurs in autumn (October–November). In desalinated aquatories, cyanobacteria (Cyanophyta) play an important role in phytoplankton composition during some periods. Cyanobacteria are the most ancient organisms among the known prokaryots and euglenoids (Euglenophyta).

Cyanobacteria show characteristics of both bacteria and plants (unicellular and colonial). They include autotrophs, heterotrophs and mixotrophs, although their functional role in aquatic ecosystems is usually close to that of autotrophic algae.

Planktonic microphytes often have organs of locomotion in water and many cyanobacteria move with the help of flagella. As a rule, euglenoids have a flagellar apparatus and include colourless forms, i.e. heterotrophs. The majority of the golden algae (Chrysophyta) have 1–4 flagella and include the ecologically important group of coccolithophora. The pelagic yellow-green algae (Xantophyta) also have flagella and can be endocellular symbionts of protozoa. Algae from the phylum Raphidophyta swim with the help of a pair of flagella. Cryptophytes (Cryptophyta) and many dinoflagellates (Dinophyta), widespread in MPs aquatories ecosystems, have the same number of flagella.

In the Black–Azov Sea basin, dinoflagellates (Dinoflagellata), or dinoflagellate algae dominate the phytoplankton composition of MPs aquatories together with diatoms (Bacillariophyta).

Among botanists, the term "dinoflagellates" includes the monades (unicellular) forms of dinophytes. In zoology, it is the same as peridinean (Peridinea). They can have one to eight flagella.

Among dinophytes, there are species using phototrophic and heterotrophic (osmo- and phagotrophic) feeding. Phototrophic species have brown chloroplasts, containing chlorophyll "a" and "c", and xanthophylls. Some tiny dinophytes are symbionts of protozoans, coelenterates and comb-jellies.

More than half of the Black Sea dinoflagellates can switch from the phototrophic to the mixotrophic (54.8%) feeding strategy and only about 12.1% of them can be considered obligate phototrophs [168].

In many MPs aquatories, water "blooms" phenomena are quite often observed and are usually caused by mass development of planktonic (from pico- to meso-planktonic) algae or protozoans. Water "blooms" are usually connected with different types of anthropogenic impact, but cases of phytoplankton mass development caused by a rare combination of completely natural causes are also known.

For example, a case of mass development of planktonic algae belonging to the genera *Gymnodinium* and *Gonyaulax* (resulting in a brown-red colour of seawater) was recorded on the southern coast of Ireland in the absence of wastewater discharges and after an extremely long period of summer calms (not even ripples). In this situation, a "red tide" phenomenon is tightly connected with lowered hydrodynamics [45].

Numerous literature data indicate that lowered water dynamics in water reservoirs along big rivers creates preconditions for the intensive development of some planktonic algae, that can then cover all the reservoir's surface. In the Azov Sea, descriptions from the 1930s report situations in which great amounts of phytoplankton accumulated in the surface water horizon, completely covering it with an intensively coloured green layer during periods of water blooms [85]. In June 2011, wide areas covered with an algal mat of the species *Nodularia spumigena* were registered in the NWBS open waters.

In both freshwater and in marine ecosystems, the prolonged natural or artificial reduction of hydrodynamics quite often causes mass development of unicellular and colonial algae. Depending on the various hydrological situations, the intensive phytoplankton development can take place in the water surface layer, in the pycnocline zone and even under the pycnocline.

In MPs aquatories, the greatest phytoplankton species diversity is observed in zones with good water exchange [40, 156, 197].

In 1976, mortality of fish, crustaceans and molluscs was registered in the region of Lower New York Bay. On that occasion, oxygen concentration near the bottom had dropped below 2 mg dm^{-3}. Occasionally, a complete absence of oxygen and the appearance of hydrogen sulphide were observed in the near-bottom water layer at a depth of 35 m. Abnormally warm and windless spring and summer resulted in the development of a stronger than usual thermocline, which separated the much lighter and warmer surface water from the heavy and cold near-bottom water. Under the thermocline, extremely strong development of phytoplankton occurred. This rarely occurring combination of natural processes caused mass mortality [45]. Similar situation could form also in MPs aquatories.

Phytoplankton usually develops more intensively in sea zones experiencing a constant pollution by domestic wastewaters than in unpolluted ones. The content of phosphates in these zones can sometimes increase so much that the planktonic algae are not able to exhaust it during the spring-summer period and "blooms" phenomena occur during most of the year [214].

In the Black Sea, several conditions are necessary for the onset of the "red tide" phenomena: 1—water temperature above 20 °C; 2—salinity in the range 11–15‰; 3—abundance of biogenic matter; 4—absence of wave action (sea force no more than 3); 5—illumination not less than 600,000 lx [12].

In darkness, the density of microalgae cultures diminishes after a couple of hours. This phenomenon is connected with the dark respiration and is called "night biomass loss". Losses during nighttime can reach 14% of initial density. Ad hoc studies showed that maximum dark loss of biomass takes place due to a decrease in carbohydrates' content, sometimes almost by a factor of two. In such situations, the content of protein slightly increases and the lipids' share reduces [170].

As it was already mentioned, many species of planktonic algae are able to shift from autotrophy to mixed feeding (mixotrophia) under certain conditions. In the Gulf of Burgas, the shift of a whole population of *Exuviella cordata* to heterotrophic nutrition was recorded during the decline period of water "blooms". The use of dissolved OM became the main source of energy supply for the cells [12].

In coastal waters and, consequently, in MPs aquatories, the heterotrophic planktonic algae are present nearly all year round and sometimes comprise more than 50% of phytoplankton biomass. The role of heterotrophic algae in the trophic dynamics of coastal waters is comparable with the role of nano- and microzooplankton [12].

A sudden die-off of the "bloom-forming" planktonic algae is often observed after the period of mass development following damage caused by viruses, protozoans, fungi, etc. The death of *Exuviaella cordata* was connected with the activity of parasitic protozoans [12].

Some algal species, in particular cyanobacteria, excrete reducing substances into the environment both after their death and during their normal functioning. Such substances react with oxygen and deplete its resources [12].

Cyanobacteria that cause water "blooms" have diurnal rhythms of vertical migration and create regular flows of matter in the water column.

Cyanobacteria and euglenoids are considered as one index of water pollution. Thus, they comprise 11.5–17.0% of total organisms' abundance in Novorossiysk and Tuapse MPs [157].

Following the death and decomposition of green algae, more or less hard detritus particles are formed along with pellets and protoplasm slimes clots populated by bacteria. The cells of various planktonic algae species have different nutritional value for their consumers. Diatoms, which quite often have hard shells with various outgrowths, are less valuable than the round-shaped cells of dinophytes with their soft membrane.

In Novorossiysk Bay, the maximal development of dinophytes takes place at the end of the period of diatoms' intensive development (May–July). The role of

Dinophyta in using the "dying bloom" dead OM can be compared only to the role of planktonic infusoria [197]. The reproduction of many fish with pelagic stages of development is also confined to the same period.

In June and August 1998, after diatoms "blooms", the dinoflagellates *Gyrodinium cornutum* and *Polykrikos schwartzii* accounted for up to 95.0% of phytoplankton biomass in Odessa MP. In September 2000, an outbreak the dinophyte *Akashiwo sanguinea* (678×10^3 cells dm^{-3} and 55.3 g m^{-3}) developed in Odessa MP aquatory and comprised 79.0% of phytoplankton quantity and 96.0% of its biomass. In August 2000, "blooms" in Odessa MP were caused by *Gymnodinium simplex* (250×10^3 cells dm^{-3}) [168]. Such mass development of these algae was closely connected with the prolonged calm weather and high water temperature.

The works of Nesterova [120–122] showed that constant accumulations of unicellular algae are often registered in the near-surface layer of the Black Sea, where they can exceed by up to threefold their quantity in the water column. Diatoms, peridinean and cyanobacteria were usually prevailing in phytoneuston composition.

In the coastal regions of the Black Sea, the foam forming on the water surface contains high numbers of bacteria, cells of planktonic algae and their cysts. In Odessa Gulf, the quantity of dinoflagellate cysts under the film of surface tension exceeds that in the subsurface water layer by twofold. Analysis of the vertical distribution of some mass species of phytoplankton showed their regular concentration at the surface of the pelagial zone [154].

In marine ecosystems, planktonic autotrophic algae, macrophytes and microphytobenthos transform inorganic components contained in marine water into organic compounds using solar light for photosynthesis. Phytoplankton is the main producer of OM in both seas and MPs aquatories. The life of organisms in the water column and in the benthal and periphytal zones frequently depends, directly or indirectly, on the OM of planktonic algae.

In MPs, the intensive development of phytoplankton changes not only the colour of water, but also sharply reduces water transparency and worsens the illumination of the deeper water layers.

In the surface sea layer, phytoplankton reaches its maximal diurnal development during daytime and evening hours. In this period, a growth of oxygen concentration is recorded because of the photosynthetic activity of planktonic algae. In the Black Sea surface water layer, the diurnal maximum and minimum values of oxygen concentration occur between 5–8 p.m. and 7–8 a.m., respectively [161].

In MPs, the hydrodynamical processes have a significant importance for the distribution and vital activity of phytoplankton and of the bacteria and protozoa trophically connected with it. Plankton diatoms with siliceous frustules develop better in conditions of active water mixing, which keep them in suspension. The diatoms most widely distributed in MPs ecosystems develop well at flow speeds of 0.7 m s^{-1} and above. For green algae, the limiting flow speed is within 0.5–0.8 m s^{-1} and flow speeds of 0.2–0.3 m s^{-1} are already unfavourable for the cyanobacteria responsible for water "blooms" [145]. In MPs aquatories, high flow speeds are absent during most of the year, thus limiting phytoplankton development. Extremely low speeds (lower than 0.01 m s^{-1}) are quite often observed [196].

For example, the phytoplankton vertical distribution in Grigorievsky Liman is unstable and depends on the hydrological situation that changes in the various seasons. In spring and summer months, the cells' number in the surface layer is up to twofold higher than in the near-bottom, while the phytoplankton accumulation occurs in the near-bottom layer during the autumn and winter months [196].

The Black Sea water column hosts both heliophilous and sciophilous species of planktonic algae, which belong to the epi- or bathyplanktonic complexes, respectively. Some peridinean have a photosensible organ and can actively move towards the source of light. Unlike planktonic algae, most planktonic animals have positive or negative phototaxis. However, they distribute in the pelagial zone according to not only illumination, but also food availability and dissolved oxygen content.

In the majority of MPs aquatories, the presence of protective HTS and of only one passage reduces, to a various degree, the possibilities of transport and dispersion of live and dead phytoplankton.

As already mentioned, many pelagic microalgae can simultaneously have autotrophic and heterotrophic feeding. Forms with a prevailing autotrophic mode of feeding inhabit the euphotic layer of Black Sea regions with relatively low DOM concentration. The big MPs aquatories do not usually belong to these regions and offer conditions for heterotrophic nutrition. At light attenuation, many microalgae easily switch from autotrophy to the heterotrophic utilization of some dissolved organic compounds. Thus, the so-called umbraticolous microalgae, usually concentrating in and below the thermocline zone, can use the heterotrophic mode of nutrition in shadowed zones and above the thermocline in MPs.

Light frequently plays a role as signal factor for selecting the best environment according to positive or negative phototaxis. For many planktonic animals, light is also essential for gonads' development and formation of vitamin A from carotenoids assimilated from phytoplankton [161].

All main algal groups present in the Black Sea plankton and inhabiting MPs (diatoms, dinoflagellates, silicoflagellates, coccolithophores, green flagellates, cyanobacteria) are able to regulate their position in the water column to a various degree by changing their specific weight by fat accumulation (diatoms), formation of air bubbles (cyanobacteria), increasing floating ability (diatoms) or owing to active movements (flagellates).

Planktonic diatoms float in the water column; their buoyancy being provided by the presence of numerous inclusions of fat in their body as well as by light-weighted tests with numerous spines, projections and spikes that increase cell surface.

Migrations of planktonic algae and their mass development in the surface layer ensure them better illumination and higher hydrodynamics and causes shading of the underlying water layers suppressing the development of non-concurrent phytoplankton species for some time.

At the initial stage of their mass development, young diatom populations increase their buoyancy by forming fatty drops inside cells. In this period, their populations remain close to the water surface under approximately optimal light conditions. The photosynthetic activity of populations reduces along with the depletion of the stocks of biogenic substances, the amount of fat in their cells decreases and the population "sinks" to a depth where their buoyancy becomes neutral. This depth is the upper limit of the thermocline.

A special flora of diatoms, adapted to low light intensities, inhabits the zone between the upper limit of the thermocline and the lower border of the photic zone. This flora is retained here owing to its ability to float and to the presence of various convective flows [161].

The "sedimentation" of microalgal populations to the depths of about 10 m, i.e. the upper border of the thermocline, is often observed in the NWBS in general and in MPs aquatories; in particular during the period of their mass development. In the thermocline zone, the concentration of microalgae can be up to twofold higher than at the surface. The mass "sedimentation" of algal populations and their die-off increase the transparency of the upper layer of the water column and favour BM turnover, thus creating conditions for the development of other phytoplankton representatives.

In MPs aquatories, a significant amount of the so-called accidentally planktonic or bentho-planktonic microalgal species, predominantly diatoms, is recorded in the water column. Thus, in the phytoplankton of Sevastopol Bay, 43 out of 110 species of diatoms belonged to bentho-planktonic species. These algae normally develop on the bottom and on periphyton and have no specific adaptations for floating. However, they can live and feed for quite a long time in the water column [161].

In MPs, diatoms provide a link between the pelagial, benthal and periphytal zones. After intensive vegetation and mass development, as well as at the onset of stress situations in water column, many diatom species sink to the surface of bottom sediments in the form of resting spores, separate vegetative cells and filamentary colonies. Spores and vegetative cells are able to return into the pelagial zone when abiotic conditions improve. Benthic diatoms can pass from benthos into plankton and vice versa in any season [99]. Such mechanism preserves the diatom populations in MPs aquatories during periods of unfavourable environmental conditions and contributes to a higher stability of MPs ecosystems. The resting stages of diatoms and other planktonic microalgae allow them to return into the cycle of matter in MP ecosystems, without passing through the stage of mineralization.

In MPs aquatories, dinoflagellates and other flagellates are able to perform active migration across the thermocline and accumulate in layers with optimal illumination and sufficient biogenic nutrition [161]. The lowered hydrodynamics of MPs facilitate flagellates in selecting the best conditions.

Some phytoplankton species in the surface layer can be grazed by the zooplankton migrating towards the surface during night hours or constantly inhabiting there. At night and during the afternoon, the growth rates of planktonic algae frequently drop [145, 161].

Diatoms (Bacillariophyta) of various species often dominate the phytoplankton composition of MPs aquatories in terms of both density and species abundance. They are an important primary source of live OM and detritus, can contribute to the aquatories' oxygen regime and serve as food for many animals of pelagial, periphytal and benthal zones. Processes of matter and energy turnover in marine ecosystems, particularly in MPs aquatories, involve their active or passive participation.

In Sevastopol Bay, diatom algae form the basis of phytoplankton. In the summer–autumn period, the contribution of pyrrophyta to the total phytoplankton quantity increases. Algae of the class chrysophyceae are registered all year round. Representatives of Cyanophyta, Euglenophyta, Cryptophyta, Flagellata, etc. are found in Sevastopol Bay at different locations. Small-sized algae dominate in terms of quantity, while big-sized ones represent most of the biomass [3].

Development of large-sized species of planktonic algae was observed in Tuapse MP during summer. Diatoms of the species *Proboscia clavata* comprised 90% of the quantity. *Gymnodinium simplex*, *Oblea rotunda* and species of the genera *Gymnodinium* and *Prorocentrum* were found among dinophytes ($\leq$5–6% of the total). The euglenophytes *Eutreptia lanovii* and *Euglena* sp., and cyanobacteria of the genus *Oscillatoria* (in total 0.5×10^6 cells m^{-3}), which are indicators of water saprobity, reached noticeable abundances. The total quantity of phytoplankton reached 84×10^6 cells m^{-3} and its biomass 453.0 mg m^{-3} [155].

In Varna Lake, oxybiont phytoplankton is observed only in the upper water layer enriched by oxygen. From the depth of 10 m down to the bottom, *Exuviella cordata* and *Prorocentrum micans* are observed in small amounts. In Varna and Beloslav Lakes, diatoms cause water "blooms" even in winter–spring months. The number of planktonic algae varies from 142×10^6 up to 838×10^6 cells m^{-3} [189].

Water "blooms" often are an ecosystem phenomenon, which testifies a surplus of BM in a marine ecosystem.

The biggest Ukrainian MPs (Yuzhny, Odessa and Ilyichevsk) are located in the desalinated NWBS and their ecosystems are variously influenced by large rivers; most importantly Dnieper and Yuzhny Bug. The waters of these rivers mix between themselves, and with seawater, in the Dnieper-Bug Liman and then spread over the saltier near-bottom mass of seawaters forming a layer up to a couple of metres thick. In MPs pelagial, an increase in the share of marine phytoplankton forms is observed from Grigorievsky Liman to the Odessa Gulf and farther towards Sukhoy Liman. Furthermore, the amount of unicellular algae, usually belonging to microphytobenthos, appears higher than in regions adjacent to these ports.

The modern structure of phytoplankton in the main Ukrainian MPs (Yuzhny, Odessa and Ilyichevsk) genetically consists in representatives of three complexes: freshwater, brackish and marine.

There are 265 species included in the general list of species and varieties of planktonic algae of Grigorievsky Liman [196]. They belong to seven systematic divisions of phytoplankton, namely: diatom (Bacillariophyta), dinophytes (Dinophyta), cryptophytes (Cryptophyceae), green algae (Chlorophyta), cyanobacteria (Cyanophyta), golden algae (Chryzophyta) and euglenophytes (Euglenophyta). More than 52% of the species are marine, brackishwater–marine and brackish water [196]. Diatoms are represented by 97 species. Phytoplankton representatives usually recorded in the NWBS are constantly observed in the liman:

the diatoms *Sceletonema costatum*, *Nitzschia closterium*, *Cyclotella caspia*, the dinophytes *Heterocarpus triquera*, *Scrippsiella trochoidea*, *Hillea fusiformis*, *Diplosalis lenticula*, the cyanobacteria *Oscillatoria kisseleva* and the green alga *Scenedesmus quadricauda*. Algal species new for the NWBS were first found in the phytoplankton composition of Grigorievsky Liman: the yellow-green algae *Amphiri epizootica*, the green algae *Pyramimonas longicauda* and the choanoflagellate *Bicosta spinifera*.

In Grigorievsky Liman, phytoplankton usually has an even vertical distribution in autumn and winter. In spring, its quantity is higher at the surface, but the biomass is evenly distributed. On the contrary, phytoplankton biomass concentrates near the surface and its quantity is evenly distributed in summer months. In Grigorievsky Liman, both quantity (741–3903 $\times$ 10^6 cells m^{-3}) and biomass (2500–34,600 mg m^{-3}) of phytoplankton exhibited considerable interseasonal and inter-annual fluctuations. Thus, phytoplankton vertical distribution in the liman is variable and depends on the hydrological situation and on season. In spring and summer, the quantity of phytoplankton in the near-surface water layer is up to twofold higher than near the bottom.

During the year, diatoms comprise 10.3–83.9% of the quantity and 22–98.0% of the biomass of phytoplankton of Grigorievsky Liman. Dinophytes algae contribute 0.1–23.3% to phytoplankton quantity and 1.6–55.5% to its biomass. In 2004–2006, a tendency in the growth of phytoplankton was observed in Grigorievsky Liman; with a continue decrease of diatoms quantity (61.3%) and biomass (34.6%) accompanied by corresponding increases (21.3 and 64.1%, respectively) of dinophytes.

Out of 29 species of microalgae that were causing “blooms” in Grigorievsky Liman, 11 were diatom representatives, 7 dinophytes, 5 cyanobacteria, 2 green algae, 1 coccolithophora, 2 euglena algae, 1 picoplankton alga (*Prochloroccocus marinus*) and a combined group of small-sized flagellate alga [196].

In Odessa MP, dinoflagellates and diatoms are the main component of plankton. In its pelagial zone, 156 species and intraspecific taxa of microalgae were registered in the 2000s: diatoms—67, dinophytes—50, green algae—22, cyanobacteria—7. Golden algae and euglenoids were represented by eight species. Among diatoms, the genus *Chaetoceros* was the most diversified with 27 species and intraspecific taxa. In 2004–2005, 82 species of dinoflagellates were found in Odessa MP, compared with 31 species recorded in Sukhoy Liman [168].

In Odessa MP, dinoflagellates quite often cause water “blooms” in autumn, During mass development in June–July 1998, the quantity and biomass of *Prorocentrum cordatum* reached 5.6 $\times$ 10^6 cells dm^{-3} and 10.6 g m^{-3}, respectively. The corresponding figures were 1.3 $\times$ 10^6 cells dm^{-3} and 13.5 g m^{-3} for *Heterocarpus triquetra* (April 1999) and 780 $\times$ 10^3 cells dm^{-3} and 70.0 g m^{-3} for *Gymnodinium sanguineum* (October–November 2000). In 2001, the highest quantities and biomasses were registered for *H. triquera* (34.3 $\times$ 10^3 cells dm^{-3}, 362.6 mg m^{-3}), *Scrippsiella trochoidea* (10.7 $\times$ 10^3 cells dm^{-3}, 124.6 mg m^{-3}), *Prorocentrum micans* (8.0 $\times$ 10^3 cells dm^{-3}, 142.1 mg m^{-3}) and *Oblea rotunda* (5.9 $\times$ 10^3 cells dm^{-3}, 23.7 mg m^{-3}) [16].

Water renewal in Odessa MP can happen quicker than in most big ports of the Black–Azov Sea basin owing to the presence of three wide and deep passages into its aquatory. In MPs, changes in phytoplankton composition are observed along with the changes of water masses. In Odessa MP, the processes of phytoplankton development are synchronized with those in the NWBS to a greater extent than in Grigorievsky and Sukhoy Limans.

In Grigorievsky Liman, the share of marine and brackishwater–marine species exceeds 52.0% of the taxa number. In Sukhoy Liman, which is much less influenced by the Dnieper-Bug Liman waters, the share of marine and brackishwater-marine species comprised 81.5% and that of freshwater and freshwater–brackish water species only 18.5% [196].

Out of the 123 species and intraspecific taxa of planktonic microalgae found in Sukhoy Liman, 43 were dinophytes, 41 diatoms, 14 cyanobacteria, 10 green algae, 9 golden algae, 4 euglena algae and 2 cryptophytes. Phytoplankton quantity varies from 0.1 to 4.6×10^6 cells dm^{-3} with an average of 1.9×10^6 cells dm^{-3}. Biomass ranges from 0.6 to 20.1 g m^{-3} with an average of 8.0 g m^{-3}. Sukhoy Liman is characterized by water "blooms" in the summer–autumn period. In June 2001, the density of *Prorocentrum cordatum* was 3.3×10^6 cells dm^{-3} with a biomass of 16.7 g m^{-3}, corresponding to 93.0% of the whole dinophytes' biomass.

In the NWBS, the phytoplankton concentration is always higher than near Crimean and Caucasian shores and everywhere is higher nearshore than offshore. In the NWBS, the highest productivity of phytoplankton is registered in limans, gulfs, bays and port aquatories. Considering a water layer 25 m deep and with a surface of 1 m^2, the daily carbon production is 0.15–0.25 g in offshore Black Sea regions and 0.2–1.0 g in the coastal zone. The diurnal carbon production reaches 1.0–2.0 g m^{-2} in Sevastopol Bay [161, 179] and it can often exceed 2 g m^{-2} in aquatories of the big MPs of the Black–Azov Sea basin during the summer–autumn period. In many cases, such high diurnal carbon production values are favoured by the entrance of river waters and various wastewaters into ports aquatories.

In MPs, the mass development of permanent planktonic algae causes self eutrophication of their ecosystems.

The total phytoplankton biomass reaches 100–150 mg m^{-3} in the open parts of the NWBS and 3–4 g m^{-3} in deeply cutting inland bays and MPs. During diatoms "blooms", especially during those caused by peridinian algae (red tides), phytoplankton biomass reaches 1 kg m^{-3} and primary production—20.0 g C m^{-3} [12].

One of the highest values of dinoflagellates density in the NWBS was registered in Odessa MP with an average of 71.8×10^3 cells dm^{-3} per year. The highest values of density and biomass of the exclusively heterotrophic dinoflagellates species *P. schwartzii* and of the mixotroph *Ceratium furca* are also connected with Odessa MP aquatory. Both species are considered indicators of high organic pollution.

During unfavourable conditions, dinoflagellates form resting cysts on the bottom. Indeed, cysts of 31 dinoflagellates species were found in Odessa MP [168].

According to Selifonova [156], the phytoplankton composition of Novorossiysk and Tuapse ports aquatories included 145 taxa: Bacillariophyceae—66,

Dinophyceae—59, Prymnesiophyceae—2, Cryptophyceae—2, Chrysophyceae—1, Dictyochophyceae—2, Euglenophyceae—4, Chlorophyceae—5, Prasinophyceae—1 and Cyanophyceae—3. The number of taxa was 131 and 104 in Novorossiysk and Tuapse port, respectively. Diatoms and dinophytes dominated in Novorossiysk Bay during summer, their total number varying between 48 and 54 taxa. In spring and autumn, the number of taxa for these algae was in the range 28–39. In Tuapse MP, the amount of diatoms increased in autumn (28) and that of dinophytes in spring and summer (23–30); while number of both groups reduced to 13–20 in winter and autumn [156, 157].

In Novorossiysk MP aquatory, the greatest contribution belonged to diatoms both in quantity (78%) and biomass (81%). In general, their total quantity was lower than in open part of the bay where the share of peridinian algae, in particular coccolithophores and dinophytes, increased. Euglenoids were registered predominantly in the port aquatory [156].

Diatoms and dinophytes made up most of the total biomass of Novorossiysk MP phytoplankton.

In the aquatory of Novorossiysk port, peaks of algal biomass and density are sometimes mismatched owing to the predominant development of small-sized phytoplankton forms. Maximal biomass of phytoplankton in Novorossiysk MP aquatory was registered in May 2006, when it reached 5.4 g m^{-3}. In the open part of Novorossiysk Bay, the same species of diatom algae as in the MP usually dominate in terms of total quantity, but among them the share of the big-sized species and also of dinophytes, silicoflagellates and golden algae (Chrysophyta) naturally increases.

In Tuapse MP, diatoms are the main contributors to both quantity (46–48%) and biomass (63–69%) of phytoplankton. Dinophytes algae significantly contribute (24–34%) only to total phytoplankton biomass. Cyanobacteria and euglenoids, which can be an index of highly eutrophic waters, are a significant component of plankton in the port (about 17% of total quantity) [156].

In Novorossiysk and in Tuapse MPs, the quantity and biomass of phytoplankton are usually 1.2–2.0-fold lower than in the adjacent open waters [156, 157].

As a rule, the share of dinophytes does not exceed 7–12% of the total phytoplankton quantity and 24–34% of its total biomass in the aquatories of Novorossiysk and Tuapse ports. Many dinophytes from the genera *Protoperidinium*, *Prorocentrum*, *Gymnodinium* and *Gyrodinium* develop in ports as subdominant species and are able to switch to heterotrophic or mixotrophic nutrition. As previously mentioned, it is commonly believed that an increase of heterotrophic phytoplankton is first of all connected with a corresponding increase in the content of dissolved and suspended dead OM in water.

In the last years, a two to fivefold decrease of phytoplankton quantity and a three to fourfold decrease of its biomass were registered in Tuapse and Novorossiysk MPs aquatories. Such tendency can be explained by the cumulative influence of various negative factors. In general, the more polluted Tuapse port is characterized by lower level of phytoplankton development [17, 157].

In Novorossiysk MP aquatory, the long-term average annual values of phytoplankton were 0.35×10^6 cells dm^{-3} for density and 1.27 g m^{-3} for biomass. The corresponding values in Tuapse MP were 0.11×10^6 cells dm^{-3} and 0.37 g m^{-3} [157].

It is important to remind that MPs eutrophicated waters typically exhibit the so-called "microplankton paradox". This phenomenon highlights that the major role in photosynthesis and OM production in MPs is often played by small planktonic algae belonging to mesoplankton net plankton and nanoplanktonic forms, especially green flagellates ranging from 2 to 20 μm in size. There is a direct relationship between the size of planktonic algal cells and the water eutrophication level [161].

According to contemporary notions, the phytoplankton in MPs aquatories not only realizes primary OM production via photosynthesis, but also uses and transforms the dead OM by mixotrophia and heterotrophia. In particular, mixotrophy is observed among many cyanobacteria and diatoms. In MPs ecosystems, the mixotrophes occupy an intermediate functional position between plants and animals and become less dependent on changing environmental conditions. Owing to their peculiarities, mixotrophes are able to maintain biotic turnover in aquatic ecosystems during critical periods.

About one-third of the Black Sea dinoflagellates are exclusively heterotrophic athecate and thecate species. In the coastal zone of NWBS (including MPs aquatories), the heterotrophic dinoflagellates comprise on average 36.36% of quantity and 48.33% of biomass [167].

2.2.2 *Bacterioneuston and Bacterioplankton*

Bacteria inhabit all biotopes of MPs aquatories and are present on the film of surface tension, in the hyponeuston layer, in the water column, at the surface of and inside bottom sediments, on HTS surfaces and on wetted surfaces of ships hulls. As it is well known, bacteria are one of the first links of trophic chains in marine ecosystems and, at the same time, they realize the mineralization of OM. They form the main food resource for mesozooplankton along with microalgae and protozoans (infusoria and zooflagellates). However, bacteria are not only food for hydrobionts, but also producers of various biologically active substances, such as vitamin B_{12} (cobalamin), in the same way as microalgae [1, 2, 145]. In seas, bacteria play a key role in primary production and metabolism of OM. They decompose dead OM, transforming the products of its decay into suitable substances for phytoplankton and phytobenthos, and are a food item for many zooplankton and zoobenthos organisms. The presence of bacteria and their activity form the basis for biological production in marine ecosystems [204, 205].

The total quantity and biomass of bacteria in the water of a basin can serve as important indices of the level of trophicity and productivity. Bacteria form the microheterotrophs group together with protozoans and, together with microalgae, can form primitive ecosystems with a close cycle of matter and energy. In marine pelagial ecosystems, bacterioplankton accounts for 60–90% of total energy flow formed by the heterotrophic part of planktonic community [161].

A significant amount of dead SOM and DOM returns into the biotic turnover owing to processes performed by heterotrophic microorganisms in the near-surface water layer. This material does not undergo total mineralization and does not sink to the bottom. This mechanism is often called the "microbial loop" [29, 55, 203].

Chemosynthesis is also observed in marine ecosystems, including MPs aquatories, but its intensity in the water column is dozens to hundred times lower than in sediments. Furthermore, chemosynthesis is considerably higher in silty than in sandy and shell-containing sediments because silts always have a higher content of dead OM and a lower oxygen content.

Several circumstances indicate the presence of bacteria throughout the marine environment. All objects immersed in seawater, in any region of the World Ocean, get covered by a slime film within a couple of days. This film is formed by various marine microorganisms, but predominantly by bacteria and diatoms. Slime film is a product of both organisms' vital activity and their habitat. However, for film-forming microorganisms, the character of the substratum plays a secondary role. The film is present on the surface of all artificial constructions and also on ships regardless of their draft. A slimy film appears even on ships' hulls treated with antifouling paints irrespective of their ability to prevent the settlement of phyto- and zoo-organisms [48]. Suspension particles, always occurring in seawater in various amounts and having different origin, are also covered by a slime film.

In MPs, besides substrata for microbial population typical of the adjacent sea zones, large areas of HTS surfaces of various form and quantity are artificially introduced into the water column. Wetted surfaces of ships hulls, performing short- and long-distance transfer, also favour the exchange of microflora between harbours of different seas.

A linear proportional dependence was established between the amount of detritus in water, bacteria and water productivity [49].

In open sea zones, bacterioplankton is responsible for up to half of the total respiration of planktonic communities. In coastal regions and bays, where MPs are located, the contribution of bacterioplankton to total respiration reaches 80–90%. The density of bacterioplankton in various water basins ranges between 10^5 and 10^7 cells cm^{-3} [94, 156].

Planktonic bacteria belong to the dimensional groups of picoplankton (0.2–0.6 μm) and nanoplankton. In water, they occur as individual free-floating cells and in aggregates. They can form microcolonies or adsorb onto organic and mineral particles suspended in water. Protozoans (representatives of nano- and microplankton) feed mainly on individual bacteria, while multicellular zooplanktonic forms, prelarvae and fish larvae feed on bacterial colonies, including those adsorbed on detritus.

Yeasts, actinomycetes and mold fungi are also included in bacterioplankton composition. In the coastal zone and in MPs aquatories, stalked and filiform bacteria attaching to dying cells of planktonic algae and detritus particles, play an important role in bacterioplankton along with bacteria that form microcolonies in the water column. In conditions of lowered hydrodynamics inside MPs aquatories, such aggregates resemble snowflakes, floating or slowly sinking to the bottom where they form a loose sediment that is easily stirred up.

Marine bacteria freely floating in water, i.e. not attached to the surface of some particles, have low metabolic speeds that sharply increase as they attach to detritus particles. With the increase in the amount of detritus, bacteria then ensure higher speeds of mineralization of dissolved and suspended OM. In the end, their activity leads to a better supply of biogenic matters for planktonic algae and microphytes [2].

The cells of unicellular bacteria are usually bigger on detritus. The average volume of bacterial cells decreases by 1.5–2.0-fold from the surface to the bottom of the water column. Maximal individual volume of bacteria is registered in periods of most intensive die-off of planktonic algae [12]. In the semiclosed aquatory of Burgas Bay, the mass development and following die-off of planktonic algae is accompanied by the appearance of filamentous bacteria in the surface water layer (0–5 m). These bacteria represent up to 50–70% of the total bacterial quantity (1–2 × 10^6 cells cm^{-3} or 150–300 mg m^{-3}) [12].

In open waters, the highest concentration of bacteria in the water column is usually found within a couple of metres below the surface. Bacterioplankton abundance gradually decreases with increasing depth.

In bays and in MPs aquatories, bacterial biomass in the upper mixed layer is equivalent, on average, to 20–40% of the phytoplankton biomass. In periods of seasonal maximums of phytoplankton development, the two biomasses are commensurable [156]. As a rule, abundance of bacterioplankton is considerably higher in spring and summer than in autumn, when the average size of bacteria also decreases.

At depths up to 20 m in intensive navigation zones within Burgas Bay, where constant water mixing occurs, the average quantity of unicellular bacteria in the water column comprises 1008 × 10^3 cells cm^{-3} and the average biomass 282 mg m^{-3} [12]. The highest bacteria concentrations are usually registered during relatively short periods of intensive phytoplankton die-off. The average size of bacterial cells and the number of bacteria aggregated on the detritus formed by decomposing algae increase at the same time.

In the open regions of the Black Sea, the total number of bacteria in the upper water layer varies from 50 up to 200,000 cells per cm^3. In coastal regions under the influence of river run-off and eutrophication (where MPs are located) the total number of bacteria increases to 300–500,000 and sometimes up to 2 million cells, and more, per 1 cm^3. In bays and gulfs in the coastal zone of the NWBS, the average bacterial biomass is in the range 100–500 mg m^{-3} [161].

The number of bacteria in various horizons of the water column is directly connected with concentrations of suspended and dissolved OM. Intensive development of bacterioneuston and bacterioplankton occurs in the presence of a high number of detritus particles. However, in the marine environment, formation of material containing organic P from dissolved OM also takes place via physico-chemical adsorption to the surface of gaseous bubbles rising up through the water column. Aggregates of OM are the best substratum for marine bacteria [173] which are the main consumers of the organic metabolites dissolved in water. The turnover speed of low molecular and directly assimilable metabolites is about 1–10 days [180]. Many planktonic algae and planktonic animals also utilize DOM to a various extent [64, 65, 145, 180].

After dying, the cells of planktonic algae become detritus and gradually sink to the bottom. Detritus is not just a form of dead SOM, but also a food supply for various bacteria and other hydrobionts. Fragments of decomposing macrophytes, fungi and animals also contribute to detritus composition. Detritus particles are quickly covered by a bacterial film. In MPs aquatories, the well-known phenomenon of mass growth of "maturing" detritus is a result of bacterial biosynthesis [161]. In marine ecosystems, including those of MPs aquatories, processes such as "rain" and "anti-rain" of dead bodies of small-sized planktonic organisms are observed [71]. These phenomena are referred to as "marine snow" [74].

Diatom algae (dead and alive), heterotrophic bacteria and protozoans (vagile and sessile) are found on detritus as well as on periphyton. These organisms are interconnected among themselves and with the marine environment. They excrete metabolites, form DOM, influence the pH of suspension and precipitate carbonates at the surface of suspended matter. Correlations between microorganisms on suspended matter and microorganisms' metabolites, DOM and pH of suspension were observed. Diatoms, bacteria and protozoans, as well as their metabolites, interact among them and with environmental factors such as pH, DOM of suspension and carbonates. In this sense, they share a common biotope represented by detritus particles. Furthermore, they respond collectively to environmental impacts. The complex of microorganisms of suspension can therefore be considered as a community [49].

In marine ecosystems, owing to plankton filtering activity, the transport of faecal pellets represents about 80–90% of the detritus flow from the water surface to the bottom. The remaining 10–20% consists of dead bodies, hard parts of planktonic organisms and mineral suspension [32]. At shallow depths, the biofiltering systems of plankton and benthos are tightly interconnected because 70–80% of benthic species have a pelagic larval stage. In fact, benthic invertebrates themselves ensure a significant part of the filtration process in the water column.

At present, a large number of studies indicates that protozoa have an important role in the functioning of aquatic ecosystems of the pelagial, periphytal and benthal zones by participating in the transformation, production and destruction of OM [31, 86, 90, 91, 93, 112, 113, 176, 177, 183, 206, 207].

There are numerous data about the selective feeding ability of infusoria. In marine ecosystems, algophagous infusoria can consume up to 38% of the phytoplankton primary production (7% directly in the near-surface layer) [93]. In the Dnieper-Bug Liman, the proportion of gross phytoplankton production consumed by ciliates varies from 30% in spring to 0.12% in summer [86]. Protozoans react very quickly to the availability of new OM resources. Introduction of protozoa into the "algae—bacteria" system stabilizes its functioning by providing a high speed of OM destruction [177].

In many coastal ecosystems, rotifers have an extremely important role [43] together with protozoans and other microzooplankton representatives [8, 32, 139, 145, 175, 201, 217].

For many heterotrophic saprophytic bacteria, detritus is the most favourable pelagial biotope and, at the same time, a source of food [187]. In marine ecosystems, there is practically no suspended and dissolved OM which cannot be used by some group or groups of bacteria. The microbial population of detritus cycles the DOM of marine water into trophic chains. Bacterioplankton (as a microflora of bottom sediments and periphytal) becomes a main source of food for filter feeders, sediment feeders and detritophages [145, 161, 181].

Bacteria are a major group of organisms, owing to which the share of primary production not directly incorporated into food chains (about two thirds of the total), is cycled back to trophic webs based on predator—prey relationships [181]. The destruction of microalgae by bacteria usually takes place within a dozen of hours [5, 145, 161, 181].

The destruction of one of the mass representatives of zooplankton copepods lasts from 4 up to 30 days depending on temperature. Regardless of their body-size, the process is completed in 5 days at a water temperature of 20 °C. The sinking speed of copepods is connected with their size and changes over time during the decomposition process [164].

In the water column, the decomposition of dead hydrobionts includes lysis, hydrolysis of high-molecular compounds to fragments of low-molecular weight and biooxidation of these compounds. Metabolites of living plants and animals also contribute to the composition of DOM. The microflora not only uses detritus and DOM, but also produces biologically active substances such as vitamins and antibiotics. Bacterioplankton, microflora of bottom sediments and periphytal, and diatoms also produce inorganic insoluble metabolites, which precipitate as carbonates. Thus, they participate in the formation of fine silt in MPs. Here, the slowed hydrodynamics and limited water exchange, the more stable hydrological and hydrochemical regimes and the high concentration of both DOM and SOM form conditions for a more intense activity of microorganisms compared with adjacent zones. These conditions also speed up the sedimentation of carbonates caused by the gravitational force. In MPs, part of carbonates precipitates because of the activity of micro- and macrophytes and animals having shells or skeletons consisting of carbonate [153, 161].

In MPs, bacteria are one of the main contributors to mineralization of SOM and DOM. In ports' aquatories, the dead OM necessary for bacterial development has autochthonous and allochthonous origin; the latter entering MPs via land run-off and discharges of different sewage waters. Conditions for OM dispersion in MP are limited and BM gradually accumulates in them.

In MPs, all the biotic components of the ecosystem participate in the formation of detritus. Plants (including micro- and macrophytes from phytobenthos and periphyton composition, phytoplankton and microphytobenthos) have an important role in detritus formation. In the macrophyte belt forming on HTS in correspondence of the upper limit of the water level, the partial die-off of plants' biomass favours the development of bacteria.

In MPs, a vertical stratification is observed in the distribution of phytoplankton and in the quantity of heterotrophic bacteria. This distribution of living organisms reflects the stratification in the distribution of particles of different origin, i.e. detritus, caused by hydrological factors [102, 145, 161]. In MPs aquatories, conditions for detritus accumulation appear in the zones of the surface tension film, in thermo- and halocline layers, on the surface of bottom sediments and in the HTS fouling.

Accumulation of microflora in the hyponeuston layer is caused by physical and chemical factors characteristic for the air–water interface (flotation, surface tension) and by the intensive reproduction of bacteria because of the increased OM content [161, 186].

In spring, summer and autumn, the quantity of heterotrophic bacteria in the near-surface micro-horizon of 0–2 cm is one to threefold higher than in the remainder of the water column [186]. The concentration of heterotrophic bacteria in the 0–2 cm layer increases sharply in summer, slightly reduces in autumn and drops to its minimum in winter. The highest peak of microbial life in the zone of surface tension film is registered in summer; along with the intensification of all living processes in the NWBS and in the whole Black Sea during this period [161, 186]. Up to 75×10^6 cells cm^{-3} are registered in neuston [145].

The amount of ammonifying and denitrifying bacteria in the 0–2 cm layer is 10–100 times higher than in the water column, which testifies the high content of OM in the surface micro-horizon. Sulphate-reducing bacteria are also registered in this well-aerated layer, so that the concentrations of sulphate-reducing bacteria in the 0–2 cm layer are comparable to those in the near-bottom horizon. The number of thiobacteria in the 0–2 cm near-surface layer is also 10–100-fold higher than in the water column [186].

According to Tsyban [186], the concentrations of heterotrophic microflora in the upper half metre horizon in the Black Sea are equal to, or higher than, in the thermocline layer. At higher depths, the vertical distribution of heterotrophic bacteria in the water column shows a maximum in the thermocline layer during all seasons in all aquatories of the NWBS [186].

In stratified water masses such as MPs water column, bacterioplankton accumulates at the upper border of the thermocline, where phytoplankton and detritus also concentrate [161].

Five ecological niches are usually distinguished with regard to the distribution of bacteria and other small-sized hydrobionts in the water column. The first ecological niche is located under the surface water film, where the micro-convective mixing of the uppermost layer (0–30 cm) of the water column takes place and the levels of oxygen saturation and nutrients are high. Here, the total quantity of bacteria ranges from 0.3 up to 4.0 milliard cells per 1 cm^3.

The second ecological niche starts at the depth of 20–50 cm and extends to the pycnocline. In this zone, phytoplankton undergoes mass development thanks to optimal illumination (photic water layer). The intensity of bacterio-, phyto- and zooplankton activity in this layer is very high owing to wind and drift currents and Langmuir vortices that ensure a good mixing.

The third ecological niche coincides with the thermocline zone where detritus particles and sinking dead plankton accumulate because of the higher water density. The density of bacterial population sharply increases in the thermocline [145].

The forth ecological niche is represented by the near-bottom water layer where the water exchange is limited and anaerobic stagnant zones occur. Under these conditions, the prevailing bacterial populations are iron reducing bacteria, thionic, methane-oxidizing and hydrogen-reducing bacteria. Sulphate-reducing bacteria and bacteria performing butyric fermentation dominate in the water zone directly in contact with the bottom.

The fifth ecological niche is connected with bottom sediments and pore waters, where bacterial density can reach dozens and even hundreds milliard of cells per 1 cm^3 of wet silt.

The lower part of the water column can be referred to as the microaerophilic zone, where dissolved oxygen is still present but reduced substances diffusing from the anaerobic zone of the near-bottom layer also appear. The anaerobic zone is located between the microaerophilic zone and bottom sediments [145].

In MPs pelagial zones, the bacterial population reflects the particular features of each ecologic niche on the one hand and directly contributes to their formation on the other.

In MPs, besides the hyponeuston layer, organic matter accumulates in the highest quantity at the border of the pycnocline. This OM includes dead phytoplankton, organic remains of animals, faecal pellets of planktonic organisms and other detritus. Detritus, populated by bacteria, gradually settles from the pycnocline layer to the bottom sediments and is partly consumed. However, in MPs, part of the sediment returns again into the water column as a result of anthropogenic activity.

Fine-dispersed sediment fractions appear in the water column during dredging and navigation of deep-draft ships. This artificially re-suspended material could then accumulate again in the pycnocline layer. Besides reducing water transparency, stirring up of bottom sediments and associated microbial population can cause other negative consequences. During dredging, the microbial population, as well as resting cysts, spores and eggs of some hydrobionts are transferred from sediments into the water column where they can contribute to the development of unwanted phytoplankton species [169]. Studies in the estuary of the river Vilaine (France) showed that silt accumulation in the estuary began after construction of a dam. After some time, the thickness of the accumulated layer of silt reached 3–4 m and brought to mass development of the toxic microalgae *Dynophysis* and *Gymnodinium*. Deposition of fine port silts dredged in ports and dumped at submerged waste deposits had similar consequences [169].

The sheltered conditions of ports aquatories lead to a significant growth of bacterioplankton in terms of quantity and biomass. The number of heterotrophic bacteria in 1 cm^3 of ports' and harbours' water ranges from a couple of thousands up to hundred thousands and even millions. Their quantity quickly reduces in open waters far from ports [48]. In Sevastopol Bay, the construction of protecting piers was one of the main causes of the sharp increase in bacterial density from 10^6 cells cm^{-3} in 1976 to 3×10^6 cells cm^{-3} in 1982 [3].

At the end of the 1990s, the total density of bacteria in the waters of Novorossiysk MP reached more than 1×10^7 cells cm^{-3} during the period of maximal water warming. In 2006–2007, the average bacterial biomass and density in the aquatory of Novorossiysk MP were 0.80–0.94 g m^{-3} and $3.8–4.7 \times 10^6$ cells cm^{-3}, respectively. The corresponding figures were 2.0–2.5-fold lower outside the port. Bacterioplankton maximal biomass reached 1.9–2.5 g m^{-3} in zones strongly polluted by coastal discharges [156].

In 2009–2010, the bacterial average biomass in Tuapse MP comprised 0.43–0.47 g m^{-3} for a total density of $2.0–2.8 \times 10^6$ cells cm^{-3} [156].

In Tuapse port aquatory, the average number of heterotrophic bacteria during the summer of 2009 reached the upper limit of 3.6×10^6 cells cm^{-3} (776 mg m^{-3}) for eutrophic waters. An extremely high level of bacterioplankton development (density of 5.2×10^6 cells cm^{-3} and biomass of 1.1 g m^{-3}) is observed in harbour inner waters, near the berths that are most distant from the passage into the port and polluted by wastewaters [155]. In MPs waters, the high quantity of heterotrophic bacteria is directly connected with increased OM content and favourable hydrological conditions.

In 1993–1996, during the period from May to September, the amount of saprophytic bacteria in Sukhoy Liman varied between 2550 and 28,700 cells cm^{-3} in the near-surface layer and in the water column and was in the range 16,200–30,000 cells cm^{-3} in the near-bottom layer. On average, the abundance of saprophytic bacteria near the bottom was 1.7-fold higher than in the superficial layer above the thermocline [154]. In June 2003, the average number of saprophytic bacteria was 3625 cells cm^{-3} in the superficial layer of Sukhoy Liman waters and 8175 cells cm^{-3} in the near-bottom layer below them [154].

Besides seasonal reorganizations and local changes, bacterioplankton composition also undergoes large-scale changes connected with the general evolution of regional ecosystems. In the summer and autumn months of 1990–2000, the absolute maximum of bacterioplankton total quantity in the NWBS was registered in the near-bottom water layer. The content of bacteria in the near-bottom waters was usually 1.2–2.0-fold lower than in the surface 0.5 m horizon. Maximal bacterial quantity in the near-bottom layer was registered following a reduction of oxygen concentrations in the near-bottom water to values between 0 and 3.87 mg dm^{-3}, i.e. under hypoxic conditions. Such conditions stimulated the development of chemolithotrophic and chemoorganotrophic bacteria [154].

In Novorossiysk MP, the quantity of bacteria included in the composition of detritus particles and aggregates was less than half of the total, while it was more than a half of the total quantity in Tuapse MP [156].

The development of bacterial populations in port aquatories generally reaches levels typical of eutrophic-hypereutrophic waters. The lower level of bacterial biomass in Tuapse port, compared with Novorossiysk MP, may be connected with better water exchange with the adjacent sea zone in the former case [156].

2.2.3 *Hyponeuston and Zooplankton*

Small-sized planktonic animals temporarily or constantly inhabiting MPs water column form two main complexes: hyponeuston and zooplankton. Hyponeuston is mainly connected with the upper sea micro-horizon (depths from 0 to 30 cm), while zooplankton inhabits the water column from the surface to the bottom. In the Black Sea coastal zones where MPs aquatories are located, merohyponeustonic and benthohyponeustonic forms are a characteristic feature of the hyponeuston, especially in summer. Merohyponeustonic organisms include species that pass into the composition of plankton, nekton or benthos after the neustonic phase (usually as eggs, larvae and juvenile) of their life cycle. Benthohyponeuston refers to the part of hyponeuston consisting in organisms alternatively staying in a benthic (during the day) or a hyponeustonic (during the night) state. Many species of amphipods, cumaceans, shrimps and polychaetes belong to this group.

Mero- and benthic-hyponeuston are tightly connected with meroplankton [71]. Holoplankton invertebrates constitute the relatively constant fraction of pelagial population in MPs.

In various regions of the Black–Azov Sea basin, the number of animal species included into plankton and hyponeuston species lists is constantly changing and being updated. This situation is connected with disappearance, introduction and discovery of new species, as well as with revision and clarification of taxonomy. In the last years, 150 taxonomic forms of animals were recorded in plankton composition in NWBS, 125 in NEBS and 162 in the coastal regions of Crimea. More than a half of them consisted in larvae of bottom (periphyton) animals [156].

Because of low, frequently almost null, flows' speeds and of the stable stratification of water masses, the ability of haloplanktonic animals and pelagic larvae of zoobenthos and periphyton to perform vertical migrations often has a fundamental importance in determining their distribution in MPs.

The microzooplankton organisms with sizes between 50 μm and 1.0 mm, the nanoplanktonic forms smaller than 50 μm and even the smallest ones, such as ultrananoplanktonic organisms that are not caught by plankton nets, are an important component of plankton composition.

In MPs aquatories, the absence of strong currents and big waves creates the conditions for the accumulation of hydrobionts from various systematic groups in the subsurface water layer. Hydrobionts are attracted to the film of surface tension for a variety of reasons.

In spring, coldwater forms dominate the hyponeuston composition and the biocenoses is characterized by overall low quantitative indices. The maximum species diversity of hyponeuston is observed in summer owing to the development of the larval stages of numerous invertebrates. In autumn, the main part of merohyponeuston disappears and the hyponeuston of the Black Sea coastal regions is poor during winter.

A reduction of zooneuston quantity has been registered over the last decades, first of all owing to a decrease in their abundance by more than one order of magnitude. The decrease mainly occurred in the previously abundant *Oithona minuta*, *Paracalanus parvus*, *Centropages ponticus* and more typical neustal copepods of the Pontellidae family. The quantity of polychaetes larvae in the neustal zone has increased because of their parental animals inhabiting on the silted bottom under the near-bottom water layer with reduced oxygen content [154].

Zooplankton is formed by many groups of small-sized animals which are not able to withstand strong currents. In marine ecosystems, zooplankton plays an important role in the formation of energy and in matter flows. On the one hand, it is a main consumer of phyto- and bacterioplankton production in the water column and, on the other one, a food item for many benthic invertebrates and commercial and non-commercial fish species at various stages of ontogenesis. In MPs aquatories, the zooplankton development directly reflects its development in the adjacent regions of the Black and Azov Seas and is controlled by the same processes. However, in MPs, there are some specific peculiarities connected with the relative isolation of aquatories, their lowered hydrodynamics and more intensive development of phyto- and bacterioplankton, and the presence of large areas of artificial HS. In MPs, the zooplankton composition depends on the geographical position and has a seasonal character.

Although zooplanktonic organisms are heterotrophic, many of them are able to accumulate and use macromolecular compounds dissolved in seawater in the same way as bacteria and microalgae. To various extents, heterotrophic feeding on DOM is common in representatives of zooplankton from various systematic groups. Concentrations of DOM nutritionally valuable components similar to natural ones can satisfy up to 30–50% of the respiratory energy requirements of some planktonic invertebrates [65, 185].

In Grigorievsky and Sukhoy Limans, a sharp reduction of zooplankton quantity is registered during years with increased continental run-off. The zooplankton quantity increases again in the following years. Zooplankton biomass can decrease sharply in years of rapid development of planktonic phytophagous animals that actively graze on phytoplankton [115]. The indices of phytoplankton can reach maximal values at low levels of zooplankton development.

In the Black Sea, the daily dynamics of phytoplankton biomass and quantity during the spring–summer period is usually described by an unimodal curve with maximum values in the evening and night hours [115]. In connection with this regular feature, zooplanktonic organisms feeding on phytoplankton also migrate to the near-surface layer during evening and night hours. Thus, in the near-surface layer, phytoplankton grazing by meroplankton occurs at nighttime.

In MPs aquatories (e.g. in Sevastopol Bay), zooplankton biomass could exceed by 5–10-fold the corresponding average biomass in open regions of the Black Sea [161]. In other MPs, zooplankton biomass is quite often a couple of times higher than in the adjacent regions [16, 154, 196].

The distribution of planktonic organisms in MPs water column is connected with the water physical characteristics and various movements, as well as with the behaviour of the planktonic organisms themselves and with their capability to move into the pelagial zone both vertically and horizontally. Hydrobionts can have passive and active methods for vertical displacement. For example, flagellates actively swim using flagella.

Copepoda, which are relatively complexly organized organisms widespread in MPs, actively control their distribution along salinity gradients. Copepoda at various stages of development, as well as Decapoda larvae, select water layers with the most favourable salinity by regulating their locomotive activity. In MPs aquatories, sharp increases in salinity lead to active displacement of planktonic animals towards the surface layer. At the opposite, a decrease in salinity causes the immersion of planktonic animals by both passive and active mechanisms [136]. Many pelagic microplanktonic infusoria and rotifers respond by active displacement to sharp changes in salinity.

Ultramicro- and microplankton (especially colourless flagellates, zooflagellates, infusoria and rotifers) have a key role in the transformation of matter in the pelagial zone of coastal ecosystems both in the Black–Azov Sea basin and in other seas [161].

As mentioned above, microplankton usually indicates a complex of heterotrophic organisms with linear dimensions in the range 20–200 μm and including dinoflagellates, infusoria, naupliar stages of copepods and some copepodit stages as well as meroplankton larvae. As a rule, heterotrophic dinoflagellates constitute the bulk (on average 63–65% of the total biomass) of marine microzooplankton [163].

Heterotrophic and colourless thecate flagellates (Peridinea) are planktonic forms in the absolute majority of cases. They occupy an intermediate position between prokaryotes and eukaryotes and belong to ancient forms of life. Their flagellum located along the longitudinal body axis allows the forward motion, while the second one, located perpendicularly to the first one, performs the rotating motion. During swimming, flagella perform a screw-like motion.

The sea sparkle *Noctiluca scintillans*, which often reaches high densities, occupies a special place among the colourless thecate flagellates or zooflagellates of the Black–Azov Sea basin. It has a spheroid body up to 2 mm in diameter and it feeds heterotrophically by swallowing and digesting different suspended particles (detritus and small-sized unicellular algae and protozoa). During the summer months, *Noctiluca* can undergo mass development in MPs and bays aquatories, reaching densities of up to 3×10^6 ind. m^{-3} and a biomass in the range of dozens of grammes per cubic metre of water. *Noctiluca* is practically not used as a food by other planktonic organisms [66].

In the pelagial zone of MPs, infusoria (Infusoria) have a high importance among protozoans. It is considered that they originated from flagellates. Infusoria have sizes from 10 μm to 2–3 mm and can reproduce asexually (by cell division) and sexually. Many of them can form cysts resting on the bottom. Variously distributed lines of cilia cover their body during their whole life or at some stages of the life cycle. Some species have bunches of cilia grouped into organs of movement called cirruses that are used for swimming and moving on a substratum. The swimming speed of infusoria reaches 1–2 mm s^{-1}. In MPs aquatories undergoing water desalination, rotifers (Rotatoria), belonging to pseudocoelomate animals, quite often reach high quantities. Their body is sacciform or spheroidal, with sizes from

10 μm up to 2.0 mm, with dimensions in the range 100–300 μm being most common. Some planktonic rotifers are the smallest representatives of multicellular animals.

Rotifers have a rotatory ciliary apparatus on the anterior end of their body. Ciliary action resembles the rotation of a wheel. In coastal zone, the large group of rotifers comprises species which can swim, crawl and temporarily attach to different substrata, including macrophytes. Numerous swimming rotifers are able to perform relatively long and sudden jumps allowing them to escape from enemies, in particular from planktonic crustaceans, larvae and juvenile fish hunting them. With regard to feeding habits, rotifers are divided into microphages, or "pacific", and predators. Microphages consume microalgae, bacteria, detritus, spores and fragments of fungal mycelia. Predators feed mainly on infusoria and other rotifers [145]. If needed, numerous predators switch to feeding on microalgae. The large majority of rotifers species is polyphagous, i.e. it is not highly selective in its food items. However, there are also monophagous species feeding only on detritus or exclusively on certain species of microalgae. On average, rotifers live from a couple of days up to 3–4 weeks. In MPs aquatories, rotifers of the genus *Synchaeta* and *Brachionus* are most often observed.

Flagellate and ciliate ways of motion are characteristic for most small-sized pelagic animals. These motion strategies are most effective only at body sizes of a fraction of millimetre and at low Reynolds numbers. Using flagella and cilia, organisms rotate around their own axis during swimming. They do not move linearly, but along a screw-like trajectory. Ciliary motion is also typical for the pelagic larvae of many benthic animals (worms, molluscs and others).

The relative speed of microplanktonic forms using ciliar and flagellar motion exceeds that of multicellular, large invertebrates. For bigger planktonic forms, the movement via body curving was more advantaged. It is typical for Annelides and Scolecida, prolarvae and fish fries. Swimming folds, fins and lever extremities, i.e. various swimming legs, appeared at a later stage in the development of motion strategies.

For the medusoid generations of small-sized pelagic coelenterates of the class Hydrozoa, a decreased or increase in salinity causes a reduction of the contraction rate of the umbrella and, thus, of motion activity. Contraction of umbrella is the main mechanism of jellyfish reactive type displacement in water and moving velocity depends, first of all, on the contraction frequency. Similar changes in the contraction rhythm are characteristic for the class of Scyphozoa. The moon jellyfish *Aurelia aurita* moves into the deeper water layers by reducing the umbrella's contractions frequency. Such behaviour of jellyfish and other zooplanktonic organisms is called "passive avoidance" [136].

Depending on the ontogenetic stage, gelatinous organisms such as hydrozoan and scyphozoan jellyfish can belong to various size groups of zooplankton, from mesoplankton to megaplankton. The moon jellyfish, saucer jelly or common jellyfish (*A. aurita*), barrel jellyfish, dustbin-lid jellyfish or the frilly mouthed jellyfish (*R. pulmo*) have planktonic and benthonic stages. While *A. aurita* performs mainly vertical displacement, *Rh. pulmo* is considered a quite good swimmer among planktonic organisms and can actually move in any direction. The diameter of *A. aurita* umbrella is usually up to 20 cm and that of *Rh. pulmo* can reach 60 cm. In MPs aquatories ecosystems, large-sized jellyfish individuals could be ascribed to megaplankton as well as to nekton.

In open waters, *A. aurita* mainly feeds on unicellular algae (32%), copepods (24%) and chaetognaths (20%). The moon jellyfish, being a food "opportunist" like the majority of gelatinous organisms, is able to consume big numbers of micro- and mesoplanktonic organisms of various groups as well as eggs and fish larvae. In coastal waters, *A. aurita* feeds on unicellular algae, bacterioplankton and microplankton. It can also consume SOM and DOM [14].

The jellyfish *A. aurita* and other gelatinous organisms mainly use the same food as many organisms from micro- and mesoplankton as well as representatives of these complexes themselves. Gelatinous organisms can achieve a high biomass, but serve as food for hydrobionts of higher trophic levels only rarely. The BM accumulated in them returns into the biological cycle mainly after decomposition by microbes.

The planula, i.e. the pelagic jellyfish larva, has cilia located in a belt-like fashion around the whole body, allowing it to actively move in the water column. However, they are not able to feed. Planulae have positive phototaxis and naturally accumulate in the near-surface layer. After some time, their attitude towards light changes and they immerse into the near-bottom layer. Here, they actively swim for a couple of days and then attach to hard substrata, including concrete surfaces of MPs HTS, by their anterior body end. Two or three days after fixation, the planula transforms into small polyps called scyphistoms. The attached scyphistoms feed on infusoria, planktonic crustaceans and others. In winter months, the quantity of scyphistoms fixed on HTS can reach dozens and even hundreds of individuals per 1 m^2 in the 0.5–2.0 m water layer. After concluding their development under attached conditions, the jellyfish larvae (ephyra) move back into life in the pelagial zone. Hydroids (Hydrozoa) resembling small-sized jellyfish are also observed in MPs plankton.

Besides jellyfish, other gelatinous organisms in the plankton composition of the Black–Azov Sea MPs aquatories are represented by three species of the phylum Ctenophora: *Pleurobrachia rhodopis* and the recent invasive species *Mnemiopsis leidyi* and *Beroe ovata*. The body length of *P. rhodopis* is about 5–7 mm, while that of *M. leidyi* and *B. ovata* is usually up to a couple of centimetres. Some *Mnemiopsis* individuals grow up to 10–11 cm. *M. leidyi* and *B. ovata* spread particularly during the period 1980–2000. As in case of jellyfishes, their biomass becomes a peculiar nutritional dead-end for a relatively short period.

The gelatinous organisms can perform not only vertical migrations, but also move horizontally in the water column. In some hydrological situations, they enter MPs aquatory together with water masses and get trapped in the zones farther away from the port's passage. The biomass of gelatinous organisms in such places is quite often calculated in kilograms and dozens of kilograms per one cubic metre of water. For a whole MP aquatory, values up to 8000–10,000 tonnes can be reached, as in Grigorievsky Liman in 2012. These organisms cannot abandon ports' aquatories and eventually die in them. Big-sized jellyfish such as *Aurelia* and, especially, *Rhizostoma* create strong water flows while moving. The influence of these flows manifest over distances of dozens of centimetres, thus mixing big volumes of water. In MPs, the artificial increase of depths facilitates the retention of gelatinous organisms in the water column, where they usually die when arriving at the bottom.

Introduction of the comb-jelly *M. leidyi* has disturbed the whole trophic structure of pelagial subsystems of the Black and Azov Seas and of big open limans for a couple of decades [154]. In eutrophic waters, *M. leidyi* can reach huge biomass. The comb-jelly *B. ovata*, which can use other species of comb-jellies as food, reaches its mass development in a basin after *M. leidyi*. By grazing *M. leidyi*, *B. ovata* not only increased its own biomass, but also reduced the quantity of the first one enough to undermine its own forage base at the same time. Breakouts of *M. leidyi* stopped after the first years following its introduction because of the pressure from *B. ovata*. At the same time, the quantity and biomass of *B. ovata* decreases because of the reduced forage base. A kind of balance between comb-jellies species was thus established in the ecosystems of the Black–Azov Sea basin.

Eight longitudinal comb rows are located on the surface of the transparent body of comb-jellies. Each row comprises swimming plates oriented transversally to the axis of the row. Along the outer edge, each plate is split into combs formed by agglutinated cilia of epithelium whose length reaches a couple of millimetres. Row lines represent the locomotor apparatus of these animals and the row plates work like oars. Comb-jellies swim slowly with their mouthpart frontward. They have pelagic larvae.

Catching tentacles with numerous branches able to elongate are located laterally. Tentacles and branches are covered by glue cells (colloblasts). In the nutrition process, small-sized planktonic organisms get stuck to the tentacles and are brought to the mouth edge. The comb-jellies are predators, also consuming eggs, larvae and early fries of fish. *B. ovata*, as already mentioned, also feeds on other smaller comb-jellies.

The comb-jellies actively graze micro- and mesoplankton in the upper layer of the MPs water column reaching high quantity and biomass (up to couple of dozens kilograms per 1 m^3), as in the case of jellyfish. Because of the presence of numerous HTS in MPs aquatories, the bodies of comb-jellies are easily damaged by rubbing and blowing against hard surfaces, even at relatively low agitation, and form a gelatinous mass in the near-wall layer. High amounts of comb-jellies are destroyed during ships navigation.

In MPs, individuals of *Oicopleura dioica*, belonging to solitary pelagic tunicates (Appendicularia) are quite often registered. They have a transparent body with a length of up to 1 mm and a tail which is nearly fourfold longer than the body. They feed on small-sized flagellates, peridinian algae, coccolithophores and diatoms.

In MPs, larvae of sedimentary ascidians or sea squirts (Ascidiae) rise up and freely swim into the water column for some time.

In MPs aquatories ecosystems, both free-living and sessile (i.e. attached to substrata) infusoria play a huge role. These organisms belong to the number of the most small-sized mass heterotrophs and mainly feed on unicellular algae and bacteria. Species that usually feed on algae quite often switch, in their absence, to bacteria consumption. Some large infusoria, the so-called macrophages, are actively and selectively able to seize food of a defined size and form. Many infusoria species can simultaneously feed on algae and bacteria. The speed of infusoria cells division, i.e. reproduction, depends on the amount of food as well as on its quality [128] and varies from a few hours to a dozen of hours.

The community of planktonic infusoria (ciliata plankton) actively participates in the production and transformation of OM and in the functioning of the planktonic trophic web. It also contributes to the foraging basis for zooplankton, pelagic larvae and fries of fish in all coastal ecosystems of the Black–Azov Sea basin, including MPs aquatories located at the open seashore as well as in the harbours of estuarine type and in limans.

The protoplasmatic body of tintinnids (Tintinnidae) infusoria is enclosed in a transparent strong lorica, consisting of OM. Different species have loricas of different forms. Extraneous inclusions could be present on the lorica surface in the form of incrustations. Tintinnids bodies usually have a conic or bell-shaped form. Ciliates tintinnids feed on phytoplanktonic organisms: diatoms, dinoflagellates and coccolithophores. In turn, they serve as a food for noctiluca, rotifers and other planktonic predators, including larvae and fish fries.

Grazing of vegetal food by infusoria in the coastal zone of Sevastopol Bay represents 3.8–9.6% of the total phytoplankton biomass. In the bay's neck, grazing is lower and does not exceed 3.2% of phytoplankton biomass [128]. This fact gives grounds to consider that phytoplankton grazing by zooplankton intensifies in protected aquatories and MP harbours.

Infusoria serve as food for some mass marine invertebrates: planktonic copepods such as *Acartia clausi*, larvae of barnacles (*B. improvisus*) and of gastropods, in particular *Rissoa splendida* [128].

In Sevastopol Bay and in the aquatories of other MPs, infusoria are represented mainly by ciliates and tintinnids. With regard to taxonomy, the composition of heterotrophic flagellates in Sevastopol Bay is considerably poorer than in the Black Sea. However, their abundance occasionally increases up to values typical for eutrophic and hypereutrophic regions, regardless of the season [3].

In microzooplankton composition of bays and gulfs (and therefore ports) of the Black Sea, various complexes of infusoria comprise, on average, about 50% of the whole biomass, but this share quite often increases up to 80–90%. Besides planktonic species of infusoria, "free-swimming" larvae of benthic and periphytic infusoria species are also found in MPs aquatories. The "free-swimming period" of the larvae (called telotroch) of attached infusoria species is short and lasts from a couple of minutes up to 2–6 h. Telotrochs stay mainly near the settlements of adult individuals and, as a rule, appear in the upper water layers only accidentally. When tearing away from parent individuals, telotrochs stay in the water column where they can be picked up even by weak flows. Attached species of infusoria are present in the fouling all year round. They are mainly represented by species of the genus *Zoothamnium*, *Vorticelia*, *Ephelota*, *Acineta* and *Follicilina*. The quick division rate of some species, equal to that of diatoms, implies that infusoria remain a constant component of the fouling community, regardless of its developmental stage. Telotrochs released from the fouling constantly replenish zooplankton so that adult infusoria, as well as their telotrochs, are used as food by various macrofoulers. It was also determined that infusoria, especially the genus *Ephelota* and *Vorticelia*, are able to utilize DOM [24].

In the Black Sea, Tuapse port aquatory is one of the most polluted with very high content of oil products. Their concentration in bottom sediments reaches 2543 μg g^{-1}, which exceeds the permissible level by 51-fold [155]. In the plankton of Tuapse port, the dominant infusoria are very small-sized species of the genus *Strombidium* (smaller than 20 μm) and big-sized loricate, phytophage infusoria *Favella ehrenbergii*. The average density and biomass of infusoria in port waters (72.8×10^3 ind. dm^{-3}, 640 mg m^{-3}) were almost half of those outside its borders [155]. Studies have showed that, in Tuapse MP ecosystem, more than 90% of the energy flow passes through the detritus food web. Its key link is the community of heterotrophic bacterioplankton. Most of phytoplankton production and rests of dead OM are consumed by and subject to bacterial destruction. In spite of the abundance of bacteria, the production of biomass by infusoria communities, which are their main consumers, is not high. It is possible to surmise that this is connected with excessive pollution [155].

Ilyichevsk and Yuzhny MPs, located in the NWBS, were built later than Sevastopol and Tuapse ports and their aquatories are less polluted. In Sukhoy Liman (Ilyichevsk port aquatory), 37 species of infusoria were identified in the water column. In the innermost, desalinated part of the basin one freshwater species of infusoria (*Strombidium viridae*) was found. In the central and southern parts of the liman, marine pelagic species predominated: *Strombidinopsis chechiri*, *S. langenula*, *Tiarina fucus*, *Pelagostrobilidium spirale* (including the form *coxliella*), *Eutintinnus lususundae* and *Metacylis mediterranea*. These species feed mainly on dinoflagellates, which agrees with the high quantity of peridinian algae in phytoplankton. Middle- and big-sized infusoria were dominating and gave relatively high biomasses (57.7–199.2 mg m^{-3}) at relatively low densities for the NWBS ($2.2–26.9 \times 10^6$ ind. m^{-3}). In the summer of 2001–2002, average values of density and biomass for the deepest southern part of Sukhoy Liman were $(3.27 \pm 1.66) \times 10^6$ ind. m^{-3} and 113.23 ± 49.82 mg m^{-3}, respectively [154].

In the plankton of Grigorievsky Liman (Yuzhny MP aquatory), 35 infusoria species were found and dominating ones were typical of non-specific complexes (eurytopic and benthic): *Euplotes balteatus*, *Euplotes* spp., *Litonotus* sp. and *Prorodon spp.* The majority of euplanktonic forms were represented by euryhaline species and species of brackish water complexes: *Askenasia stelaris*, *Rimostrombidium caudatum*, *R. velox*, *Cyclotrichium sphaericum*, *Strombidium vestitum* and tintinnids. Quantity and biomass reduced during the spring–summer period: from 4.0×10^6 up to 0.012×10^6 ind. m^{-3} and from 78.5–134.9 mg m^{-3} up to 0.87 mg m^{-3}, respectively. A high density of predatory infusoria was registered in Grigorievsky Liman. A considerable part of algophagous infusoria was represented by diatomophagous ones [98, 154].

As a rule, no more than 10 infusoria species are constantly present in the NWBS pelagial during one year. In total, about 50 species belong to mass ones and develop with various intensities in different seasons. The majority of the registered infusoria belongs to widespread species that are common in the pelagial zone of seas, oceans and estuaries. The ratio of marine, brackish water, euryhaline and freshwater species is relatively stable during the year, although the share of freshwater species somewhat increases in spring because of the influence of river floods. A total of 184 taxa of infusoria were identified in the NWBS, while the corresponding figures for Sukhoy and Grigorievsky Limans were 37 and 35 species, respectively [98].

In the NWBS, 57.5% of the occurring species are euplanktonic. In Grigorievsky and Sukhoy Limans, euplanktonic species represent 81.1–94.6% of the total [98]. Apparently, such differences are mainly determined by a more marked stratification of the water masses inside the limans. In the water column of MPs aquatories, the presence of infusoria non-specific for planktonic complexes (as well as of benthic unicellular algae) could be connected with their presence in the fouling community on the lateral surfaces of various HTS. These structures accomplish the function of a vertical bottom for infusoria. Active mixing of the whole water column during strong storm winds and during the passage of deep-draft ships can also play a role in this process.

In the NWBS, the size distribution of planktonic infusoria varies from nanoplanktonic forms with a cell volume around 600 μm^3 to mesoplanktonic ones with volumes of 6–12 $\times$ 10^5 μm^3. The most common size class (75.0% of the total) is represented by species with volumes of 1–50 $\times$ 10^3 μm^3 (more than 50.0% from them having cells volumes up to 2.5 $\times$ 10^4 μm^3) usually belonging to "small-sized" groups. These are the mass species *Myrionecta rubra, S. vestitum, S. emergens*, *Urotricha* spp., *Rimostrombidium* spp. and *Holophrya pelagica.* With few exceptions, the indices of quantitative development of infusoria in the surface water layer of 0–0.5 m and in the near-bottom layer (3.5–4.0 m overlying the bottom) are higher than in other regions near the surface. In the surface layer, small-sized species prevail [98].

Infusoria are divided into four main trophic groups: 1—microphages (ration basis: picoplankton, detritus, dissolved and suspended OM), 2—algophages (food basis represented by various microalgae), 3—predators (mainly feeding on other infusoria), 4—mixotrophs (having photo-endosymbionts). Some obligatory mixotrophs species energetically depend on photo-endosymbionts, while others are facultative mixotrophs using photo-endosymbionts only when food is lacking. Overall, mixotrophs comprise about 13.0% of all infusoria species registered in the NWBS [98]. However, it could be considered that mixotrophs have a stabilizing role in the trophodynamic processes of coastal ecosystems.

Algophages and microphages are the most numerous groups over the year. In spring, the ratio among trophic groups changes due to a sharp increase in the quantity of mixotroph species which reach more than 90.0% of the whole infusoria quantity. The maximal development of phytotrophic infusoria occurs in winter (December–March). Diatomophagous infusoria prevail from December until the beginning of February, leaving room for consumers of flagellate plankton in February–March [98].

In the NWBS, the algophagous groups represent 62.3% of all infusoria species. Microphages (8.6% of all species) develop predominantly in summer, while the maximal diversity of mixotrophes (4.7% of all species) is registered in autumn and winter at temperatures not higher than 18 °C. As a rule, predators (4.4% of all species) are not abundant. The ratio among trophic groups in Sukhoy and Grigorievsky Limans is similar to that in the NWBS open zones. In general, the trophic structure of Odessa Gulf (including Odessa MP) is characterized by the prevalence of algophages and mixotrophs in terms of density and biomass.

In Grigorievsky and Sukhoy Limans, algophagous infusoria dominate in density and biomass. A positive correlation between the quantitative development of planktonic infusoria and a complex of three factors (temperature, salinity and dissolved oxygen content) was found in these MPs aquatories, with dissolved oxygen showing the highest correlation. In MPS aquatories located in limans, a reduction in the quantity and biomass indices of planktonic infusoria is registered in the near-bottom layer of the deepwater zones [98].

In the NWBS, the ratio among the biomass of various trophic groups of infusoria is shifted towards dominance of algophages during the year. In comparison with phytophages, infusoria-microphages are much smaller and do not reach a considerable biomass despite their high quantity. In general, the contribution of mixotroph forms to quantity as well as to biomass remains relatively constant during the year. Compared with their quantity, the contribution of predatory infusoria to biomass is more noticeable owing to their bigger sizes [98].

For those infusoria which are obligatory mixotrophs, chlorophyll content and photosynthesis level are similar to those of microplanktonic algae [98].

It was determined that Odessa MP aquatory is characterized by a low yearly variability of infusoria trophic structure compared with the adjacent waters. The level of phytoplankton consumption by infusoria is usually much higher in limans and MPs aquatories than in the open zones of the sea. The degree of connection with the sea largely influences the size and species composition of the infusoria community in aquatories. The annual dynamics of planktonic infusoria quantitative development shows two regular maximums in spring and summer. The first one corresponds to the spring "bloom" of phytoplankton and the second one to the warmest period and OM mass destruction [98].

During the vegetative period (May–October), infusoria of the NWBS plankton daily consume from 31.3 to 190.3 mg m^{-3} of algae and 15.1–39.9 mg m^{-3} of bacterial biomass [98].

In the Odessa Gulf and other regions of the NWBS, phytophages and mixotrophs infusoria consume up to 1677 mg of algae and up to 456 mg of bacterial biomass per day and per 1 m^3 of water. Predatory infusoria graze up to 406 mg of "pacific" species [98].

Production indices of planktonic infusoria vary from negative values, when predators prevail in the community, up to 585 mg m^{-3} day^{-1} in Odessa Gulf and up to 364.1–2299.6 mg m^{-3} day^{-1} in Sukhoy and Grigorievsky Limans. Most of the infusoria production (82.0–88.0%) is formed by species belonging to the size range of food objects for mesozooplankton and fish larvae. In Odessa Gulf and in

Sukhoy and Grigorievsky Limans, the maximal values of production are characteristic for the summer period [98].

In coastal marine ecosystems (including MPs aquatories), infusoria play a leading role in the energy transfer to all heterotrophic plankton components. Compared with invertebrates, the average indices of planktonic infusoria production during the vegetative period (May–November) are 1.4-fold higher and the metabolic expenditures 7.0-fold higher, in spite of the much lower biomass of infusoria. The contribution of planktonic infusoria to heterotrophic and total destruction is 0.84 and 0.82%, respectively. The food rations of infusoria comprise 1.0% of bacterial production and 6.5% of net primary production of phytoplankton corresponding to 0.5 and 18.0% of their biomass [98].

In the NWBS open regions and limans, the expenditure of infusoria on metabolism has similar values and does not exceed 10.3% of the consumed food energy. The transmission efficiency of energy to higher levels is low and, on average, only 12.5–14.1% of the ingested food transforms into production. The majority of consumed food returns into the environment in a transformed form and renews the OM stock circulating in the system DOM + OM → bacteria → protozoa → DOM + SOM [98], i.e. the so-called "microbial loop".

The relatively low levels of resources used by ciliatoplankton in coastal marine ecosystems of the NWBS, particularly in MPs aquatories, may be explained by the high absolute quantity of phyto- and bacterioplankton from one side and by pollution and anthropogenic transformation of MPs aquatories from another. The latter changes environmental conditions and affects the natural structure of hydrobionts communities [9, 40, 156].

Dominance of big-sized diatom algae in phytoplankton is considered to reduce its total forage value for zoo- and ichthyoplankton. The speed of phytoplankton utilization by zooplankton generally depends on temperature and biochemical composition of food, as well as from its concentration in the water column.

In MPs aquatories located in desalinated regions of the Black and Azov Seas, water fleas (Cladocera) with the size up to 1 mm quite often play a noticeable role in meroplankton. These crustaceans move by short leaps mainly with the help of the second pair of antennae that they can swing abruptly. They feed on small algae and detritus particles that they filter out of water using thoracic, leaf-like appendages equipped with numerous hairs or setae. These structures collectively form the filtration apparatus. Cladocerans usually cannot sort edible particles from inedible ones. They are themselves a valuable food supply for fries and juvenile fish and play an important role in zooplankton composition (especially forage) of the Black Sea in general and of MPs aquatories in particular.

Small-sized cladocera, especially copepods, are able not only to stay in a defined horizon, but also to actively move in the water column and usually spend their whole life in the pelagial zone. Over a day, each population of planktonic copepods has a well-defined living depths [158]. Without compensation of gravitational settling by active motion, many cladocera and copepods would sink deeper than their optimal living depths, or even to the bottom, owing to their negative buoyancy [164]. As already mentioned, oxygen deficit and appearance of hydrogen sulphide quite often occur in the near-bottom layer of MPs. Such conditions harm the generally oxyphilous planktonic crustaceans to a various degree.

Among pelagic copepods, eggs carried by adults hatch into nauplius larvae. There are 11–12 larvae stages, out of which 5–6 are naupliar and 6 copepodid. The last copepodid stage is sexually mature. Copepods larvae stay in the near-surface layer. Naupliar and younger copepodid stages usually do not perform vertical diurnal migrations. In the case of calanids (Calanida), the swimming speeds during vertical migrations are in the range 10–30 cm min^{-1} [135, 136].

In MPs aquatories of the Black–Azov Sea basin, the most common groups of copepods include representatives of the genera *Acartia*, *Oithona* and *Paracalanus*. For example, *Acartia* accounts for 81.0–91.0% of the zooplankton composition in Novorossiysk MP, while it contribution ranged from 34.6 to 63.0% in open part of the bay [156].

In MPs ecosystems, copepods are one of the main links along the food chain. They ensure the transfer of matter and energy from the phyto- and bacterioplankton to the higher trophic levels. Copepods are important food objects for fish larvae and juveniles. Representatives of the genus *Acartia* are the main copepod group accomplishing this function in the coastal zone and in MPs aquatories. In the Black Sea, *A. clausi* is a typical inhabitant of gulfs, bays and limans. Its eggs rest on bottom sediments.

During the last years, the relatively recently introduced copepod *Acartia tonsa* occupies a prominent place in MPs ecosystems. In Novorossiysk MP, *A. tonsa* populations are present in plankton from June until November. They appear at temperatures around 18.0–18.5 °C and disappear at temperatures lower than 15.0–18.0 °C. From the end of June, the abundance of *A. tonsa* can even exceed that of the eurybiontic *A. clausi*. Its development peak occurs in July–September with 24–26 × 10^3 ind. m^{-3} in the port and 6–8 × 10^3 ind. m^{-3} in the open part of Novorossiysk Bay [156].

Free and colonial forms of diatoms do not have a high forage value for copepods or other zooplanktonic organisms. Copepods feed mainly on nanoplanktonic dinophytes which are a more suitable food owing to their soft, round-shaped and small-sized cells which are free from the hard theca typical of diatoms [136, 156].

Three types of predators are distinguished among copepods consuming animal food: ambush, examiners and suspension feeders. The "examiner strategy" is common to the majority of calanids. They have an evident selectivity towards preys of a defined size. In the majority of cases, the body length of preys is within 4.0–40.0% of the predator's body length and 0.1–8.0% of its mass. However, there are exceptions in which the predator–prey body ratio approaches one [136].

Copepods feed by using appendages to create water flows directed toward their mouth apparatus. These appendages work at low Reynolds numbers (0.01–0.001) and practically create laminar flows. Studies performed over the last decades showed that feeding on vegetal food by passive filtration is an exception, if it exists, because it is not possible in a water mass with Reynolds number <0.1. Algal cells are trapped separately and, in principle, copepods have a discontinuous, discrete feeding and not one based on filtration. The feeding process is based on active regulation processes, i.e. individuals distinguish density gradients, size, form, algal physiological state and biochemical composition of metabolites surrounding the cells. Some copepod species show a distinct avoidance reaction towards algal species containing toxins such as *Gonyaulax grindleyi* [136].

Experimental studies showed that copepods move by leaping, consisting in a proper leap followed by inertial motion. The lengths of a proper leap and associated inertial motion are about 0.88 and 0.52 cm, respectively, giving a full leap length of about 1.40 cm. Depending on the situation, leaps could be short and long. The last ones are characteristic during avoidance of dangers and during diurnal vertical migrations [136].

Besides intermittent motions, copepods are capable to actively slide and float. In the pauses between movements, individuals with a slightly negative buoyancy sink passively with speeds varying from 0.3 mm s^{-1} for *A. clausi*, 0.6 mm s^{-1} for *Paracalanua parvus* up to 1.0–2.5 mm s^{-1} for bigger calanoid copepods. For the majority of the studied species, the passive phase exceeds the active one both in the presence and in the absence of food in the environment [136].

It is important to emphasize that among the copepods of the genus *Acartia*, most frequently registered in MPs aquatories, the speed of passive immersion is relatively low and allows them to stay in the more desalinated epipelagial layer. Different movement types such as abrupt leaps, short leaps and sliding are characteristic for *Acartia*, as well as for other copepods. All of them are performed by sharp movements of the second pair of antennas. Short, frequent leaps (with a speed of 0.02–0.03 mm s^{-1}) and abrupt leaps (0.4–0.8 mm s^{-1}) prevail among copepods; while smooth gliding is rarely observed [130].

Acartia are not able to escape to a significant distance from light and are forced to constantly stay in the well-illuminated layer, accomplishing short migrations to accessible depths [130, 132].

In MPs aquatories, both live and dead *Acartia* are an important food supply. In the case of *A. clausi*, body density decreases from 1.0497 to 1.0256 g cm^{-3} during decomposition, which favours the accumulation of their dead bodies in the near-surface layer [164].

The lower the aquatory depth, the higher the energy that planktonic organisms must spend to remain in a given horizon [164]. When alternation of passive immersion and active emersion periods becomes too frequent, crustaceans die from inanition or sink to the bottom where they either die or are grazed. In MPs aquatories, dredging ameliorates the living conditions of many planktonic crustaceans and other hydrobionts.

It is well known that emerging in warmer water requires relatively low energetic expenses [164]. Apparently, this is also one reason why, in MPs aquatories, small-sized species of planktonic crustaceans have a propensity for the epipelagial zone.

Big-sized calanoid copepods, which are usually registered in MPs aquatories during winter months, could cover a distance of a couple of dozens metres (and up to 71 m) during 1 h. In this time, individuals of the coldwater species *Calanus helgolandicus* were moving by leaps for 5.9 min and were sliding for 7.4 min, i.e.

they were spending 13.3 min, or less than 0.25 h, on active movements [164]. Periods of passive immersion are always considerably longer than time of active motion as for other pelagic copepods.

Marine arrow worms (*Sagitta*) have a transparent body with a length of 10–20 mm (resembling plumage arrow) and are registered in the plankton composition of many MPs. They have a caudal fin at the body end opposite to the mouth as well as anterior and posterior tail fins on the lateral parts of their body. With the help of fins, arrow worms can perform rush leaps exceeding their own length by several fold.

During feeding, marine arrows are able to seize preys which are bigger than their own body. As a rule, they feed at night and eat mainly crustaceans, fish larvae and their own juveniles. Juvenile individuals feed on bacteria and unicellular algae.

In MPs, isopods (Isopoda) from the genus *Idotea*, *Sphaeroma*, *Jaera* and *Eurydice* are often observed in the near-wall layer of HTS. Isopods specifically feed on macrophytes and diatoms fouling them, bacteria, protozoans and other microscopic organisms.

In MPs, shrimps from the genus *Palaemon*, opossum shrimps (Mysidacea) and hooded shrimps (Cumacea) move from benthal and periphytal zones into the near-surface water layer at nighttime. Shrimps can crawl and swim by pushing themselves with the help of pleopods. During such movements, antennas and thoracic (walking) legs are pressed against the body and the abdominal segments are straightened. At the approach of a danger, shrimps accomplish leaps frontwards by using the posterior part of their body. Abdominal segments are sharply bent and the telson and the uropods are used to create the thrust. Shrimps are mainly predators, but can feed on dead hydrobionts.

Shrimps and crabs larvae are quite good swimmers and pass 4–5 weeks in the water column during their development process. During the zoaea stage, larvae use thoracic appendages for swimming and zoaea can cover a distance of 0.5–1.0 m per minute. During the subsequent stages, pleopods become the organs of motion. Equipped with pleopods, larvae swim with their anterior body end frontward. Decapods larvae feed on planktonic organisms such as algae and animals.

The semi-transparent mysids, being similar in appearance to small-sized shrimps, have lengths of 10–20 mm. At daytime, they stay near the bottom and near HTS walls. They swim with the help of thoracic legs whose anterior pair presents maxillipedes. Mysids mainly feed by filtration or by selectively seizing suspended particles and small-sized animals. They are also able to consume fragments of macrophytes and crustaceans' dead bodies.

Opossum shrimps perform regular vertical (diurnal) and horizontal (seasonal) migrations during which they form shoals of various sizes. Mysids stay in the shaded near-bottom or near-wall water layers during the day and rise up towards the surface for feeding at dusk.

Cumaceans usually spend day time burrowed into the silty-sand or sandy sediment; going out from sediment and swimming in the water column during evening hours and at night. Females swim using external branches of thoracic legs and males also by swimmerets. Swimming is also favoured by bending and straightening of the abdominal segments. Individuals with a length of about 3 mm can move to surface with a speed of 1 m s^{-1} [67].

In MPs, and particularly in Sevastopol Bay, some representatives of meiobenthos, including harpacticoids and nematodes regularly move from the bottom into the water column during night hours [70]. This behaviour could be connected with both nutrition and nocturnal oxygen deficit at the bottom. Sorokin [161] surmised that nocturnal migrations of some bottom organisms into the thermocline layer are caused by the presence of increased concentrations of dead OM, i.e. food, in this zone. We have observed amphipods migrations towards the water surface in early morning hours, which was undoubtedly caused by oxygen deficit.

Excessive enrichment of ecosystems with food supplies decreases species diversity, increases the abundance of few species and quite often causes a reduction in the size of the individuals of such species [41, 47]. These phenomena manifest most clearly in heavily polluted MPs aquatories.

In MPs aquatories, the observed mass mortality of bottom invertebrates and other organisms is caused by lack of oxygen in the bathypelagial and benthal zones. Oxygen depletion is often connected with phytoplankton and macrophytobenthos hyperproduction in the periphytal zone. Mortality of bottom invertebrates is compensated by an increased survival of their typically pelagic larvae which find themselves in improved trophic conditions [5].

The pelagic larvae of bottom invertebrates and HTS foulers play an important role in the life of many plankton, benthos and periphyton representatives of MPs owing to their mass character and role in food chains. Due to their participation in diverse biological links and food chains with numerous components, they not only ensure species survival, but also contribute to OM turnover between the benthal and pelagial, periphytal and pelagial, and benthal and periphytal zones.

In MPs, the mass reproduction of benthic invertebrates with pelagic development usually occurs during spring and summer in plankton and is directly connected with intensive development of phytoplankton. Diurnal migrations of zooplankton are also synchronized with development of unicellular algae in the near-surface sea horizon.

In the afternoon, when the intensity of illumination decreases, zooplankton starts migrating towards the water surface. The maximal quantity of zooplankton is observed in the surface 10 cm layer at nighttime around 10 p.m. The reverse redistribution of zooplankton from the surface layer into deeper horizons takes place later at night [161].

Being in plankton and subject to phenomena of "local transfer" near the coast, larvae of bottom invertebrates serve as a potential reserve for reestablishing the normal structure of bottom communities in coastal zones affected by pollution after amelioration of living conditions [116]. Being in protected MPs aquatories, pelagic larvae of bottom invertebrates quickly settle on recently installed artificial HS and restore damaged settlements.

Pelagic larvae of bottom invertebrates have tight and diversified relations with benthos, plankton, neuston and nekton and participate in the system of biotic links among them, in particular via food relationships. In this way, pelagic larvae of bottom invertebrates contribute to the constant turnover of OM between the water column and the bottom [116]. In MPs, additional links with other groups exist thanks to the presence of a periphytal zone and its population, which is recruited mainly from benthos composition.

Pelagic larvae of bottom invertebrates significantly influence the biological cycle and structure of plankton, neuston and nekton communities inhabiting the water column, where they temporarily live during the pelagic stage of their development [116].

The Black Sea decapods' larvae, as well as adult individuals of some species, lead a benthic life and perform vertical diurnal migrations.

Adult shrimps *Crangon crangon, Palaemon elegans* and *P. adspersus* are registered at night in the hyponeuston layer at the water surface. In some cases, adults (both females and males individuals) of the mud shrimp *Upogebia pusilla* have been reported to rise towards the surface light of electric lamps at nighttime. While adult shrimps' individuals rise up towards the surface only in some cases, their larvae stay in the surface layer at any time of the day in the same way as the larvae of the majority of decapods' species of the Black Sea. The majority of decapods larvae (zoaea, megalope) exhibits a positive phototaxis. Decapods larvae have special adaptations for remaining in the hyponeuston layer [109]. Shrimps accomplish mass seasonal migrations from MPs aquatories into the deeper sea regions, thus eliminating a part of BM from MPs ecosystems.

Vagile invertebrates inhabiting the MPs periphytal (amphipods, isopods, cumaceans, mysids and shrimps) move along the lateral surfaces of HTS during both day and night and can move away from HTS at distances ranging from a couple of centimetres to couple of metres. Because a large quantity of land-based nocturnal sources of light is present around MPs and in MPs aquatories, various insects concentrate and die there in greater number than in adjacent sea zones. Insects get into port aquatories at some meteorological situations and during mass migrations period. They accumulate mainly at the water surface in the apex zones of aquatories. Some of them are eaten by fish, but most of them end up enriching MP ecosystems in dead OM. In MPs, the night sources of light attract various species of vagile hydrobionts to the surface layer.

In MPs of the Black–Azov Sea basin, constant (holoplankton forms) as well as temporal (meroplankton forms) inhabitants of the pelagial zone are registered in the zooplankton composition during all seasons. The presence of inflowing rivers has a significant influence on the zooplankton horizontal distribution in MPs; especially during the high water season. Freshwater zooplankton organisms can be locally found in MPs. In the NWBS, the river influence is so high that salinity in the aquatory of Odessa port sometimes decreases to 3.5–4.0‰. In the case of Ilyichevsk and Yuzhny MPs, the composition of fauna is influenced by their location inside limans and by their geographic proximity to the main source of desalinated water in the eastern part of the NWBS: the Dnieper-Bug Liman.

In the aquatories of Grigorievsky Liman and Odessa Gulf, many species of planktonic animals and fish, which cannot survive for a long time in seawater, appear during periods of intense desalinization. Cape Bolshoy Fontan diverts seawards the desalinated waters coming from the Dnieper-Bug Liman, thus decreasing the possibility for such organisms to enter the aquatory of Sukhoy Liman [16, 154, 196].

Quantitative and qualitative composition of Yuzhny, Odessa and Ilyichevsk MPs zooplankton is significantly enriched in benthic animals larvae compared with the open sea. The same peculiarity is also typical for other MPs, irrespectively of their location in limans (as for Yuzhny, Ilyichevsk and Varna-West MPs), bays of the estuarine type (as Sevastopol in Sevastopol Bay and Sevastopol fishing port in Kamishovaya Bay) or on open seashore (as Odessa, Mariupol, Novorossiysk, Tuapse, Sochi and others) [16, 154, 196].

With respect to species composition, the population of the pelagial zone in Sukhoy and Grigorievsky Limans and in Varna and Beloslav Lakes became richer after the construction of shipping channels, but the importance of freshwater and brackish water species has reduced. Genetically the modern composition of zooplankton in these water basins comprises representatives of three complexes: freshwater, brackish water and marine [7, 54, 117].

In the 1990s, the average long-term zooplankton quantity in Grigorievsky Liman comprised 35,347 ind. m^{-3} for a biomass of 292.14 mg m^{-3} (excluding the comb-jellyfish *Mnemiopsis*). The highest density of zooplankton (53,693 ind. m^{-3}) was registered in spring and was followed by a progressive decrease during the other seasons (30,893 ind. m^{-3} in summer, 12,437 ind. m^{-3} in autumn and 4436 ind. m^{-3} in winter). Zooplankton biomass increased from spring (198.79 mg m^{-3}) towards summer (376.65 mg m^{-3}), reaching its maximum in autumn (10,282.70 mg m^{-3}). The winter period was characterized by the lowest biomass (103.81 mg m^{-3}). In spring, representatives of the genus *Synchaeta* prevailed in terms of quantity and biomass. Their quantity remained also high in summer, when the role of the sea sparkle *N. scintillans* and of meroplankton, especially *B. improvisus*, was increasing. In autumn, the role of *N. scintillans* decreased, larvae of barnacles remained present in significant amounts and the role of the tunicate *O. dioica* markedly increased. In early winter, the sea sparkle dominated plankton in terms of quantity and biomass.

In Grigorievsky Liman, representatives of Cladocera (11 species), Copepoda (11) and Harpacticoida (11) were registered in the composition of forage zooplankton. In 2003, besides *A. clausi*, the calanoid fauna copepod *Acartia tonsa* (new for the Black Sea) was registered in the southern part of the liman. In the 1990s and 2000s, Rotatoria and Copepoda composed the main structure of the zooplankton in the basin. Species of the genus *Brachionus* dominated among rotifers and *Synchaeta* and *Calanoida* among copepods. The holoplanktonic forms (77 taxa) comprised 73.8% of the total number of taxa and meroplanktonic ones (27 taxa) 26.2%. Among holoplanktonic forms 4.8% are Caspian relicts (*Centropages pengoi*, *Cornigerius maeoticus*, *Podonevadne trigona*, *Heterocope caspia* and *Calanipeda aquae-dulcis*).

The number of zooplankton species in Grigorievsky Liman has increased threefold between the 1960s, when there was no shipping channel, and the 2000s. As a consequence of spring floods from the Dnieper-Bug Liman, the spring zooplankton of Grigorievsky Liman became the most abundant group as to taxa number and was substantially represented by rotifers. In the 1970s, after construction of the channel, the role of *N. scintillans* and *A. clausi* in the zooplankton composition of Grigorievsky Liman increased and these two species became dominant. The quantity and biomass of meroplankton increased significantly. The zooplankton development in Grigorievsky Liman reflects its development in the adjacent part of the Black Sea [196].

Unlike Yuzhny MP, which is located in Grigorievsky Liman, Odessa MP is situated on the open coast of Odessa Gulf and is connected with the sea by three passages having a total width of more than 900 m. Fifty-three taxa, excluding meroplanktonic forms, were identified in the zooplankton structure of Odessa port: larvae of Polychaeta, Cirripedia, Bivalvia and Gastropoda. The most abundant were copepods (30%) and cladocera (9%). Representatives of *A. clausi* and meroplankton forms (larvae of Polychaeta, Cirripedia, Bivalvia and Gastropoda) were always present. A slightly lower percentage of occurrence frequency was registered for *Sagitta setosa* (79%), *Pleopis polyphaemoides* (79%), *O. dioica* (77%) and *N. scintillans* (75%). Larvae of *B. improvisus* (9805 ind. m^{-3}, 58.3 mg m^{-3}) and *N. scintillans* (1295 ind. m^{-3}, 61.38 mg m^{-3}) were the most abundant in terms of quantity and biomass. Rather high densities were observed for *P. polyphaemoides* (876 ind. m^{-3}), polychaetes larvae (840 ind. m^{-3}), bivalves larvae (627 ind. m^{-3}) and *Synchaeta baltica* (420 ind. m^{-3}). During August–December 2001, 53 taxa were registered in the zooplankton composition of Odessa port aquatory, compared with only 16 taxa in the adjacent coastal region [16].

Sukhoy Liman was quickly colonized by new zooplankton organisms after its junction with the sea. The number of taxa increased from 26 in 1967 to 48 after 2 years [20]. In the 1990s, the zooplankton of Sukhoy Liman was represented by 24 species and subspecies during the various seasons. The highest quantity and biomass were registered during the summer months: 73,286 ind. m^{-3} and 306.70 mg m^{-3}, respectively. These indices were decreasing towards autumn (17,965 ind. m^{-3}, 65.54 mg m^{-3}) and reached their lowest values in early winter (2178 ind. m^{-3}, 22.54 mg m^{-3}). *A. clausi* and *N. scintillans* were the dominant species in the 1960s and 1970s, but in the 1990s, the role of *Noctiluca* considerably reduced and the role of barnacles larvae increased in all seasons and especially in summer. The highest indexes of zooplankton were registered for the southern and central parts of Sukhoy Liman. In September 2002, mass development of *Favella ehrenbergii* (56,283 ind. m^{-3}, 258.90 mg m^{-3}) occurred. The average quantity of zooplankton (without comb-jellies) comprised 679,594 ind. m^{-3} for a biomass of 477.75 mg m^{-3} [154].

In Odessa MP, representatives of coldwater complexes such as *N. scintillans*, ephyra stages of the scyphomeduse *A. aurita*, copepods *Pseudocalanus elongatus*, and comb-jellies *Pleurobrachia pileus* were found in compensatory waters after offshore winds in spring, late autumn and in summer.

It is evident that the increase in the importance of meroplanktonic forms in MPs is directly connected with the increase in the underwater surface area of berths, piers, breakwaters, floating docks and ships.

In Sevastopol Bay, 59 species of the classes Polychaeta, Gastropoda, Bivalvia and of the order Decapoda were found in the composition of meroplankton during the period 1987–1997. Compared with previous observations (1950–1970s), pelagic larvae of Bivalvia had decreased by 60% and species diversity of other three big taxa was also reduced by one-third on average.

In all zones of Sevastopol Bay aquatory, the meroplankton is an important component of zooplankton all year round. The highest biomass was registered for *B. improvisus* larvae, followed by Polychaeta [3].

In Sevastopol Bay, Copepoda (mainly *A. clausi* and larvae of benthic animals) form most of the biomass of summer mesozooplankton. The total zooplankton biomass has reduced after the construction of moles that restrict water exchange between the bay and the open sea. Larvae of the polychaetes *Neanthes succinea, Polydora cornuta* and *Capitella capitata*, which are resistant to anthropogenic pollution, were found in the plankton composition [3].

In Sevastopol Bay, organisms from benthic and fouling communities were found in the zooplankton composition: larvae of barnacles, mussels, pearlworts and *Botrillus* ascidians. Nauplii and cypris of barnacles are the most numerous, while veligers of mussels occur in lower numbers. The role of fouling hydrobionts larvae in the zooplankton composition of the bay changes following reproductive cycles as well as peculiarities of the hydrological regime. Larvae of fouling hydrobionts occur all year round in plankton, but their mass development is observed from May to December, with some reduction in July–August. Because of the vertical migrations peculiar to the majority of species-foulers larvae, their quantity in the upper water horizons is twice as much at nighttime than in morning hours. The average quantity of larvae in the near-bottom water layer (depth 4 m) is relatively similar during the day- and nighttime, but considerably lower than the quantity observed at night in the upper horizon. In Sevastopol Bay, during the period of abundant development, larvae of fouling organisms comprise 20–43% of the total zooplankton mass [25].

Zooplankton quantitative composition in the NWBS coastal zone is similar from Novorossiysk to Sochi [50]. In strongly polluted areas of MPs, plankton is impoverished both quantitatively and qualitatively. Cladocera are particularly sensitive to pollution [52].

In 2004–2010, 71 taxonomic forms were found in the holoplankton composition of Novorossiysk Bay including 56 Copepoda, 4 Cladocera, 6 Porifera, 2 Ctenophora, 1 Dinophycea (*N. scintillans*), 1 Chaetognata and 1 Appendicularia (*O. dioica*). Copepods composition consisted in 15 Black Sea species and 41 alien species.

In Novorossiysk port aquatory, holoplankton comprised 66.0–77.0% of the total zooplankton quantity compared with 73.0–88.0% in the open part of the bay. The maximal holoplankton biomass reached 0.4–0.8 and 0.8–2.0 g m^{-3} in the open part of the port during summer and at the beginning of autumn. In most cases, the amplitude of biomass fluctuations in port waters was lower than in the open part of the bay.

Oligotrophic forms of crustaceans, predators and others prevail in the planktonic communities of the open sea. On the other hand, organisms resistant to high OM concentrations such as the neritic forms of crustaceans, rotifers and meroplankton are most common in port waters rich in detritus. In Novorossiysk MP, the genus

Acartia accounted for 81.0–91.0% of the Copepoda abundance, compared with values of 34.0–63.0% in the open part of the sea. The total density of *Acartia* in the port aquatory was 1.5-fold higher than in the open part of the sea for similar values of biomass. In the apex part of the port aquatory, nauplii and copepodites were observed more often in the structure of *Acartia* populations. Reproduction of *A. clausi* occurs during the whole year, with exception of one or two months (August–September).

Mass development of cladocerans is registered in summer. In the open part of Novorossiysk Bay, their average long-term density is 7.8 × 10^3 ind. m^{-3} and their biomass 195.9 mg m^{-3}, which is two to fivefold higher than in the port aquatory. The small-sized *P. polyphaemoides*, which feeds mostly on detritus, was the most abundant species in the port aquatory.

In the port aquatory, the average long-term abundance and biomass of the sea sparkle (*N. scintillans*) was 1.7-fold higher (6.7 × 10^3 ind. m^{-3}, 0.4 g m^{-3}) in comparison with the open part of the harbour.

In Tuapse port, the holoplankton included 40 taxonomic forms, among which 26 Copepoda, 4 Cladocera, 5 Porifera, 2 Ctenophora, 1 Dinophycea (*N. scintillans*), 1 Chaetognata and 1 Appendicularia. Copepods composition comprised 14 invasive alien species and 12 Black Sea species. In Tuapse port aquatory, the average long-term density of forage holoplankton was 10.6 × 10^3 ind. m^{-3} for a biomass of 137.6 mg m^{-3}, which is twofold lower than in Novorossiysk port. Holoplankton organisms comprised 80.0% of the total zooplankton quantity in the coastal zone adjacent to port and 65.0% in the port aquatory.

The long-term average annual density of the genus *Acartia* were 5.1–54 × 10^3 ind. m^{-3}. In ports aquatories, the indices of biomass (0.08 g m^{-3}) were 1.5-fold higher than in the open waters. The maximal density of populations of the neritic copepod species *A. tonsa* was 2.2 × 10^3 ind. m^{-3} and 0.9 × 10^3 ind. m^{-3} in port aquatory and in open part of the sea, respectively. The meroplankton, i.e. larvae of bottom invertebrates, is an important component of neritic zooplankton in various regions and MPs of the Black and Azov Seas. In Novorossiysk Bay, larvae of bottom invertebrates were present in plankton during all study periods.

In 2004–2010, the meroplankton of Novorossiysk Bay and Tuapse port was represented by 68 taxonomic forms out of which 23 belonged to Polychaeta, 4 to Cirripedia, 1 to Phoronida, 20 to Decapoda, 8 to Bivalvia, 10 to Gastropoda and 2 to Hydrozoa. The highest number of taxonomic forms (67) was registered in Novorossiysk Bay, while only 46 were present in Tuapse port.

In bays and ports of the NEBS, the season of mass reproduction of bottom invertebrates occurs from May to September. On average, meroplankton accounts for 23.0–26.0% of the total zooplankton quantity (15.0% in the central part of Tuapse port aquatory), reaching 50.0% and more during spawning peaks. In Novorossiysk Bay and Tuapse port, the complex of meroplankton dominant species comprised larvae of *B. improvisus*, *B. reticulatum*, *M. lineatus* and *Polydora* ssp. which are tolerant to pollution.

Larvae of the bivalves *Anadara inaequivalvis* and *Chamelea gallina* as well as of decapods played a certain role in the meroplankton structure of Tuapse and Novorossiysk ports.

Non-uniform distribution of meroplankton in MPs aquatories is explained by the peculiarities of waters circulation and the formation of larvae accumulation in the apical stagnant zones of aquatories.

Meroplankton of the NEBS port aquatories is characterized by high densities and species diversity. Spatial and temporal changes in the composition of dominant species and their abundance are determined by periods of benthic animals spawning, circulation of water masses, and eutrophication levels of water and bottom sediments.

Observations in the estuarine zone of Tsemes river, in the north-western part of Novorossiysk port, showed, that rotifers (40,800 ind. m^{-3}) were the mass representatives of plankton in April at a temperature of 10.6 °C. Other organisms that developed in considerable amounts were *O. minuta* (1108 ind. m^{-3}) and *A. clausi* (11,887 ind. m^{-3}) among Copepoda and *B. improvisus* (9960 ind. m^{-3}) among larvae of benthic organisms. In May, at a temperature having increased to 19.5 °C, the most abundant zooplankton organisms were infusoria (1600 ind. m^{-3}), rotifers (444,400 ind. m^{-3}) and acartia (17,886 ind. m^{-3}). In plankton, the number of *O. minuta* (2984 ind. m^{-3}) and *P. parvus* (1372 ind. m^{-3}) significantly increased. Among benthic organisms' larvae, Polychaeta (3916 ind. m^{-3}) reached the maximum development, while the quantity of *B. improvisus* larvae, albeit decreasing, remained significantly high (3376 ind. m^{-3}). Thus, infusoria, rotifers and benthic organisms' larvae develop in mass quantity in the aquatory of Novorossiysk Bay during spring [51].

In Novorossiysk port, summer (June–September) is characterized by the most diverse species composition of zooplankton. The peak of its development is registered in June, at a water temperature of 21.6 °C, and a couple of mass developing species constitute the bulk of zooplankton quantity. In June, the maximum quantity of acartia (38,620 ind. m^{-3}) occurs and that of *O. minuta* increases by more than twofold to attain 6200 ind. m^{-3}. Among the other representatives of Copepoda, *P. parvus* (1174 ind. m^{-3}) is the most numerous. In June, the number of rotifers reduces by 10-fold. Barnacles larvae (4760 ind. m^{-3}) and Gastropoda (3448 ind. m^{-3}) are found in great quantity, while other species do not significantly contribute to zooplankton composition. In Novorossiysk port, the period July–September is the most unfavourable for zooplankton development, possibly because of the high water temperature (25.0–26.0 °C). In October, with water temperature lowering to 16.2 °C, the zooplankton species composition acquires the characteristics typical for the autumn period. The number of thermophilic forms reduces, the development cycle of all Cladocera concludes and the cryophile form *Oithona similis* appears again [51].

In the aquatory of Novorossiysk port, the species composition of zooplankton changes not only seasonally, but also across years. Two dominant species, *A. clausi* and *O. minuta*, account for most of the zooplankton quantity. With regard to benthic organisms, larvae of *B. improvisus* are recorded in mass quantity in April–

June and September–October. The development peak of gastropods larvae is registered in June and August and that of bivalves in October [51].

In Tuapse port, the abundance of holo- and meroplankton is 1.5–2.0-fold lower (40,000 ind. m^{-3}, 371.7 mg m^{-3}) in the heart of harbours, near the berths, than in the central part of the aquatory and beyond its boundaries. Copepods from the genera *Acartia* and *Centropages*, predatory *Sagitta* and larvae of gastropod *Bittium reticulatum* represent most of the biomass [155]. An increased role of protozoans, infusoria, rotifers and larvae of some small-sized polychaetes has been observed in the zooplankton composition of MPs aquatories ecosystems over the last decades. This change is accompanied by a simplification of the zooplankton structure owing, first of all, to pelagic Copepoda, Cladocera and Decapoda larvae.

2.2.4 Ichthyoneuston and Ichthyoplankton

Fish have an extremely important role in aquatic ecosystems. Frequently, fish occupy the highest trophic level in food chains and transform the phytoplankton, invertebrates and small-sized fish biomass into valuable food resources for humans. The long-lived, large-sized predatory fish preserve BM in their biomass for a prolonged period of time, thus supporting the normal, stable functioning of ecosystems. Replacement of long-lived species by short-lived ones, as observed in the Black Sea during the last decades, testifies about excessive water eutrophication. The absolute majority of Black Sea fish has pelagic development stages, i.e. for some period of time fish become components of ichthyoneuston and ichthyoplankton.

For many fishes, MPs aquatories ecosystems represent a sufficiently nutrient-rich environment for the various stages of their ontogenesis. While there is no active fishery in MPs, they are at the same time eutrophicated and frequently polluted by various toxic substances.

In MPs aquatories, the vertical and horizontal distribution of ichtyoplankton is mainly connected with water hydrodynamics, especially with the surface microconvection, Langmuir circulations, wind surface currents, and water density at various horizons and in the different harbours.

It is known that the survival ability of pelagic fish larvae significantly depends on how effectively individuals find food and graze zooplankton micro-accumulations, particularly copepod ones. Fish larvae with a length of 10 mm are able to detect their preys at a distance of about 10 mm [219]. It is important to consider that elements of gregarious behaviour appear in zooplankton organisms' accumulations. Individuals displace as a whole while preserving defined (1–2 cm) inter-individual distances. Aggregation "responds" to danger imminence, shading and other factors [136]. Fish larvae spend energy in search of food micro-accumulation because search of and feeding on single preys could be energetically insufficient and lead to death.

The majority of studied Copepoda form monospecific assemblages. In coastal zone, such assemblages often have an irregular spherical form and sizes of centimetres to dozens of centimetres [136]. In aggregations, the copepods density reaches 24×10^3 ind. m^{-3}. Copepods are the main food object of early larval stages of fish in the pelagial zone.

The basis of the Black Sea ichthyofauna is formed by more or less thermophilic fish of Mediterranean origin. They usually reproduce from the end of spring until mid autumn. The group of cryophile species, with pelagic eggs and usually spawning in March–April, comprises sprat *Sprattus phalericus,* whiting *Merlangius euxinus*, shore rockling *Gaidropsarus mediterraneus* and flounder *Platichthys lucsus.*

The changes in ichthyofauna composition, which took place during the last decades in the entire Black Sea and in its specific regions, also reflected in ichthyoneuston and ichthyoplankton composition of MPs aquatories. Many species have almost completely disappeared and others stopped to live and to spawn in the coastal zones of the whole NWBS.

Eggs, larvae and fries of different fish species can be found in the water column of MPs aquatories. Here pelagic spawn of fish reproducing directly in ports are recorded along with pelagic spawn entering port aquatories together with water masses. Prelarvae and larvae hatch from eggs which remain in the water column for some time. Here larvae and fries developing from demersal spawn are present together with spawn of ovoviviparous fish (Syngnathidae).

In the Black Sea coastal zones, the ichthyoneuston species composition is diverse due to eggs and larvae of the bottom and near-bottom fish. In the open regions of the sea, eggs, prelarvae, larvae and juveniles of fish mass species (khamsa, sprat, horse mackerel, goatfish, striped mullet, golden mullet, leaping grey mullet and other) are predominantly observed. The ichthyoneuston species composition in the eastern and southern part of the Black Sea and near the Crimean coasts is richer than in the NWBS. Studies showed that, at the early stages of development, some fish species reside in the near-surface sea layer over long time spans, while others rise up to the film of water surface tension only for a couple of hours. However, the neustonic stages of development are vitally important for all fish species. In the near-surface layer of the Black Sea, fish are present both during passive stage of development (spawn) and as active vagile forms (prolarvae, larvae and young fish) [38, 39].

The buoyancy of the eggs developing at the water surface depends on the relative densities of the eggs and of seawater. The important hydrostatic adaptation of marine pelagic eggs is their high content of low-density water and, often, the presence of large-sized fatty drops (striped mullet, golden grey mullet, leaping mullet, common bluefish, red mullet, horse mackerel, greater weever fish, annular seabream and others).

Observations of eggs from various fish species near the surface showed that the embryo occupies a quite defined position inside one egg, depending on the ratio between the densities of yolk and embryo and on the location of fat drop(s) (if present). In highly buoyant eggs, the embryo is located under the yolk-sac. In pelagic eggs of marine fish, the fatty drops occupy a position opposite to that of the embryo and are always located above it, which favours the quick return of the egg to a stationary position in case of disturbances. In the near-surface layer, the conservation of a stationary position by fish eggs, exposed to intensive illumination by sun rays, apparently allows them to regulate irradiation with the help of pigment-containing cells located near the vegetative pole [38, 39].

Prelarvae of fish rise up from various horizons to the surface in variety of ways. Embryos developing from eggs laid on the bottom (as in the case of sand smelt, garpike, picarel, gobies and others) rise up with active body movements using the primal fin fold and pectoral fins. Prelarvae developed from floating eggs use the primal fin fold as a thruster, but do not yet possess the horizontal rudders represented by pectoral fins. Prelarvae move by undulations of body and fin fold, alternating periods of activity and rest. They swim up during periods of activity and slowly sink down while resting. Periods of activity and rest have approximately the same duration, but prelarvae sink slower than they rise up. This results in a gradual rising up of the prelarvae in the water column. After reaching the water surface, prelarvae accumulate there and periodically perform vertical displacements. Positive phototaxis undoubtedly favours the rising up of prelarvae and larvae which develop from bottom eggs.

The yolk sack, fatty drop and subdermal cavity of prelarvae developing from floating eggs favour their upward movement. At first, prelarvae hatching from pelagic eggs swim with their yolk sack upward. They turn into the back-up position only after the formation of the subdermal cavity and the almost complete resorption of yolk. Prelarvae of different fish species swim differently in the near-surface sea layer depending on the position of fatty drop(s) in the yolk sack. The red mullet, Black Sea horse mackerel, greater weever fish and others have a fatty drop located in the anterior part of the yolk sack. During activity and rest periods they are oriented with their head towards the water surface and move along steep vertical spirals. However, they cannot swim horizontally because of the insufficient development of pectoral fins that would function as horizontal rudders. The prelarvae of the striped mullet, golden grey mullet and Black Sea turbot have a fatty drop located in the middle part of yolk sack. They move along an almost straight line, directed at a slight angle towards the film of water surface tension, or emerge along a wide flat spiral rotating around body axis.

The subdermal cavity forms in the dorsal part of the prelarval body, mainly over the head, and takes the place of the yolk sack as the hydrostatic organ of the embryo.

Until the development of pectoral fins, the yolk sack, fatty drops and subdermal cavity of fish prelarvae and larvae are special organs of orientation and steering that lighten the anterior part of the prelarval body and assist it in floating up to the surface. Thus, the character of fish prelarvae movements, the precise direction of movements and the presence of organs of orientation and steering indicate that many the Black Sea fish prelarvae have an "inborn" ability to swim up.

It is known that the majority of sea bony fish larvae and fries have an air bladder, even if they lose it in adult state.

The primary function of an air bladder is to save energy while remaining in a defined sea horizon. It was determined that the initial filling up of the air bladder of larvae and fries of physostomous and physoclistous fish takes place with atmospheric air. Air enters into the air bladder via an air passage (ductus pneumaticus), which, in physoclistous fish, is occluded after a couple of days. During subsequent developmental stages, gas regulation within the air bladder takes place via the circulatory system. If the primary intake of atmospheric air does not take place, then the gas regulation in the air bladder fails. Fish larvae with empty air bladders develop abnormally, experience delayed growth and die [38].

In MPs aquatories, fish eggs with a too high density and abnormally developing larvae and fish fries sink down to the pycnocline zone and become food for various hydrobionts. Heavier larvae and fries sink to the bottom.

Spawning of 17 species was registered in Odessa port aquatory: European anchovy *Engraulis encrasicolus*, sea mouse *Callionymus risso*, sand sole *Pegusa lascaris*, red mullet *Mullus ponticus*, horse mackerel *Trachurus ponticus*, sprat *S. phalericus*, greater weever fish *Trachinus draco*, the Black Sea turbot *Psetta maeotica*, shore rockling *G. mediterraneus*, striped mullet *Mugil cephalus*, golden grey mullet *Liza aurata*, sharpnose mullet *Liza saliens*, haarder *Liza haemotocheilus*, the European black scorpionfish *Scorpaena porcus*, stargazer *Uranoscopus scaber* and the Black Sea flounder *P. lucsus*. Larvae of grey mullet, leaping mullet, atherines, the Black Sea flounder, goby (Gobiidae), combtooth blennies (Blenniidae) and sea needles (Syngnathidae) were also caught [16].

In the 1950s, eggs of 16 fish species were reported to occur in the ichtyoplankton of Sevastopol Bay. This number had decreased to only 6 species in 1998.

In Sevastopol coastal zone, larvae of commercial fish with pelagic eggs comprised 80.0% of the total quantity of ichthyofauna in the 1960s. In 1988–2008, this type of larvae was sporadic or absent and the average percentage of larvae of fish with demersal eggs had increased up to 80.0% of the total, reaching 100% in some years [35].

Eggs, larvae and fries of various fish species were found in Novorossiysk and Tuapse ports' aquatories. At the end of the 1960s, eggs and larvae of 13 fish species were found in the plankton composition of Tuapse Bay during summer. Early stages of development of 13 fish species were also present in the ichthyoplankton of Sochi port aquatory. The qualitative composition of ichthyoplankton was similar in both ports [97]. Eggs of 10 fish species were found in the aquatory of Novorossiysk MP in the 1970s and at the beginning of the 1980s [23].

Anchovy eggs constituted 86.2% of the Tuapse port ichthyoplankton in August 2010. The average density of ichthyoplankton in the port was 2.5-fold lower than outside its boundaries (21 ind. m^{-2} in vertical catches, 163 ind. 100 m^{-3} in horizontal catches). The overall quantity varied from 0 to 104 ind. m^{-3} [155].

During the summer period, the majority of Black Sea ports aquatories is characterized by the presence of eggs and larvae of the European anchovy *E. encrasicolus.* Findings of eggs of the annular seabream *Diplodus annularis* are common in the eastern and southern part of the Black Sea ports aquatories and at Crimean shores.

Eggs of the Black Sea flounder *P. lucsus* are found in the aquatories of Sukhoy and Grigorievsky Liman in March–April. Larvae and low-pigmented fries of the Black Sea flounder stay in the surface layer and, after metamorphosis, approach the shallow waters and shift to the near-bottom way of life. Fries of the Black Sea turbot *P. maeotica* and of the sand sole *P. lascaris*, that develop from pelagic eggs in the spring-summer period, behave in the same way.

In Sukhoy and Grigorievsky Liman, *Atherina pontica*, which deposit eggs on water vegetation, spawns in April–May. Its hatched larvae and, at a later stage, fries stay in schools in the upper water layer with the thickness of 0–10 cm, feeding in the coastal zone or near various HTS.

In the NWBS, the abundant fries of striped, golden grey and leaping mullets develop from pelagic eggs and, during the first 2 to 3 weeks of their life, remain under the surface tension film and feed on plankton. Juvenile fish subsequently migrate from the open sea towards the coast and enter into all shallow-watered gulfs, bays and limans; including MPs aquatories. After some time, their behaviour change and they start to feed mainly on periphyton. Moving alongshore in the Odessa region of the NWBS, the fries of haarder get in Odessa, Ilyichevsk and Yuzhny MPs aquatories. In Sukhoy and Grigorievsky Limans, shores zones not yet affected by hydrotechnical construction and suitable for their growing period are still preserved. The fries of haarder stay in MPs aquatories until late autumn. In Odessa MP, there are no conditions for grey mullets to switch from feeding on plankton to feeding on periphyton on the bottom surface. In all three MPs, 1-year-old grey mullets and grey mullets of older age live both in preserved shallow waters and near HTS lateral surfaces from spring until late autumn. The highest mass concentrations of leaping mullet and striped mullet fries, counting million of individuals, were registered in Odessa, Ilyichevsk, Khorly and Skadovsk MPs in winter. Until ice formation, the dense schools of juvenile individuals were concentrating in the upper water horizon along the port berths and under ships' hulls [149]. In winter, during the ice regime, mass concentrations of fries of mullets are found near the berths of Mariupol, Berdyansk, Henichesk and Temryuk ports [150].

In last years, fries of the recently acclimatized haarder *L. haemotocheilus* have acquired a significant role in the ichthyoplankton composition of Odessa port and Sukhoy and Grigorievsky Limans. At the end of summer and beginning of autumn, juveniles of the garfish *Belone euxini* are often observed in the near-surface water layer (0–10 cm) of the aquatories of all three ports in the Odessa region.

It is natural that fries of the black goby *Gobius niger* and of gobies from the genus *Pomatoschistus*, developing from demersal eggs, rise up to the surface water horizon for the first infill of the air bladder by atmospheric air. After a couple of days, they settle to the bottom. In MPs aquatories, some benthic fish with demersal eggs, as for example the tubenose goby *Proterorhinus marmoratus*, spawn in cracks and cavities of HTS as well as in cavities between the fouling of mussels' druses. The hatched fries of tubenose goby are registered in the water column near piers, breakwaters and berths. The combtooth blennies from the family Blenniidae and wrasses from the family Labridae behave in the same way.

In general, the larvae and fish fries species composition in ichthyoneuston and ichtyoplankton reflect the specific state of an MP ecosystem.

In Novorossiysk Bay, eggs and larvae from 31 taxonomic groups (14—migrating, 17—non-migratory) were found in the ichthyoplankton composition. In 2006, eggs and larvae of 27 fish species were found. In 2006, the average quantity of ichthyoplankton in the aquatory of Novorossiysk MP comprised 9.2 ind. m^{-2} in vertical catches and 170 ind. 100 m^{-3} in horizontal ones. In July, spawns of anchovy (80.0%) and red mullet dominated. In July, the average quantity in the port decreased to 2.0 ind. m^{-2} and was 26.0 ind. m^{-2} outside it. These changes were likely connected with grazing of eggs and fish larvae by the zoophagous comb-jelly *M. leidyi* [156].

In 2007, eggs and fish larvae from 15 taxonomic forms were found, with eggs of anchovy (80.0%), horse mackerel and annular seabream being dominant. According to data of summer vertical catches, the average quantity of ichtyoplankton comprised 90.0 ind. m^{-2} in the port aquatory and 121.0 ind. m^{-2} outside the port aquatory. The ichtyoplankton density exceeded 100.0 ind. m^{-2}, which is similar to open waters indices, in most part of the port aquatory. However, the proportion of dead organisms and ichthyoplankton individuals with developmental anomalies was almost twofold higher (60.0%) in the port aquatory than outside its borders [156].

In 2008, 2009 and 2010, ichthyoplankton was represented by eggs and larvae of 11, 15 and 11 fish species, respectively [156].

In Tuapse MP, the species composition of eggs and larvae was poorer than in Novorossiysk MP and was represented by 17 species; 11 migratory ones and 6 non-migratory ones. The eggs of anchovy (70.0%), red mullet, horse mackerels and annular seabream prevailed [156]. Fifteen and 9 taxonomic forms were found in the port in 2009 and 2010, respectively [156].

In Tuapse port, the average quantity of ichthyoplankton was two to threefold lower and species composition poorer than in Novorossiysk MP in spite of the better water exchange [156].

The presence and abundance of forage zooplankton for larvae, juvenile and adult fish species in many respects determines the composition of ichthyofauna in MPs aquatories ecosystems. In MPs, the zooplankton biomass (mainly holoplankton) and its species composition fluctuate significantly over the years because of various causes, while species composition and concentrations of meroplankton are more stable.

Embryos, prelarvae, larvae and fries of fish are sensitive to different pollutants. In MPs, the amount of abnormal and nonviable prelarvae and larvae in ichtyoplankton composition increases because of constant or periodical pollution.

2.2.5 Nekton

In MPs aquatories, fish form the main mass of nekton; i.e. organisms able to swim actively in the water column, withstand water flow and cover considerable distances. Pelagic fish are typical nekton organisms of MPs. However, all fish species inhabiting on the bottom, pass some time in the water column. Many of them have pelagic stage of development and their larvae belong to nekton [4, 39, 40].

In marine ecosystems, feeding relationships are conditioned by the relative sizes of organisms. Although the ratio between the masses of predators and preys differs among species, the value of this ratio usually slightly exceeds 100. The corresponding ratio for predator vs. prey dimensions is approximately equal to 10. As both growth and mortality depend on age, the availability of food objects with the optimal target size decreases as the size of the predators increases. Big predators quite often encounter conditions in which they have to feed on preys with less than optimal sizes [103].

In MPs aquatories ecosystems, fish, like other big vagile organisms, are in constant search of food, while most of the mass forage planktonic or benthonic species are sedentary or completely motionless. Fish move to ecosystems zones with high densities of food, switching again to other regions when food density decreases below a certain threshold. This behaviour also allows the population of food objects to restore its biomass. In this scenario, fish and other big predators control aquatic ecosystems "from above" [8, 32].

Changes in the ichthyofauna composition of the Black and Azov Seas as a whole and also in their separate regions inevitably affect the biota of MPs aquatories. In the main Ukrainian MPs aquatories (Odessa, Ilyichevsk, Yuzhny), the quantitative and qualitative composition of ichthyofauna is tightly connected with processes occurring in the NWBS ecosystem (in particular in the estuaries and the near-estuaries zones of inflowing rivers) and directly in the Odessa region [39].

The peculiarities of underwater landscapes and hydrodynamics of MPs make them more attractive for fish of Mediterranean origin, inhabiting in stony and rocky regions that are not ecologically connected with the bottom. They mainly feed in the periphytal developing on HTS and include representatives of wrasses (Labridae), seabreams (Sparidae) and other families.

In MPs, fish species usually living on or near the bottom avoid hypoxia and anoxia by staying in the periphytal zone of HTS, where they feed and find shelter in the macrophyte belt and among bivalve druses.

In MPs, the summer season temperature is, as a rule, a couple of degrees lower at the depths of more than 5–10 m than at the surface. This circumstance allows coldwater fish such as the Mediterranean sprat *S. phalericus*, whiting *M. euxinus* and shore rockling *G. mediterraneus* to remain in deep layers of MPs water column.

We studied the ichthyofauna of MPs aquatories of the Black–Azov Sea basin in three ports: Yuzhny, Odessa and Ilyichevsk. These MPs are located in the Odessa region of the NWBS, between the mouths of Grigorievsky and Sukhoy Limans. Fish species composition in this region is determined by regional peculiarities of the ecosystems as well as by general seasonal and long-term processes, of both natural and anthropogenic origin, taking place in whole basin. About 150 fish species and subspecies were registered in the ichthyofauna of the NWBS. A couple of dozens of them have become extremely rare, or have completely disappeared, in the region. At the same time, there are data about the entrance of several other species in these waters. The presence of species found only in single specimen likely has a random character. More than 120 fish species directly inhabit in the NWBS [154].

Theoretically, all of them could be found in the aquatories of Odessa, Ilyichevsk and Yuzhny MPs.

Because of the run-off of the big rivers such as the Danube, Dniester and Dnieper with Yuzhny Bug into the NWBS, freshwater and diadromous fish can be found in the sea. Proper freshwater fish are represented by the families Cyprinidae (commonly called the carp family) and Cobitidae (commonly called the true loaches). The semi-anadromous fish could be ascribed to a special subgroup whose importance increases in desalinated waters and whose occurrence is delimited by the 10–12‰ isohaline. During strong floods, the freshwater diadromous and semidiadromous cyprinoids can reach the Odessa Gulf from the Dnieper-Bug Liman. Representatives of such fish include the following: vimba bream *Vimba vimba*, shemaya *Alburnus sarmaticus*, common bream *Abramis brama*, common carp *Cyprinus carpio*, sichel *Pelecus cultratus*, taran *Rutilus rutilus heckelii*, sheatfish *Silurus glanis*, pike *Esox lucius*, zander *Lucioperca lucioperca*, crucian carp *Carassius carassius* and other. In Yuzhny and Odessa MPs aquatories, they can survive for a couple of days to a couple of weeks.

Usually, up to 65–70% of fish of Mediterranean origin, the so-called Mediterranean immigrants, are present in the NWBS ichthyofauna composition. Their quantity decreases in desalinated regions.

Alien fish species occupy a special place in the ichthyofauna composition of the region. Prominent among them are the haarder *L. haemotocheilus*, especially brought from the Far East of Russia and perfectly acclimatized to the Black and Azov Seas, as well as the pumpkinseed sunfish *Lepomis gibbosus* which, apparently, was accidentally introduced.

The ichthyofauna of the NWBS has various reproductive strategies: (a) viviparous species (spiny dogfish *Squalus acanthias*, common stingray *Dasyatis pastinaca*); (b) fish spawning big capsules from which hatch formed fries with a length of about 12 cm (thornback ray *Raja clavata*); (c) fish producing bottom attached and protected (Gobiidae, Blenniidae, Gasterosteidae and Gobiesocidae) or unprotected (Belonidae and Atherinidae) eggs; (d) fish producing bottom non-attached and guarded (Labridae) or unguarded (*Gimnammodytes cicerellus*) eggs; (e) fish carrying eggs on their body (Syngnathidae); and (f) fish producing pelagic eggs (Clupeidae, Engraulidae, Mugilidae, Mullidae, Sparidae, Scophthalmidae and others). Fish with pelagic eggs comprise about 30% of the entire ichthyofauna composition and usually reproduce far away from the shore [39].

According to their feeding strategy and food objects, fish present in the NWBS, can be divided into benthophagous (about 65%), plankton feeders (about 21%), predatory ichthyophagous fish (about 8%) and fish feeding on zoobenthos and other fish (about 7%). Species that feed on fouling and detritus (e.g. Mugilidae, Mullidae and others) are considered benthophagous.

In the NWBS, proper brackish water species constitute an important group. Among them there are typical species such as the Black Sea kilka *Clupeonella cultriventris*, the common percarina *Percarina demidoffi* and the large group of gobies (Gobiidae).

Marine temperate water fish are represented by eight species of the so-called Boreal-Atlantic relicts which are apparently remains of the Ice Age fauna. These are the spiny dogfish *S. acanthias*, the thornback ray *R. clavata*, the sprat *S. phalericus*, the Black Sea flounder *P. lucsus*, the haddock *M. euxinus*, the three-spined stickleback *Gasterosteus aculeatus*, the salmon *Salmo labrax* and the European eel *Anguilla anguilla*.

Marine thermophilic fish are the mass group of fish in the region. Among them there are highly euryhaline forms that enter even freshwaters, such as some sea needles (Syngnathidae) and grey mullets (Mugilidae). In general, these are Mediterranean immigrants. Part of them has evolved into Black Sea local subspecies, differing from initial forms by a series of morphological characteristics and with regard to size.

The gobies, herrings and sturgeons are most characteristic for the NWBS. The following species are recorded practically everywhere: the Black Sea sand smelt *A. pontica*, the Black Sea grey mullets (striped mullet, golden grey mullet, leaping mullet), haarder, red mullet *M. ponticus*, greater weever *T. draco*, stargazer *U. scaber*, the Black Sea flounder *P. lucsus*, spiny dogfish *S. acanthias*, anchovy *Engraulis ponticus*, sprat *S. phalericus*, whiting *M. euxinus* and others.

In the NWBS, the qualitative composition of ichthyofauna and its distribution depend on seasonal temperature changes which control spawning and feeding migrations. In autumn, the absolute majority of species moves away from the shore to return back to it in spring.

During the last decades, a significant reorganization in the quantitative and qualitative composition of the NWBS ichthyofauna took place because of different causes. The stocks of long-lived, large-sized commercial fish (sturgeons, the Black Sea turbot *P. maeotica*, large-size herrings, big-sized gobies) have sharply reduced and the Atlantic bonito *Sarda sarda*, the common bluefish *Pomatomus saltator* and the Atlantic mackerel *Scomber scombrus* have almost stopped to approach the shore.

Seventy-one fish species, belonging to 34 families, are indicated in the composition of the NWBS coastal ichthyofauna, including marine Mediterranean and Boreal species, Caspian relicts and acclimatized species. Near-bottom benthic forms prevail everywhere. Forty-one species of fish were found near Odessa city shores. In the aquatory of Grigorievsky Liman, subject to the strong influence of the Dnieper-Bug Liman waters, marine Mediterranean migrants comprise 59% of the total species [154, 184].

In various years and seasons, the following 48 fish species were found in various parts of Grigorievsky Liman after its junction with the sea: spiny dogfish *S. acanthias*, the Russian sturgeon *Acipenser guldenstadtii*, stellate sturgeon *A. stellatus*, the Black Sea sprat *C. cultriventris*, sprat *S. phalericus*, the Black Sea herring *Alosa pontica*, the Black Sea shad *A. tanaica*, European anchovy *E. encrasicolus*, the Black Sea salmon *S. labrax*, European eel *A. anguilla*, garpike *B. euxini*, rockling *G. mediterraneus*, haddock *M. euxinus*, the three-spined stickleback *G. aculeatus*, stickleback *Pungitius platygaster*, sea horse *Hippocampus guttulatus*, straightnose pipefish *Nerophis teres*, black-striped pipefish *Syngnathus nigrolineatus*, pipe fish *S. argentatus*, flathead grey mullet *M. cephalus*, haarder

L. haemotocheilus, golden grey mullet *L. aurata*, leaping mullet *L. saliens*, the Black Sea sand smelt *A. pontica*, horse mackerel *Trachurus mediterraneus*, picarel *Spicara flexuosa*, red mullet *M. ponticus*, grey wrasse *Symphodus cinereus*, ocellated wrasse *S. ocellatus*, rusty blenny *Parablennius sanguinolentus*, tentacled blenny *P. tentacularis*, sphinx blenny *Aidablennius sphynx*, transparent goby *Aphia minuta*, grass goby *Gobius ophiocephalus*, black goby *G. niger*, toad goby *Mesogobius batrachocephalus*, ginger goby *Neogobius cephalarges*, monkey goby *N. fluviatilis*, round goby *N. melanostomus*, ratan goby *N. ratan*, racer goby *N. gymnotrachelus*, Caucasian dwarf goby *Knipowitschia caucasica*, sand goby *Pomatoschistus minutus*, marbled goby *P. marmoratus*, tubenose goby *P. marmoratus*, the Black Sea turbot *P. maeotica*, the Black Sea flounder *P. lucsus*, and sand sole *P. lascaris* [39].

In Odessa Gulf, other species rather exotic for the region were also observed besides those found in Grigorievsky Liman, including: common stingray *Dasyatis pastinaca*, sea scorpion *S. porcus*, painted comber *Serranus scriba*, annular seabream *D. annularis*, shi drum *Umbrina cirrosa*, sea mouse *C. risso*, European barracuda *Sphyraena sphyraena*, St. Pierre fish *Zeus faber*, flying gurnard *Cephalacanthus volitans* and beardless tadpole goby *Benthophiloides brauneri* [154].

In the apical part of Sukhoy Liman, which experiences significant freshening influence from the Bolshoy Dalnik River, up to 11 species of freshwater fish have been recorded [162]. In Sukhoy Liman, marine fish of Mediterranean and Caspian origin and two species of sticklebacks constitute the basis of ichthyofauna.

The annular seabream occurring in Odessa MP aquatory can survive in the salinity range 8.0–37.5‰, but spawns only at salinities higher than 16.0‰. It is a typical inhabitant of the *Cystoseira* biocenosis [151] and it probably enters into Odessa Gulf with the cyclonic water circulation. It feeds on microfouling and, apparently, finds the food it needs in the periphytal zone of Odessa MP.

The transparent goby has a certain importance in the ecosystems of Odessa, Ilyichevsk and Yuzhny MPs. Schools of this nearly transparent fish, with a length of 3–4 cm, are found in the water column close to the lateral surfaces of HTS. Here also live fish from wrasses family (Labridae), needlefishes and sea horse (Syngnathidae). In MPs, combtooth blennies (Blenniidae) and some gobies (Gobiidae) are often observed. Most fish encountered in the water layer adjacent to HTS are big-sized grey mullets such as the striped mullet and the golden grey mullet.

In MPs, the silts contaminated by hydrogen sulphide are avoided even by bottom and near-bottom-dwelling fish that are resistant to lowered oxygen concentrations.

Fifty-eight species of marine fish are present in the composition of Odessa Gulf ichthyofauna. In Sukhoy Liman and in the adjacent part of the Black Sea, representatives of 58 fish species were also found during 40 years [154]. The ichthyofauna of Grigorievsky Liman comprises 48 fish species [196].

The aquatory of Grigorievsky Liman is characterized by the nearly constant presence of whiting which usually occurs in water masses with a temperature range from 5 to 15–16 °C. This euryhaline fish spawns in the upper water layer in winter.

Juvenile whiting lead pelagic life and feed on plankton. Adult individuals stay in the near-bottom water layer and are near-bottom pelagic predators. A significant number of whiting individuals is eaten by spiny dogfish, turbot and dolphins.

The whiting is an intermediate predator along the food chain. In the early stages of its development, it connects plankton with small-sized fish (sprat, anchovy, horse mackerel, gobies) and invertebrates (shrimps, crabs, etc.) and, in its adult stage, with big-sized predators (spiny dogfish, rays, turbot, dolphins) [30].

In Odessa, Ilyichevsk and Yuzhny MPs, the sand smelt and grey mullets are the most common mass nekton representatives. Mass entries of anchovy, sprat and horse mackerel are occasionally observed. Spiny dogfish and striped mullet are the biggest nekton organisms among fish of these ports.

In MPs aquatories, big-sized specimen of the jellyfish *A. aurita* and of the root-mouthed jellyfish *R. pulmo* can also be ascribed to nekton. The dice snake *Natrix tesselata*, which feeds mainly on gobies and creeps out on land with its prey, can be quite often observed in Sukhoy and Grigorievsky Limans. Small schools of the short-beaked common dolphin *Delphinus delphis* sometimes enter Odessa, Ilyichevsk and Yuzhny MPs aquatories in the spring-autumn period.

Nekton organisms actively participate in BM uptake from MPs ecosystems.

Chapter 3
Periphytal Zone Subsystem of the Marine Ports Aquatories

E. Hentschel was one of the first who used the quantitative methods of calculation for the study of fouling or periphytal community in 1916 in Hamburg MP [104].

In periphytal communities, macroforms first attract the attention of scientists. In MPs, the attached organisms are the basis of macroperiphyton, forming its spatial structure and creating specific conditions in the adjacent water layer. In the Black Sea macroperiphyton communities, the biomass of mobile forms reaches only 1–6 g m^{-2} and comprises 5–38% of the quantity [15]. Attached invertebrates (filter feeders) form the basis of fouling community and their favourite habitat is the zone located below the depth of the water turbidity maximum [114].

Academician Zernov [50] noted that rocks and immobile stones in marine environment gave shelter to special life forms, i.e. fouling hydrobionts, and other species from various systematic groups connected with them. He considered the fouling of berths and buoys to be analogous to the fouling of rocks. Already in 1913, S.A. Zernov pointed to the resemblance of the various Black Sea MPs fouling and underlined its resemblance with MPs of the Mediterranean Sea. According to him, geographic position, depth, light intensity and level of water pollution were the main factors conditioning fouling types. This approach is still valid [4, 79–84, 114].

In 1924, A.L. Bening introduced the term "periphyton" (from the Greek peri—all around, and phyton—plant), meaning the organisms community overgrowing hydroengineering structures. He distinguished fixed substrata from independently moving boats and ships. After Bening [10], the terms "periphyton" and "fouling" are considered to have the same meaning by many other scientists. At the same time, the term "fouling" is quite often used for designing animal and plant complexes settling on artificial structures, i.e. substrata of anthropogenic origin, in the scientific literature. In contrast, those complexes of organisms that are present on rocks, stones and other natural objects are considered as benthos [44, 74, 90].

According to Romanenko [110], the hydrobionts settlements on the surface of hard objects submerged into water belong to periphyton. In his detailed definition, periphyton is a complex of organisms forming on any HS surface regardless of its origin (natural, artificial, experimental) and in more dynamic conditions than on

A.K. Vinogradov et al., *Ecology of Marine Ports of the Black and Azov Sea Basin*, DOI 10.1007/978-3-319-63062-5_3

basins bottom. Periphyton includes auto- and heterotrophic organisms, as well as OM of various origin and different degrees of processing. The taxonomic specificity of periphyton is determined by dominance of the attached forms of organisms [59].

In its composition are distinguished bacterioperiphyton, phytoperiphyton and zooperiphyton. Periphyton is characterized by the constant presence of planktonic and benthic forms. The presence of plankton and of attached and freely moving benthic organisms in the periphyton composition determines the use of such terms like "euperiphyton" and "pseudoperiphyton" species [59, 104, 106].

Marine fouling or periphyton origin and development are controlled by several abiotic and biotic factors. Abiotic factors include salinity, temperature, pH, content of dissolved oxygen and hydrogen sulphide, water dynamics (flows' velocity and characteristics of water circulation), ice conditions, illumination, pollution, depth, season, region, size and peculiarities of aquatory configuration and features of the aquatory connection with the adjacent sea zone. Biotic factors include concentration in water of floating stages of fouling organisms, intra- and interspecies relations; presence, accessibility and excess of food. Anthropogenic factors include material, sizes and peculiarities of constructions and structures.

Marine organisms adapted to life in conditions present at the surface of artificial constructions could pass into the fouling type of life. Such passage is especially easy to accomplish for species attaching to natural substrata submerged in water. The characteristics of fouled surfaces change progressively with the development of these organisms and create suitable shelters for numerous vagile forms [20, 22]. At this stage of community development, the number of forms inhabiting fouling increases significantly and comprises species that are unable to attach to surfaces [90].

Usually, marine fouling is divided into coastal versus oceanic [48] and near-bottom versus pelagic [101]. The HTS fouling in MPs of the Black–Azov Sea basin should be considered of the coastal and near-bottom types, i.e. genetically connected with the benthos communities predominant on hard substrata (stones and rocks) of the coastal regions.

The marine fouling can reach very high biomass values. In coastal waters of Southern Vietnam, at favourable conditions, the fouling biomass could exceed 300 kg m^{-2} [12/48]. At stationary meshed and latticed objects in the Black Sea it reaches 100–150 kg m^{-2} [56].

By analogy with the terms "pelagial" and "benthal", we use the term "periphytal" to facilitate further discussion of MPs ecosystems structures. "Periphytal" means the ensemble of wetted surfaces of immobile and mobile HS together with their population (Fig. 3.1).

Fouling biotope is, first of all, an interface of solid and liquid phases, i.e. the periphytal [15, 102]. About 41% of introduced species (14 species not counting fish) entered the Black Sea with ships because their life cycle is closely connected with HS. For this reason, the coastal biocenoses where HS are located becomes the habitat of introduced organisms [4]. These species find favourable conditions for settling in MPs where there are various arrangements and combinations of HS and corresponding biocenoses. At present, the basis of the Azov Sea (including MPs)

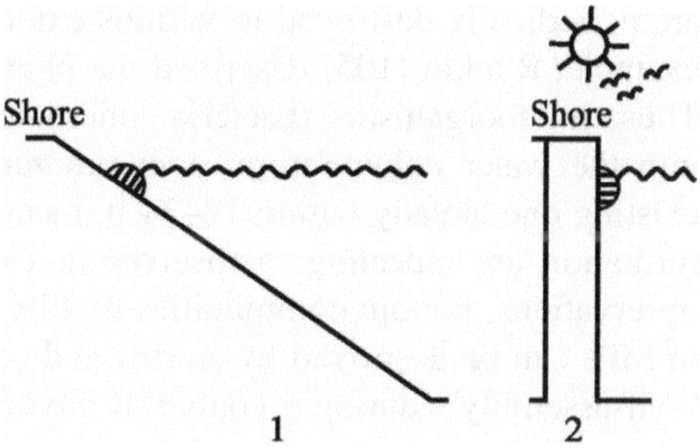

Fig. 3.1 General scheme of contact zones: *1* natural coast—water—atmosphere, *2* hydrotechnical structure—water—atmosphere

fouling is formed by invasive alien species which, on average, comprise more than 90% of the total biomass [4].

There are reasons to surmise that the role of introduced species in fouling cenoses of marine ports of the Black–Azov Sea basin could also increase in future due to advancement of already acclimatized species as well as the appearance of new ones.

Species settling in fouling of artificial HS usually reach a higher (often one- to twofold) quantitative development than in natural biotopes. Thus, the biomass of fouling can reach 100 kg m^{-2} for mussels in Sevastopol Bay, 40 kg m^{-2} for barnacles in Kerch Strait and 15 kg m^{-2} for hydroid in brackish Taganrog Bay [128].

The majority of animal species inhabiting fouling have a pelagic larval stage. Their reproduction and sedimentation on substrata take place in different periods and at different temperatures, so that the various species appear in sequence in the fouling composition and a succession of fouling biocenoses and seasonal changes in the process of their formation are observed [47].

The qualitative composition of the fouling of any artificial substratum is usually determined by the flora and fauna composition of surrounding regions and the species number of fouling is always lower than the number of benthic species registered in the region. The total species number in macrofouling communities in the Black, Azov and Caspian Seas usually does not exceed 10. In these seas, quite well-structured fouling biocenoses are very often formed by 3–5 species (2–3 attached species and 1–2 free-living ones) and, generally, the quantitative development of one of the sessile species significantly exceeds the development of other forms [128]. Microperiphyton and meiobenthos, for which macrozoobenthos serves as substratum and habitat-forming factor, are characterized by much higher species numbers.

In this way, the structure of the fouling community composition is always characterized by sharp domination of a small number of species despite its seasonal changes. In a fouling community, the dominating species usually belong to groups far apart as to systematic and ecological aspects. In such communities, competition between dominating species is absent and their negative impact on each other is minimized [128]. In fouling communities, dominant species fulfil various ecological functions complementing each other.

The upper part of the periphytal zone as well as the littoral benthal zone is the most exposed to storm influence and the hydrobionts communities inhabiting them are periodically destroyed to various extents. Using periphyton communities as an example, Rialkin [105] described the phenomenon of community "self-assembly". Thus, microorganisms (bacteria, microalgae, protozoans) washed from substrata into the water column form a community structurally similar to the previously existing one already within 12–24 h, meaning that all sequences of the community formation are repeating themselves quicker. In the same way, according to our observations, bottom communities of silty, silty-sandy and silty-coquina sediments in MPs can be destroyed by storms and complete a more or less total recovery, or "self-assembly", during a couple of days after termination of wave influence. The most mobile invertebrates and fish occupy their place in communities later than other components. This peculiarity of a community to self-recover provides relatively stable functioning of the specific MPs aquatories ecosystems.

3.1 Main Abiotic Peculiarities

In MPs, the periphytal consists of permanent, immovable, man-made components such as piers, berths, docks, breakwaters and mobile components such as incoming and outgoing ships.

Laminar and turbulent flows can originate in the water column of MPs aquatories, but laminar flows inevitably become turbulent after a certain period. Various experiments showed that, depending on flow speed and walls asperity, laminar water movement near artificial HS walls sooner or later acquires a vortex-like character. The walls of HTS in MPs always create flow turbulence and generate a mixed boundary layer near them [1].

Light intensity and direction play an important role for fouling communities. Studies in the NWBS testify that westward-oriented sides of sunken ships foul more than eastward-directed sides [112]. In the Black Sea, the limiting values of solar energy intensity reduction for macrophytes occur at the vertical surfaces oriented northward [86].

Organisms inhabiting the surface of substrata receive food from both the water column (detritus, plankton) and the hard surface on which form detritus–bacterial, bacterio-algal films [104] and bacteria–algae–fungi settlements.

In particular, in fouling and benthos communities, fungi play the main role in OM decomposition together with bacteria accelerating biogeochemical cycles and increasing their biomass.

3.1.1 Periphytal of the Hydrotechnical Constructions

For fouling hydrobionts, the most important characteristic of substrata is their hardness [103].

In the sea coastal zone, microbial communities of microperiphyton and detritus are among the most active and widespread participants in the ecological metabolism [131, 132].

In the sea, the quantity of fouling accumulating on HTS having different purposes depends on their form and construction material [90]. Many authors indicate that the physical and chemical properties of HS influence the formation of fouling communities [4, 103]. However, the HTS location and relative position in the marine environment have hydrological and hydrochemical consequences that also impact fouling. For instance, an important factor such as illumination intensity depends not only on the form, but also on the location of a construction.

In the White Sea, fouling of asbestos cement plates suspended at the depths of 0.5, 2.5 and 5.0 m highlighted minimal similarity between the upper and lower plates. Foraminifera, nematodes, harpacticoids, ostracodes, marine mites, tardigrades and turbellarian worms were recorded among the representatives of meiobenthos. Juvenile forms of oligochaetes, polychaetes, gastropods, bivalve molluscs, isopods, amphipods and insects larvae were present in pseudomeiobenthos. The less constant species composition was on the plates recovered from the depth of 0.5 m [23].

Extended berths, moles, separately located breakwaters and other HTS in MPs create conditions not only for the settlement of attached and vagile hydrobionts, but also facilitate the retention of some fish species near them. At various meteorological conditions, they can freely migrate along these constructions choosing the most favourable areas.

According to some data, the number of fouling species in the Black Sea exceeds 130. In the Black Sea, fouling biocenoses are mainly constituted by vagile molluscs, polychaetes, crustaceans and fish [48]. The main fouling species are mussels (*Mytilus galloprovincialis*), acorn shells or barnacles (*Balanus improvisus* and *Balanus eburneus*), bryozoans (*Lepralia pallasiana*), tunicates (*Botryllus schlosseri*) and hydroids (*Obelia loveni*) [15, 48].

Following current practice, artificial reefs (AR) are used in some seas for improving environmental conditions, attracting hydrobionts and ameliorating the quality of aquatic environment [4, 42]. The main mechanism of AR influence on aquatic environment is connected with intensive colonization of artificial HS by hydrobionts and with functioning of the new fouling or periphyton biocenoses [4].

During construction of MPs, HTS are not supposed to fulfil AR functions, but lots of them eventually meet the AR definition. AR are anthropogenic substrata laying above the bottom or buoyancies in the water column. They are built in a water basin with the aim to ameliorate its aquatory by creating specific combination of abiotic and biotic characteristics, distinct from the surrounding environment of hydrobionts [4, 42]. Breakwaters, moles, berths, pile foundations, pier towers, flyover supports and cargo handling terminals not only rise over the bottom, but extend through the whole water thickness and have above-water parts. Concrete and metal floating docks and constructions immersed into the water at various depths occupy a special place.

HTS of MPs have many characteristics of AR and, considering their extension and area of underwater surfaces, they variously influence biota and ecological processes in MPs aquatories and in each zone adjacent to a specific port. External and internal surfaces of HTS protecting MPs experience different conditions and have their own ecological influence on the adjacent sea zones.

Most zones of the NWBS coast, the Dnieper-Bug Liman and the Azov Sea have a deficit of natural HS. The surfaces of port HTS frequently and effectively replace missing HS of natural origin for many hydrobiont species, which leads to an increase in bioproduction.

Stones of natural origin have been one of the main materials for construction of marine ports over millennia. In the twentieth century, they continued to be used for hydrotechnical constructions in MPs, but mainly for coastal protection and breakwaters construction. The extent of their use can be illustrated by the example of Rostock port (Germany) whose Eastern mole required 60,000 stones for its construction [133]. In the Black Sea, the protective breakwaters in Skadovsk MP and in Sevastopol Bay were built from stones. Stones have also been used for protection of the Danube–Black Sea Shipping channel via Bystroe arm and in other cases.

The Danube–Black Sea Shipping channel dam of 1.2 km length is made of prism-shaped granite blocks dumped on a gravel layer. A right bank dam with a similar rectilinear profile was also initially planned to protect the southern part of the channel, but it was not constructed.

Concrete, commonly used in port construction from the end of the nineteenth century, turns out to have characteristics closer to limestone than to granite as HS for marine fouling hydrobionts. In the zone of coastal protecting structures in Odessa region, the average species number belonging to micro- and macrophytobenthos, meiobenthos and macrozoobenthos comprised 39 species on limestone and 36 on concrete [4].

The modern breakwaters are made of concrete blocks having the shape of equilateral trigonal pyramids. In the 1980s, concrete elements having a tetrapod ("stabilopod") shape, a total mass of 4.5 t each and a height of 2.25 m were used for constructing the protective moles of Constanza MP. Breakwaters of any size can be constructed using such elements. They are stable against wave influence and their irregular laying with numerous cavities creates favourable conditions for hydrobionts. Frequently, moles are constructed from concrete masses casted on the shore and weighing 50–100 t each or from floating reinforced concrete cases installed on prepared places and filled up by stones later.

The wide use of concrete and reinforced concrete in hydrological construction became possible after the elaboration of the method of concrete industrial production in 1844 and of ferro-concrete in 1867. Construction of the first iron ships started at the beginning of the nineteenth century and steel ships appeared at its end. Practical aspects of shipbuilding, ship navigation and hydrotechnical construction in MPs immediately showed that these new solid and relatively durable materials could be densely colonized by marine hydrobionts. First observations already confirmed that the concrete blocks were usually populated by aquatic fouling organisms even more densely than traditional wooden piles and constructions [90].

It turned out that fouling develops slowly in ports located in river estuaries. In strongly polluted MPs aquatories, zoofouling by macroforms can be completely absent.

Information about fouling peculiarities, forming on concrete and metallic HTS in different seas are reported in various publications [4, 9, 14, 45, 46, 55, 99, 105, 152]. Five years after completion of wave-protecting mole construction in Constanza MP, consisting of concrete elements, the average biomass of fouling on its surface comprised 26.6 kg m^{-2} and was dominated by attached filter-feeding animals [140].

It is known that marine drilling organisms generally attack rocks of sedimentary origin, but several cases of some molluscs drilling volcanic rocks are also registered. The concrete HTS were also subject to attacks of some drilling "rock borers" organisms and, in particular, algae [125]. Sedimentary and volcanic rocks are used as filler material in concrete. Microbiological destruction of concrete occurs in both seawater and freshwater.

Concrete is prepared by mixing cementitious materials with a defined inert admixture (sand, crushed stone, gravel, etc.) and water. Portland cement, or its mixtures with silicates or other admixtures, serves as the bonding substance. Portland cement composition comprises up to 65% of calcium oxide. After hardening on air, a thin (usually 0.2–0.3 mm) shell that plays a certain protective role forms on the concrete surface. This shell confers mechanical, chemical and water resistance. The softer mass containing calcium hydroxide $Ca(OH)_2$, which easily react with seawater, is located under the shell, so that damage to the surface shell speeds up the destruction of concrete constructions. In modern concrete mixtures designed for HTS, various admixtures are used to hamper water intrusion.

In seawater, concrete and metallic constructions undergo damage not only in the water column but also within the sediments thickness. Despite anaerobic conditions in the silt zone, corrosion speed increases as a result of the impact of metabolic by-products (sulphides, nitrites, ammonia and other) released via the vital function of sulphate-reducing and other bacteria.

High numbers of bacteria were registered on concrete HTS in Odessa, Sevastopol, Novorossiysk and Batumi MPs both on surfaces fouled by high organisms and on concrete "free" from macrofouling. In the latter case, many diatoms settle on concrete a couple of days after bacteria [125]. The cumulative action of bacterial, vegetal and animal population on concrete surface is usually observed in the sea.

A case of a big dry, reinforced concrete dock being put out of order owing exclusively to bacterial activity was registered. The dock's bottom was at a depth exceeding 10 m and its lateral walls and bottom were in contact with the silt layer. Constant cement leaching took place via reaction with the hydrogen sulphide contained in silt. Acting as a weak acid, H_2S dissolved concrete via formation of calcium sulphide (CaS) that was washed out by water containing new amounts of hydrogen sulphide [125].

At MP berths construction, wooden piles and concrete monolith blocks were replaced by sheeting and pile rows. The use of piles with groove-and-tongue joints allows to obtain a solid wall. In pile rows with a 1.68 m pace, the reinforced concrete piles of 40 × 40 cm, steel pipe piles with a diameter of 62 cm and box piles are used. At a pace of 2.52 m, the steel pipe piles of 122 cm diameter, steel double-T piles, reinforced concrete prestressed shell piles with the diameter of 160 cm and Gambia piles in situ reinforced concrete piles with 120 cm diameter are used [130]. The use of pile foundations under moles and berths influences hydrodynamics in the water column and in the near-bottom layer to a lesser degree. However, illumination under such berths is very low.

Comparison of lateral underwater surfaces area near the berths of equal length, but constructed by different ways, shows that berths constructed by regularly laid monolithic blocks have the minimal area. When piles with a square and rectangular cross section are disposed at a given distance from one another, all their four sides can theoretically foul to various extents. The area of piles having a square section will exceed that of a solid wall from monolith blocks by nearly fourfold, provided that piles are located in one row.

When using round piles, their surface area is nearly threefold the area of a continuous mooring wall of specified length made from solid blocks and excess value depends on the distance between piles. The use of a pile platform or a pile field with piles arranged in rows, or in a chessboard wise fashion, increases the area of MPs underwater surfaces by many folds. Fouling can also occur on vertical constructions of mooring walls, which are fastened on piles and partly rise over the water surface and partly remain submerged at a certain depth. Storm walls or screens are fastened in a similar way on breakwaters with pile foundation.

One round pile with a diameter of 1.6 m and a height of 10 m has an area of about 50.0 m^2.

It is accepted that the minimal size of AR structural components should be around 100–250 m^3, that of separate AR of 800–1000 m^3 and those of groups of reef and reef complexes of 5000–10,000 m^3 and 80,000–100,000 m^3, respectively [143, 147, 148]. Following this classification, the breakwaters in Odessa, Berdyansk and Mariupol MPs have sizes corresponding to reefs and all the HTS of big MPs can actually be considered as reef complexes. In Japan, artificial-reinforced concrete reefs, resembling multi-storeyed buildings, have already been built and can be up to 27 m high. They are placed at depths up to 60 m [127]. In the Black Sea, the deepwater berths are rising for more than 20 m over the bottom.

From the hydrological and hydrobiological point of view, separately located breakwaters are narrow, vastly elongated sandbanks or islands on one side and AR on the other. In MPs, separate harbours function as AR together with berths and moles. Even a whole bay could be totally transformed into an AR by various HTS and permanently moored ships, as in the case of the Southern Bay adjacent to Sevastopol Bay.

However, not all the components of the MPs periphytal zone constitute an AR. The main characteristic of ARs is their stationarity, while various ships in service cannot be considered as AR because of their mobility. At the same time, ships that remained moored for a long time (years), sink as a consequence of catastrophes, or are submerged on purpose actually become AR.

The information about the total length of berths, moles, breakwaters and depths near them does not fully reflect the actual area of the HTS underwater surfaces, especially when they have pile foundations and various irregularities in their shape. When correctly laid, square or cubic monolith constructions create uniform ecological conditions. Structures built using regularly spaced blocks with a rectangular section or round piles have a higher total surface for an equal construction length. Furthermore, variable conditions as to illumination, hydrodynamics, etc. originate around each pile. Pile foundations cumulatively create an entire complex of various ecological microniches. In Odessa MP, 1300 piles were used during the construction of the new berth for a container terminal at Karantinny mole, giving a total breakwater length of about 1 km.

In MPs, the hard surfaces of HTS and the wetted parts of ships' hulls covered by hydrobionts are washed by water whose hydrochemistry is determined by both the ecosystem of the adjacent sea zone and the entire MP ecosystem. However, the role of periphyton can be clearly observed in the near-wall layer where, during daylight, the photosynthesizing micro- and macrophytes increase the oxygen content. An alkaline environment usually forms in the near-wall water below the fouling vegetation. Fouling by animals, especially by molluscs, is accompanied by the formation of an acidic environment in the near-wall water layer following the release of carbon dioxide. The microperiphyton plays an important role in the production of vitamins, hormones and ferments, accumulating them in the near-wall layer. In the near-wall boundary area, filter feeders extract SOM from water during their vital activity and excrete their metabolites into the water enriching it with DOM [20, 22, 28, 131, 132].

As a general trend, fouling lowers the SOM concentration due to the filter-feeding strategy of invertebrates in the fouling community and increases the total nitrogen (N_{total}) content as result of its intensive excretion during the nutrition process of the organisms. Increase in the total surface of HS in MPs favours the development of new fouling and the reduction of total phosphorus (P_{total}) concentrations.

Besides living organisms (bacteria, diatoms, etc.), organic and inorganic detritus, sand, silt and other particles suspended in seawater are included in the mucilaginous film of primary fouling which captures suspended OM particles and ensures further development of bacteria.

In the near-wall water layer, filter feeding by barnacles and molluscs creates microflows that propagate to distances between a couple of centimetres and one metre, thus creating conditions for its mixing. In MPs aquatories, gravitation completes and substitutes, to some extent, the vertical water exchange for fouling organisms. Owing to gravity, the faeces and pseudofaeces from the fouling biocenoses deposit onto bottom sediments.

In MPs, illumination of HTS vertical walls is about 75–85% of the normal light intensity [38, 58]. At the same time, specially constructed AR create shadowed zones that attract many fishes [127]. In MPs, large shadowed zones appear under the berths and moles with pile foundations. The vertical position of surfaces and artificial HS is not optimal for macrophytes [115].

In HTS fouling, cracks and hollows having various sizes form between the single molluscs and their druzes. These spaces serve as a shelter for mobile invertebrates and fish. Intensively growing macrophytes shade the underlying water layer where some invertebrates and fish can also find shelter.

In MPs, water layer fluctuations occur for various reasons and lead not only to a decrease in MPs total depth, but also to a temporary drying of the upper part of HTS lateral surfaces. Many algae and animals (barnacles, bivalves and other) attached to the HTS upper parts can survive exposure to air during a couple of days.

To correctly understand the ecosystems peculiarities of MPs aquatories, it is important to examine two extreme cases. On the one hand, we find MPs located on an open part of the shore with aquatories without protecting HTS. Here, the water and matter exchange is not hampered and occurs freely. All hydrological, hydrochemical and hydrobiological indices will clearly be identical to those in the adjacent coastal regions. On the other hand, we can imagine an MP aquatory consisting in a sea zone protected by a continuous mole. In this closed aquatory, hydrological and hydrochemical changes, as well as biological reorganizations, take place after some time. Although the aquatories described in the above examples cannot be considered real ports, the existing MPs occupy an intermediate position between the two extreme situations, usually exhibiting characteristics closer to those of the latter case. The ports of the Black–Azov Sea basin usually have only one passage leading into the aquatory, the presence of two or three passages being a rare feature.

In MPs, each HTS represents an artificial HS, intentionally placed into aquatic environment, independently from its size and its construction material. All constructions built at shallow beach shores create conditions for the development of fouling organisms, i.e. for an increase in bioproduction and biological diversity. Analogous constructions located on rocky shores influence biotopical diversity to a smaller extent, but still increase HS area and bioproduction.

The following approximate figures document the scale of anthropogenic impact onto the near-coastal ecosystems. The underwater surfaces area of HTS is 190,000 m^2 in Sevastopol Bay, 170,000 m^2 in Odessa MP, 100,000 m^2 in Novorossiysk MP, 90,000 m^2 in Yuzhny MP, 80,000 m^2 in Ilyichevsk MP, 55,000 m^2 in Tuapse MP and 10,000 m^2 in Sochi MP. The total area of the underwater surfaces in Constanza MP exceeds 400,000 m^2.

The near-surface horizon (uppermost 0.5–0.6 m) of HTS is a less favourable environment for many fouling organisms because it is influenced by wave action, water level fluctuations and ice formation. The lower part of HTS at a depth of more than 8–10 m and at a distance of 0.5 m (or more) from the bottom is also quite often unsuitable for stable fouling formation. The reasons for this include chronic oxygen deficiency and periodical or constant presence of hydrogen sulphide because of redoxcline rise up [9, 115, 139]. According to Ukrainian standards, estuarine waters with a BOD_5 of more than 1.6 mgO_2 dm^{-3} are classified as polluted. In seawater, BOD_5 should not exceed 3.0 mgO_2 dm^{-3} [111, 113].

3.1.2 *Periphytal of Ships' Hulls*

Ancient seafarers already faced the problem of hulls' fouling. Mechanical cleaning was eventually replaced by planking of the underwater parts by lead and copper plates as a method to prevent fouling. Attempts to replace the metal coating with a special paint-and-lacquer material (PLM) or coating (PLC) were already made during the times of wooden sailing fleets. However, their effect was insignificant. In the nineteenth century, with the advent of iron and, later, steel ships the situation further changed and became more complex. The hulls of metal ships not only foul, but are also subject to various forms of corrosion. Marine corrosion happens to be closely interrelated with fouling [57, 61, 90].

Marine and oceanic fouling of various objects could cause significant damage during their exploitation. Fouling reduces ships' speed by up to 50% and increases their fuel consumption (up to 40%) and weight (by 20% or even more) [56]. During docking, up to 200 t of fouling can be removed from a ship's hull [90]. Experimental observations showed that a mucilaginous film of fouling formed by bacteria and diatoms can increase friction resistance by 11–13% already after 10 days [90].

Comparing with barnacles communities, mussel fouling develops much slower and usually appears only when ships stay in the port during a couple of days. The specific weight of barnacle and mussel fouling is about 1.2–1.3 g cm^{-3}. On ships, the surfaces exposed to fouling always preserve an equal distance from the draught line and, in this case, the fouling distribution is directly connected with the immersion depth [90]. The degree of fouling on laying up ship's hulls and on floating light vessels does not differ much [124].

Macrophytes most often present in the ships' fouling include the green algae from the genus *Cladophora*, *Chaetomorpha*, *Enteromorpha* and *Ulva*; the brown algae of the genus *Ascophyllum* and *Fucus*; and the red algae of the genus *Callithamnion* and *Ceramium*. The green algae most often settle near the draught line, while the brown and red algae settle at progressively increasing depths below it. If algae do not belong to calcareous ones, they easily drop off after dying. Various macrophytes species were registered in ships' fouling during various seasons and the ship's mobility has an influence on the morphology of algae settling on their hulls [124].

In the Black Sea MPs, ships' hulls could be immersed into water up to depths of 19.5 m. At such depths, illumination is usually not sufficient for settlement and growth of Black Sea macrophytes. Because of the construction features, the underwater part of the majority of ships' hull has a more or less marked curvature towards the keel. This feature leads to a nonuniform illumination of the various parts of the ships' hull. The ships' streamlining in different parts varies significantly depending on the hull's shape. All these factors affect the distribution of fouling on the underwater parts of ships. The ship operation also has a certain influence. After discharge operations, a significant part of the cargo ship hull dries out and is again submerged into water after loading. At prolonged drying, part of attached hydrobionts could die off and detach from the hull so that a new fouling can be formed after loading.

The fouling organisms of ships in operation form the so-called physically (cleaning) and chemically (antifouling paints) controlled communities. Similar natural communities, typical for the littoral zone and estuaries, form during 1–3 years and quite easily return to their original state after stress impact [97, 98].

Because of the regular hull cleaning, the fouling communities of ships' hulls seldom reach climax condition and remain at some succession stages corresponding to relatively stable "fouling types"—hydroids communities, barnacles, mussels and others [91].

The distribution of total, plant and most animal biomass on sea ships has a distinct patchy character which, first of all, concerns the dominant and subdominant species. Algae (*Ulva*, *Enteromorpha*, *Ectocarpus* and others) are usually concentrated in the upper part of the ships' hull, below the waterline, while animals are irregularly distributed on its remaining parts. In terms of settlement density, the majority of individual, attached species of invertebrates has a rather aggregated distribution, while mobile forms are more randomly distributed. The total biomass of foulers and animals' biomass (providing 55–100% of total biomass) quite often significantly increase along the hull from the fore end towards the stern section. In the case of algae, their biomass reduces from the draught line towards the ship's bottom [91].

It is important to note that separate patches of fouling settlements form particular belts at different depths.

Chloride ion is a strong corrosion activator in seawater, so that corrosion speed in the ship's hull splash zone is three- to fivefold higher than in its permanently immersed zone [51]. This peculiarity requires the use of different protective paintings.

The important components of primer paints are toxic metals such as chromium, lead, zinc and sometimes cadmium. Recently, tin and copper have become the main toxic metals in the composition of anticorrosive fouling. Zinc is also used for protecting steel hulls against corrosion.

Until now, the most effective method of simultaneously preventing ships' corrosion and fouling is the use of special poisonous PLC. It is considered that the most effective PLM could totally prevent organisms' attachment for prolonged periods. At present, the antifouling PLC used for ships' protection serve from 1 to 5 years and more. As it was already mentioned, the antifouling paints usually contained zinc, copper, mercury, arsenic and tin compounds in different combinations depending on the historical period. The specific toxicity of these elements prevents fouling. Over time, toxic substances from PLC leach away and paint consumption occurs. PLC thickness, total poison (biocide) content and dissolution speed determine the duration of its action [90].

The slimy film forming on the paint surface during the first stage of fouling absorbs part of the poison leaching from PLC. The presence of slime could slow down the speed of biocide leaching from the paint's surface, thus weakening its antifouling action.

Several stages, and four generations of PLC, can be distinguished in the history of antifouling coatings development. In the 1940s, the first-generation enamels contained inorganic compounds of copper, zinc and mercury as active components. Second-generation enamels appeared in the 1940s–1950s and their biocidal action was obtained by using cuprous oxide. Third-generation enamels were developed in the 1960s and contained a low-molecular organotin compound as a biocide. At the end of the 1960s—beginning of the 1970s—fourth-generation enamels appeared. They consisted of acrylic-type polymers containing organotin radicals included into polymeric chain and released into the marine environment following hydrolysis reactions. The so-called self-polishing coatings were developed on the basis of these fourth-generation enamels. These products release the biocide at constant speed both during navigation and moorage. Constant washout of the coating leads to self-cleaning at berth and self-polishing during motion [56].

Marine corrosion processes are influenced by many biotic and abiotic factors such as oxygen concentration, salinity, temperature, conductivity, pH, water dynamics and impact of micro- and macroorganisms on hulls surface [56]. Fouling is tightly connected with metals corrosion and, with some exceptions (see below), increases its speed and degree of damage by many folds. The MPs constructions (berths' pile foundations, flyovers, pylons of various constructions in the sea and ships' underwater parts) are all subject to corrosion. Fouling can not only speed up corrosion of metallic surfaces but also alter PLC designated for metal protection from corrosion.

Microorganisms are the first stage of fouling and they can both increase or suppress metal corrosion [30, 60]. The corrosion speed of carbon steel increases by 2.5-fold in the presence of algal fouling [5].

Fouling can : 1—mechanically disturb antifouling coatings and thus contribute to the corrosion process; 2—cause changes in the chemical and physical characteristics of the near-wall water layer which also intensifies corrosion; 3—at sparse settlement of fouling organisms, form potential differences between the zones occupied by organisms and zones free from their presence leading to local corrosion phenomena; 4—in some cases, isolate the metal surface from water and protect it from corrosion [56].

At the same time, modern PLC should protect ships' hulls from fouling and corrosion. Depending on application, PLCs used for ships' painting are subdivided into paintings for underwater hulls parts and for the line of alternating watering. PLCs for protection of the ships' underwater parts against corrosion are based on copolymers of chlorovinyl and vinyl acetate, chlorinated rubbers and epoxide (pure and modified) foam formers. This covering can last between 4 and 6 years [56].

Due to their widespread application, biocides PLC simplify the biota present in semi-closed MPs aquatories and favour the selection of species that are resistant towards their active components. The ecological danger of an antifouling method is determined by the biocide type and quantity present in a PLC. By leaching into the environment, biocides prevent sedimentation and attachment of fouling organisms in MPs, but at the same time may accumulate in these ecosystems and cause unwanted impacts.

In the presence of a high number of ships treated with antifouling paints, the poisonous components which leach from the ships' coatings accumulate in ports at levels exceeding the existing norms of maximum permissible concentrations (MPC). During the last decades, the substances and methods used for protecting ships' hulls and HTS against fouling and corrosion already led to significant visible negative consequences in numerous coastal aquatories ecosystems [56].

Sea-going ships are quite often the main cause of increased content of toxic metals in molluscs filtrators. In Newport harbour (United Kingdom), the copper content of *Mytilus edulis* is ninefold higher than in molluscs fished in the open coastal zones. Increased concentrations of cadmium, chromium, lead, tin and zinc in mussels were also registered [24].

There is an actual need to develop modern and ecological methods to protect ships and port constructions from fouling and corrosion [52, 56]. In the last years, research has been carried out on "ecologically safe" organic biocides, repellents and other specific mixtures. Biocides coating with a slippery and especially smooth surface to prevent settlement of sea organisms have also been developed. Coatings bases on silicone polymers have also come into use because fouling is easily removed from them under the action of seawater during the ship motion [56].

Theoretically, the control of fouling and corrosion should not cause large-scale, negative consequences for aquatic ecosystems. In open sea waters, biocides leaching from PLC are easily diluted to concentrations that are considered safe, but such diffusion process is sharply limited in MPs. Elaboration of ecologically safe PLC is the main problem in the optimization of the management of natural resources in the marine environment, especially in MPs ecosystems.

Until now, alternative physical methods such as cathode protection and ultra-hydraulic strike do not provide sufficient protection from fouling, are technically complex and require high power expenditure. These methods sometimes cause damage of protecting surfaces, ships or constructions [66].

The wetted surfaces of the hulls of big ships are temporary component of MPs periphytal, constitute HS and often have an area reaching 10,000 m^2 and more, which should be taken into consideration.

3.2 Main Biotic Peculiarities

The conclusion of academician Zernov [50] about the strong resemblance of fouling in various MPs of the Black Sea has found confirmation in studies of other authors [17, 83, 84, 89].

It was determined that, in the Black Sea, bacterial film and diatoms, fungi, protozoans, attached plants and invertebrates (barnacles, bivalves, hydroids, bryozoans, sponges and others) form the basis of marine fouling. Various worms, gastropods and crustaceans are the main mobile hydrobionts found in fouling. The free-living settling stages such as spores and plankton larvae of animals represent the majority of fouling organisms. After completion of the pelagic stage, they settle down and attach to substrata [3, 15, 48]. Small-sized animals can find shelter in fouling, i.e. barnacles' shells, empty polychaetes' tubes, molluscs' shells, etc.

3.2.1 Main Mechanisms of Formation of the Fouling Biocenose

In MPs, the settlements of various marine organisms, starting from bacteria up to large invertebrates, on various HS, are called marine foulers and periphyton [15, 46, 90, 124]. In fact, the two terms are synonyms.

Depending on their position relative to the shore, depths, hydrodynamics and illumination, HTS underwater lateral surfaces, apparently uniform at first sight, actually provide various habitat combinations (microniches) to the hydrobionts foulers organisms connected with them. In Black Sea MPs, the molluscs *M. galloprovincialis* and *Mytilaster lineatus*, the acorn barnacles *B. improvisus* and *B. eburneus* and some other micro- and macroforms are major constituents of fouling in terms of quantity and biomass. Not all horizons of HTS are equally favourable to all fouling organisms which show a visible vertical distribution. The surfaces situated at depths between 1 and 7 m are usually the most densely inhabited. For example, in 2001, the average fouling biomass at the depth of 3 m in Odessa MP amounted to 30.0 kg m^{-2}, 75% of which was represented by mussels [3].

In all regions of the World Ocean, including the Black Sea, population of artificial HS starts with the formation of a slimy film [15, 26, 151]. As already mentioned, such film includes bacteria and diatoms with the slime being their habitat and metabolite. Detritus particles of organic and non-organic origin enter the slime layer and are retained therein. Particles containing OM serve as a vital substratum for bacteria which reproduce quickly.

Diatoms, including pennate diatoms (Pennatophyceae), also participate in the formation of the slime film. The body structure of pennate diatoms allows them to live in various ecological zones of water basins. The presence of longitudinal grooves, called raphes, on the diatoms' valve helps them to move actively on HS and, occasionally, in the water column. Their gliding movement is realized by flowing of cytoplasm in the fissures of raphes or by flowing of water inside their cavity. Unlike cyanobacteria, which intensively develop in zones with strongly reduced hydrodynamics, diatoms are attracted towards hydrodynamically active zones. Horizontal flows or vertical water mixing favour the sustention of these algae, burdened by silica-based skeletons, in the water column [110]. Owing to such adaptations, pennate diatoms easily pass from the pelagial into the periphytal zone and vice versa.

In Artelleriyskaya Bay, the diatoms density was 53.02×10^8 ind. m^{-2} on mussels' shells inhabiting the pier vertical wall and 364.40×10^8 ind. m^{-2} on stones. The number of species was 33 and 41, respectively, and species resistant to strong pollution (*Navicula pennata var. pontica*, *Niebla ramosissima*, *Grammatophora marina*, *Cocconeis scutellum*, *Amphora coffeaeformis*, *Bacillaria paradoxa* and *Caloneis liber*) were the dominant ones [78].

Different species of infusoria and rotifers appear in the slime film after its formation. Many protozoans are less demanding in terms of oxygen water saturation compared with more highly organized hydrobionts. In zones of Odessa shore characterized by the presence of protective structures, the biomass of infusoria is commensurable with that of other zooplankton and their food ration exceeds 2.5-fold that of zooplankton [3].

Rotifers (Rotatoria) play a noticeable role in the formation of primary fouling and are divided into microphages, which feed on microalgae, detritus and bacteria, and predators. The life cycle of rotifers lasts from 5 to 24 days.

At a later stage, macrophytes prothalli and barnacles' shells become visible in primary fouling of the well-illuminated horizon adjacent to the water surface. Settlements of *M. galloprovincialis* and *M. lineatus*, represented by young specimens of 1–2 mm length, establish themselves at higher depths. Barnacles and molluscs settlements initially have a patchy character, but form continuous belts of rather uniformly sized individuals after some time.

Compared with the bottom, the artificial HS located in MPs aquatories provide better conditions to fouling animals owing to lowered hydrodynamics and food abundance. As already mentioned, the growth of bivalves, especially mussels, living in the water column on anthropogenic HS is 120–240% faster than on the bottom. As a consequence, the form of their shells resembles that of molluscs inhabiting the benthal zone in conditions of lowered wave activity and the molluscs mass on HS is two- to sixfold higher than in the benthos [115].

In coastal ecosystems, including MPs aquatories, the main function of invertebrates is the consumption and destruction of OM. Among the bottom invertebrates that move into the periphytal zone, 90% of food ration is consumed by filter feeders which filter SOM from the water. In the periphytal zone, the phyto-detritophages also have an important place.

Formation of fouling biocenoses in MPs of the Black Sea basin has a regular character and many of its aspects follow a specific algorithm. The basis of biocenosis comprises representatives of bacterio-, phyto-, microzooplankton and benthic organisms, developing from algal spores and larval stages of invertebrates that settle onto artificial HS from the water column and survive in the periphytal zone until maturity [19, 47, 80, 83].

Fouling communities develop according to the general laws of ecological succession, with their species structure and overall functioning changing over time. Successions take place as a result of modifications of the physical environment by the fouling community and of specific interrelations among populations. The most important features of a succession are its irreversibility and directionality towards a higher control of the environment by the community of living organisms. While the pioneer stages of a fouling community are entirely dependent on the external environmental conditions, climax communities control the surrounding environment to various extents [95, 109].

When developing on new HS, the vegetation passes through several successional stages. Benthic and planktonic species of diatoms establish onto HS within 24 h. However, the planktonic forms disappear from the community composition within a week. The first macrophytes appear after 9–12 days and all of them are highly productive and quickly growing opportunists with a filamentous or tubular thallus, such as the green algae *Ulothrix* and *Enteromorpha* sp. and the red algae *Ceramium* sp. and *Polysiphonia*. Vertical colonies of diatoms (*Navicula*) also develop. This successional stage lasts for 30–120 days after the beginning of the settlement and it is also known as the stage of *Ulva* algae because of their dominance in the community. In the transition period between the first and second phases, the diatoms species diversity drops, their distribution on HS becomes patchy and the early colonizing species gradually disappear. The third phase, which starts approximately after 3 months, is characterized by the appearance and development of long-lived forms with slower growth rates [40].

Field observations showed that, during the second and third stages of development and change of vegetation on HS in MPs, the macrophytes thallus become itself a substratum for the formation of diatom fouling and the settlement of other epiphytic forms. It is possible to consider that this fouling is an attractive source of food for some invertebrates and fish.

Some general features of fouling development on HS within the photic zone could be identified based on averaged data. The ratio of total production and destruction (respiration) of a community is a proxy of fouling stability [4]. Based on measurements of P/R ratios, B.G. Aleksandrov defined three stages of fouling successions [4]:

1. the balanced initial stage, or stage of stable microcommunity (P/R = 0.5–1.3);
2. the autotrophic transitional stage (P/R = 10.3–19.0);
3. the final equilibrium stage, or stage of stable macrocommunity (P/R = 1.5–5.2).

Depending on different factors (type of substratum, distance from the shore, depth, season and others), the duration of the various stages in the Black Sea is up to 1 month for the first one, 2–4 months for the second and more than 1 year for the third. During annual succession, the P/R value is determined by the balance between productive and destructive processes directly occurring within the fouling community. The interrelation of the organisms living in the attached community with the pelagic producers increases with the volume of water required by the former for their vital activity.

In the first stage of a succession, the microalgae and microheterotrophs attached to the HS surface equally contribute to create conditions for the formation of a mature community.

In the second stage of a succession, the metabolic excretions of macrophytes provoke the suppression of microalgae fouling HS. According to available data, macrophytes attachment precedes the settling and development of macrozoobenthos mass forms in the well-illuminated zone of HS. Antagonistic interactions are possible between macrophytes and invertebrates [4].

The third stage of a fouling succession is characterized by the development of edificator species (*Balanus*, *Mytilus*, *Mytilaster* and others). The shift in the size structure of invertebrates from meiobenthos to macrozoobenthos takes places owing to the appearance of older stages of the same organisms and requires more time than for macrophytes development. The share of pseudomeiobenthos exceeds 50% of the total number of organisms in the zone of the shore protecting structures (made of limestone, concrete and granite) of Odessa city [4, 19]. Consequently, at the second stage of fouling development, the presence of small-sized animals becomes more significant than that of microphytobenthos. In particular, they contribute 4–50% of the total invertebrate biomass and 3–20% of the respiration intensity. At the stage of mature community, the importance of meiobenthos is comparable with that of microphytes and does not exceed 0.01% for all indices. It is not by coincidence that the macrophytes production reaches its average maximum during the second stage of a succession, when the biomass of invertebrates is onefold lower than their average value in mature community [4].

A shift in the main source of primary production, necessary for existence of the fouling community, takes place at the third stage of a succession. In this period, most invertebrates in the community are not trophically connected with macrophytes, but with filter feeders such as mussels, i.e. the proportion of phytophages decreases [4].

Approximately, 89% of the total production of autotrophs involved in the metabolism of the mature fouling community via the water filtration process belongs to pelagic microphytes (phytoplankton) and only 11% to macrophytobenthos. During this period, macrophytes' development is inhibited until the next exodynamic succession connected with the detachment of fouling from the surface of substratum [4].

The biological diversity of the species forming fouling communities increases with their development. Studies in the zone of the coastal protective structures of Odessa city found a total of 83 representatives from different taxonomic groups: 31 microphytobenthos species, 12 microzoobenthos species, 9 taxonomic groups of meiobenthos and 31 macrozoobenthos species [4].

In Sevastopol Bay, fouling organisms' larvae comprise 20.0–43.0% of the total plankton biomass [56].

In the NWBS, all mass species of meroplankton such as *M. galloprovincialis*, *M. lineatus*, *Mya arenaria*, *Polydora cornuta*, *Neanthes succinea*, *B. improvisus* and others are also part of fouling communities and represent up to 76.0% of the organisms' quantity and up to 99.0% of the community biomass [4].

In the Black Sea coastal waters, fouling communities forming on HS play a significant role in the process of matter and energy transformation [49]. In the sea coastal zone, up to 74% of the primary production and about 90% of OM destruction are related with fouling. Based on the general features of fouling community formation and functioning, their role in the specific conditions of MPs aquatories could be assumed to be even more significant. When estimating the ecological consequences of the construction of new MPs or when assessing the environmental impact of new HTS in already existing MPs, it is important to consider both the impact on MPs ecosystems as well as the overall impact of MPs ecosystems on the adjacent sea zones.

On MPs artificial HS, the fouling composition is subject to natural seasonal changes that are characteristic of each particular region of the Black and Azov Seas.

In MPs, the increase of primary and secondary production in fouling biocenoses intensifies the flow of OM sedimentation to the bottom.

Fig. 3.2 Scheme of the location of coastal (1) and near-surface macrophytes belts (2) in marine ports

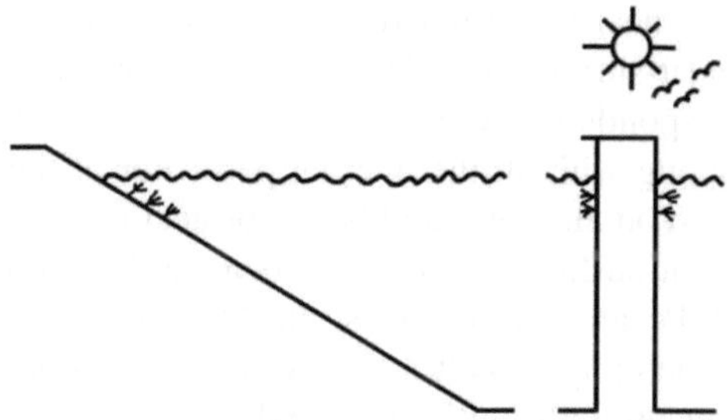

3.2.2 Near-Surface Macrophytes Belt

In MPs aquatories, macrophytes are the second autotrophic link by importance and, although significantly less abundant than phytoplankton, play a rather important role in the functioning of port ecosystems.

A littoral macrophyte belt develops near all the shores of the Black and Azov Seas to various extents. Under favourable conditions, many macrophyte species settle on all suitable HS and biotopes owing to the presence of pelagic stages. Green and brown algae have actively swimming zoospores, while red algae have immobile zoospores and their dispersal takes place at the zygote stage. Eggs are fertilized by sperm in the water column.

A littoral macrophyte belt directly or indirectly serves as a food resource for many invertebrates and fish species. The productivity of bottom communities is frequently determined by macrophytes that form an underwater blanket. Planktonic larvae of gastropods and bivalves settle on the macrophytes of the coastal belt where mites, crustaceans, polychaetes and others live and reproduce. In MPs aquatories, macrophytes of shallow waters are natural reservoirs of fouling organisms that attach to HTS and ships.

In MPs, a near-surface macrophyte belt exists along with the macrophyte coastal belt. The coastal belt is preserved, in various forms, in places that are not affected by hydrotechnical constructions, while the near-surface belt is formed by macrophytes that inhabit artificial HS (Fig. 3.2).

Macrophytic algae and higher aquatic plants are present in the composition of the coastal water belt. For example, the fennel pondweed *Potamogeton pectinatus* grows on silty sediments at depths between 0.4–0.5 m and 1.5–2.0 m in Sukhoy and Grigorievsky Limans where it forms ramified sprouts up to 1.0 m long. A great number of bottom invertebrates and some fish are connected with pondweed bushes. The seaweeds *Zostera marina* and *Zostera minor* are found on silty sediments in the same places and in similar conditions between depths of 0.4–0.5 m and 2.0–2.5 m. *Z. marina* grows on silty sediments, while *Z. minor* grows on silty-sand. The bushes of seagrass also attract many invertebrates and fish. However, MPs artificial HS do not offer conditions for fastening and growth of high aquatic vegetation which is absent in the near-surface macrophytes belt. In MPs aquatories, the near-shore macrophytes belt is either entirely replaced or accompanied by the near-surface one. On a natural shore, the extension of the macrophytes belt corresponds to the length of the shore stretch. In MPs, the macrophytes belt can form along with all the internal perimeter of the port (excluding passages into aquatory) and on all moles and berths projecting seawards. In slightly different conditions, the near-surface macrophytes belt can form on the external surfaces of HTS protecting MPs and in passages. In MPs, the shallow waters and the associated coastal macrophytes belt and coastal biocenoses are destroyed during the construction of quays with berths and moles, because deep waters (up to 10–21 m) are created directly close to the shore. Part of the invertebrates and fish usually found in the coastal zone move to live among the macrophytes of the HTS fouling or simply disappear from biota composition. In MPs, a near-surface macrophytes belt develops on berths, moles and breakwaters, which constitute artificial capes and AR projecting into the sea. This macrophytes belt is analogue to coastal ones in many aspects, except that it is mainly located on vertical surfaces. Inside aquatories and on HTS external surfaces, macrophytes experience various hydrodynamic influences. In MPs, the near-surface belt of macrophytes can form both on HS, lowered or immersed into water at 0.5–1.0 m, and on constructions located in the water column from bottom up to the film of the surface tension and raising over it.

On HS, the process of the near-surface macrophytes belt formation is the same regardless of the immersion depth of the HS. Development of the near-surface and coastal belt of macrophytes in MP is actually limited by light intensity and hydrodynamics.

A specific water line formed by a belt of green algae, among which prevail species of the genus *Enteromorpha* and *Cladophora*, can be observed on all HTS of MPs. This line rarely extends deeper than 0.3 m from the surface, even if individual green algae can be observed much deeper. Brown and red algae prevail below the green algae line and can be present nearly down to the bottom in some zones of MPs. The depths up to which they can grow depend on the water transparency [90].

In the NWBS, the number of macrophytes species sharply increases near the natural coasts between 0 and 1 m. This increase continues, albeit more gradually, up to depths of 2–3 m owing to red and brown algae, while a decreasing trend is observed moving to higher depths because of the disappearance of green algae. In aquatories with low water transparency, as it is the case in many MPs, the decrease in floristic composition is observed starting from depths of 1 m. In Odessa region, the biomass peak of the most widespread macrophytes occurs at depths of about 0.2–1.0 m, decreasing with a further increase in depth [37].

In the basins of the coastal protecting zone of Odessa city, the vegetative fouling on HS is represented by 11 species belonging to the divisions of the green algae (Chlorophyta: *Bryopsis plumosa*, *Cladophora vagabunda*, *Enteromorpha intestinalis*, *Enteromorpha linza*, *Enteromorpha flexuosa* and *Chaetomorpha linum*), red algae (Rhodophyta: *Ceramium elegans*, *Ceramium rubrum*, *Polysiphonia denudata*), cyanobacteria (Cyanophyta: *Spirulina tenuissima*) and diatoms (Bacillariophyta: *Berkeleya rutilans*). The composition of the vegetative fouling reduces to 7 species during the algal decomposition period [4].

In closed concrete basins of the coast-protecting structures of Odessa city, the composition of fouling community includes typical multicellular algae and is characterized by the presence of the diatom alga *Amphipleura* sp. Its slimy colonies, with a height up to 3 cm, have the form of branched tubes attached to HS and their morphology resembles that of the brown alga *Ectocarpus confervoides*. In the MPs of the region, conditions are favourable to the development of this diatom alga whose biomass can reach 250.0 g m^{-2} [4].

In Odessa region, a zonal distribution of macrophytes (green, brown, red algae) is generally observed on HS in the coastal zone (supra-, pseudo- and upper part of sublittoral). However, due to various ecological reasons, some macrophyte belts may be absent or the borders between them not well defined. Such distribution is already visible at depths up to 1.0 m and it becomes more distinct in deeper waters [11]. In MPs aquatories, similar features are observed on HTS surfaces.

In the zone of Odessa city, macrophytes developing on the concrete coastal protecting structures can contribute up to 43% of the total photosynthetic activity. From a functional point of view, the key species belong to red and green algae such as *C. elegans*, *C. vagabunda*, *C. rubrum*, *E. intestinalis* and *E. linza* which, all together, account for 89% of macrophyte production [4, 115].

In the course of evolution, the macrophytes playing an important role in MP fouling have elaborated special mechanisms to ensure their life on HS surface located in various layers of the water column and at different illumination intensities. During ontogenesis, the photosynthetic apparatus of marine macroalgae easily adapts to light, temperature and other environmental conditions. A change in light conditions influences the growth of macrophytes adult individuals and leads to significant structural and functional changes in the photosynthetic apparatus. A decrease in the intensity of illumination leads to pigments accumulation and to a change in their ratios, which results in an increased photosynthetic potential. The plasticity of the photosynthetic apparatus manifests in diurnal variations of its structure and functioning. The optical density of the thallus increases around midday and drops towards evening. During a year, the anatomy and ultrastructure of chloroplasts, their position in cells, the content and ratio of pigments as well as temperature dependence and photosynthetic potential change significantly [126].

In the first half of the twentieth century, the brown algae *Cystoseira barbata*, with which a whole complex of invertebrates and fish is connected, had a significant role in the composition of phytofouling of MPs HTS. In the 1960s, deterioration of aquatic environment quality led to the disappearance of *Cystoseira* in the NWBS and other regions. The state of *Cystoseira* bushes serves as an index of various MPs ecosystem well-being.

In Odessa MP, nine Rhodophyta species (*Ceramium diaphanum*, *C. elegans*, *C. rubrum*, *Acrochaetium savianum*, *Callithamnion corymbosum*, *Anthamnion cruciatum*, *Kylinia secundata*, *P. denudata* and *Lomentaria clavelosa*), eight Chlorophyta species (*Enteromorpha prolifera*, *E. intestinalis*, *Cladophora sericea*, *C. vagabunda*, *B. plumosa*, *B. hypnoides*, *C. linum* and *Ulothrix implexa*) and three Cyanophyta species (*S. tenuissima*, *Oscillatoria corallina* and *Oscillatoria* sp.) were observed in 2001–2002.

The red algae belonging to the genus *Ceramium*, which attach mainly to mussels shells, were recorded at nearly all stations. *C. elegans*, *C. rubrum* and *C. diaphanum* were used, in turn, as basiphyte by the epiphytic microscopic red algae *A. savianum* and *K. secundata*.

In Odessa MP, two species of siphonous algae from the genus *Bryopsis* dominated in the fouling of concrete constructions and molluscs in terms of biomass and density in 2001 [9].

Higher aquatic vegetation was not observed in Odessa MP aquatory where species of the genus *Enteromorpha* and *Cladophora* are less frequent than in Sukhoy and Grigorievsky Limans [9].

In Sukhoy Liman, 38 macrophytes species were recorded: Rhodophyta—8, Chlorophyta—16, Phaeophyta—4, Cyanophyta—5 and Thalassiophyta—5 [115].

In Grigorievsky Liman, the microphytobenthos composition included 46 species of bottom vegetation: Rhodophyta—12, Chlorophyta—20, Phaeophyta—6, Cyanophyta—3 and Thalassiophyta—5. Green algae of the genera *Cladophora* and *Enteromorpha* were specially developed. On HS, the average and maximal biomasses of *Cladophora albida* were 1158.6 and 2777.1 g m^{-2}, respectively. The corresponding values for *E. prolifera* were 1356.8 and 3581.8 g m^{-2} [115].

In MPs aquatories, substitution of natural HS by concrete surfaces sometimes reduced macroalgal species diversity by many folds and allowed the development of a new community dominated by green algae [92, 123].

On smooth-walled vertical HTS, macrophytes biomass frequently comprises no more than 10–50% of the corresponding values typical of the same species on natural HS [42].

The decrease in species diversity and the change of algal communities in MPs are connected with the spatial orientation of the substratum in water [115, 123]. In the ports of the Black and Azov Seas, the external side of breakwaters is characterized by maximal species diversity. In particular, the highest values of biomass and species diversity are observed on concrete blocks, placed more than 15 years ago and located slantwise (35°–75°) towards the bottom [123].

Living conditions for macrophytes are markedly different in the near-surface and coastal belts compared with those on HTS. In particular, the energy of a wave is immediately suppressed when breaking against the lateral surfaces of HTS, while it decreases gradually on a flat shore. Waves striking external and internal surfaces of MPs HTS have markedly different sizes. Fouling biomass is always higher in zones protected from the direct impact of breaker waves, but the intensive development of algae–macrophytes is characteristic for surf zones [46, 47]. In MPs, macrophytes grow better on the external surfaces of HTS which, however, host less animals than the macrophytes growing on the internal sides of protective moles [76].

Visual observations in Odessa region show that, on concrete surfaces of MPs HTS-oriented south-eastward, southward and south-westward, the upper layer of macrophyte belt is formed by green algae with red and brown algae growing in lower layers. On surfaces oriented north-eastward, northward and north-westward, green algae are less abundant, or totally absent, and red algae prevail.

In Sevastopol Bay, Kalugina-Gutnik [54] reported 184 macrophyte species out of which 49 were green algae, 44 brown algae and 73 red algae. In Novorossiysk Bay, 161 species were registered: 40 belonging to green algae, 48 to brown algae and 73 to red algae.

In the NWBS region (Ilyichevsk—Cape Severny Odessa, including Odessa Gulf) 84 species were found: 30 green algae, 12 brown algae, 28 red algae and 14 cyanobacteria [9]. The smaller number of macrophyte species in the NWBS compared with Crimean and Caucasian coasts is explained by several factors such as the more severe climate, lower salinity, higher OM content, better development of phytoplankton and reduced water transparency.

In Novorossiysk port aquatory, the green algae prevail in the upper sublittoral zone (0.5–0.8 m) [31]. With increasing depths, the green algae are substituted by brown ones and, at even higher depths, by red algae. In Novorossiysk port, liquid silt devoid of any vegetation is found near moles at depths of 8.0–10.0 m and below. In Novorossiysk MP, the highest values of biomass and species composition of algae are registered at the depths between 0.5 and 5.0 m. Macrophyte species typical of relatively deep water, which are registered at depths of 10.0–15.0 m near open shores, can be found at depths of 1.0–5.0 m in ports [53]. The macrophytes composition of Novorossiysk MP and Novorossiysk Bay waters changed and impoverished along with increasing pollution.

In 1927–1930, 37 macrophytes species were found in Novorossiysk MP: 18 species of green algae, 8 of brown algae and 11 of red algae. In the 1960s, the number of species was 35 (14 green algae, 8 brown algae and 13 red algae). In 1968, the floristic composition was represented by 24 species including 12 green algae and 12 red algae. The brown algae, which are more sensitive to pollution, had disappeared [53].

In Novorossiysk Bay, simple communities dominated by the green algae *Ulva* and *Enteromorpha* prevail in the zone of sublittoral border up to the depth of 0.5–0.8 m. As a rule, associations dominated by *C. barbata* appear starting from the depth of 2.0 m and develop up to 6.0–7.0 m [35].

Kalugina-Gutnik [54] marked the resemblance of ecological conditions in Sevastopol and Novorossiysk Bays. She also noted the leading role of green and red algae, accompanied by the loss of brown ones, in these ecosystems, which testifies to the presence of strong anthropogenic pollution. The flora of Sevastopol Bay significantly differs from that of the adjacent regions of the south-western Crimea. However, the index of species similarity of Sevastopol and Novorossiysk Bays algal flora is very high (C = 71), which confirms the proximity of their ecological conditions. The macrophytes index of species similarity of Sevastopol Bay with the flora of Odessa coast is C = 30 [54].

Milovidova and Tsymbal [83, 84] have studied vegetation on the hydrotechnical structures in Sevastopol and Kamishovaya Bays. In the apex parts of the bays, macrophytes are present only as single specimens at the water surface. When advancing towards the exit from the bays, green (4 species) and red (2 species) algae appeared. However, these algae were found only in mussel fouling on the middle part and near the water surface of berths. The lower parts of berths were usually deprived of mussels and vegetation. The dominating alga was *Ulva rigida* whose biomass reached 1260.0 g m^{-2} on one of Kamishovaya Bay berths which was most isolated from the sea. *E. intestinalis*, *Cladophora laetevirens* and *C. rubrum*, typical for natural HS of both bays, were present in low amounts on the near-surface layer of berths.

In Sevastopol and Kamishovaya Bays, the winter form *B. plumosa* was observed during the summer season. The same phenomenon was previously described for the moles of Novorossiysk MP [87, 89]. N.V. Morozova-Vodyanitskaya explained such phenomenon by the presence of specific conditions, i.e. increased shadowiness on the vertically located substratum. Brown algae were not registered on berths [84].

In 1977, 24 algal species (9 green, 3 brown and 12 red) were found on natural rocky substrata in Sevastopol Bay, while only five species were found on berths. Three of them belonged to polysaprobs and two to mesosaprobs. The same phytosaprobic composition was registered on the berths of Kamishovaya Bay and oligosaprobic algal species were not found in bays [84].

The plant fouling was richer on the mole enclosing Kamishovaya Bay than on berths. Two species of green algae, two of red algae and four of brown algae were collected on the internal side of the mole. Fouling of the entire internal side of the mole was characterized by a significant dominance of brown algae, accounting for 64.0% of total biomass. Red and green algae were present with roughly equal biomass [84].

On the internal side of the mole, the highest biomass and species number are registered in its terminal part, near the passage into the aquatory. The species number decreases progressively moving from the passage into the aquatory and towards the shore. In the middle part of the mole, the surface layer has the richest algal species composition. Biomass in the near-surface layer is sevenfold higher than in the underlying middle layer. The algae *Nereia filiformis* and *Cladostephus verticillatus* grew in the latter layer. Green algae were not found on the internal part of the mole adjoining the shore where *C. verticillatus* reached its maximum development.

In Kamishovaya Bay, the highest diversity of macrophytes species was observed on the external side of the mole where their biomass exceeded by 3.5-fold that on the internal side. Thirteen macrophyte species (three green, six red and four brown algae) were found. The middle part of the mole and the external side of its coastal part are the richest in vegetation. Phytosaprobiont algal composition of the protective mole significantly differs from that of berths and oligosaprobic species prevail on both its sides. Among mesosaprobic species, only *U. rigida* was found in significant amounts. Polysaprobic species were represented by only one species of *C. corymbosum*. *C. barbata* was the predominant species on the mole as well as on natural HS of Kamishovaya Bay. The example of the mole clearly shows that, at sufficient hydrodynamics, a rich phytocenoses, with a composition similar to the phytocenoses of natural rocky substrata, settles on HTS [84].

In the near-bottom horizon, at the depth of 9.0–14.0 m, phytofouling of the internal side of the Kamishovaya Bay mole was not registered [84].

Studies in the relatively clean Laspi Bay (Crimea) showed that red algae species diversity was 1.5–2.0-fold higher on a protected mole than in benthos on natural HS; while little differences were observed for green and brown algae. The total species number in phytocenoses on the western and eastern sides of the mole differed only slightly, 57 and 59 species, respectively. During the year, phytomass varied from 113.9 g m^{-2} in February up to 1911.0 g m^{-2} in August, with an average of 523.0 ± 201.0 g m^{-2} [36].

As mentioned above, macrophytes are capable of organotrophy, i.e. to absorb DOM from water [131, 132]. The opportunistic species have higher absorbing capacity than other species and, for this reason, they have an advantage in the eutrophicated waters of MPs aquatories.

Macrophytes react to excessive eutrophication not only by changes in species composition but also by increasing the specific surface of thalli structural elements and of populations. At the community level, the overall result of these processes manifests in a radical morphofunctional change of phytocenoses: small-sized, short-cycled, slimy branched, cylindrical and highly productive forms substitute large-sized, perennial, coarsely ramified laminar species with low specific production. During ecosystem eutrophication, this process transforms phytobenthos from an OM bioaccumulator into an object that intensifies the process of secondary eutrophication [85]. It is important to note that pluriannual, coarsely ramified algal species with a low specific production are usually adapted to waters with low BM and OM content and active hydrodynamics. That is why they are first to disappear from MPs phytocenoses.

In MPs aquatories, the vegetative fouling of ships hulls adds to the coastal the near-surface belts of macrophytes, even if active measures are taken to contrast its development.

The most frequent macrophytes in ships' fouling are green algae from the genus *Cladophora*, *Chaetomorpha*, *Enteromorpha* and *Ulva*; brown algae *Ascophyllum* and *Fucus* and red algae *Callithamnion* and *Ceramium.* Green algae are most often found near the waterline with brown and red algae settling at progressively increasing depths. Such distribution is determined by the illumination conditions. However, this scheme is further complicated by three factors: the presence of line algae (developing in illumination conditions unusual for their division), the actual ships' contours (which determine illumination near the sides and bottom of the ship) and water turbidity which is relatively higher in ports than in the open sea. The ships' speed affects the morphology of algae inhabiting their hulls [124]. When a ship passes from waters with high salinity into strongly desalinated waters, part of fouling algae and animals dies away [65, 90, 124].

As a rule, fouling is less pronounced for ships with a small draft in which case fouling is slowed down by various factors. Roughly speaking, green algae and barnacles form a belt about one metre wide immediately below the draught line, although the width of this belt sometimes reaches 3 m [65, 90, 124].

The periphyton of ships' hulls contributes to increasing SOM and DOM in MPs, thus favouring their further eutrophication.

In MPs, the macrophyte belt attracts many vagile invertebrates some of which perform regular night migrations into it. The amphipod *Stenogammarus* can collect food from the sediment surface and by grazing fouling. Corophiidae, which mainly feed by filtration, can also nibble various macrophytes [32].

Macrophytes frequently act as a substratum for smaller organisms and are populated by unicellular microalgae as well as by some colonial, multicellular algae. In the Black Sea, the average quantity and biomass of macrophytes increase three- to fourfold from May to August. The quantitative indices of attached microphytes and their distribution in the coastal and near-surface belts vary accordingly.

In the Black Sea, macrophytes–basiphytes offer a habitat to epiphytic unicellular algae and to some multicellular epiphytes (e.g. *Ectocarpus*, *Cladophora*, *Ceramium*, *Chaetomorpha*, *Sphacelaria* and others) whose surface is in turn colonized by microalgae. Quite often, macroepiphytes are more densely populated by microalgae than macrophyte–basiphyte which, being the basic substratum, have the main load of fouling by epiphytic animals and plants. Within the epiphyton of macrophytes, diatoms form multilayered communities in which the small-sized species inhabit on the surface of bigger ones. During the period of intensive reproduction, epibenthic diatoms occupy the upper layer of the settlement, forming colonies of various types [74]. Various direct and indirect links form in multispecies settlements of fouling organisms.

Different factors are important for attached unicellular algae: the morphology and mechanical properties of substrata (perforations, irregularities, roughness and thallus ramification), their physiological condition (age and life cycle stage), and seasonality and habitat conditions (depth, temperature, illumination and water transparency, BM content) [74].

The epiphyton of 46 macrophyte species widespread in the Black Sea and of two species of seaweeds contained 254 microalgae taxa, including 237 diatom species. The quantity of diatoms in the epiphyton of brown, red and green algae follows the ratio of 20:8:1, in agreement with changes in the morphological characteristics of macrophyte–basiphyte from ramified axial structures towards flat thalli [74]. The epiphyton of algal macrophytes included 184 taxa of microalgae and the epiphyton of 2 species of seaweeds consisted of 42 species with a prevalence of diatoms [74].

3.2.3 Zoofouling

In MPs of the Black–Azov Sea basin, the underwater parts of HTS function as an AR onto which larvae and juveniles of different taxa of hydrobionts settle down. These taxa include important commercial species, edificatory species that play a key role in fouling development and species present in biocenoses only episodically. The absolute majority of organisms inhabiting the coastal benthal zone can be found in periphyton. The introduction of artificial HS in regions with a deficit of natural HS is especially noteworthy. Development of additional areas of artificial HS always leads to an increase in the bioproductivity of the adjacent sea zone. Artificial reefs (AR) are anthropogenic structures that sharply change the natural environment in the place where they are positioned [25]. Piers, berths, breakwaters and other HTS of MPs perform the functions of an AR, but in conditions of lowered hydrodynamics and weakened water exchange. Port structures are durable and constructed from materials relatively favourable to fouling formation.

In MPs, HTS fouling mainly develops owing to representatives of fauna and flora having already settled on hard sediments in the adjacent sea zones. Some migrant species, whose usual habitats are located at considerable distances from MPs, can complete fouling composition. The well-defined zonal distribution of foulers is observed on the HTS vertical surfaces for both macrophytes and animals. One or two species, forming the basis of the periphyton, usually dominate in each belt [33, 34, 88].

Animals feeding mainly on deposited detritus predominate on MPs silty sediments, while filter feeders, consuming seston and suspended detritus, and predators are the dominant groups on HS periphyton. When choosing its dwelling horizon(s) on HTS surfaces, each fouler forms the highest biomass on the surface of the HTS horizon offering the most favourable balance among the necessary (abiotic and biotic) living conditions, the amount of food and the energetic expense for its procurement. The HTS of MPs offer a wide choice of such microniches.

Because of the large amplitude of changes in temperature, salinity and oxygen content, eurybiont forms dominate the fouling composition in the different horizons of the HTS near-wall water layer of MPs. Microorganisms play a very important role in the periphyton composition of MPs aquatories on new and pluriannual substrata. Bacteria and diatoms are the main components of primary fouling on newly constructed artificial HS. Microorganisms have a different influence on both plant and animal organisms and, in general, on all fouling at various stages of succession [26].

In the same way as microphytobenthos, microperiphyton constitutes a multi-species community of unicellular microscopic algae (diatoms, cyanobacteria, dinophytes, green algae, golden algae and others) which can settle on loose sediments and on living and non-living substrata, or spend prolonged periods in the water column. Macrophytes growing on artificial substrata are an additional surface for microalgae settlements with which they make a significant contribution to the oxygen balance and primary production of coastal ecosystems, including MPs aquatories [78].

Cyanobacteria, diatoms, golden and green algae most actively develop in microphytobenthos and in microfouling in summer, at seawater temperatures higher than 12–15 °C. Many microalgae exhibit mixotrophic feeding and most actively develop in polluted bottom sediments and silted fouling of artificial substrata. Under certain specific conditions, cyanobacteria, diatoms and other microalgae switch from autotrophic feeding to the assimilation of different organic compounds, thus showing heterotrophic or photoheterotrophic form of nutrition or combining them with photosynthesis.

Bays and MPs aquatories polluted by various waste products are the major ecological niche for microalgae with increased heterotrophic capabilities. Some of the forms inhabiting the water column switch to a sessile life using the increased OM concentrations which are found in the water layers situated at the boundary between the HS and the water column. Pennate diatoms (Pennatophycea) inhabiting silted aquatories typically show such behaviour and many of them grow even in total darkness by assimilating different organic compounds. Some diatom species reproduce in sediments in the absence of light, conditions that are found in some zones of the deepwatered berths. According to Khaylov [132], forms with increased heterotrophic requirements contribute particularly to the composition of fouling on HTS surface, macrophytes and surface of detritus particles. From an energetic point of view of OM utilization, the heterotrophic feeding by SOM and DOM is more favourable than autotrophic feeding. Some microalgae alternate auto- and heterotrophic feeding during day and night.

Algal "dark" respiration is in many respects similar to the metabolism of heterotrophic organisms. In the course of a year, their photosynthetic intensity correlates with the water temperature [134]. As already mentioned, the water transparency in MPs aquatories of the NWBS and the Azov Sea is often limited to 0.5–1.5 m, especially in summer. Illumination thus becomes a main limiting factor at depths higher than 5.0 m.

For some diatom algae, more favourable conditions are frequently encountered in MPs aquatories than in the adjacent zones. In the winter period of 1994–1995, 52 diatom species were found in HS fouling in the region of Cape Bolshoy Fontan (Odessa) and 65 species in Oil harbour of Odessa MP [107]. The number of diatom species registered on the HS of Ilyichevsk MP was higher than in the region of Ilyichevsk city municipal beach, located on open seashore (50 vs. 44 species). In Ilyichevsk MP, the diatoms' species number and diversity were higher and the biomass was lower than on the open seashore due to the development of small-celled forms resistant to pollution [108].

After their formation, the slimy films formed on HS by bacteria and algae become densely populated by ciliates. These protozoans are constantly present in the aquatic environment and reproduce very quickly. The number of cellular divisions for the species present in the fouling composition varies from 1.5 up to 5.6 per day. Infusoria are important and labile component of the fouling. In Sevastopol Bay, the density of infusoria settlements during the first days and weeks of fouling exceeds the quantity of other fouling groups by a hundred- and, sometimes, a thousand-fold. Small-sized infusoria and their telotrochs constitute an important food resource for larvae and juveniles of different vagile organisms finding themselves on artificial substrata. Infusoria, together with bacteria and diatoms, most actively develop on HS between January and April. The highest infusoria quantity occurs at depths up to 10 m [15]. Hydroids, barnacles, bivalves, nematodes, turbellarians, polychaetes, harpacticoids and other groups of hydrobionts appear in fouling only after the formation of a thin slime film due to the presence of bacteria, diatoms and protozoans and of their metabolic products [15, 26].

In fouling communities, vagile protozoans correlate with heterotrophic and rod-shaped bacteria, dead diatoms, coccoid bacteria and pH values [27].

The attached protozoans forms most distinctly correlate with alive and dead diatoms, coccoid bacteria, pH value, DOM, heterotrophic rod-shaped bacteria and carbonates. These correlations testify their more important role in fouling communities compared with mobile protozoans [27].

Proteins, carbohydrates and OM of the communities of periphyton and suspension microorganisms closely correlate with each other [27].

During their vital processes, microorganisms of periphyton and suspension communities deposit quartz, while biogenic silica is deposited from the shells of dead diatoms in the same communities. It was determined that quartz deposition is mainly influenced by calcite (which is always deposited together with quartz), pH, water temperature and DOM originating from the communities of periphyton organisms [27].

In April, hydroids (Hydrozoa) acquire a special place in fouling of the Black Sea, together with infusoria. Their colonies on substrata, underpopulated by other microorganisms, quickly expand and cover all the available area. In such situations, the probability of settling for other species is very limited, because hydroids scare them away with their tentacles covered with explosive cells (cnidocytes). The period of hydroids predominance in fouling cenoses lasts about 1 month. At the end of April–beginning of May barnacles (Cirripedia) determine the beginning of a community transition towards a new succession phase. At this stage, single individuals of barnacles, larval stages of bryozoans (Bryozoa), newly settled *B. schlosseri* (Tunicata) and, sporadically, mussels appear in the fouling [15].

Barnacles and hydroids coexist for a certain period and then the importance of *Botryllus* and bryozoans in fouling increases. *Botryllus* is able to replace barnacles by forming a continuous cover over them, thus worsening their feeding and respiratory conditions, lowering their activity and preventing their reproduction. Up to 85% of barnacles die after 35–40 days. On the other hand, the interactions of *Botryllus* and bryozoans have the character of commensalisms, with *Botryllus* protecting bryozoan zooids from clogging by detritus and natural enemies. The polychaete *Pomatoceros triqueter* can be found in fouling along with *Botryllus* and bryozoans [15].

At the beginning of June or, rarely, July, empty barnacles shells comprise 40–60% of the total amount. The empty shells are favourable substrata for the settlement of all fouling species. The relative importance of bryozoans, *Botryllus* and mussels increases and the polychaete worm *P. cornuta* (*P. ciliata limicola*) appears in great numbers [15].

During a period of 1 or 2 months after the beginning of mass sedimentation, a continuous layer of mussels could cover a substratum. The predominance of mussels in a community with all the complex of organisms associated with it (bryozoans, polychaetes, corophids, nematodes, decapods, harpacticoids, ascidians and others) indicates a dynamic equilibrium state in that community [15]. In the region of Sevastopol, about 50 taxa belong to the vagile organisms regularly found in the fouling community of artificial substrata (Nematoda and Harpacticoida were not determined until species). At that stage, a sharp increase of biomass takes place, especially upon reaching sexual maturity by molluscs. Species diversity in the community increases due to the appearance of vagile forms inhabiting various microniches [4, 15, 138].

A fouling community can remain in a quasi-stationary state for a couple of years [15]. In the Wight Sea, as well as in the Black Sea, perennial fouling communities in the coastal regions up to the depth of 10 m are characterized by a dominance of bivalves. Considering a zone with an overall depth of 25 m, the fouling reaches its climax conditions in 3–4 months at depths of 1–2 m and in 15–18 months at the depth of 5 m, while the fouling structure does not stabilize during 4 years at the depth of 10 m [96]. In MPs of the Black–Azov Sea basin, perennial fouling settlements at the depths of 5.0–10.0 m are mainly formed by bivalves and associated sessile and vagile organisms.

In Mariupol MP (Azov Sea), six mass species were found in the fouling biocenoses developing on HTS of a metallurgic plant: the hydropolyp *Perigonimus megas*, mainly feeding on copepods of the species *Calanipeda aquae dulcis*; the barnacle *B. improvisus*, feeding on phytoplankton and small-sized zooplankton; the epibionts peritrich ciliate *Zoothamnium* sp. and pearlwort *Bowerbankia imbricata*, feeding on microplankton and small-sized detritus particles; the predatory nudibranchiate mollusc *Tenellia adspersa* and the omnivorous Dutch crab *Rhithropanopeus harrisi tridentata* [128].

The complete system of identified simphysiological links of the named fouling community counts about 40 direct and indirect topical and trophic links. The majority of them manifests in the spring–summer period when larvae settle down, macrofouling organisms grow and, in general, the community forms its structure. The different connections within this system have unequal duration, impact and intensity on the populations of the interrelated species [128].

According to our observations, the formation time and the period of prolonged functioning of climax fouling community increase in MPs along with the weakening of wave action (moving from the MP entrance farther into the aquatory) and with the depth. Autumn–winter storms or catastrophic storms "rejuvenate" the MPs fouling cenoses and their whole ecosystems. Ice has the same influence in the near-surface horizons of MPs experiencing freezing.

Below we will examine the roles of the representatives of some main taxonomic groups in fouling of the Black–Azov basin MPs.

In the Black Sea, hydroid polyps Hydrozoa settle on natural and artificial HS, including HTS of MPs and ships. Sessile hydropolyps of *Eudendrium* and *Obelia* resemble bushes and reproduce via the formation of buds from which develop a young jellyfish leading a pelagic life. The colonial polyp *Aglaophaenia* does not form free-swimming jellyfish. The hydroids colonies easily attach to wood, metal and concrete and can form stable settlements on ships' hulls. They are characterized by a quick growth and reach a height of 5.0–7.0 cm in 1 month. The hydropolyps are able to seize planktonic animals such as infusoria, crustaceans, larvae of molluscs and polychaetes with the help of their tentacles.

Hydroids are not afraid of breakers and wave action. Young polyps of many species are protected by a special capsule, called hydrotheca, and their colonies bend according to the waves' direction so that they do not break.

The hydropolyp *Obelia longissima* is registered in the fouling composition of Odessa MP and Grigorievsky and Sukhoy Limans [9, 139]. It is a common mass species near all the coasts of the Black Sea.

Studies of fouling of Constanza MP showed that the initial stage of cenosis development is carried out by hydroids and then by barnacles. The maximum density of hydroids and barnacles settlements was reached in July, with more than 229,000 and 215,000 ind. m^{-2}, respectively [141].

It could be stated that, in the Black Sea, hydropolyps prefer the well-illuminated and hydrodynamically active near-surface horizon of MPs HTS. The polyp of the species *Aglaophaenia pluma* inhabits the MPs HS near the Crimean shores located in the eastern and southern parts of the Black Sea. Single specimens and small colonies of the polyp *Moeresia (Odessia) maeotica*, preferring brackish waters, are registered in the ports of Odessa region in the near-surface horizon of HTS (up to 1.5–2.0 m). In the Azov Sea, the brackish water polyp *Blackfordia virginica* inhabits MPs fouling, predominantly the macrophytes belt. It is also present in the NWBS.

The role of tunicates, particularly ascidians, in the fouling of Black Sea ports was already mentioned above. These bottom single-celled and colonial animals lead a sessile life. The species *Ciona intestinalis*, widespread in other seas, reaches sizes of 6 cm. Ascidiacea, commonly known as ascidians or sea squirts, are hermaphroditic and have sexual and asexual reproduction. Ascidian larvae lead a planktonic life for periods ranging from a couple of hours up to 5 days. Settling on a substratum, the larvae attach with the help of adhesive papillae and become sessile. Ascidians have the characteristic ability to regenerate lost body parts [39]. The sea squirt *Molgula euprocta* is registered in Odessa MP and in Sukhoy Liman.

An important representative of the fouling biocenosis of the Black Sea is the sea squirt *B. schlosseri* [15]. It is not registered in Constanza MP [141], but it is present in Odessa MP [9]. It is also absent from the desalinated regions of the NWBS and in Kerch pre-strait region. However, colonial ascidians can form large settlements on artificial substrata located in waters within the normal Black Sea salinity range, in particular in Batumi MP and Sevastopol Bay. They feed on detritus and on microorganisms which they collect by filtering water.

The sea anemones (Actinia) deserve attention among the coelenterates found in perennial fouling of the Black Sea MPs. They are typical marine benthic animals with both sexual and asexual reproduction. However, they have pelagic stages of development and their swimming larva, called planula, stays in the plankton for up to 7–8 days. They feed on small-sized organisms which they seize with tentacles. They can attach to HS of natural and artificial origin and, when necessary, are able to slowly move on the substratum. Four sea anemones species were found in the Black Sea and one species in the Azov Sea [39]. The sea anemones *Actinia equina* and *Actinothoe clavata* occur in Odessa MP, in Sukhoy and Grigorievsky Limans, in Sevastopol Bay and near the Caucasian coast. *A. equina* is registered in small numbers in mature communities formed mainly by mussels and macrophytes and it avoids illuminated zones. On the other hand, *A. clavata* reaches densities of 400 ind. m^{-2} in the 0.1–2.0 m horizon inside MPs.

Sponges (Porifera) are present in significant amount in mature communities of MPs located in the less polluted regions of the Black Sea, but are less numerous in the desalinated and eutrophic waters of the NWBS. The species *Suberites carnosus*, *Suberites prototipus* and *Spongia* g. sp. were found in Odessa MP [9].

Bryozoans (Bryozoa) are sessile colonial organisms feeding on detritus and small-sized forms of phyto- and zooplankton. They have sexual and asexual reproductions and are mainly hermaphroditic animals.

In the Black and Azov Seas, bryozoans are found in ships' and ports' HTS fouling together with other sessile organisms. About 30 species of bryozoans inhabit the Black Sea, while only seven species are known for the Azov Sea. Bryozoans feed using tentacles covered by cilia. Cilia constantly move and create a water flow going towards the mouth. This mechanism filters various planktonic microorganisms and detritus out of water. Indeed, bryozoans are typical filter feeders [39]. The bryozoan's larvae inhabit the water column for some time.

Bryozoans together with *Mytilus*, *Mytilaster*, *Dreissena* and barnacles play a leading role in fouling of MPs of the Black–Azov Sea basin. They can settle on various substrata and be found in the composition of all Black Sea biocenoses, including phaseoline ooze. They are present in the aquatories of many MPs, particularly in Odessa and Batumi MPs, and in Sevastopol Bay. For bryozoans, a substratum allows them to be retained in a specific water horizon. Their larvae have positive phototaxis. They are selective towards substrata, but are able to settle on any surface in the absence of appropriate ones. Closely located bryozoan colonies start to depress each other, especially if they are settled on a limited space [13].

In order to settle, bryozoans larvae more often move to the bottom, but can also approach vertical surfaces and slide along them with the sensitive cilia of their glandiform organ. After having crawled on substrata, trochophores of bryozoans stop in some place, but can start swimming again when finding themselves in unsuitable conditions. Settled individuals tightly attach to the chosen surface with the help of an adhesive suck and undergo further metamorphosis. Substrata roughness plays an important role for settling of bryozoans' larvae. The concrete surfaces possess suitable characteristics. The positive reaction of larvae to light is so strong that it suppresses the ability to choose substratum, so that larvae settle on the illuminated side of a surface independently of its characteristics. Phototaxis is connected with water temperature. When temperature rises up to 25.0 °C, the positive phototaxis of part of the larvae is changed to a negative one [13]. Owing to these mechanisms, bryozoans survive in the critical conditions of each particular MP.

The colonies, up to 25 cm thick, of the bryozoan *Membranipora* were found in Batumi MP in the presence of strong oil pollution [124]. *Membranipora denticulata* commonly occurs in the Black Sea and inhabits Constanza MP [141]. *Membranipora crustalenta* withstands significant salinity fluctuations (up to the freshwater conditions) and is present in both the Black and Azov Seas. In the

Caspian Sea, *Membranipora* fouls HTS and ships covering them with a dense calcareous crust. Representatives of the genus *Victorella* and *Bowerbankia* cover port structures in the oil-polluted waters of Baku and Krasnovodsk (presently Turkmenbashi) MPs. In Odessa MP, a sunken ship was covered by a complete crust of bryozoans over a period of 3–4 months [39]. Bryozoans, mainly from the genus *Bowerbankia* and the species *Conopeum seurati* and *L. pallasiana*, are common fouling constituents in Batumi Bay [15]. In Odessa MP, the bryozoans *L. pallasiana*, *C. seurati*, *Electra pilosa*, *B. imbricata*, *Bouteloua gracilis* and *Brassia caudata* can be found in fouling [9].

The bryozoan *L. pallasiana*, also inhabiting MPs aquatories, reproduces even in the period between December and March, when most hydrobionts do not reproduce, but the hydrochemical conditions are improving.

Among barnacles, *B. improvisus*, *B. eburneus*, *Chthamalus stellatus*, *Ch. depressus* and *Verruca spengleri* are known in the Black Sea. These species inhabit and spawn in the Black Sea. *B. improvisus* is resistant to water desalination and inhabits the MPs of the Azov Sea. *Balanus amphitrite* is sometimes present in various ports of the Black Sea, particularly in Odessa and Burgas MPs, but it cannot survive temperatures below 7.0 °C [9].

In Odessa MP, the highest density (81,360 ind. m^{-2}) and biomass (5884.0 g m^{-2}) of *B. improvisus* were registered at depths of less than 1 m [9]. In Odessa MP, isolated individuals of *B. eburneus* were registered at the depth of 3 m in zoocenoses of mussel's fouling. In the northern part of Sukhoy Liman, *B. eburneus* becomes a mass species in mussels zoocenosis with a biomass of 5.5 kg m^{-2} at the depth of 1 m. The biomass and density of *B. eburneus* reach 712.0 g m^{-2} and 340 ind. m^{-2}, respectively [138].

Barnacles with a population density of more than 500 ind. m^{-2} are considered dominant in a community [15].

B. improvisus is the most widespread barnacle in MPs of the Black–Azov Sea basin and belongs to major animals foulers, forming dense settlements. Its pelagic larvae, released into the water column, are eaten by larvae and fries of many fish species. Except for their larval stages, barnacles lead a sessile life. Barnacles shells are built from a couple of calcareous laminae. *B. improvisus* is a very eurybiontic species. It reaches sexual maturity 3 months after completing larval development and its life span is up to 3–4 years or more. In the Black Sea, its larvae are sometimes found during the whole year, but their number is not significant during winter.

Barnacle settlements usually die because of extreme pollution and some natural causes, but not as a consequence of overpopulation or ageing. It is considered that mussel fouling represents a serious danger for barnacles settlements [90].

Barnacles have six pairs of thoracic branched (biramous) limbs. The numerous appendages of their limbs are called cirri. During feeding, cirri are stretched from the shells, resembling a fan, and are used to drive water with detritus into the shell cavity. The strokes' frequency of cirri depends on flow speed, concentration of suspended particles and water temperature. At flow speeds above 10 cm s^{-1}, barnacles stop feeding and close their shells. Water renewal inside the shell, respiration and removal of excretory wastes are performed via the strokes of thoracic limbs. Water colouring near barnacle settlements showed that the simultaneous joint motions of numerous individuals create noticeable water flows at distances of 10–15 cm from HS.

Within settlements, barnacle shells are located close to each other. This distribution is determined by the peculiarities of their cross-fertilization which occurs via a tube-like, stretchable penis entering into the mantle cavity of neighbouring individuals. Barnacle settlements initially have a patchy character and interstices are gradually filled.

After hatching from eggs, the larvae (nauplius) of barnacles leap out into the water column and lead a free-swimming life for some time. After a couple of moults, larvae develop into metanaupliuses and, at the last moult, into bivalve cyprid larvae which will later settle on a substratum, attach to it and develop into sessile individuals. Larval settling is not possible at flow speeds above 50 cm s^{-1}. Larvae settle especially readily on the shells of other individuals belonging to their own species. The contact with shells further stimulates attachment. Once settled on a HS, larvae move in such a way to distribute evenly on a given surface. Usually, larvae actively move against the flow and settle down when the flow speed is equal to the moving velocity [39]. Apparently, this is one of the mechanisms to select locations with a sufficient hydrodynamics to guarantee survival.

Cyprids initially exhibit a positive phototaxis. However, they manifest a negative phototropic behaviour at the moment of attaching to a substratum when they move to deeper horizons and settle down with their heads directed towards the source of illumination [90]. Dense fouling belts of *B. improvisus* usually form at 20–30 cm below the water surface and extend up to depths of 60–90 cm. Some groups and individuals of barnacles occur at higher depths, but mainly on bivalve shells. Once they have settled, larvae of barnacles grow very quickly and completely cover a surface in a short time. They are also able to remain attached to moving ships.

During the 4 warm months, the weight of fouling by *B. improvisus* and accompanying species reached 20.0 kg m^{-2} per year in Sevastopol Bay and 12.0 kg m^{-2} in Novorossiysk MP [9].

In the Black Sea coastal zones, *Mytilus* is the most common and abundant genus of the Mytilidae family.

M. galloprovincialis forms the basis of zoofouling in all MPs of the Black Sea. Following salinization, the Black Sea *Mytilus* also diffused into the most saline zones of the Azov Sea over the last decades. Its larvae occupy a noticeable place in the Black Sea plankton. In Odessa MP, specimens of mussels belonging to morphological types common for *M. edulis* and *Mytilus trossulus* are also found along with *M. galloprovincialis.* This indicates the possible presence in the Black Sea of mixed populations of native *M. galloprovincialis* and other introduced species of the genus *Mytilus*, as well as of hybrid forms [115, 136].

Shurova [137] recommends to consider them as *edulis*- and *trossulus*-like forms of *M. galloprovincialis* until there is no other conclusive genetic-biochemical analysis. The growth speed of the Black Sea *M. galloprovincialis* is determined not only by the ecological and environmental conditions, but also by the genetic and phenotypic peculiarities of the molluscs. An oxygen deficit typically leads to a reduction in growth speed. The growth speed decreases by 2.5-fold and 27-fold at oxygen water saturation of 40 and 20%, respectively. Under typical conditions, about 2.5 years are needed to reach a shell length of 40 mm and more than 3.1 years are necessary for conditions of periodical temperature stratification [137].

The Black Sea mussel *M. galloprovincialis* forms populations on silts and rocks and occurs at salinities higher than 10.0‰. In the MPs located in the most desalinated regions of the Black and Azov Seas, *Mytilaster* and *Dreissena* take the place of *Mytilus* in the composition of HTS fouling. In the Black Sea, the maximal registered age of mussels is 28 years. In the last decades, the increased eutrophication of various zones of the Black Sea has led to a decrease in *Mytilus* total stocks and to a reduction of their age. Increase of *Mytilus* mortality in the NWBS, resulting from hypoxia of the near-bottom waters, led to a decrease of two- to threefold in their quantity and three- to sixfold in their biomass during 1989–1990 in comparison with the 1960s. *Mytilus* settlements respond to a reduction of eutrophication by an increased quantity, biomass and longevity [136].

It must be mentioned that eutrophication mostly harmed, mussels' settlements on silted sands and, to a lesser degree, those on rocky substrates. These molluscs withstand relatively well muddy and polluted waters as well as significant fluctuations in temperature and salinity. They could survive up to 3–4 weeks in conditions of oxygen deficit and in the presence of hydrogen sulphide.

Their gametes are released into water where eggs fertilization occurs. In natural conditions, the release of gametes is stimulated by sudden drops in temperature and salinity. Two peaks of mussels' larval quantity are registered in the Black Sea plankton: the first one in April–May and the second one in September–October. The quantity of larvae is relatively low for the rest of the year.

Larvae actively move in the water column and, after 10 days, already have a quite visible shell comprising two valves which are smaller than the larval body and do not close. Larvae undergo metamorphosis after about 4 weeks in the pelagial zone and complete it before they start to attach. In this period, they can still swim freely, but are already protected by a quite well-formed bivalve shell. By the time of settlement, shells are pigmented and have a length of 0.7–0.8 mm [90, 122].

Mytilus larvae are present all year round in the plankton composition of Sevastopol Bay. The maximum number of larvae is observed from July to September. During the first month of attached life, the mussels' growth speeds depend on where larvae have attached [119].

After settling down, juveniles of *Mytilus* can sometimes return back into the water column, continuing to choose a substratum. They are able to move on the substratum with the help of their foot. After attachment by the byssus, the foot slowly loses its function. The ability of young mussels to attach tightly to objects, including artificial substrata, is used in aquaculture. The size of a 1-month mussel varies between 0.5 and 6.8 mm. The fastest (2.0–6.8 mm) and slowest (0.5–1.0 mm) growth during the first month of attached life are registered for *Mytilus* settling down on a substratum between June and October and between November and March, respectively [119].

In natural conditions, the maximal quantity and biomass of juvenile and adult molluscs are registered on coquina and silted sand at depths up to 10.0 m. In the Black Sea, *Mytilus* are somewhat bigger in the zones with higher salinity than in the desalinated ones. An important role in mussels feeding belongs to plankton algae (first of all diatoms), protozoans and detritus. According to some estimation, one middle-sized mussel could catch up to 100,000 microscopic organisms during one day [122].

The fouling mussels actively participate in the formation of bottom silt by excretion of faeces and pseudofaeces. The silt layer formed by the mussels' vital activity can reach a thickness of a couple of metres [39]. Mussel's faeces have the form of compact ribbons. Pseudofaeces form only at concentrations of 1–6 mg (dry weight) of suspension per litre of water [41]. According to some estimations, up to 80.0% of the matter assimilated by mussels can be consumed in faeces and pseudofaeces production. Assimilation efficiency of fresh detritus by *Mytilus* and *Mytilaster* is around 60.0–80.0% [4, 41]. DOM concentration can increase three- to fourfold in zones of mussels' mass settlements. Excretion of ammonia by mussels favours high primary phytoplanktonic production in the near-wall water layer [20, 21].

"Rocky" mussels, settling on artificial HS in MPs, survive better in zones having moderate hydrodynamics and wave activity where a sufficient amount of food is present.

In perennial fouling, mussels themselves quite often serve as a HS to which attach small-sized sessile organisms and plants such as protozoans, sponges, coelenterates, worms, bryozoans, crustaceans, smaller molluscs and algae. The sponge *Cliona vastifica*, *Rapana*, some polychaete worms, drilling algae, viruses and fungi are mussels' pests.

Studies of sinking peculiarities and behaviour of young-of-the-year mussels (based on the example of *Mytilus grayanus* whose biology is similar to that of the Black Sea "rocky" *Mytilus*) showed that the ratio between the number of young-of-the-year and of other age groups was 0.5:1 on silty-sand sediment; 3:1 on sandy sediments; and 10:1 on rocky substrata. The total immobility of rocky bottom substrata, as well as MPs HTS, creates conditions for the formation of big druses, with a dense and clean web of byssus filaments. Sinking larvae easily enter into this web. They attach on byssus filaments and grow under the shelter of adult individuals while freely receiving food and oxygen [116].

In their settlements, the Black Sea *Mytilus* and *Mytilaster* form aggregations of three main types: 1—one-layered dense aggregations where individuals are oriented in a similar fashion; 2—one-layered aggregations of variously oriented individuals, i.e. "simple druses"; 3—multilayered aggregations of oppositely oriented individuals, i.e. "complex druses". In all types of settlements, molluscs are connected into a complex by byssus filaments. Sinking onto druses, larvae initially settle in equal quantities on shells of adult molluscs and on their byssus filaments, but then migrate to the lower part of druses [100]. Further studies of the Black Sea *Mytilus* and *Mytilaster* confirmed that their larvae sink near older individuals in notably bigger amounts than elsewhere. The *Mytilaster* larvae immediately settle on byssus filaments.

In places with strong break waves, *Mytilus* and *Mytilaster* create settlements having the form of dense continuous layers, i.e. "complex druses" protected from the influence of storms. The big druses periodically break under their own weight and under wave action. In settlements located on HTS, the biggest *Mytilus* appear at depths of 1.5–2.0 m (or deeper), while molluscs size decreases towards the surface. At low hydrodynamical activity, all the area of artificial HS in the water layer between 0 and 10.0 m can be considered as optimal for supporting "rocky" mussel. Reinforcement of hydrodynamical activity displaces the upper border of the optimal zone to depths of 2.0–3.0 m [41].

As already mentioned, besides settlements formed by single species the Black Sea *Mytilus* and *Mytilaster* also form mixed settlements. In the region of Sevastopol, *Mytilaster* comprises on average about 5.0% of Mytilidae on piles and concrete constructions inhabited by *Mytilus* [41].

On stones dumped close to Sevastopol Bay Southern mole, the distribution of *Mytilus* and *Mytilaster* at different depths is mainly determined by the number of larvae in the water column rather than by hydrodynamics. In general, stones (marble-like limestone) experience less fouling than concrete tetrapods. Mussels at 1.7 and 15 m depths were represented by individuals with sizes up to 10 mm. Mussels quantity increased with depth and wave influence was not observed below depths of 4–5 m [121].

In fouling of Constanza MP, *Mytilus* had already become predominant 2–3 month after the beginning of pier colonization. In September, the mussels quantity comprised 24,063 ind. m^{-2}. A high number of bivalve veliconchs was found in fouling of concrete surfaces. Fouling biomass increased with community age. In 1979, the maximal biomass of fouling exceeded 35.0 kg m^{-2} from April to December. In the port aquatory, numerous vagile forms were found in the fouling cenoses. The presence of the mobile forms was especially characteristic for the closed zones of aquatories. Turbellarians, nemertines, nematodes, polychaetes, copepods, amphipods and isopods were recorded among the vagile forms [141].

Regular diurnal migrations of vagile fauna (mainly crustaceans and gastropods) are observed in fouling biocenoses. This phenomenon takes place in Sevastopol Bay and in the aquatories of many ports. It is supposed that oxygen deficit at HTS foundations and in the deepwater layers is one of the main causes of these migrations. Consequently, many animals migrate towards the water surface layer where oxygen diffuses from the atmosphere. In the surface water layer of Sevastopol Bay, catching of invertebrates by nets is twofold higher at night than during daytime. Opossum shrimps (Mysidacea) appear at the surface only at night and are caught into nets 18–19-fold more often at nighttime. Gastropods crawl up the macrophytes and, at night, they are present near the water surface film at levels tenfold higher than in daytime. Similar migrations were not registered at depths of 2–5 m [75].

Conditions for the formation of foulers settlements of second and even third order originate in the near-wall water layer adjacent to settlements of *Mytilus* mussels. Second- and third-order foulers use molluscs shells and macrophytes thalli attached to shells as substratum. Settlements of microfoulers (bacteria, diatoms, protozoans and others so-called epiphytes) form on macrofoulers.

In the NWBS, the number of macrozoobenthos taxa inhabiting a given and equal area of HTS is 35–40 on Pontian limestone (coquina), 30–35 on concrete and 5–8 on granite. The corresponding macrozoobenthos biomasses are in the ranges 8.0–10.0 kg m^{-2} on coquina, 10.0–15.0 kg m^{-2} on concrete and 0.002–0.005 g m^{-2} on granite [6].

Mytilus settlements on soft sediments and on HS included an average of 6.84 ± 0.62 and 13.33 ± 0.44 taxa, respectively. As a rule, the species number in a zoocenosis increases with mussels' biomass. Accordingly, no more than nine invertebrates species are present in mussels' settlements with biomass between 0.1 and 2.6 kg m^{-2}, while the total number of macrozoobenthos species increases up to 20 in mussels zoocenosis with a biomass of 37.8 kg m^{-2}. In fouling of Odessa MP, at depths of 0–7 m, the mussels' biomass varies from 0.16 to 21.40 kg m^{-2} and the species number in mussels' zoocenosis changes from 10 to 21 in the various harbours [138].

The mussels' settlements in MPs of various seas are functioning in a similar way, maintaining relatively stable conditions close to them. For this reason, invasive alien species frequently acclimatize in MPs mussels' settlements or near them. In 2000–2001, the following old and new invasive alien species were found in the mussels zoocenosis of Odessa port and in the adjacent area of Odessa Gulf: the polychaetes *Polydora limicola* (*P. cornuta*) and *Mercierella enigmatica*, the oligochaete *Peloscolex (Tubificoides) benedeni*, the barnacles *B. improvisus, B. eburneus* and *B. amphitrite*, the crab *Rh. harrisi tridentata*, the gastropod mollusc *Rapana venosa*, the sea slugs *Doridella obscura* and *Ercolania funereal*, and the bivalves *Anadara inaequivalvis* and *M. arenaria* [9].

In MP aquatories, the distribution of nematodes differs between benthos and fouling where *Mytilus* plays the leading role. Thus, in Odessa MP, 24 nematodes species were registered on silts and 7 species in the fouling, but nematodes' quantity in the fouling (from 142,500 to 312,500 ind. m^{-2}) was one order of magnitude higher than in silts. Within mussels' settlements, the quantity of marine mites (Halacaridae) in 1997–2000 comprised 2250 ind. m^{-2} on the coast-protecting structures and 4140 ind. m^{-2} in the fouling of Odessa MP [9].

The marine mussel *M. lineatus* is found in fouling along all shores of the Black Sea and is usually the second most abundant species, after *Mytilus*, among the molluscs of the Mytilidae family. *M. lineatus* forms both monospecific settlements as well as colonies inside *Mytilus* settlements and has a competitive advantage in more desalinated zones.

Mytilaster reaches its sexual maturity in less than 1 year. In the region of Sevastopol city, in Kazachya Bay, sexually mature *Mytilaster* starts to appear among individuals with a shell length of 5.5–6.0 mm. All molluscs with lengths of 6.5 mm and above have reached their sexual maturity. In other Sevastopol harbours, the share of sexually mature individuals exceeded 50.0% in the size group of 7.0 mm and above [41]. *Mytilaster* life span is 5–7 years and fertilization is external as in *Mytilus*. Eggs develop into free-swimming larvae which, after some time, sink onto a substratum, crawl, attach with the byssus and switch to a mode of life similar to that of adult individuals. At the moment of sedimentation onto substratum, *Mytilaster* larvae have a length of about 0.4 mm and their colour depends on the composition of grazed algae. *Mytilaster* larvae do not metamorphose on silt, sand or HS not covered by a bacterial–algal film. Larvae specifically settle on filiform substrata, among which they prefer *Cystoseira* beds. In the Black Sea, *Mytilaster* larvae in plankton are numerous during the warm months. *Mytilaster* is a euryhaline species adapted to a salinity range of 7.0–30.0‰. It inhabits the water column from the surface layer to a depth of 20 m, although its most numerous settlements are usually located below the belt of macrophytes and barnacles (1–7 m).

In the coastal zone, *Mytilaster* forms clumps settlements (dense and loose) and, rarely, druses inside mussels "complex druses". *Mytilaster* juvenile individuals are concentrated inside clumps on the byssus filaments of adult individuals. The actual settlement type is determined by environmental conditions [41]. Dense clumps appear in zones often influenced by waves. In dense settlements (clumps), the number of *Mytilaster* mature individuals can range from a couple of thousands up to hundreds of thousands of specimen per 1 m^2 [122].

The Yuzhny (Southern) protective mole of Sevastopol Bay is populated by both *Mytilus* and *Mytilaster*. At the depths of 1.7 and 15 m, *Mytilaster* individuals had sizes of 1–20 mm. The molluscs quantity decreased with depth and the influence of roughness was not observable below depths of 4–5 m [121].

Settlements of *Mytilus* and *Mytilaster* are registered on various HS in Odessa MP, Sukhoy and Grigorievsky Limans. On the HTS of Odessa MP, *Mytilaster* was registered at depths of up to 7 m. The maximum age of molluscs was 5 years in the majority of samples, with 6-year-old individuals being present in only two samples. Two- and 1-year-old individuals were most abundant [115]. Studies have shown that there is no considerable antagonism between the young of the current year *Mytilus* and *Mytilaster* [41]. In many regions of the Black Sea, *Mytilaster* colonies inside *Mytilus* settlements constitute specific reserves, which maintain a rather constant presence of their larvae in plankton and allow the formation of new settlements at the onset of favourable conditions.

In highly hydrodynamic conditions, *Mytilaster* forms one-layer dense clumps, in which molluscs' density reaches 100,000–700,000 ind. m^{-2} depending on the size and age structure of the settlement [41]. In Odessa MP, average density and biomass of *Mytilaster* were 5594 ind. m^{-2} and 1624.0 g m^{-2}, respectively, in the 3 m horizon and 3988 ind. m^{-2} and 1333.0 g m^{-2} in the 7 m horizon [9]. In the 7 m horizon *Mytilaster* individuals were bigger than those at the depth of 3 m and near the water surface.

In Odessa MP, *Mytilus*, *Mytilaster* and barnacles comprise 99.5% of fouling animal biomass. At the depths of 3 m, where conditions are more stable, the qualitative composition of macrofauna was twofold richer than at the surface and the average density and biomass 10- and 243-fold higher, respectively [118].

In the absolute majority of cases, the quantity and biomass of fouling organisms is dozens of times higher than that of benthos on loose sediments [48]. In the NWBS, these differences can be over 400-fold [55].

Unlike *Mytilus*, *Mytilaster* does not bear well muddy and polluted waters. In particular, *Mytilaster* is sensitive to silting of settlements and its larvae do not attach to ships' hulls in muddy and polluted port aquatories. *Mytilaster* prefers zones with moderate water hydrodynamics and reaches the highest densities in zones sheltered from strong currents.

In MPs of the Azov Sea, *Mytilaster* is one of the main fouling organisms, with depths up to 6 m being the most favourable to its settlements. In the autumn period, the density and biomass of *Mytilaster* settlements in the Azov Sea reach 11,000 ind. m^{-2} at 660.0 g m^{-2}, respectively [122].

Mytilaster is resistant to oxygen deficit and to the presence of hydrogen sulphide. In anoxic environments and at hydrogen sulphide concentrations between 0.1 and 1.0 mg dm^{-3}, it can close its valves and stay alive for 18–19 days. If oxygen reappears in the water column during this time span, the mollusc opens its valves to replenish its oxygen reserve and can continue living for another 12–14 days in an anoxic environment [122].

Because of differences in shell shape, the force required (by e.g. wave action) to tear individuals of *Mytilus* from substrata is 2.4-fold stronger than for *Mytilaster* [41]. This difference is important for settlements on loose substrata such as limestone and coquina, while settlements of *Mytilus* and *Mytilaster* (in clumps) are equally resistant to wave action on concrete and metallic constructions.

Dense, single-layered *Mytilaster* and *Mytilus* settlements have greater resistance to wave action, than separate individuals, small molluscs' groups or druses. The damaging action of waves especially manifests near the water surface and along the settlements' edges and completely disappears with depth. Owing to the better wave resistance of isolated *Mytilus* individuals, and to their bigger size compared with *Mytilaster*, the latter usually forms colonies within *Mytilus* settlements, while the opposite situation is not observed.

Besides *Mytilus* and *Mytilaster*, the oyster *Ostrea edulis* and zebra mussels *Dreissena polymorpha* and *Dreissena bugensis* can be present in the composition of fouling biocenoses in MPs of the Black–Azov Sea basin. In the Black Sea, oysters are found in regions of high salinity such as the Crimean and Caucasian coasts, while zebra mussels inhabit brackish waters. Under favourable conditions, both oysters and zebra mussels can form dense monospecies settlements on HS of anthropogenic origin.

The oyster *O. edulis* "cements" to the substratum by growing with one of its valves tightly attached to it, so that it is very difficult to separate one organism from its substratum. Larvae develop from fertilized eggs. The development time from the egg until the moment of sinking and attachment lasts 13–16 days. Twenty hours after egg fertilization larvae are already covered by a shell, swim, and feed with the help of a ciliated organ, the velum. The development of pelagic larvae lasts 6–14 days. Veligers leave the female when reaching a size of 0.15–0.17 mm. Immediately before sinking, larvae develop a strong foot for crawling and secrete a sticky cement-like substance that helps them to attach to HS. After settling on a substratum, larvae crawl for some time to choose a place for attachment. After permanent adhesion to a substratum, the larval leg atrophies. Oysters are filter feeders and mainly consume detritus. In the Black Sea, they survive in anoxic conditions for up to 5 days and at hydrogen sulphide concentrations up to 5.6 mg dm^{-3}. The presence of relatively strong currents is important for oysters [122]. Because of overexploitation and pollution of coastal waters, the total reserves of oysters in the Black Sea have reduced catastrophically and their contribution to fouling is almost negligible in comparison with the first part of the twentieth century.

Zebra mussels resemble *Mytilus* and *Mytilaster* and, in the same way, attach to HS with the help of byssus filaments. Settlements of *D. bugensis* and freshwater *D. polymorpha* can form in MPs located in desalted regions of the Black–Azov Sea basin (e.g. the Dnieper-Bug Liman, Taganrog Bay).

Zebra mussels live up to 18–19 years. *Dreissena* inhabits at depths up to 10 m and its density and biomass reach 10,000 ind. m^{-2} and 7.0 kg m^{-2}, respectively. *Dreissena* reproduces during the warm season between April and October. Eggs are fertilized in water and develop into a veliger which leads planktonic life for 8 days. After this period, the veliger sinks onto a substratum and transforms into a tiny crawling bivalve with a relatively long leg. Young and adult individuals live attached to a substratum; however, if necessary, they can detach the byssus from the body and creep over distances of 10–11 cm per day using their leg. *Dreissena* can form big settlements, continuous layers and "druses", comprising hundreds individuals. *Dreissena* filters its food out of water and, depending on the habitat, feeds on phytoplankton and detritus. These molluscs prefer non-silted substrata and well-aerated zones. *Dreissena* dies at salinities above 10.0–11.0‰, but is able to survive in anoxic conditions for about 4–5 days at 18.0 °C. In winter, *Dreissena* falls into an anabiotic state [122].

In the last decades, the bivalve *M. arenaria* has been registered in fouling of MPs aquatories. It is a species that has been introduced relatively recently into the Black Sea (end of the 1960s—beginning of the 1970s) and that is present in small amounts in mussels' druses. *Mya* usually burrows deep within silty-sand sediments. However, *Mya* is a typical filter feeder, thanks to the presence of a siphon that is extended up to the sediments surface. Its free-swimming veliger settles on mussels' druses, macrophytes or artificial HS. Sunken young individuals are able to crawl with the help of a leg. *Mya* individuals living in fouling do not have the possibility to dig into the substratum. *Mytilus* individuals involuntary retain *Mya* in their settlements by attaching to their shells with byssus filaments. As a rule, *Mya* in HTS fouling composition is represented by small-size individuals with lengths up to 20 mm. Furthermore, part of the individuals have deformed shells.

In 2000–2002, a few specimen of the bivalves *Abra ovata* and *Cerastoderma glaucum* were registered in fouling biocenoses of Odessa MP. In the same way as *Mya*, these species are typical infaunal filter feeders of porous soft sediments. Their pelagic larvae having settled on *Mytilus* settlements are retained there by the growth of *Mytilus* byssus filaments.

The nudibranch sea slug *D. obscura*, another invasive species for the Black Sea, was discovered comparatively recently (end of the 1980s—beginning of the 90s) in *Mytilus* fouling of Odessa MP. Until now, it has been found near the Crimean shores, Odessa Gulf and Kerch Strait, as well as near Bulgarian and Romanian coasts [117]. Bryozoans are the only food of this nudibranch [149] which reaches its maximal development in semi-closed aquatories. This peculiarity explains why the quantities of *Doridella* are much higher in MPs than in the adjacent sea zones.

Doridella is a common species in the periphyton of Odessa MP and its average density and biomass are 50 ind. m^{-2} and 0.17 g m^{-2}, respectively. Maximum, occasional values of 1360 ind. m^{-2} and 4.32 g m^{-2} were found immediately below the water surface. However, typical average values below the water surface were 102 ind. m^{-2} for density and 0.34 g m^{-2} for biomass. The corresponding figures were 38 ind. m^{-2} and 0.17 g m^{-2} for the 3 m horizon and 10 ind. m^{-2} and 0.01 g m^{-2} for the 7 m horizon. Outside the port aquatory, the average quantity of *D. obscura* was 3 ind. m^{-2} for an average biomass of 0.01 g m^{-2} [9].

The tanaid *Tanais cavolini* (Anisopoda), constructing tubes and feeding on detritus and microalgae, was registered in Constanza [141] and Odessa MPs fouling [9].

Turbellarian worms, nemertines and polychaetes are observed in fouling of MPs located on various coasts of the Black Sea.

Turbellarians are a common taxon in the periphyton of Odessa MP. Their occurrence frequency comprised 100.0% at 27 stations. The average density and biomass of turbellarian worms were 289 ind. m^{-2} and 3.59 g m^{-2}, respectively. The highest average density (363 ind. m^{-2}) and biomass (4.39 g m^{-2}) are observed at depths around 3 m. The maximal density observed in one sample was 1220 ind. m^{-2} and the corresponding biomass 21.80 g m^{-2}. The average density (16 ind. m^{-2}) and biomass (0.22 g m^{-2}) of turbellarians in the benthos of the port aquatory were much lower than in fouling [9].

In Odessa MP periphyton, nemertines are a usual taxon (100.0% of occurrence frequency), although they are not numerous with an average density and biomass of 16 ind. m^{-2} and 0.11 g m^{-2}, respectively. The highest average density (23 ind. m^{-2}) and biomass (0.16 g m^{-2}) were registered near the surface horizon. In the port benthos, the indices of nemertines quantitative development were one order of magnitude lower: average density—2 ind. m^{-2} and average biomass—0.02 g m^{-2} [9].

Free-living polychaetes species occur in fouling of the Black Sea MPs along with species having a chitinous or lime exoskeleton in the form of straight or spiral tubes. Many species use mussels' shells as substrata. Representatives of the genus *Spirorbis* settle both on HS and on phyto- and zoofouling. The polychaetes *N. succinea* and *P. cornuta* were found in the periphyton of Odessa MP [9]. Mass numbers of *P. cornuta* and *Nereis zonata* were found in the fouling of Constanza MP [141].

The polychaetes *Vermiliopsis infundibulum*, *M. enigmatica* and *P. triqueter* played a significant role in the fouling biocenoses of Poti MP and Batumi Bay [15, 93, 124].

Some polychaetes species inhabiting MPs mussel fouling are able to live in nearly anaerobic conditions. Thus, representatives of the family Capitellidae survive for a long time in environments containing hydrogen sulphide and *Hediste*

(Nereis) diversicolor and *Harmothoe imbricata* are able to withstand the absence of oxygen for up to 7 days [18, 71]. In Odessa MP periphyton, detritophages (*N. succinea*, *P. cornuta*, *Heteromastus filiformis*) dominate among polychaetes in terms of quantity and biomass. Predator species (*H. imbricata*, *Harmothoe reticulate* and *Exogone gemmifera*) occur in far smaller amounts, while sestonophages (*Fabricia sabella*, *M. enigmatica*) and phyto-detritophages (*Platynereis dumerilii*) are represented by isolated individuals.

Table 3.1 presents the qualitative composition of macrofauna in HS fouling community of Odessa MP. Regardless of the region under study, differences in species composition and in the intensity of fouling development are regularly observed in the vertical distribution of fouling [15].

In established fouling MP cenoses having reached a steady state, inhabit edificatory species capable of switching to anaerobic metabolism and of surviving in hypoxic and anoxic conditions, as well as in the presence of hydrogen sulphide, for at least a couple of days.

Numerous authors note that, in the case of barnacles larvae, the selection of optimal substratum is stimulated by the presence, or traces, of previously settled barnacles [13, 90, 124]. When fixing to HS, the larvae of *Mytilus* and *Mytilaster* prefer byssus filaments of adult molluscs and creep into the inner parts of existing settlements [41, 100]. However, the passage from the stage of larva to the juvenile one has a random character for the majority of benthic forms despite the preference of some species for some specific substrata and the ability of some others to slow down their sinking toward the bottom [15].

In MPs aquatories, the ability of sessile animals larvae to inhabit both natural and artificial HS increases the ecosystems bioproduction and eutrophication level.

A significant part of the larvae produced in ports either sinks onto a suitable substratum or dies on the bottom of the aquatory or following grazing by other hydrobionts. These phenomena occur because of the enclosed nature of MPs, the weakened water exchange, the restricted possibility for bottom invertebrates pelagic larvae to escape from the periphyton composition, the presence of adaptations for buoyancy and vertical displacement and the nearly total absence of adaptations for active horizontal displacement. These phenomena collectively preserve, to a certain extent, the foulers populations inhabiting MPs aquatories. However, renewal of existing settlements and formation of new ones by invasive species can be supported by survival of very small fractions of their gametes (hundredth or thousandth of the total).

In the Black–Azov Sea basin, functioning of fouling cenoses is determined by attached (sessile) species regardless of their developmental stage. The mobile (vagile) forms do not have a significant influence on the interaction processes within the community [15]. However, mobile species do ensure a more complete use of food resources in fouling cenoses and thus maintain favourable hydrochemical conditions.

In MPs, the adult stages of all dominant species in the communities of epifaunal filter feeders are either tightly attached to HS or bottom dwellers. The mass filter-feeding organisms of the fouling produce faeces and pseudofaeces that

Table 3.1 Qualitative composition of the macrofauna in the fouling community of Odessa MP hydrotechnical constructions [9]

Taxonomic group	Taxa
Porifera	*Suberites carnosus, S. prototipus,* Spongia g. sp.
Coelenterata	*Obelia longissima, O. exigua, Actinia equina, Actinothoe clavata*
Turbellaria	Turbellaria g. sp.
Nemertini	Nemertini g. sp.
Nematoda	*Paralinhomoeus filiformis, Cylindrotheristus maeoticus, Prochromadorella mediterranea, Cyatholaimus gracilis, Cyatholaimus* sp., *Viscosia glabra, V. minor, Oncholaimus dujardini, Oncholaimus* sp., *O. compilocercoides*
Polychaeta	*Harmothoe imbricata, H. reticulata, Exogone gemmifera, Neanthes succinea, Platynereis dumerilii, Polydora cornuta, Heteromastus filiformis, Fabricia sabella, Mercierella enigmatica*
Oligochaeta	Oligochaeta g. sp., *Peloscolex (Tubificoides) benedeni*
Bryozoa	*Conopeum seurati, Electra pilosa, Lepralia pallasiana, Bowerbankia imbricata, B. gracilis, B. caudata*
Gastropoda	*Mohrensternia lineolata, Setia valvatoides, Hydrobia acuta, Doridella obscura*
Bivalvia	*Anadara inaequivalvis, Mytilaster lineatus, Mytilus galloprovincialis, M. edulis, M. trossulus, Cerastoderma glaucum, C. lamarcki lamarcki, Abra ovata, Mya arenaria*
Cirripedia	*Balanus improvisus, B. eburneus*
Decapoda	*Athanas nitescens, Palaemon elegans, P. adspersus, Pisidia longimana, Carcinus mediterraneus, Pilumnus hirtellus, Xantho poressa, Rhithropanopeus harrisi tridentata, Pachygrapsus marmoratus*
Mysidacea	*Mesopodopsis slabberi*
Cumacea	*Iphinoe tenella, Cumella limicola*
Anisopoda	*Tanais cavolini*
Isopoda	*Sphaeroma pulchellum, Idotea baltica basteri*
Amphipoda	*Stenothoe monoculoides, Gammarus subtypicus, G. insensibilis, Marinogammarus olivii, Melita palmata, Dexamine spinosa, Hyale pontica, H. perieri, Microdeutopus gryllotalpa, Amphithoe vaillanti, Corophium bonelli*
Halacaridae	*Copidognatus brachiostomus, C. magnipalpus, Rhombognathus tonops, Agaue chevreuxi, Rhombognathides pascens*
Tunicata	*Eugyra adriatica, Molgula euprocta, Botryllus schlosseri*
Insecta, Diptera larvae	*Thalassomyia frauenfeldi*

eventually lead to the formation of black silts enriched in OM. A constant accumulation of detritus and silt takes place at the foundations of HTS where molluscs' shells, barnacle houses and exo- and endoskeletons of different hydrobionts accumulate. In MPs aquatories, the highest values of live and dead OM are found in the fouling of artificial HS and in bottom sediments, respectively.

MPs aquatories are ecosystems characterized by an increased production of SOM, high amounts of HS and intensive sediment accumulation. Because of these characteristics, communities dominated by epifaunal filter feeders develop in zones with a sufficient hydrodynamic activity. In such communities, the epifauna representatives comprise up to 84.0% of the total biomass, the infauna—2.0%, the collectors—2.0% and the deposit feeders—4.0% [63]. Detritophages and deposit feeders acquire a leading role in zones with lowered hydrodynamics where sediment accumulation occurs [40, 63]. In MPs, these zones are located at 3.0–5.0 m distance from HTS.

Barnacles and mussels are the most important species in MP fouling communities. Both live in conditions of sufficient hydrodynamics and are filter feeders, but have a different apparatus for seizing food. At sufficient water velocity, barnacles simply put out and orientate their thoracic legs towards the movement of the flow, periodically eliminating the captured suspension. The ecological efficiency of barnacles *B. improvisus* (K_1) is quite high (about 30.0%) compared with other invertebrate species whose K_1 is 5.0–15.0%. This peculiarity indicates that barnacles play a key role in energy transformation in marine ecosystems [64].

Mussels direct water flows inside the mantle cavity by changing the opening of shell valves and the alignment of the ruffled edges of the siphons. Too strong flows reduce the use of thoracic legs by barnacles, so that suspension is washed away from them while the thoracic legs are kept inside the shell. In the same conditions, mussels leave only a narrow opening between shell valves, thus decreasing the water flow inside the mantle cavity. At low hydrodynamics, barnacles must actively move their food-catching apparatus (cirri), while mussels must create a water flow through the mantle cavity, both of which require additional energy [77].

The larval stages of barnacles, bivalves and polychaetes from the fouling cenoses mainly feed on unicellular pico- and nanoalgae (with sizes of 3.0–5.0 μm), bacteria, SOM and DOM [62, 144–146].

In MPs, the juveniles of molluscs foulers developing on artificial HS are characterized by a metabolism higher than that of their benthic variants inhabiting on bottom. This difference is directly connected with the link between intensity of the energetic metabolic processes and growth. On artificial HS, animals develop faster because of the more favourable life conditions. Such conditions have less influence on metabolism intensity of adult molluscs [120].

Epifauna filter feeders of a mature fouling community consume SOM 160 times more intensively than epiphyton [2, 4].

It was determined that cenoses of HS fouling and of benthos of loose sediments react in opposite ways to eutrophication. A significant increase in the quantity and biomass of fouling organisms takes place at high phytoplankton production and mud accumulation on the bottom. This situation, in turn, favours the increase of benthos total reproductive potential as it is documented in MP aquatories and in the NWBS in general. In this part of the sea, the quantity of meroplankton in the coastal zone has increased between 1973 and 1991 due to the spawning activity of the sessile organisms community. The spawning activity increased by threefold for bivalves, 15-fold for gastropods, 17-fold for barnacles and 29-fold for polychaetes. Before eutrophication of the NWBS, the reproductive output of fouling was 3.5–11.0-fold higher than that of the benthos living on soft sediments and occupying the main area of the shelf. At present, this difference exceeds 21-fold [2, 4].

Mass edificator species of fouling cenoses and the whole cenoses themselves are not only influenced by the environmental conditions present in the near-wall water layer of MPs, but also actively contribute to create conditions suitable for long-term and stable functioning of the entire community. Extensive, vertical settlements of epifauna filter feeders:

1. brighten the near-wall water layer;
2. create microflows that remove their metabolites from the environment immediately surrounding them;
3. enrich the near-wall layer in biogenic matter, DOM and SOM.

Conditions for the intensive development of micro- and macrophytes are thus created in the near-wall water layer. As a rule, in MP fouling cenoses, animal settlements are located under the macrophytes belt and not the other way round.

Owing to constant contact and interaction with the fouling community, the near-wall water in the periphytal zone, the near-bottom overlying and interstitial waters, and the water layer under the film of surface tension have different properties compared with the main water masses of MPs.

Periphyton organisms produce vitamin B_{12} [28], ferments, hormones, antibiotics and other biologically active substances. Blooms of phyto- and bacterioplankton occur in zones of bivalves mass development and change the gas content and pH of these zones. Mussels' settlements feed on phyto- and bacterioplankton, but the mussels' metabolic processes favour further development of these planktonic organisms, thus intensifying water eutrophication [29]. Owing to their reproduction via pelagic larval stages, fouling cenoses are one of the main sources of meroplankton replenishment in the pelagial zone. At the same time, the holoplanktonic forms [4, 135] and fish at different stages of development concentrate near the fouling cenoses [16, 115, 139, 150].

In dense fouling settlements, organisms do not behave only as separate structural units, but their presence as a whole creates various and useful group effects and characteristics. Formation of the primary bacterial and algal slime film, covering all surfaces of artificial HS, is an obligatory step in periphyton formation.

Owing to tentacles movements and to the presence of stinging cells, layers of hydroids collectively protect themselves and prevent the penetration of individuals belonging to other species into their settlement. Being attached to the lower part of their body, hydroids can not only bend and wriggle according to water flows, but can also bend randomly, thus increasing the water volume available for catching preys. The tentacles of the entire hydroids' settlements thus become a huge weapon for fishing.

After mass sinking, barnacles tightly attach to HS and to each other. They subsequently react to wave action as a single, solid and well-streamlined formation.

More or less synchronized movements of thoracic legs transform barnacle settlements into a big web for capturing suspended nutritive particles. At the end of the mass development stage, the presence of empty shells and traces of their previous attachment to a substratum stimulate settling of barnacles and other hydrobiont species in such places [15, 43, 90].

In big settlements of barnacles, bivalves, bryozoans and hydroids, the common activity of all organisms puts large water masses into motion in different directions, thus facilitating ventilation of the near-wall water layer, its enrichment by oxygen and pH stability. With respect to this phenomenon, all organisms are functioning as a single entity.

All dense settlements of foulers (barnacles, bivalves, bryozoans and hydroids) increase the survival chances of an individual organism by protecting them from grazing, especially when they are still part of the meiobenthos in terms of size. Excluding some situations of *R. venosa* mass attack, grazing occurs, if at all, only at the settlements' periphery.

For MPs sessile fouling organisms attached to a substratum, the water flows with respect to their settlements are of vital importance, because they are the equivalent of the movements of animals themselves in the aquatic environment.

Resistance of fouling organisms towards wave action is achieved in various ways:

1. macrophytic algae and hydroids, being attached by their body end to the substratum, easily bend and twist according to water flows;
2. sessile organisms (barnacles, oysters and polychaetes) attach firmly to a substratum with the help of cementing agents and remain immobile;
3. some sessile species (e.g. mussels) have byssus filaments and are able to change their position relatively to the substratum under certain conditions;
4. a number of species attach to substrata only temporarily by using suckers and different adhesive substances.

At the strengthening of roughness, some molluscs living in settlements react by changing the distance between their shell valves, thus synchronously responding to the changing conditions. All other manifestations of their life activity also change in the same synchronous way.

The strength of bivalves and barnacles soft tissues is much less than that of their shells. For this reason, strengthening of sea roughness above certain dangerous levels leads to shells closing. The feeding and growing conditions of molluscs and barnacles therefore worsen during periods of strong wave action. Strong roughness also resuspends silty particles and sand grains in the water column, which also negatively influences the animals' respiratory apparatus. On HTS, the zones of foulers' delayed growth are confined to the near-surface horizon and to some other zones particularly exposed to wave hazard.

All internal angles present on MPs HTS surfaces and with edges of more than 4–5 cm are usually colonized by molluscs foulers forming simple and complex druses. On HTS external surfaces, in conditions of much stronger and more frequent wave action, molluscs' accumulations in the form of druses are preserved in fissures and big cavities. On the surfaces adjacent to them, molluscs are usually distributed in one continuous layer and druses can appear only at depths of more than 3 m.

The process of fastening one layer of molluscs foulers onto another one has its natural limits, which are determined by the strength of byssus filaments in calm water and by their resistance to rupture during storms. Animals in the lower layer could also die of food or oxygen deficiency. Under the influence of gravity or wave action, such dead individuals could detach from the substratum along with a whole mass of dead and alive foulers. Detachment of the first layer of molluscs thus uncovers parts of artificial HS which can be readily populated by sinking juvenile individuals of *Mytilus* and *Mytilaster*. A permanent "rejuvenation" of molluscs' old settlements goes on in this way.

For HS mass foulers, it is important to estimate the resistance towards hydrodynamic influences not only of single individuals, but of the organisms' groups and settlements as a whole. Mussels individually attached to a vertical surface by their flexible byssus filaments are affected by gravity, directed downwards, and by lateral flows bending them in various ways. Similar changes of position relative to the substratum and to other individuals are limited or absent for mussels growing in dense settlements. In the first case, separate individuals are pushed out from their layer or their shells are deformed.

Field studies showed that the settlements most resistant to wave action are the dense, continuous solid layers formed by juveniles of *Mytilus* and *Mytilaster* and raising 0.5–0.7 cm over the substrata. Such settlements are usually free of epiphytes. The less wave-resistant settlements are those consisting in "loose druses" of *Mytilus* comprising molluscs of a couple of years of age and about 2 cm long. "Loose druses" are usually confined to harbours' zones with limited water exchange at depths above 3–4 m. These molluscs are frequently covered by epiphytes.

Sessile and vagile forms of meio- and macrozoobenthos representatives are found in MP fouling. Macrozoobenthos larvae sinking onto a substratum spend their meiobenthos dimensional life period in periphyton. These larvae belong to the pseudomeiobenthos cenoses and pass into macrozoobenthos relatively quickly. Species ascribed to eumeiobenthos are also present in MP fouling and include Turbellaria, Nemertini, Nematoda, Polychaeta, Harpacticoida, Halacaridae and others. Species with pelagic larvae are almost totally absent from the eumeiobenthos composition of periphyton. In general, these are crawling organisms and, as a rule, their development takes place in fouling.

In MP aquatories, meiobenthos forms constitute the basis of populations in silts and silts with coquina. At the bottom, close to HTS foundation, the meiobenthos biotope constantly renews owing to the steady flux of detritus and dead molluscs' shells from the fouling cenosis.

When inhabiting in the fouling of MPs HTS, hydrobionts with heavy shells, skeleton or testa fall onto the bottom, which is covered with black semi-liquid silt. Here, they either immediately drown in it or gradually die because of oxygen deficit, the presence of hydrogen sulphide and low pH values [30].

At the basis of HTS foundations, ecotone zones hosting inhabitants of the periphytal and benthal zones originate in places with sufficient dissolved oxygen content. An extension of the fouling cenosis sometimes forms on the bottom at a distance of up to couple of metres from the concrete HTS surface.

It is well known that polychaetes play an important role in both benthos of soft sediments and fouling of HTS MP. In 2001, the ecosystem of Odessa MP aquatory was studied in the framework of international project "GloBallast". In particular, a comparative analysis was carried out with regard to species composition and quantitative indices of polychaetes in the fouling and benthos of Odessa MP [73]. The biodiversity of polychaetes was studied using various indices (Simpson, Shannon and Pielou) [72].

Nineteen polychaetes species were found in Odessa MP aquatory, with the number of species in benthos being nearly twice than in the fouling (17 and 9, respectively). However, a high number of species does not always translate into a high ecological diversity. This information can be obtained by applying an index approach. The basic data used for calculating the diversity indices (in terms of both quantity and biomass) of polychaetes in fouling and in benthos are presented in Table 3.2.

Polychaetes species diversity in the benthos of Odessa MP was much higher than in HTS fouling, especially in terms of biomass. Furthermore, diversity indices calculated on the basis of density were somewhat, or even several folds, higher than the corresponding ones calculated on the basis of biomass for both benthos and fouling (Table 3.3). On the contrary, the Berger–Parker dominance index was significantly higher in fouling (0.811) than in benthos (0.377).

Besides species diversity, the ecological diversity of polychaetes was also studied in the fouling and in the benthos of the port. Ecological diversity was determined as the diversity of living forms, the ratio between species number of different living forms and their relative abundance in the community [67]. Five main living forms had been previously distinguished among the Black Sea polychaetes [68], but mass species such as *H. diversicolor*, *N. succinea* and *P. cornuta* and some rarer species were not considered. Such rare species, that could also inhabit on porous and hard substrata, were ascribed to intra-epibionts life form.

All six main living forms of polychaetes were present in Odessa MP aquatory. According to the number of species, the vagile epibionts and intra-epibionts dominated in the fouling, while fossorial intrabionts polyphages and intra-epibionts were the dominant forms in the benthos. In terms of density and biomass, intra-epibionts were dominant in both fouling and benthos (Table 3.4).

The diversity of polychaetes life forms is higher in benthos than in fouling. This difference is caused by the fact that life forms of intrabionts are nearly completely absent in the fouling biotope. On the other hand, all life forms of intrabionts, as well as epibionths, occur in benthos of loose sediments (occupying nearly all the port aquatory) owing to the presence of the admixture of coquina near moles and breakwaters.

Indices of polychaetes ecological diversity were calculated using the average biomass and density of species complexes belonging to the various living forms (Table 3.4).

Table 3.2 Species composition, quantitative indexes (*N*—average quantity, ind. m^{-2}, *B*—average biomass, g m^{-2}, *P*—frequency of occurrence, %) and ecological characteristics of polychaetes in the fouling and benthos of Odessa MP in 2001

Species	Fouling			Benthos			Life form[a]
	N	*B*	*P*	*N*	*B*	*P*	
Phyllodoce mucosa Oersted	–	–	–	0.1	0.001	0.8	IV
Eteone picta Quatrefages	–	–	–	0.1	0.001	0.8	IV
Harmothoe imbricata (L.)	26.9	0.271	49.1	47.9	0.344	56.1	I
H. reticulata Claparede	2.1	0.011	7.7	0.4	0.001	4.1	I
Exogone gemmifera Pagenstecheri	0.1	0.001	0.5	0.4	0.001	1.6	VI
Hediste diversicolor (Muller)	–	–	–	7.4	0.389	8.1	VI
Neanthes succinea (Frey et Leuch)	1329.0	20.129	100.0	173.9	3.706	82.9	VI
Platynereis dumerilii (Audouin et M.-Edwards)	10.5	0.102	33.8	4.6	0.087	4.9	I
Nephtys hombergii Savigny	–	–	–	0.3	0.046	1.6	IV
Nephtys cirrosa Ehlers	–	–	–	0.9	0.088	4.9	IV
Spio filicornis (O.F. Muller)	–	–	–	16.5	0.012	43.1	III
Polydora cornuta Bosc	262.0	0.094	76.6	284.8	0.134	81.3	VI
Prionospio cirrifera Wiren	–	–	–	1.2	0.001	5.7	III
Heteromastus filiformis Claparede	0.4	0.001	2.7	210.7	0.982	30.1	V
Capitella capitata (Fabricius)	–	–	–	4.6	0.008	17.9	V
Capitomastus minimus (Langerhans)	–	–	–	0.1	0.001	0.8	V
Melinna palmata Grube	–	–	–	0.2	0.002	2.4	III
Fabricia sabella (Ehrenberg)	0.3	0.001	2.3	–	–	–	II
Mercierella enigmatica Fauvel	0.1	0.001	0.9	–	–	–	II
Total	1631.4	20.611	–	754.1	5.804	–	–

[a]I—vagile epibionts, II—sessile epibionts, III—semi-vagile intrabionts, IV—fossorial—polyphagous intrabionts, V—fossorial—detritophages intrabionts, VI—intra-epibionts

Table 3.3 Indexes of polychaetes species diversity in the fouling and benthos of Odessa MP in 2001

Index	Fouling					Benthos				
	d_1	*D*	*H* (ln)	*H* ($\log_2$)	*e*	d_1	*D*	*H* (ln)	*H* ($\log_2$)	*e*
Quantity	1.088	0.310	0.574	0.828	0.261	2.415	0.721	1.464	2.112	0.516
Biomass	–	0.043	0.136	0.197	0.062	–	0.554	1.221	1.762	0.431

Margalef—d_1, Simpson—*D*, Shannon—*H*, Pielou—*e*

Diversity indices for polychaetes living forms are nearly always one order of magnitude higher in the benthos than in the fouling. However, indices based on density and biomass show only small differences in both fouling and benthos (Table 3.5). The Berger–Parker dominance index was 0.975 in fouling and 0.618 in benthos. It was established that polychaetes diversity indices in benthos of

Table 3.4 Composition and quantitative indexes (*n*—species number, *N*—average quantity, ind. m^{-2}, *B*—average biomass, g m^{-2}) of polychaete living forms in fouling and benthos of Odessa MP in 2001

Living forms	Fouling			Benthos		
	n	*N*	*B*	*n*	*N*	*B*
Vagile epibionts	3	39.5	0.384	3	52.9	0.432
Sessile epibionts	2	0.4	0.002	–	–	–
Restrictedly semi-vagile intrabionts	–	–	–	3	17.9	0.015
Burrowing intrabionts, polyphages	–	–	–	4	1.4	0.136
Burrowing intrabionts, detritophages	1	0.4	0.001	3	215.4	0.991
Intra-epibionts	3	1591.1	20.224	4	466.5	4.230
Total	9	1631.4	20.611	17	754.1	5.804

Table 3.5 Polychaetes indices of ecological diversity in fouling and benthos of Odessa port in 2001 (Simpson—*D*, Shannon—*H*, Pielou—*e*)

Index	Fouling				Benthos			
	D	*H* (ln)	*H* (log_2)	*e*	*D*	*H* (ln)	*H* (log_2)	*e*
Quantity	0.058	0.118	0.171	0.085	0.544	0.941	1.359	0.585
Biomass	0.037	0.094	0.135	0.068	0.435	0.829	1.196	0.515

Odessa MP are much higher than in fouling and their species diversity is higher than ecological diversity (Tables 3.3 and 3.5). Furthermore, in the port aquatory, polychaetes have a higher species diversity in terms of density than of biomass.

The r-strategist *N. succinea* quantitatively dominates, especially in HTS fouling where it comprises 81.5% of the average quantity and 97.7% of average biomass of all polychaetes. In the benthos, this species dominates only in terms of biomass (63.9%), while *P. cornuta* and *H. filiformis* dominate in terms of quantity (37.8 and 27.9% of total average, respectively). The latter species are tolerant with respect to pollution [69, 70]. *N. succinea* is the third most quantitatively abundant species representing 23.1% of the total average. Mass development of these three species and their dominance in the polychaetes taxocene testify about the presence of organic pollution in the aquatory of Odessa MP.

Very low indices of polychaetes species diversity (less than 0.05 according to Simpson and less than 1.0 according to Shannon) are registered only in HTS fouling. Simpson indices of about 0.05–0.20 are characteristic of ecosystems under stress [129] and Shannon indices below 1.0 indicate an unfavourable state of marine bottom sediments [142]. Simpson indices of species diversity for polychaetes living in port benthos (0.72 on quantity and 0.55 on biomass) are comparable with those for the benthos of Odessa region outside the port, where, in 1998, their numerical values were 0.73 on coquina and 0.63 on silt [69, 70].

Being considered as a proxy of a species "significance", the density of a species is usually employed for calculating diversity indices. However, the use of other indices of "significance", e.g. biomass and productivity, is not excluded [94]. It is known that biomass is related to the energy flow in communities. When calculating diversity indices on the basis of density, individuals of a given species in the community are assumed to be comparable with regard to size and metabolism [129]. Nonetheless, in the Black Sea, the size of individuals of various species within the polychaetes taxocen ranges from 1–2 up to 100 mm. As a consequence, the small-sized detritophagous polychaetes (species of the families Spionidae and Capitellidae) give only a tiny contribution in terms of biomass even during their mass development in the presence of eutrophic conditions. For this reason, the role of these polychaetes in a community is largely overestimated when density is used to estimate the "importance" of their species. Indices of diversity calculated on the basis of biomass are therefore more informative than those on the basis of density, at least in some cases. In favour of this approach, it can be observed that the indices of polychaetes species diversity calculated using biomass are very close to the indices of their ecological diversity calculated using biomass and quantity.

Attention was previously drawn to the big difference in the functioning conditions of fouling cenoses on the internal and external surfaces of MPs HTS.

The fouling of the external vertical concrete surfaces of Karantinniy and Reydovy moles of Odessa MP was studied in July 2006. Representatives of 45 macrofauna taxa of euryhaline marine complex were registered, including 9 worms, 12 molluscs, 21 crustaceans, 2 coelenterates and 1 chironomid [118]. Apart from 6 species occurring in low numbers, all observed hydrobionts had been previously registered in the fouling of Odessa MP HTS [9]. The average density and biomass of the macrofauna were 31,396 ind. m^{-2} and 8303.9 g m^{-2}, respectively.

Three epifaunal species of sessile filter feeders accounted for 80.6% of the quantity and 99.5% of the biomass: *M. galloprovincialis*, *M. lineatus* and *B. improvisus*. *Mytilus* was the dominant species of the fouling cenoses with an average density and biomass of 11,243 ind. m^{-2} (35.8%) and 7619.2 g m^{-2} (91.8%), respectively. A high-density value (11,648 ind. m^{-2}) was also registered for *Mytilaster*.

Crustaceans were the prevailing group in terms of taxa number (21), while molluscs were predominant with respect to density (73.2%) and biomass (97.9%). In terms of quantity and biomass, macrofaunal invertebrates with a pelagic stage significantly dominated over species exhibiting hypogenesis (i.e. direct development). Among the six observed trophic groups, detritophages were dominant in terms of taxa number (15), while filter feeders represented 80.7% of quantity and 99.5% of biomass. The index of uniformity of the food structure was 0.99.

Representatives of epifauna were dominating according to taxa number (30), quantity (99.1%) and biomass (99.9%). Vagile forms had the highest number of taxa (40), but sessile forms represented 80.6% of total organism quantity and 99.5% of the corresponding biomass.

Six species of self-acclimatizing organisms were registered (*P. cornuta, D. obscura, A. inaequivalvis, M. arenaria, B. improvisus* and *Rh. harrisi tridentata*), comprising in total 7.8% of organisms' quantity and 1.6% of their biomass. With the exception of crustaceans, these species appeared quite recently (1962–1989) in the Black Sea.

Table 3.6 Comparative characteristics of the quantitative indices of macrofauna fouling of Karantinniy and Reydovy piers of Odessa marine port for different depths in 2006

Parameter	Depth, m		
	0	1.5	3.0
Taxa number, total	20	19	38
Including worms	2	4	8
• Molluscs	3	2	10
• Crustacean	14	13	19
• Other groups	1	–	1
Average quantity of meiofauna, ind m^{-2}	7022	8687	71,923
Including *Mytilus galloprovincialis*	293	3167	30,053
• *Mytilaster lineatus*	105	3210	27,336
• *Balanus improvisus*	454	543	6197
Average biomass of macrofauna, g m^{-2}	88.63	2611.95	21,533.50
Including *Mytilus galloprovincialis*	12.58	2426.88	20,034.88
• *Mytilaster lineatus*	5.72	139.24	1114.13
• *Balanus improvisus*	11.94	33.90	337.91
Biomass of forage benthos, g m^{-2}	88.63	444.19	2617.00
Biomass of forage benthos, % from total	100.00	17.01	12.15
Trophic groups, quantity, %			
• Sestonophages	12.14	79.66	88.50
• Detritophages	9.23	7.22	7.44
• Phyto-detritophages	68.34	6.76	3.56
• Other groups	10.29	6.36	0.5
Trophic groups, biomass, %			
• Sestonophages	34.13	99.54	99.79
• Detritophages	4.10	0.05	0.07
• Phyto-detritophages	34.47	0.11	0.06
• Other groups	27.3	0.30	0.08
Index of uniformity of food structure	0.14	0.99	0.99

With the exception of coelenterates and mussels with shell lengths above 20 mm, all hydrobionts belong to the forage fauna component for fish [11]. The average biomass of forage benthos comprised 1150.7 g m^{-2}, which amounted to 13.9% of total. Molluscs, especially *Mytilus*, formed the bulk (85.1%) of the forage component biomass. Molluscs with length less than 20 mm represented only a minor part (466 g m^{-2} or 6.1%) of the total average biomass of *Mytilus* (7619.2 g m^{-2}).

The indices of macrofauna quantitative development varied depending on water depth (Table 3.6).

Crustaceans formed the bulk of macrofauna quantity (94.1%) and biomass (79.2%) immediately below the water surface. At the depths of 1.5 and 3.0 m, this role belonged to *Mytilus*, *Mytilaster* and barnacles which, overall, represented 79.7–88.4% of quantity and 99.5–99.8% of biomass.

The average biomass of macrofauna and of its forage component sharply increased between the 0 and 3 m depth. At the opposite, the share of the forage component (expressed as a percentage of the total macrofauna biomass) decreased from 100.0% on the 0 m depth to 12.2% on the 3.0 m horizon (Table 3.6). This change reflects the increase of large-sized individuals in the mussels population of the 3 m horizon.

Molluscs accounted for most of the forage component biomass (89.7 and 85.4%) on the 1.5 and 3.0 m depths, while this role belonged to crustaceans (79.2% of forage biomass) on the 0 m depths.

Changes in the quantitative indices of macrofauna mass foulers are mainly determined by hydrodynamics, which depends on depth, and by their feeding spectrum. The quantity and biomass of vagile detritophages and sessile filter feeders increase one- to twofold between 0 and 3 m. On the contrary, the quantity and biomass of phyto-detritophages and phytophagous forms, represented by crustaceans, reduce by one- to twofold between 0 and 3 m, which is explained by the mass development of macrophytes at the water surface (Fig. 3.3).

In the end, the overall state of the macrofauna fouling the external concrete vertical surfaces (0–3 m) of Karantinny and Reid piers of Odessa MP in July 2006 could be considered as satisfactory.

The total extension of MP HTS in the Black–Azov Sea basin exceeds 300 km and their total area is over 2×10^6 m^2. The periphytal of MPs aquatories located at flat shores constitute outposts and oases of rocky life among uniform, loose, coastal bottom sediments. The MP periphytal, especially the fouling biocenosis formed by *Mytilus*, usually provides shelter to the first generation of auto-acclimatizants in newly colonized habitats.

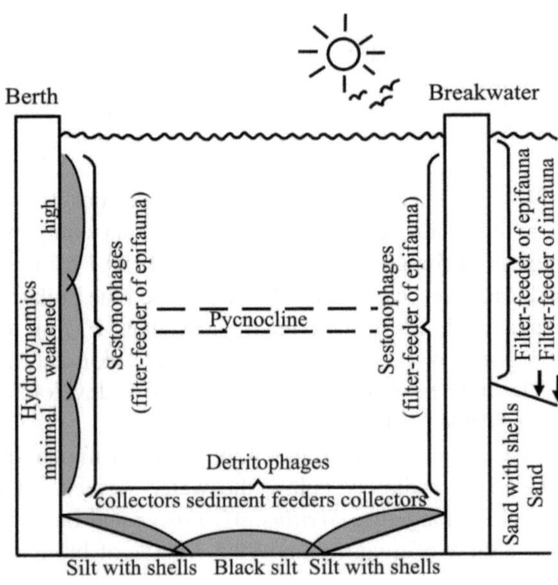

Fig. 3.3 Links between trophic groups of benthic invertebrates and hydrodynamics in marine ports

In theory, mechanical withdrawal of fouling from MPs aquatories and, ideally, its further use could be an effective method for bioproduction control. However, the problem remains difficult to solve in practice.

The simplest methods for limiting periphyton production in MPs could be:

1. reduction of illumination;
2. use of material less prone to fouling than concrete and steel, and of non-fouling, ecologically safe materials and coatings;
3. construction of hydrotechnical structures with the smallest area of underwater surfaces.

As already mentioned, *Mytilus* and *Mytilaster* settlements quickly clarify MPs waters owing to their filtration activity which, in turn, accelerates the sedimentation of suspended material. This process is often presented as an example of "self-cleaning" of marine ecosystems. The term "self-cleaning" is often used to indicate the combination of physical, chemical and biological processes leading to elimination of toxic substances and excess nutrients. However, a true "self-cleaning" of a given environment is achieved only by chemical and microbiological transformation of contaminating substances into the final non-toxic products. Processes of mass transfer and redistribution of contaminating substances from the water column towards bottom sediments only create an illusion of "self-cleaning". Furthermore, diffusion, redistribution among biotic components and export of contaminants into the adjacent ecosystems can also occur [7].

Products of *Mytilus* and *Mytilaster* metabolism serve as "fertilizer" and stimulate plants' growth, which again produces OM and creates an environment for the development of bacteria and protozoans. Molluscs filtrators concentrate bacterial cells, including coliform bacteria [7]. *Mytilus* settlements favour the formation of a thick silt layer on the bottom. This layer contains increased concentrations of OM whose degradation under anaerobic conditions leads to the formation of hydrogen sulphide. Properly speaking, the process of "self-cleaning" of the MPs water column by mussels actually is a mechanism of "self-pollution" for the MPs benthal zone.

Hard substrata, artificially introduced in MPs aquatories in the form of gravel, rubble stone, irregularly dumped concrete blocks and metallic constructions, initially fulfil the functions of a periphytal zone. Such substrata can become partially or entirely covered by a detritus layer with time. Cavities of various sizes (where detritus, faecal particles, algal fragments, molluscs valves, etc. also accumulate) form on vertical surfaces fouled by macrophytes and mussels druses. The periphytal can thus become suitable for the settlement of typical benthic forms from various systematic groups over different periods. During strong storms, the accumulated sediment and many hydrobionts can be washed away and the conditions for the settlement of foulers performing primary succession on hard surfaces appear again. In many MPs, transition zones where the periphytal and benthal communities closely interact form at HTS basement.

Chapter 4
Benthal Zone Subsystem of Marine Port Aquatories

The benthal zone of marine ports comprises the bottom sediments, overlying waters and interstitial waters of the harbours' seabed and access channel. It is populated by microorganisms, fungi, plants and animals inhabiting the surface or the inner layers of the sediments. Benthos is divided into bacteriobenthos, micro- and macrophythobenthos, micro-, meio- and macrozoobenthos according to the type and size of the organisms. Some benthic organisms inhabit the surface of sediments (epifauna), while others live within their different layers (infauna). Among them there are mobile (vagile) and non-motile (sessile) species. Bacteriobenthos sustains the continuous functioning of marine ecosystems, including those of marine ports [16, 42, 103, 106]. The bacteriobenthos regenerates biogenic matter in the bottom communities and its metabolic activity is over 20% higher than the primary production and four- to eightfold higher than the metabolic activity of benthic animals [8].

In terms of covered area, the seabed is the biggest biotope of MPs and is mostly formed by silts. Various sediments serve as ecotopes for different benthic forms and the type of substratum is tightly linked to its population [16, 42, 103, 106]. Sediments are formed by particles of different sizes and active surface area, creating a variety of living conditions for hydrobionts. Notably, a cubic particle with an edge of 1 cm has a surface of 6 cm^2, while the cumulative surface of the particles with an edge of 0.01 mm ideally contained in the same cube amounts to 6000 cm^2 [58]. Granulometric and mineral composition of the bottom sediments are one of the main ecological factors for the benthic organisms [15, 16, 42, 80, 103, 138, 141, 147].

The vertical distribution of benthic organisms in the sediments' layers is mainly determined by hydrodynamic conditions, feeding habits and oxygen requirements. Ninety-five percentage of all meiobenthos organisms could be present in the upper 2 mm of sediments [64].

In aquatic bodies, the upper layer of silty sediments is characterized by a certain microzoning of habitat conditions and associated microorganisms distribution. Eight niches inhabited by different organisms have been identified in the upper 2 mm layer of silty sediments. Diatoms dominate in the upper microzone of photosynthesis above the second microzone—microzone of iron oxidation—inhabited by

A.K. Vinogradov et al., *Ecology of Marine Ports of the Black and Azov Sea Basin*, DOI 10.1007/978-3-319-63062-5_4

attached forms of iron bacteria. In the third microzone, located below, lives the predatory bacterium *Dyctiobacter rapax*, which feeds mostly on iron bacteria. In the fourth microzone develop bacteria that are related to *Azotobacter*. The fifth microzone is inhabited by the saprophytic filamentous algae similar to bacteria, while filaments, rings and loops of the predatory bacterium *Cyclobacter* are registered in the sixth microzone. The seventh zone is formed by the two interlaced cell chains of *Liskeella* and the eighth microzone, characterized by the presence of dissolved sulphides and precipitated black iron sulphide, is inhabited by the mobile bacterium *Thiospira* [100].

The depth of foraging of the organisms of macro- and meiobenthos is determined by the sediment density and oxygen content and by the amount of detritus settling down from the water column.

The water depth in MPs is a primary factor for shipping and efficiency of cargo transportation, but it is not an ecological factor for benthic organisms. On the other hand, hydrodynamics influences the biotopes and composition of the benthos at depths up to 20 m that are characteristic for the aquatories of marine ports. In MPs, as the pressure of the water column increases with depth, the influence of wave movements decreases and the conditions at the sea bottom become more stable. At the same time, the character of sediments accumulation changes and vertical mixing and illumination weaken, creating the preconditions for oxygen deficit. On the whole, in MP aquatories, the number of benthic species reduces with depth.

Human economical activity connected with the construction and functioning of MPs influences the coastal flows of drifts and causes their sorting and accumulation in some zones of the coast and their removal in others. During the construction of hydrotechnical structures, changes in the movement of the drifts along, from and towards the coast may become apparent. Furthermore, the construction of shipping channels (SC) and the deepening of the marine ports basins cause the displacement of great masses of bottom sediments. At present, near the MPs and in their aquatories, the bottom sediments are often of altered and mixed composition because of their constant or periodical dredging, nourishment, dislocation and mixing.

Different measures are taken to compensate the destruction of sandy beaches and coastal habitats caused by the changes induced by the MPs HTS in the alongshore flows of drifts. Thus, in MP Yevpatoriya, the sand extracted during the regular dredging of the shipping channel is dumped into the sea south-east from it, but the extension of the beaches continue to reduce [33]. Piers, dikes, sea walls, breakwaters, berms and fill dams can therefore be built to prevent coastal degradation. However, at present, the most ecological option consists in the artificial replenishment of the alongshore drifts. To this end, soil-derived material is deliberately introduced into the sea coastal zone. This material is transported and mixed by the waves so that the coastal zone gradually regains its natural appearance. Creation of wide beaches ensures suppression of the waves energy by beach deposits, thus eliminating the need to use metal and cement during the construction of HTS. Figure 4.1 shows the general scheme of bottom substrata distribution as a function of depth in Odessa region.

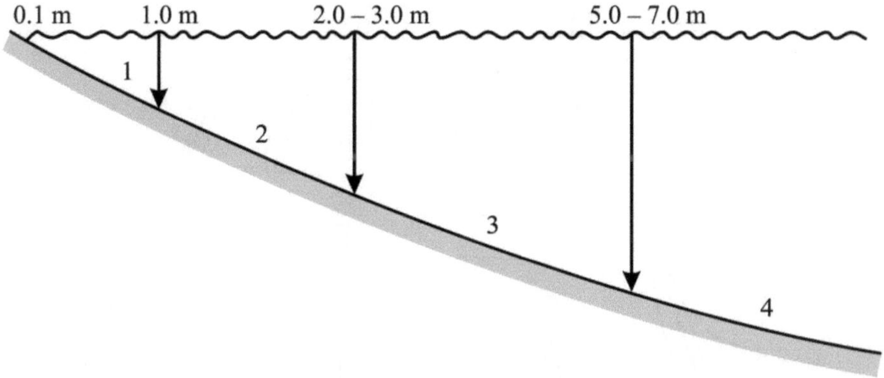

Fig. 4.1 General scheme of bottom substrata distribution as a function of depth in the coastal zone of Odessa region at sites non impacted by the construction of hydrotechnical structures: *1*—zone of boulders, pebbles, broken shells, coarse washed sand; *2*—zone of middle- and fine-grained sand; *3*—zone of silted and oozy sand; *4*—zone of silt and clayey silt

4.1 Main Abiotic Peculiarities

In marine environments, the characteristics of bottom sediments, overlying waters and interstitial waters determine the distribution of benthic hydrobionts. In MPs, the bottom sediments could be represented by: the native bedrock, deposits of abrasive, aeolian, biological (autochthonous) origin and drifts entering the MPs with the longshore current. In this last case, sorting of the drifts takes place during their movement.

Depending on the predominance of particles of a certain chemical composition and size, sediments can be divided into silts of different origin and colour and sands of different grain sizes. Gravel, pebbles, stones and rocks are rather rare in MPs of the Black–Azov Sea basin, while silts and sands are present in different combinations. In some zones of MP bottom sediments, shells of subfossil and modern molluscs occur. All biotic components of MP ecosystems take part in the formation of the autochthonous sediments; from the microbial population and diatoms up to pelagic fish, aquatic and nearshore birds and dolphins.

In natural conditions, the breaking of waves near the coastal zone determines periodical fluctuations in the water level, causes the appearance of currents and has variable impacts on the displacement of the drifts [111]. In MPs, the hydrodynamic activity and, above all, the nearshore waves' activity are artificially reduced. Areas of sediment accumulation and shallow-water zones can form in different MPs as a result of dislocation of the drifts and their interaction with the hydrotechnical structures (Fig. 4.2).

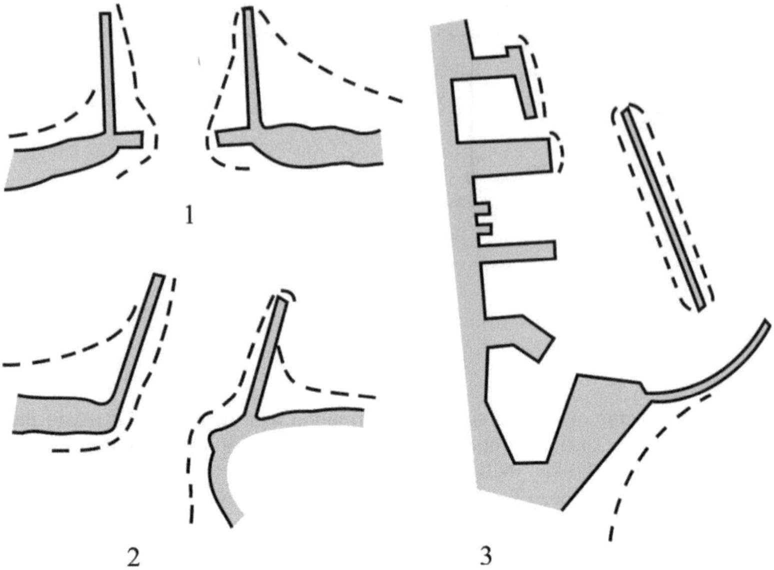

Fig. 4.2 Schematic representation of the shallow-water zones in the main marine ports of Odessa region: *1*—passage into Ilyichevsk MP; *2*—passage into Yuzhny MP; *3*—south-eastern part of Odessa MP. *Grey areas* represent the port hydrotechnical structures, *dotted lines* indicate the extension of shallow-water areas

4.1.1 Coastal Shallow Waters

To properly understand the functioning of MP ecosystems, it is important to consider their newly acquired characteristics, as well as the properties that they lose, compared with shallow coastal zones after the construction of HTS and the deepening of their basins. During dredging operations and berths construction, the extraction of bottom material, along with the associated biocenosis, takes place from the shoreline up to depths of 20–21 m, as in the ports of Constanza (Romania) and Yuzhny (Ukraine). The coastal shallow zones are destroyed together with the complex of organisms of the supra-, pseudo- and upper sublittoral zones [38] and conditions typical of rocky steep shores are created.

On gently sloping shallow coastal zones composed of sand, sand with shells (fragments) and sand with silt, such as the north-western part of the Black Sea (NWBS), the Dneprobugsky Liman and the Azov Sea, the waves' energy starts to gradually reduce far away from the coast. This process shapes the different biological features of the coastal biocenoses. When the MPs HTS are built, waves cannot anymore fulfil their ecological function of constantly shifting the border between water and land as they do at shallow marine coastal areas. As a consequence, specific biotopes develop in MP areas (Fig. 4.3).

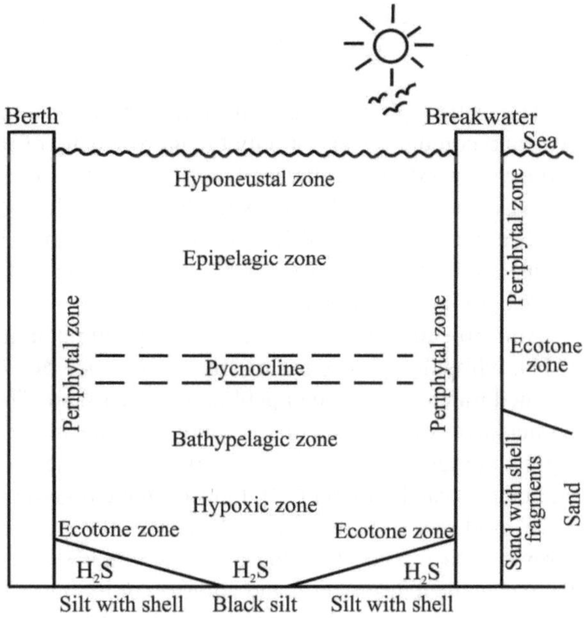

Fig. 4.3 General scheme of the main biotopes usually found in marine ports of the classical type

The sea coastal zone is a natural interface for the discharge of mechanical energy from sea waves of different types (mainly wind waves) and from sediment matter flows. The quantity of mechanical energy in this zone is so high that, during some periods, it suppresses the influence of other types of energy (light, thermal, chemical, gravitational) that also play an important role for organisms and biocenoses. Biotic components of the coastal ecosystems are thus adapted to function in a range of mechanical energy conditions and levels of mobility of the various parts of the coastal zone. Degradation and destruction of the coastal biocenoses could be caused by both disastrous storm waves and hurricanes and by the absence of water roughness near the shore [133].

During sea roughness, compensatory, discontinuous and proper wave flows develop in the coastal zone and contribute to the displacement of drifts along the shores. When the wave crest falls down on the natural sandy shallow shore, it captures air and transforms it into the foam masses. The major part of the energy accumulated in the zone of storm formation is progressively consumed over a few seconds while reaching the shallow shore. During this time, the turbulent layer reaches the upper border of the swash zone where it partly vanishes in the sand sediments and partly returns back into the sea taking away sediments. All these processes consume most of the waves' energy.

In porous sediments, the wave-induced water infiltration plays an important biological role by transporting nutrients and oxygen into their thickness. At the same time, the coastal sands are washed out up to a certain depth. The content of

OM (organic matter) in these sands is usually several times less than in the muddy sediments. Frequently, all these mechanisms are completely absent in MP aquatories.

The active physical, chemical and biological processes taking place at the sea-sand contour affect not only the narrow band of psammocontour, but also zones farther seaward and landward [26, 112]. In the scientific literature, the concept of the wave-break coastal zone as the biggest mechanochemical "enterprise" on Earth has become firmly established. In particular, abrasion of pebbles, sand and mollusc shells constantly takes place in the coastal zone and continuously creates new and chemically active particles [12].

Abrasion affects 5–20% of the corresponding annual input in the case of pebbles [111], while the corresponding figure can reach 50–70% for mollusc shells that are crushed much quicker than pebbles or silica sand. The presence of mollusc shells is particularly conspicuous on the sandy beaches of the Black–Azov Sea basin, where a great number of subfossil shell molluscs (mussels, *Cardium*, *Venus*, oysters, *Solen* and others) can be found [12]. Indeed, in the region of Odessa city, tens of tonnes of alive molluscs (mussels, *Mya*, *Rapana*, etc.) per km of shore can sometimes be thrown onto the coast during autumn–winter storms. The fragments created from the mechanical destruction and abrasion of molluscs' shells in shallow waters play a significant role in the formation of the beach body. In the swash zone, the layers of sand formed by shell fragments can attain 3–4 m of width and 15–30 cm of thickness. When milled to the size of the sand fraction, particles originating from mollusc shells have a form which is more similar to a plate than to a sphere. In comparison with ordinary sand, packed sandy-shell layers and banks are therefore less affected by wave movements and stabilize the shore line [110]. However, the shells of molluscs and barnacles can become more vulnerable to mechanical destruction by wave action following the activity of various drilling algae that are widespread in the sea coastal zone. These algae participate in the processes of calcium turnover by corroding and dissolving lime-containing matter, including the shells of molluscs and barnacles [124].

The shell-based material mostly comes from the sea bottom and plays an important role in the drifts' balance of the coastal zone. It represents from 5–15% to almost 100% of the material forming beaches and bottom sediments along all the coasts of the Black and Azov Seas. The partial abrasion of the shells takes place under the influence of the rip flows. The percentage mass of molluscs lost by wave abrasion ranges from 30 to 70% (average 25–30%) per annum [111]. Mussels' shells undergo two and three times less abrasion than *Cardium* and *Mya*, respectively. The shells of *Abra* and *Monodactna* are the least resistant [47]. The main characteristic of the shell component of the coastal zone drifts is its constant renewal by benthos. Even at 30–50% of shell abrasion per annum, accumulation of shell sand can take place owing to the renewal of molluscs population. In the coastal zone of the NWBS (Tendra spit, Dzharylgach Bay), the layer of the shell sand and fragments originating from the crashed molluscs' shells can reach 3–4 m. The shells are smashed to pieces by the action of currents and waves until they are reduced to a final particle size approximating that of siltstone and pelite grains [98].

During shell abrasion, a mineral suspension is formed and then carried away by currents from the coastal zone to the great depths, where it forms the carbonate component of bottom sediments [135]. In the aquatory of MPs, the accumulation of molluscs' shells takes place, but their destruction is usually not possible.

Both in summer and winter, mechanical wave energy and temperature are the leading factors that determine seasonal changes in the qualitative and quantitative characteristics of the coastal biocenoses up to the depth of 10 m along all coasts of the Black and Azov Seas. In this region, many important biocenoses are especially concentrated in the coastal zone between the depths of 0 and 5–6 m. A community of specialized interstitial organisms is connected with the sandy beaches above and below the water edge [26].

In the coastal zone of the NWBS, sandbanks are present from the Danube delta to Ochakov city and farther east up to Crimean shores. Sandy beaches are also present along numerous areas of the Crimean peninsula. Many shores of the Azov Sea constitute sandy formations extended over hundreds of kilometres and vast sandy beaches also form the coasts of Turkey, Romania and Bulgaria. The position of the coastal line results from the interaction of different natural forces subject to seasonal changes. During winter time, because of the fierce storms, the width of the beaches is reduced, while it increases again during the summer period. In summer, when storms are weaker, the dead organic matter (OM) is stocked as detritus in natural, shallow-water zones. In the absence of anthropogenic influence, these sediments are washed out by the following autumn–winter storms, maintaining favourable conditions for hydrobionts. In fact, the shallow coasts usually have a wide wave-break zone. This zone is completely absent in areas where quays and berths of marine ports have been constructed.

In shallow coastal waters, numerous animals are sustained by the abundance of food resulting from the development of a great mass of attached and free-floating vegetation, as well as from the entrance of OM of plant and animal origin in the form of storm casts. Part of the OM comes here from the mainland. On the littoral zone, OM is transformed by mechanical, physico-chemical and biochemical processes. A variety of organisms (microbes, protozoa, aquatic invertebrates, larvae of terrestrial insects, fish, birds and mammals) take part in the transformation and utilization of OM that can thus be removed from the aquatic ecosystems. Another mechanism that reduces organic and biogenic matter stocks in the coastal waters is the formation of foam and its transfer on land by the winds [129].

Partial or total destruction of the coastal shallow waters in MPs markedly worsens or prevents the dwelling of hydrobionts that are adapted to live in aerated waters. At the same time, conditions for fouling hydrobionts and hydrobionts inhabiting the depths of 10–20 m far from the coastal zone are artificially created or ameliorated. Furthermore, construction of the ports' HTS and berths prevents the erosion and abrasion of the shores, weakens the land sliding processes and stabilizes and simplifies the coastal line. Unlike at natural shores, the detritus accumulates on the bottom of MP harbours and takes part in the formation of silted sediments.

4.1.2 *Marine Ports and Longshore Drift Flow*

On an annual basis, the rivers of the Black–Azov Sea basin carry millions of tonnes of sediment run-off to the sea and initiate the main longshore drift flows. This input of solid matter ensures the stability of the coasts and provides substrata for the normal existence of various benthic organisms. However, the same input can interfere with the normal functioning of MPs by causing siltation. In some MPs of the Black–Azov Sea basin situated in the lower reaches of rivers (e.g. MP Sulina, Kiliya, Izmail, Reni, Dzurdzuleshti, Nikolaev, Kherson), the silting of aquatories is rather limited. Except during years with low run-off, the drifts flow by without accumulating and periodic strong floods make a "general wash-out" of drifts deposits. At the opposite, drift accumulation is observed in MPs situated in or near river deltas (e.g. MP Poti, Temryuk, Bartin, Ust-Dunaysk) where the sediment matter is discharged to the sea. Drift deposition is slower at MPs situated on coasts with a sharp continental slope or near underwater canyons (e.g. MP Yalta and Batumi). Overall, the choice of location, operation and modernization of MPs is connected with the different mechanisms of displacement and distribution of bottom sediments. Sediment displacements can be caused by near-bottom water movement, anthropogenic activity or a combination thereof and influence the population of the benthal zone.

Benthic organisms inhabit the relatively thin water layer (1–10 cm) overlying fixed substrata or inside the first 10 cm of such substrata. The thickness of the overlying water layer is mainly connected with the roughness of bottom sediments (defined as the average or modal size of grains) and influences water movement [42, 95, 103]. In the aquatories of MP, the water speed and friction near the muddy bottom can approach zero; a situation frequently observed in shallow zones and in various pits.

The interaction between the rough seabed and the water in turbulent movement forms microvortexes that enter the near-bottom zone. This phenomenon increases the possibility to come in contact with the substrata for the larvae of benthic organisms. At the same time, it provides food to attached animals and tube-dwelling organisms by increasing the possibility of "predator–prey" interactions. Many animals are settling near the zone influenced by microvortexes because they also contribute to the removal of metabolites. Microvortexes therefore are a main factor in determining the general pattern of the distribution of bottom organisms [42, 95, 106].

In the near-bottom layer of MP aquatories, the water movement has a three-dimensional character and depends on the entrance of energy via the vortexes formed in the upper water layers, as well as on the changes in water flow caused by the roughness and other irregularities of the sea bottom [42, 95, 103, 106]. The seabed, especially in marine ports, plays an important role in transforming and recycling matter and energy between the sea bottom and the water column. Geothermal heat can be transferred through the sea bottom at a specific flux corresponding to about 0.03% of the solar heat flow. This quantity of geothermal heat

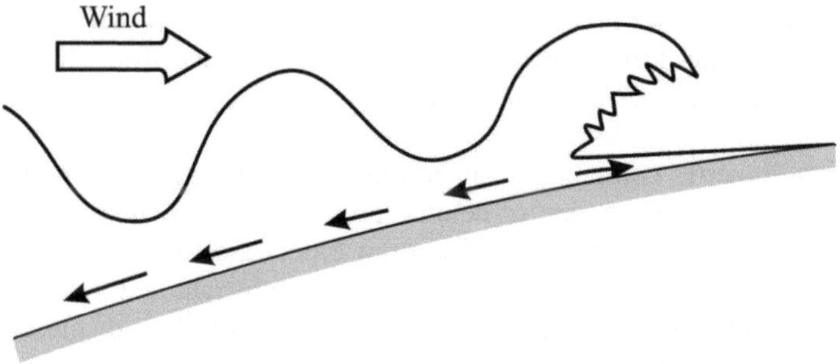

Fig. 4.4 Main scheme of wave formation and currents at shallow-water coast

is sufficient to bring the near-bottom water into motion providing it with a weak, but constant, regenerative mixing. The average figures of the geothermal flow for the World Ocean and for the mainland are similar (3×10^{-11} W m^{-2}) and, at the sea bottom, the convection has a pulsatile character [2, 66].

The displacement speed of bottom sediments is mainly connected with the speed of the water flow and differs between shallow and deepwater coasts. During waves' deformation at shallow-water coasts, the elliptic orbits of water particles turn open-ended; which creates faster velocity of transportation, comparing to deepwater coasts. The fastest water transportation velocity is reached at the moment of maximal wave deformation, when waves become unstable and drop down. Accumulation of water at the water edge, formation of the inclined shoreline and flow-out, or compensational displacement of water take place at the same time. At shallow coasts, the water movement gradually weakens during the gradual dissipation of energy that follows the break process at the shallow coast [66] (Fig. 4.4).

Protection of sea coastal areas by piers or breakwaters always leads to the formation of a closed bay shore. The waves coming from the sea do not reach the areas behind the coastal protection structures or reach them in a weakened form. Here, as well as in most marine ports, only small local waves are active. However, when waves break onto the external surfaces of HTS, the power of water movement quickly increases and the process ends with an intense discharge of energy. Cases of displacement of concrete constructions weighing tens of tonnes are known [58].

In Tuapse MP (Russia), the strongest and most stable currents are registered in the zone of the access channel (AC) where the water flow concentrates and intensifies. Thus, the central part of the port is the most aerated and well mixed [114]. A similar situation is characteristic of many other marine ports [18]. In the zone of contact with the bottom, a value of wave energy from 24 up to 47 GJ was shown to negatively influence the psammophyle fauna [105]. These phenomena are observed not only in big lakes, but also in MP aquatories where a wide spectrum of depths exists. This leads to a predominance of attached and deeply borrowing forms in the composition of benthos.

In the coastal zone, the main mass of the bottom sediments and drifts originates from the shore. Numerous observations and studies showed that the surface run-off and atmogenic processes are more intense in the aquatories of MPs located within shore bays, gulfs and limans (which are surrounded by land from all sides) than in MPs constructed at open sea shores and protected by artificial hydrotechnical structures.

The granulometric composition of bottom sediments and its temporal stability play a key role in the formation of benthic populations in each specific area of the sea coastal zone. The distribution of bottom biocenoses at natural and open sea shores is connected with the zonal distribution of substrata and biotopes and with the migration velocity and differentiation of drifts during their displacement.

In the coastal zone, there is dissipation of the energy of waves, wave flows, run-up flows and of some other types of coastal currents that are not directly connected with the wind waves and sea ripples. The coastal accumulation of the drift material takes place in the zones of energy dissipation [94]. Drifts in the coastal zone can move in different directions. The drifts movement from the shore takes place in the form of rip and compensational currents. During the negative water set-ups, the more mobile drifts accumulated near the gangways at MP aquatories enter the MPs because of compensational flows.

Under certain wave regimes and flows, the displacement of drifts can occur in longshore currents having opposite directions. Such bidirectional flows of drifts were registered, for example, in Odessa region southward the Cape Bolshoy Fontan [94].

The resuspension of sediment particles depends on the flow speed, while their sedimentation speed depends on size and density. At each flow speed, the smaller and lighter particles are carried away, while the bigger and heavier ones are deposited. It is more difficult to set the bigger sand grains into motion than to maintain them in motion because, while colliding among themselves, the moving sand grains contribute to the general displacement of the flow. When sediments consist of small muddy particles (aleurite, pelite), the bottom roughness is low and the flow remains laminar. The transition from laminar to turbulent flow starts at particles' size of 0.1–1.0 mm [17]. Silt and clay silt, due to smoothness of their surface, are washed out less. However, when being suspended, these particles are transferred across longer distances and remain suspended in the water column over longer periods. With the increase of the sediments' particle size and of the water movement speed the laminar flow is disturbed, vortexes appear and the water movement becomes turbulent, transmitting lots of energy to the sediment particles. In the case of relatively heavy quartz grains, the transition from laminar to turbulent flow is registered at grain sizes of 0.1–2.0 mm. The turbulent water movement hampers the quick sedimentation of the suspended particles to the bottom.

For the erosion of rough sand sediments, less current velocity is needed than for the particles of other sizes. In the absence of turbulence, the sediments remain immobile.

The sediments that are mainly composed of pelite and silt particles have less tendency to create turbulent currents. However, these sediments, exposed to constant or periodical bioturbation, formation or disappearance of hills and burrows along with the accumulation of faecal pellets, considerably activate the formation of

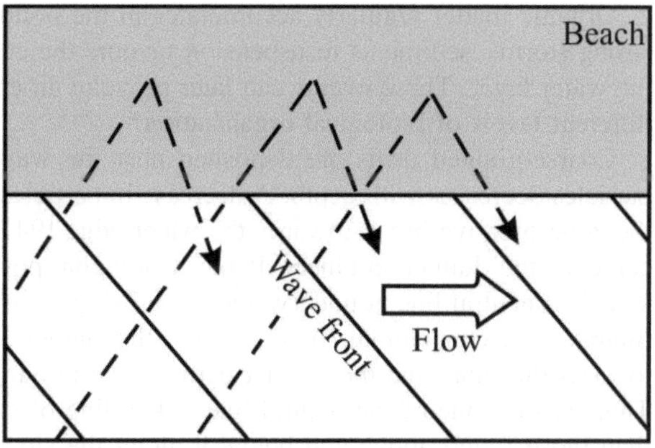

Fig. 4.5 Scheme of waves energy consumption at the slanting approach to the shallow sand shore and formation of the longshore flow of drifts

micro turbulences. Clay particles have diameters less than 0.01 mm, muddy particles from 0.01 up to 0.1 mm and sandy particles from 0.1 up to 1.0 mm. For fine-grained sand, the shear velocity of water flow is about 22 cm s^{-1} [94].

In situ measurements of the flows in Kalamita Gulf, a region adjacent to Yevpatoriya MP, showed that the shear velocity did not exceed 10 cm s^{-1} in 97% of the observations during a period of 3 years. The velocity in the near-bottom layer was nearly two times less than in the surface layer. During moderate and strong storms, the drifts along the coast reached 2 kg m^{-1} s^{-1}. The main flows were located in the coastal zone of the gulf and reached 0.2–0.4 kg m^{-1} s^{-1} for sands with an average grain size of 0.5 mm [34]. It is known that several coastal biocenoces are connected with the beach zone (psammocontour) of the sea. The term "beach" implies an unstable and mobile sand layer. The characteristics of each beach are determined by the equilibrium between the forces providing the input of hard material and the forces (drift) washing out this material. Usually, in autumn, the increase in the wave energy transports the sand offshore to higher depths and the width of the beach can noticeably diminish. In spring, as the wave energy decreases, the beach width is increased, i.e. beaches rehabilitate (Fig. 4.5).

At coastal beaches, the wave energy, besides causing the drifts displacement, is spent on particles abrasion and filtration. The loss of wave energy because of filtration becomes visible only at particles size of about 1 mm and above [94]. During abrasion, particles acquire different forms depending on their origin: globular for fragments of igneous rocks, flat oviform for fragments of sedimentary rocks and lamellar for fragments of broken shells. The abrasion of drifts' particles is called abrasive deterioration and indicates the loss of volume and weight of individual particles. The coastal zone where the pounding and abrasion of the drifts take place is named "breakers mill" [46].

In the confined environments of MPs, the formation of sediments is determined by factors other than wave activity in most parts of the aquatory bottom. All types of sediments can be found at the bottom of different MPs, providing different types of substrata for various hydrobionts. Texture and grain size distribution, colour, composition of OM, concentration of dissolved oxygen in interstitial waters are all indicators of the hydrodynamic activity in the near-bottom layer.

Organic matter regularly accumulates in the bottom sediments of MPs. During strong storms, sediments resuspension favours the entrance of biogenic matter into the water layer. These events can later provoke an entire chain of consequences at different levels of biological organization.

Coarse-grained drifts are deposited near the wave-break zone and the size of particles decreases with depth. A decrease in particles size is usually observed from the zone of wave-break towards the water edge [94]. At the end of the nineteenth century, the Italian engineer Paolo Cornaglia postulated the existence of the so-called neutral line at underwater coastal slopes, where the drifts of a certain size migrate and concentrate. In this zone, the bigger sand particles are moving up towards the shore and the smaller particles are moving down, away from the shore. Thus, at the zone of the neutral line (a.k.a. line of silt) conditions are more stable than those found above and below it along the slope. Thus, Cornaglia's line represents a special ecotone zone for hydrobionts inhabiting silt and sand.

At open, extensive sand coasts, one to three bar crests and trenches are usually formed at some distance from the shore. The sandbars usually start to form along the shore at a depth of 0.5–0.6 m and are parallel to the shore and to each other. As it was already mentioned, their number is usually two or three. The last one, as a rule, is situated at a depth of 2–3 m. The depth at bar hollows is 40–50 cm higher than over the bar crest, but the differences can be up to 1 m. Heterogeneous conditions for marine fauna are formed in the troughs and crests of the coastal zone. During calm weather, the troughs are filled up with lighter detritus particles. Rip currents originate periodically during the breaking of waves at the shallow water and sandbars. In this case, water returning back to the sea creates a narrow channel in the sandbar. Strong rip currents create new, additional negative forms of relief located perpendicularly to the coast. In settling down after a storm, shrimps, jellyfish and *Cumacea* (hooded shrimps) are usually concentrated in rip current channels, while grass crabs are found on their slopes.

In the coastal zone, at weak roughness, the refraction effects start to appear and create complexes of microrelief forms at the surface of sandy and silty-sandy sediments located at depths up to 2–3 m. An alternation of small crests (couple of centimetres) and dividing narrows, called ripple marks, is observed [94]. During ripples formation, microheterogeneous conditions appear on the crests and in the narrows. Because of the very weak water movements, the bigger particles (molluscs shells fragments, macrophytes shreds, coarse sand particles) are concentrated in the narrows. The relatively light detritus particles are carried over the crests either in suspension or by rolling. These circumstances are important in creating a patchy distribution of the organisms of micro- and meiobenthos.

In aquatories of big MP, the wave energy is often not sufficient for the formation of sandbars and channels of rip currents, also the formation of small underwater sandbars and ripple zones occurs in shallow-water areas. Ripples at the surface of bottom sediments increase roughness and favour turbulence.

All MPs of the Black–Azov Sea basin located directly at the sea shore in gulfs, bays and limans are under a certain influence of the longshore drifts flow. The construction of MPs HTS inevitably causes changes in drifts displacement and into the habitat conditions of benthic communities. Some biotopes can be destroyed, but new ones are created nearby at the same time. The functioning of the marine ports HTS usually causes the appearance of new accumulative forms in the port aquatory and at a considerable distance from them. Retreat and extension of the shore could amount to kilometres. Quite often, the changes in the bottom relief could include an area a couple of kilometres wide and up to the depth of 6–10 m and more.

As shown in Fig. 4.5, the longitudinal flow of the drifts is created by roughness. By hitting the shore obliquely, the waves shift the sediment particles not only up and down the coastal slope but also longshore. Since the predominant roughness has a defined direction in each specific region, the drifts are also moving in a defined direction. The flow speed undergoes changes and the highest registered speeds reach hundreds of metres per day. The extension of the flow strongly depends on the shore configuration. Along straight shores, flows sometimes stretch for dozens or even hundreds kilometres. However, each natural or anthropogenic irregularity in the coast morphology interrupts the drifts flow or changes its direction [111]. Most of the large and heavy sand particles in the bottom drifts are moved as bed load, practically without being torn off from the sediment. Organic particles and most small mineral particles transported by advection and turbulence move in suspension and can reach the areas more distant from the MP entrance.

In the main part of the MP aquatory, the separation of sediments and drifts practically does not occur or, if it does, it happens very slowly. On the other hand, in the adjacent areas of the swash zone the drifts move selectively. Seawards of the swash zone, the net resultant of the sand displacement is always directed towards the shore, although the drifts could move from the shore into the sea at certain periods.

During fierce storms, at sandy beaches, sand layers located at depths above 40–50 cm are displaced and redeposited from one site to another. Severe storms cause the destructions of bottom habitats as witnessed by the transportation on the shore of deep borrowing species such as *Mya*, *Upogebia* and *Nereis* from sediments located at depths of 4–5 m.

Drifts usually accumulate at zones where the shore is protected from sea roughness by breakwaters, piers, berths and other HTS. In Odessa MP, the drifts are most intensively accumulated near Stariy, Noviy and Zavodskoy breakwaters. Here, the heavier and bigger sand particles settle to the bottom and remain there where the wave energy is reduced and the currents fade away. At these areas of Odessa MP, water depth is often reduced to 3–5 m. More or less marked sandbanks, sandbars and tombolos are formed in the wave shadows of the breakwaters. The sands are also accumulating on the outer part of the Karantine, Raid and Oil piers and at their wave shadows. The same type of accumulative forms appears in the ports of Mariupol and Berdyansk (Ukraine), Varna East (Bulgaria), Constanza (Romania) and others (Fig. 4.6).

On the inner side of the protecting structures, the shells and druses of molluscs fall from the piers and berths onto the silty bottom and sink into it. On the external side, because of the strong flow and shallow depth (3–5 m), part of the shells is mechanically destroyed into shell fragments, while another part rolls over the sandy

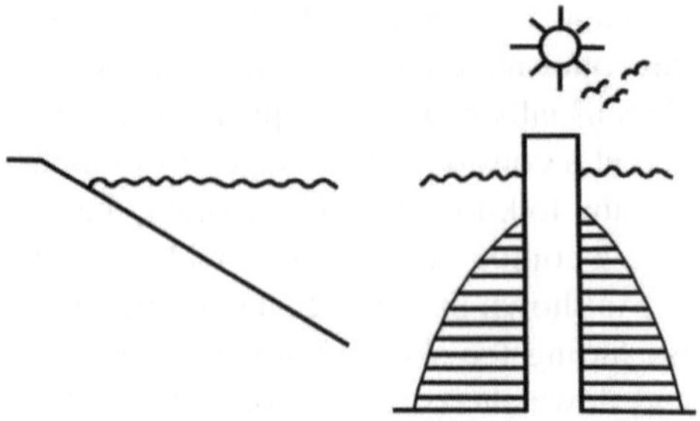

Fig. 4.6 Scheme of the distribution of shallow-water areas formed by the longshore drifts flows in: natural coastal zones (*1*) and around the hydrotechnical structures of marine ports (*2*)

bottom and is transported at some distance. Part of these molluscs preserve their vitality and form a zone which is an extension of the fouling biocenoses. The intact mollusc shells provide hard substrata (HS) for hydroids, barnacles, pearlweed *Sagina*, sponges, polychaeta with lime and soft tubes and some macrophyte species. Some species of small bullheads (*Gobiidae*), combtooth blennies (*Blenniidae*) and wrasses (*Labridae*) spawn in this biotope.

Opposite phenomena causing the washing out of existing drifts deposited by the longshore current happen during storm winds with unusual directions and when the replenishment of the flow is interrupted for some reasons. These reasons include the presence of the protecting HTS of MPs, that block up the shore areas, and also the ACs that trap part of the drifts [76].

The majority of marine channels located at the coastal sandbanks of non-tidal seas are subject to filling because of the storms. Only about 3% of the fine fractions of the drifts are retained in the channels. On the other hand, large fractions are totally caught, but their percent content in the mass of the moving drifts is not high [10]. Such sorting of the drifts favours the formation of different biotopes and biocenoses.

For particles with a diameter of about 1 mm and at a flow thickness of 15 cm, the displacement of the bottom drifts during head winds and fair winds are 5–13 times and 1–4 times higher than the displacement at the lack of wind, respectively [111]. In the marine channel of Mariupol, during strong storms, the concentration of suspended material in water reaches 5.0–6.0 kg m^{-3} [10]. The sand drifts are usually entering the MP aquatory in the form of tongues, splitting according to the predominant currents. The fine fractions of the drifts are entering and depositing in MPs because of the low wave energy. The fine granulometric composition of the sediments often plays an important role in the formation of the bottom biocenoses. The lower the current speed, the smaller the fractions in the drifts and the higher their concentrations of detritus and OM [98].

At the bottom of the majority of MPs of the Black–Azov Sea basin, layers of soft sediments prevail and grow not only because of longshore transfer and sedimentation of the suspended particles from the water column, but also due to the filtration activity of the living components of plankton and periphyton. Via the production of faeces and pseudofaeces, these organisms transform dead and alive suspended organic matter (SOM) into sediments together with the mineral particles (seston). Part of the dead plankton, organisms of nekton and periphyton also sinks to the bottom sediments. The accumulation of these residues (detritus) is particularly visible near the foundation of berths, piers and breakwaters in the zones not affected by the entrance of the coarse fractions of the drifts. In MPs aquatories, the tendency of accumulation of OM in sediments, predominantly represented by black silts, is observed. In the NWBS, the maximal concentrations of OM in bottom sediments are registered for black silts. Only in access channels to the aquatory, there are areas with a sandy bottom.

In Odessa MP, the drifts can enter from the costal area that has an extension of dozens kilometres. The longshore flow of the drifts in Odessa Gulf is several kilometres wide and it can pass around artificial obstacles farther into the sea. The sources of drifts in the aquatory of Sukhoy and Grigorievsky Limans are mainly located in their drainage areas and shores. At these sites, the width of the longshore drifts flow, where it is preserved, is no more than a couple of metres. In the aquatories of Sukhoy and Grigorievsky Limans, stirring up, displacement and redistribution of the sediments take place during dredging and construction of HTS. During this activity, the settlements of bottom organisms could be totally silted, while previously buried sands could be uncovered.

After construction of the access channel and of two jetties in Ilyichevsk MP, the sandy beaches north-east of the channel completely disappeared in a few years as far as 5 km from the site. The sea bottom became exposed, losing its sand layer up to the depth of 5–6 m. Grigorievsky Liman, which is the aquatory of Yuzhny MP, is located in the zone between the capes Severny Odessa and Adzhiyask. This zone is a sea coastal lithodynamical cell (system) having unified longshore drift flows where all morphodynamic and lithodynamic processes are interconnected. It is known that disturbance in one of the areas of a given cell provokes a chain of changes in the other zones of the same cell. Changes take place not only along the coast, but also landward and at the underwater slope [134]. After construction of the access channel and jetties of Yuzhny MP north-east of Grigorievsky Liman, the processes of abrasion increased twofold and the destructive phenomena at the underwater slope extended from the water edge down to the depth of 5–6 m [136]. In the region from Grigorievsky to Sukhoy Liman, the depth of 6–7 m could be taken as the external limit of the longshore drifts flow. Here, the maximum of the wave impact is reached at the depth of 0–2 m [134].

The absolute majority of MPs of the Black–Azov Sea basin is characterized by an increase in the silt layer on the bottom of the port area. The average data on the granulometric composition testifies that the fractions smaller than 0.25 mm, including pelite particles, predominate in the bottom sediments of Odessa MP [14, 40]. In Sukhoy and Grigorievsky Limans, the water-rich black silts are exposed mainly during dredging. These silts also predominate in the aquatory of Odessa MP.

In the Odessa region, at the coastal zone located between Cape Odessa North and Cape Adzhiyask, the drifts flow moves southwards. From Cape Bolshoy Fontan, the drifts are directed northwards towards the Gulf of Odessa and south-west up to Zabrianskaya Bay. However, bilateral drifts flows dominate between Sukhoy and Dnestrovsky Limans. The drift regime in the coastal zone of Cape Bolshoy Fontan–Sukhoy Liman is characterized by strong annual fluctuations. Its maximum is registered in summer, the least stormy period, and its minimum during the period of intense storms in autumn–winter. During periods of increasing wave action, the beach zone undergoes erosion, while beach accretion takes place during periods of reduced wave action [89].

4.1.3 Bottom Sediments

About 90–95% of the bottom area in the deepwater parts of MPs aquatories of the Black and Azov Seas is covered by silts, mainly black silts. The mineralogical and granulometric composition of the sediments reflects the hydrodynamic conditions. The presence of silt and clayey silt testifies the very quiet hydrodynamics of these zones. Silts, and most of all black silts, belong to the so-called soft sediments because of their high water content (50.0–55.0%). Water-rich silts are not able to sustain the weight of hydrobionts with heavy shells or testa, which sink into the sediment. Such silts are also unsuitable for the V-shaped passages of polychaetes, in particular *Nereis*. At the bottom of MPs, silts form combinations with sands of various sizes and with shells. Among sand particles, there are always free spaces and passages that offer possibilities for the existence of a specialized community of interstitial organisms. These cavities are absent in the finest silt particles (0.005–0.100 mm), so that the higher the number of pelite and aleurite particles in silty-sand, the less favourable are the conditions for the interstitial fauna.

Since the 1970s, two biocenoses are distinguished at the bottom of Odessa MP. The biocenosis of the sand, with shells containing live mussels, stretches as a narrow line along the Stariy breakwater as far as 10 m away from it. The biocenosis of the black silt, with nematodes, occupies the rest of the aquatory [40]. At the beginning of the 2000s, sediments consisting of fine-grained black silts (pelite–argillaceous particles with size less than 0.005 mm and with an odour of hydrogen sulphide) covered nearly all the bottom of Odessa MP [11].

The bottom of Sevastopol Bay, having depths between 3 and 20 m, is covered by sediments of various composition and thickness. Most of the bottom area, including the Yuzhnaya and Artilleriyskaya branches along with Kilen-Bay, is also occupied by fine-grained black silts. Other parts are covered by coarse-grained sand and sands with silted coquina, while grey silts are found in Karantinnaya Bay [4, 49, 87, 131].

In the aquatory of Novorossiysk MP, the uppermost 15 cm of sediments consist of black silts with a greasy shine and a strong smell of hydrogen sulphide. These sediments are loosely packed and have a high water content (from 38.9 up to 53.3%) compared with the underlying silty sediments (between depths of 15–30 cm) that are more compact and less water-rich (32.3–36.9%) [50]. Outside the zone of Novorossiysk port basin, grey silts with a yellow thin coating were observed over the upper 11 cm of sediments. As in the port aquatory, this sediment is loosely packed, but has only a weak odour of hydrogen sulphide. This difference indicates that the processes of oxidation of the surface and subsurface layers of bottom sediments are much slower in ports' areas than outside them [50, 85]. The sediments of Tuapse MP are formed by black silts with an odour of hydrogen sulphide [49].

Black silts are also uncovered in Sukhoy and Grigorievsky Limans during dredging and the intensive formation of fresh black silts takes place in the artificially deepened areas [112, 137]. The freshly uncovered silts are often non-sufficiently oxidized. Dredging at the aquatory of an MP forms trenches and other irregularities on the bottom. Under the influence of gravity, currents and ships' propellers, the semi-liquid fluid silt gradually evens the seabed and stabilizes the position of the channels and of the deepened areas.

The bottom sediments include autochthonous (of local origin) and allochthonous (originating from outside) mineral and organic matter. In fact, mineral particles in the water column could consist of sand grains, silt, clay and salts of various metals. These particles enter water as the result of wash-out of shores and bottom sediments, terrestrial run-off, and inputs from the atmosphere and dust-forming cargoes in MPs. Depending on their physical properties, these particles have various sedimentation speeds. At temperatures higher than 4 °C, their sedimentation speeds up because water viscosity decreases. In stratified water masses, particles could be retained for some time in the pycnocline zone. The sediment formation and accumulation in MPs depends on their trophic status. In eutrophic waters (to which the port aquatories belong), the speed of sedimentation is higher than in the adjacent ecosystems because of the artificial increase in bioproduction [106]. At high input rates, most of the OM is moved out from the biotic turnover and is buried by other, constantly depositing OM before it can be mineralized or used by hydrobionts.

The amount of OM in the silts of Odessa MP is 4–5 times higher than in the open part of Odessa Gulf [11]. In 2001, the OM content in the bottom sediments of Odessa MP ranged from 1.15 to 20.35% directly near the berths and from 1.42 to 15.44% at a distance of 50 m from them [9]. The average content of OM in various harbours and different zones of Odessa MP is shown in Figs. 4.7 and 4.8. For comparison, the content of OM is usually 1–2% in sand and 5–10% in oxidized aerated silts [106, 111].

In Novorossiysk Bay, the quantity of OM in sediments increases from 11.39% near the port entrance to 24.36% near the berths [49]. The OM content of silts at the sediment–water interface is about twice than at 20 cm below the sediment surface, but it does not vary at higher depths within the sediments. This situation is observed regardless of the OM content at the sediments' surface [111]. Studies carried out in one of the Sevastopol bays (water depth 1.5–15 m) showed that the amount of settling detritus increased with depth, with the sedimentation speed changing accordingly. The relative content of OM in detritus at the depth of 3.5 m is 2.5 times less than at the depth of 1.5 m and is practically the same as at the depths of 10.0 and 15.0 m [131].

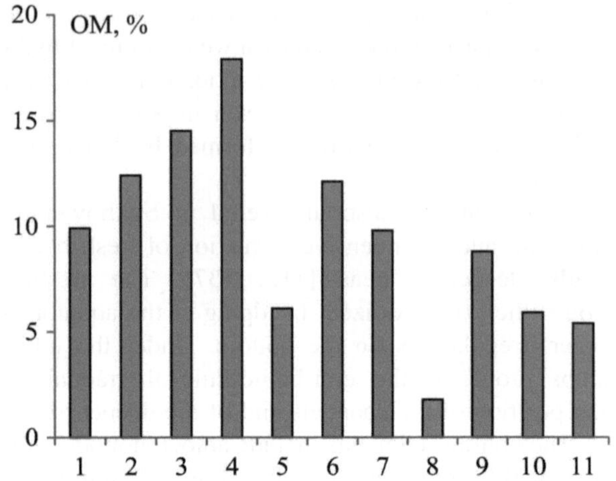

Fig. 4.7 Average content of organic matter (OM) in bottom sediments of Odessa MP harbours: *1*—Karantinnaya, *2*—Novaya, *3*—Kabotazhnaya, *4*—Prakticheskaya, *5*—Zavodskaya, *6*—Grain, *7*—Oil, *8*—Container terminal, *9*—Stariy breakwater, *10*—Noviy breakwater, *11*—access shipping channel

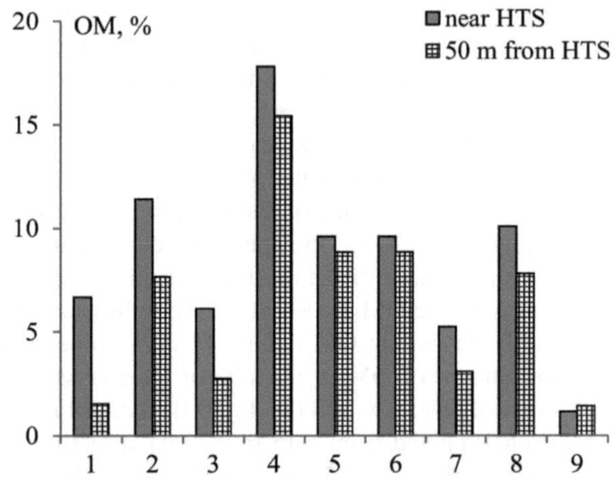

Fig. 4.8 Average content of organic matter (OM) in bottom sediments of Odessa MP near the hydrotechnical structures (HTS) and at a distance of 50 m from them. Harbours: *1*—Novaya, *2*—Kabotazhnaya, *3*—Zavodskaya, *4*—Grain, *5*—Oil, *6*—open berth of Oil terminal, *7*—Zavodskaya breakwater, *8*—access shipping channel, *9*—Container terminal

The MP aquatories belong to the zones of the sea with a high sedimentation rate. Here, the mineralization processes do not have time to oxidize all the incoming OM; a significant part of which is buried under the thickness of bottom sediments. Mass mortality in MP is mainly connected with the increasing of organic production that causes sulphate reduction and accumulation of sulphides and hydrogen sulphide in bottom sediments.

Intensive accumulation of sulphides was registered even in sediments of the relatively clean Gelendzhik Bay at a depth of 8–10 m [120]. High concentrations of labile acid-soluble sulphides (>600–940 mgS dm^{-3}) are present in the upper layer of Tuapse MP bottom sediments, which confirms their strong pollution by OM. Such situations inevitably lead to an oxygen deficit (hypoxia and anoxia) [113].

Sulphate reduction occurs more intensively at the surface of bottom sediments because of the constant input of easily degradable OM from the water column. The speed of sulphate reduction at the depth of 10–15 cm within the silt thickness decreases, on average, by more than one order of magnitude. More than 90% of the hydrogen sulphide forming in the upper sediment layer of the oxygenated zone of the Black Sea is oxidized by the oxygen dissolved in water. The reserve of hydrogen sulphide in the upper layer of sediments has a turnover time of 0.5–2 months [125]. The formation of hydrogen sulphide in the black silts of MPs is a constant phenomenon that has its maximum development during the warm months characterized by prolonged calm weather.

Because diffusion of dissolved oxygen into the sediments thickness is very limited, an anaerobic zone containing hydrogen sulphide forms very close to the sediments surface. At the border between the aerobic and anaerobic zones, the value of redox potential decreases sharply and a redoxcline is formed. The redoxcline separates the upper layer of the bottom sediments, where the substances are in an oxidized form, from the underlying layer with reducing conditions. Rather often, the redoxcline is located at the sediment–water interface and can even rise into the overlying water.

In bottom sediments of Sevastopol Bay, the pH ranges from 7.43 up to 7.79 [4]. In the upper 10–15 mm layer of black silts, the pH can vary from 8.2 down to 6.0; a value which can severely affect most species of bottom invertebrates [58, 106].

The presence of black silts in the benthal zone of MPs periodically causes mass mortality of the bottom fauna. Wind processes and upwelling phenomena can transport to the water surface (and release to the atmosphere) the toxic hydrogen sulphide, methane and mercaptans gases formed during decomposition of OM under the reducing conditions found in black silts. In Odessa MP and in the adjacent waters of Odessa Gulf, this phenomenon is often observed in August–September.

Mineral and organic particles (detritus) depositing at the bottom undergo a complex series of physico-chemical, physiological, biochemical and bacteriological transformations. Different types of sediments can form depending on the size and composition of these particles, with silts being the main sediment type in MPs. The quality and even the colour of bottom sediments can change during their formation. In well-aerated areas of MPs (e.g. in the channels), the bottom sediments have a light colour. Under predominantly anaerobic conditions and oxygen deficit, silts containing a lot of OM are saturated by sulphide compounds and the sediment has a black colour.

The presence of iron, which enters the sea together with river run-off, influences the colour of sediments in the MPs aquatories situated near river mouths. During the mixing of marine and river waters, dissolved iron precipitates from solution. Iron sedimentation is closely connected with formation of siderite or iron spar ($FeCO_3$) in the sediments. Siderite has a yellow or brown colour in contrast to iron sulphide (FeS) which has a black colour. The presence of FeS indicates the occurrence of sulphur precipitation in the sediments.

When enough oxygen is present in MP waters, the sediments containing iron hydroxides have a yellow-grey or brown colour. The colloidal-sized iron oxides (Fe_2O_3) accumulate at the surface of siltstone grains or distribute in the pelitic fraction. The thickness of the oxidized layer depends on the extent of sediment aeration, which is connected with granulometric composition, porosity, sedimentation rate and bioturbation. It ranges from a couple of millimetres up to 10–30 mm in silt, but it reaches a couple of centimetres in sludgy sand and even more in sand [111].

Under the hypoxic and anoxic conditions encountered in the sludgy sediments of MP aquatories, the sulphate-reducing bacteria produce hydrogen sulphide that reacts with iron oxides to form FeS. The iron sulphide hydrotrollite ($FeS \cdot nH_2O$) gives sediments their black colour in the reduction (monosulphide) zone. The thickness of the monosulphide layer could reach many metres, as in Sukhoy and Grigorievsky Limans, Sevastopol Bay and other MPs.

The quantitative and qualitative composition of the MPs bottom sediments directly depends on the quantity and composition of seston and detritus in the overlying water column. The sedimentation of detritus takes place under the influence of gravity and hydrobionts activity (biosedimentation). Seston is mainly formed by OM and its content of mineral matter is 4.0–6.0% [118].

In the coastal zone of the NWBS, contributions to the formation of detritus are as follows: phytoplankton—2%, zooplankton—0.1%, macrophytes—31%, invertebrates of porous sediments—5%, invertebrates of hard substrata—21%, fishes—6% and outside sources—35% [6]. It is assumed that 85% of the macrophytes mass is decomposed during 1 month following their death, while 15% of the mass remains in the form of detritus. During the decomposition of microalgae, 40% of undecomposed OM is incorporated into the detritus [6]. In general, biosedimentation could prevail over gravitational sedimentation. In aquatories with low flow speed, the zooplankton filtrates from 5 up to 90% of the total water volume each day [6, 106]. A settlement of 1 m^2 of mussels, the most important filter feeders in MP aquatories, daily transforms 8–47 g of the filtered suspension into bottom sediments [52].

While consuming the fine suspension of various origin, filter feeders of the pelagial and periphytal zones agglomerate it into faecal pellets with diameters of 0.5–3.0 mm (up to 5.0 mm). These organisms play an important role in suspension sedimentation and speed up this process by eightfold or more compared with the usual speed of sedimentation due to gravity. The faecal pellets of some hydrobionts are sedimenting through the water column hundreds of times faster than the original source material [111], with speeds of 30–150 m per day for faecal pellets with a size up to 2×10^{-4} m [3].

Up to 80% of the matter filtered by mussels is transformed into faeces [6]. When consuming 1 kg of macrophytes, the amphipods convert 370 g of it into faeces. These faeces are enriched by bacterial OM and serve as food for various detritophagous animals, including some amphipods species, along with the rest of detritus [36].

Because of their characteristics, the detritus particles are good sorbents and participate in the transportation of various contaminants from the water column to the sediments, where pollutants and toxicants accumulate. In Sevastopol, Novorossiysk and Tuapse Bays, the sediments are polluted by oil products and the majority of silt bottom sediments in the Sevastopol Bay smells of fuel and sulphur. In Novorossiysk MP and in its adjacent areas, both strongly reduced and slightly oxidized conditions were registered in the sediments (silts, silted sands, silt with pebbles and shells). The prevalence of the bitumization of oil products over their ulmification indicates that, in these areas, the transformation of the organic part of bottom sediments is slow [49].

The bottom of various MPs is polluted by oil products at various concentrations, especially in old ports with reduced water exchange. The sediments of Novorossiysk, Tuapse, Odessa, Ilyichevsk MPs and Sevastopol Bay are the most polluted. The content of oil products in the bottom sediments of Tuapse port reached 2543 mg kg^{-1}, which is 51 times higher than the maximum allowable concentration [113]. In some areas of Sevastopol Bay, concentrations of oil products reached 24,000 mg kg^{-1} dry weight, corresponding to the maximum level for the Black Sea. The maximal concentrations of oil products were 6000 mg kg^{-1} in Odessa and Ilyichevsk MPs and about 1900 mg kg^{-1} in the more recently build Yuzhny port [39].

It was experimentally proved that, even under conditions of constant water exchange, the oxidation of the black silts containing oil products happens very slowly. An oxidized layer with a thickness of about 1 mm is formed over a period of 1 month. During 1 year, its thickness reaches 5–12 mm [87]. The hydrobionts inhabiting the silt contribute to oxygen penetration into the sediment and the passages dug by polychaetes inside the silt have the same light colour as the upper, oxidized layer [87].

Oil products, especially their heavy fractions, are preserved at the surface of and within sandy, silt-sandy and silty sediments over the years. The high concentrations of oil products at the MP bottom lead to the nearly total degradation of their benthic ecosystem. Baku Bay (Caspian Sea) is a typical example of extreme degradation of the bottom community. The maximal depth of the bay is 14 m and its sediments consist mainly of oiled, black sludgy sand where even amphipods are absent. The shortest distance from the shore of the bay where benthic organisms (*Mytilaster lineatus*, *Abra ovata*, *Cerastoderma glaucum* and *Balanus improvisus*) were found was about 7 km [7].

In MPs, the presence of silts is mainly the result of the vital functions of all biotic components of their ecosystems. Infusoria, sponges, worms, molluscs and crustaceans are forming the pelite microparticles. The sparingly soluble organic and mineral compounds are turned into dissolved forms of nitrogen, phosphorus, silicon and other biogenic elements by the activity of bacteria, infusoria, nematodes, polychaetes and other benthic organisms. The maximal levels of sums of provitamins D, methostenol and cholesterol are registered in grey silts, while the minimal amounts of sterins are found in black silts [48]. For many species inhabiting MP saprobiotical zones, the black silts are at the same time a substratum and a source of food. The detritus particles and faecal pellets, depositing at the sediment surface from the pelagial and periphytal zones, are enriched by bacteria and become food for detritivores of various taxonomic groups.

Feeding on the silt substratum, the bottom worms of the meio- and macrozoobenthos make it more structured. The zoobenthos organisms gradually change the chemical composition of the upper layer of the bottom sediments and facilitate the circulation of matter and energy among the benthal, pelagial and periphytal zones. In MPs, there is a constant exchange of biogenic elements between the bottom sediments and the water column. This exchange ensures a more or less stable functioning of MP ecosystems. The return of BM into the water column after OM mineralization in oxidized or reduced conditions takes place thanks to the physical processes of wave-induced mixing and upwelling. Inside MPs, upwelling is caused by the offshore–onshore wind phenomena and compensational flows.

The content of BM in the bottom sediments is one order of magnitude higher inside Odessa MP than outside it [5]. In the bottom sediments of Sukhoy and Grigorievsky Limans, the content of BM is much higher than in the adjacent aquatories [112, 137].

In Novorossiysk MP, the pore waters of the bottom sediments contained 42–66% more total nitrogen and nearly twice as much ammonia nitrogen than the open parts of Novorossiysk Bay [50].

In Sevastopol Bay, pore waters contained twice as much phosphates as overlying waters; while the concentrations of the other biogenic elements were one order of magnitude higher in pore waters. The flux of phosphates from the sediments is determined mainly by the processes in the upper 1 cm layer. However, the phosphates transformation continues in the upper 10 cm layer and stabilization of the phosphorus content occurs deeper in the sediments [59].

Bottom sediments, and particularly silts, are zones of BM accumulation. The easily stirred up fresh detritus silt, which is composed also of faecal pellets, is accumulated in great quantity in the deepest areas of the MP aquatories following the particularities of their hydrological regime. Thus, for example, the main amount (93–98%) of mobile BM in the pore waters of the bottom sediments of Grigorievsky Liman is found in their upper 10 cm layer. In pore waters of the bottom sediments of Sukhoy Liman, accumulation of organic phosphorus and nitrogen was three and two times higher, respectively, than in adjacent areas. Following biological activity, sediments can become a secondary source of BM to MP waters. This process is relatively slow, but intensifies during natural cataclysms, dredging and construction of hydrotechnical structures.

In MPs, the organic particles that are formed during the hydrobionts vital activity and after their death enter the sediments. Some organisms constantly create suspended matter, while others are filter the suspension from the water column and convert it to faecal pellets or fix, transform and displace the bottom sediments. These processes have a seasonal dynamics. The structure of bottom sediments plays a key role in controlling the settlement, quantity and biomass of benthic organisms. The semi-fluid silts are usually avoided by the majority of meio- and macrozoobenthos, except for some small worms.

As in the majority of ecosystems, the annual balance of the primary production and overall decomposition in MPs of the Black–Azov Sea basin is negative and OM accumulates. In the pelagial subsystem of MP, the largest part of the energy flow is conveyed via the detritus chain [18, 20, 114, 120]. Direct consumption of phytobenthos (mainly microphytobenthos) represents only 10% of the zoobenthos diet. The main diet of benthic and periphytic fauna consists of microplankton, bacteria, protozoa and microflora of bottom sediments and detritus (suspended organic matter—SOM, dissolved organic matter—DOM) [6, 20, 114, 120].

4.2 Main Biotic Peculiarities

In MPs benthal zone, three main subzones can be defined according to hydrodynamics intensity, peculiarities of granulometric and chemical composition of the bottom sediments and dissolved oxygen content in the near-bottom water layer and in the upper sediment layer: shallow waters, intermediate depths and bottom deepwater, stagnant zones. For each one, a specific complex of hydrobionts species commonly dominates during most of the year. In MPs, the size of the benthic fauna reduces with increasing depths. Small organisms of meio- and macrozoobenthos (deposit feeders and detritophages) adapted to life in saprobiotic conditions prevail in the deepwater zones of the bottom. Meiobenthos organisms from various taxonomic groups and numerous small-sized organisms of macrozoobenthos (polychaetes, molluscs, crustaceans) with size up to 1–2 cm constitute the basis of the fauna in the coastal zone of the pseudolittoral and up to depths of 1–2 m. The most large-sized zoobenthos representatives (nereid polychaetes, flat-browed mud shrimp *Upogebia*, bivalve *Mya* and others) inhabit the intermediate zone and populate the sediment's surface as well as its upper 20 cm.

In shallow waters of MPs, a key role belongs to macrophytes among autotrophs and to burrowing species of meio- and macrofauna among animals. In the intermediate zone, the autotrophic processes are mainly connected with the microphytobenthos, while macro-, meio- and microbenthos have a significant role in fauna composition. In the deepwater zones of the bottom, autotrophs are represented by microphytobenthos and its resting stages, while micro- and meiobenthos organisms form the animal biomass. The presence of macrozoobenthos hydrobionts is sporadic.

During meiobenthos studies in the Wight Sea, based on the example of nematodes, three areas with a high species similarity coefficient were found. The first one is delimited by the littoral zone and the zone of upper sublittoral up to 1–2 m depth, the second is located at 5–12 m depth and the third one starts from the depth of 18 m [29]. In aquatories of the Black Sea MPs, the second and third areas, analogue to those of the Wight Sea, are slightly shifted towards the shore because of the lowered hydrodynamics. Under the conditions of the Black–Azov Sea basin MPs, the mentioned areas correspond to shallow, intermediate and deepwater zones. In the White Sea, the highest number of meiobenthos organisms is registered at depths of 3–10 m during all seasons [28, 29]. The maximal values of species number and indexes of species abundance, for example harpacticoids in Grigorievsky Liman, are registered for depths of 3.0–9.5 m and the minimal ones at 10–17 m [114].

For the analysis of the MPs biota peculiarities, we define the shallow-water zone between depths of 0–2 m depths, the intermediate zone between 3–6 m and the deepwater zone starting from 8 m to the maximal depth of about 20–21 m [114]. At the same time, it is important to consider that there are no sharp borders, connected only with depths, between the zones. Such subdivision permits a better understanding of MPs aquatories ecosystems peculiarities of functioning and of their influence on the adjacent bottom zones.

Unlike those in the water column, the bottom biotopes have more or less marked gradient according to a series of factors such as the hydrodynamic water action, illumination, the character of bottom sediments, the velocity and peculiarities of sedimentation, the hydrochemical regime and others. Near natural shallow shores with a great number of harbours, rocks and stones, the biotopes are quite often very complex. In MPs aquatories, this mosaic structure is observed only when a part of the coastal shallow waters has been preserved in pristine condition. Dredging and construction of berths, moles and breakwaters inevitably lead to a simplification of the bottom biotopes structure to more or less uniform silty or silty-sand bottom, but, on the other hand, create a diversity of artificial HS which function as impoverished analogues of the bottom biotopes.

In the shallow-water zones of MPs, the bottom substrata are represented by stone, different grain-sized sand, sand with shells, silt with coquina and argillaceous silts. In the intermediate zone sand, silty-sand, silts with coquina, grey and black silts without hydrogen sulphide odour are present. The presence of grey silts testifies about the prevalence of oxidative processes over reductive ones. In MP aquatories, freshly deposited and water-rich silts with coquina whose formation is promoted by HTS foulers (molluscs filtrators and barnacles) prevail directly near the berths and up to a distance of about 5–10 m from them. At a distance of a dozen metres from berths and piers, less water-rich and denser black and grey silts are usually present; sometimes with admixture of coquina and sand [9, 21, 22, 40, 84, 92, 113, 116, 117].

At the bottom and on vertical surfaces, the so-called mussels' silt forms during the functioning of mussels' settlements. Leonov [67] noted that it has a dark colour in both wet and dry condition because of the high OM content. The mussels' silt quite often has a hydrogen sulphide odour. Because of the near-bottom hypoxia, anaerobic bacteria decompose the OM of the mussels' silt considerably slower than aerobic ones and form several compounds with unpleasant odour. A high content of SiO_2 and Al_2O_3 is characteristic for mussels' silt.

Historically, people discharged organic wastes and faeces into the sea during cities construction in the coastal zone, construction of ports and navigation. Concentration of population on a small territory always creates various negative consequences on natural ecosystems. Thus, at San Clemente Island, located at 76 km distance from the California coast, was located a garrison of only 300 marine soldiers, which resulted in an annual discharge of about 100 m^3 of sewage waters into the bay. During a short period, the macrophytes from the genus *Egregia*, *Hydrolithon*, *Halidrys*, *Phyllospadix* and *Sargassum*, which are indicators of the clean waters, completely disappeared within a distance of 100 m from the discharge. Near the discharge pipe, they were replaced by species from the genera *Ulva* and *Gelidium* and by cyanobacteria [31].

In various seas, flora and fauna are impoverished on the shelf near the coastal cities and in ports. For example, in the Adriatic Sea and in a number of Black Sea regions, *Cystoseira* algae were previously abundant on stony and rocky zones, but disappeared mostly because of organic pollution along with the whole biocenoses of *Cystoseira*'s beds.

The big-sized macrophytic algal species are usually substituted by *Cladophora*, *Enteromorpha*, *Ectocarpus*, *Polisiphonia* and others. It was established that domestic-grey wastewaters, entering into relatively closed aquatories, led to excessive input of BM and reorganization in macrophytes phytocenoses. Degradation of the conditions for some macrophyte species, and creation of new favourable conditions for other forms of vegetation, better adapted to the new conditions, are initially accompanied by a growth in biomass and productivity. Near the discharge point of sewage into the sea, a more abundant development of phytoplankton and microphytobenthos is periodically or constantly registered.

Wastewaters can influence macrophytes cenoses directly via BM input and indirectly via the development of epiphytic micro- and multicellular algae, bacteria, protozoa and animals of various systematic groups on the macrophyte thalli.

The river and sewage waters carrying a large quantity of BM and organics play an important role in the formation of ecosystems biota in the MPs aquatories of the NWBS, NEBS, at Crimean coast and in the Azov Sea. The estuarine zone of the Tsemes River serves as the aquatory of Novorossiysk MP. In Tuapse MP, the mouths of rivers Pauk (north-westward) and Tuapse (south-eastward) adjoin the external sides of the ports' protective structures. During south and south-eastward winds, superficial waters polluted by the river run-off enter into the port from the open sea, which intensifies the eutrophication of the port aquatories [114]. The same situation occurs in the port of Sochi, whose aquatory receives waters polluted by the run-off of the river Sochinki and the sewage waters of Sochi city. For example, in the NEBS, the ports of Novorossiysk and Tuapse have the biggest cargo turnover.

Eutrophication of MPs aquatories occurs because the increasing anthropogenic impact leads to suppression of some macrophytes species. These species are replaced by more resistant and productive once. Eutrophication also stimulates development of epiphytic algae, growth of drifting algal biomass and formation of green algae and cyanobacteria communities. Bottom macrophytes could totally disappear in hypertrophic MPs aquatories.

In the marine coastal ecosystems, the increasing oxygen deficit caused by excessive eutrophication initially changes the behaviour of meio- and macrozoobenthos organisms and then causes a gradual decline in species abundance, total quantity and biomass of bottom communities.

In the zones of strong organic pollution, MPs ecosystems are characterized by a low quantity of bottom inhabitants with domination of small-sized polychaete worms and nematodes and lack of filter feeders. One of the main causes of such structural–functional transformation of ecosystems is the accumulation, in the upper layer of bottom sediments, of labile sulphides which are toxic for many bottom inhabitants [114, 119].

In bottom sediments, accumulation of labile sulphides inevitably leads to degradation of the port ecosystems, impoverishment and changes in species composition of the bottom fauna, worsening of water quality and reduction of the self-purifying capability of ecosystems [114].

In ports' ecosystems, the main part of the energy flow is spent on formation of labile sulphides [119], which can be considered as an important self-polluting mechanism of the ports' benthal zone.

At high concentrations of OM, silts acquire a black colour and the anaerobic process of sulphates reduction by sulphate-reducing bacteria causes the formation of sulphides. This process is accompanied by the release of free hydrogen sulphide into the overlying water layer. At concentration of sulphides above 300–500 mgS per dm^3 of sediment, labile sulphides are the main cause of quick hypoxia development in the near-bottom water layer; even at relatively short-term stratification of the water column. Furthermore, labile sulphides are a strong cytochrome poison [145, 151].

Some organisms of the bottom fauna are able to adapt to the presence of labile sulphides in the bottom sediments and withstand concentrations up to 200–400 mgS dm^{-3} of wet silt. However, a reduction in the total biomass of zoobenthos and an impoverishment of its species diversity are observed at the same time [144, 151]. Organisms that are more resistant to the high sulphide level of contamination of bottom sediments become dominant [75, 150].

However, studies in Helgoland Bight showed that, following an increase in OM entrance and the appearance of hydrogen sulphide, a significant increase in the quantity of the molluscs *Abra alba* and polychaetes *Pectinaria* can occur in benthos [140].

Organisms tolerant to the presence of sulphides are among the mass representatives of the Black Sea bottom fauna. They include polychaete worms of the genus *Nereis*, *Polydora*, *Capitella*, *Melinna*, *Nephthys*, gastropods of the genus *Hydrobia*, *Tritia* and bivalves of the genus *Abra*, *Cerastoderma*, *Mya* and *Mytilaster*. Molluscs of the genus *Pitar*, *Modiolus*, *Mytilus*, *Lucinella*, *Chamelea* (Bivalvia) and *Rapana* (Gastropoda) could withstand moderate sulphide contamination [122].

In MPs aquatories, colonization of the bottom areas mainly by bivalves adapted to sulphide pollution such as *Mya*, *Abra*, *Tapes*, *Cerastoderma* and *Mytilaster* maintains high sulphides concentrations [148]. These molluscs filter suspended material and deposit it in the form of faeces and pseudofaeces, further stimulating sulphate reduction in sediments.

The high concentration of labile sulphides in the upper layer of sediments should be considered among the main causes of degradation of coastal marine ecosystems subject to anthropogenic pollution [114, 121, 142].

Oil and oil products are one of the main organic pollutants of the marine environment in general and of MPs aquatories in particular. Up to 2000 different molecules can be registered in oil composition. During destruction of oil patches and films by waves, oil globules of various sizes form in seawater. They can remain suspended in the water column, adhere to HTS surface and fouling organisms and settle down onto the bottom together with detritus, thus forming a layer of oiled sediments. Oil globes are perceived as food particles by some hydrobionts with phagocytosis type of feeding. More often, oil globes enter into the digestive tract of filter feeders. In general, accumulation of oil hydrocarbons in food chains occurs following pollution of aquatories by oil products [18, 86, 88, 97, 101, 114, 119].

At the same time, it was determined that the average amount of oil hydrocarbons in faeces of *Rissoa splendida* (a mass bivalve which feeds on microperiphyton) was about 93%. This shows that, unlike bivalve filter feeders, they practically do not accumulate oil hydrocarbons. However, by binding them into the faecal pellets, they increase the intensity of hydrocarbons degradation by microorganisms (bacteria and fungi) [86]. Other small-sized gastropods feeding on periphyton apparently fulfil analogous functions in MPs ecosystems.

As known, oil oxidizing bacteria, yeasts and fungi are able to decompose and utilize oil hydrocarbons in the marine environment. In turn, they become a food resource for the following links of trophic chains, among which an important role belong to protozoan and especially infusorians. In periphyton communities, many invertebrates feed on infusorians.

Popova [101] showed that periphyton infusorians (cilioperyphyton) are consuming not only the bacteria and fungi that destroy oil products, but also directly the oil globes. By becoming a food source for other hydrobionts, they pass oil hydrocarbons and products of their decomposition and transformation to higher trophic levels.

Infusoria also have an important role in the composition of microbenthos where they are quite often represented by the same species as in the periphyton. It is therefore possible to assume that they also utilize oil products in benthos.

Unlike other pollutants of the marine environment such as toxic metals, oil hydrocarbons could increase the productivity of some groups of hydrobionts belonging to low trophic levels. However, their accumulation along food chains is an extremely negative process at the ecosystem scale.

4.2.1 General Biotic Peculiarities of the Benthal Zone

Micro- and macrophytes play an important role in the primary production of OM and in establishing common patterns of ecosystems functioning in MPs aquatories. Significant amount of allochthonous OM also enter the aquatories of most MPs.

In MPs, saprobiotic conditions (excess of dead OM, absence or deficit of oxygen, presence of hydrogen sulphide) are formed in the black silts, i.e. areas of dead OM accumulation. Here, only organisms with specific adaptations can survive either permanently or temporarily. The fauna of the black silts is composed by representatives of micro- and meiobenthos along with the smallest sized forms of macrozoobenthos. The presence of big quantities of dead OM explains the prevalence of detritophages and mud feeders from various taxonomic groups on silty sediments. Bacteria and protozoans passing through mesh with 0.1 × 0.1 mm size belong to microbenthos, while protozoans, small-sized invertebrates and juvenile forms of macrozoobenthos passing through mesh with 1.0 × 1.0 mm size belong to meiobenthos [25, 29].

In the deepwater parts of MPs, the weak hydrodynamics and low mobility, or total immobility, of silty sediment create stable conditions for micro- and meiobenthos. At the same time, this situation can cause the redoxcline to rise up from the sediment—water interface into the overlying water and into the water column.

In MPs, micro- and meiobenthos occur in very high quantity in the upper layer of silty sediments. Although they do not yield high biomass, they favour the processes of OM total mineralization together with bacteria and Phycomycetes.

Colonization of detritus particles by bacteria already starts in the water column [13]. In MPs aquatories, detritus is mainly of phytogenous origin. The filter feeders inhabiting the water column and the HTS walls play an important role in transferring OM to the MPs bottom. From all the food of autotroph origin and used by a mature fouling community for its metabolism during the water filtration process, 88.8% originate from phytoplankton and 11.2% from macrophytobenthos [6]. On the one side, this testifies the important role of filtrators in maintaining the quality of the water column and, on the other, their leading role in OM transport to the bottom in the form of faeces and pseudofaeces.

As already mentioned, MPs sediments reflect all the processes occurring in aquatories. The fouling community is the main component of MPs ecosystems via which passes the main part of transformed matter and energy in the same way as in other, similar coastal ecosystems. Fouling community accounts for 88 ± 9% of total destruction of all pelagic and benthic invertebrates [6]. The share of bottom hydrobionts of micro- and meiobenthos, the black silts inhabitants, seem to be insignificant at the first sight. However, in terms of specific production, the functional activity of meiobenthos exceeds that of the large-sized bottom invertebrates by 10–15-fold [6].

In MPs, the presence of a high quantity of artificial HS increases the amount of produced OM compared with the adjacent waters, but it also lowers the possibility of OM elimination outside aquatories. The constant transfer of SOM and DOM into detritus followed by removal to sediments has a primary importance in determining the water quality and life conditions at the bottom.

In addition to sedimentation driven by gravity, the so-called "ecological pump" operates in some MPs aquatories and causes SOM to sediment quicker to the bottom. The "ecological pump" is based on biosedimentation performed mainly by phytophagous organisms, detritivores and filter feeders, inhabiting the water column and the HTS walls fouling.

After reaching the MPs bottom, the OM can: 1—mineralize in oxidizing conditions and return back into the biological turnover via the autotrophic link; 2—mineralize in reducing conditions and be included into the biological turnover via the autotrophic link in the form of BM; 3—be buried in sediments; 4—be incorporated into the biomass of micro- and meiobenthos; 5—become the initial link of heterotrophic food chains by being ingested by larvae and adult individuals of some mass micro- and meiobenthos representatives which, in turn, became food for mobile macroinvertebrates and fish. In the last case, the mass representatives of micro- and meiobenthos in silty sediments form a flow of live OM and energy directed from the bottom to the water column.

In MPs, the most unfavourable living conditions for the majority of meio- and macrozoobenthos representatives form in the benthal deepwater zones. At the same time, the presence of high values of OM in all seasons creates the possibility for the mass development of some detritivores and deposit feeders (nematodes, polychaetes, harpacticoids, foraminifera and others) adapted to life at lowered oxygen concentrations. Situations when hydrobionts either die or leave the polluted zone form on the bottom during some periods. In the latter case, hydrobionts actively move into ecotone zones and biotopes with more active water dynamics (shallow waters, HTS surfaces), use pelagic dispersal stages, or form resting stages (Fig. 4.9).

Despite their small size, meiobenthos organisms can form settlements with densities of more than 5×10^6 ind. m^{-2}. They are very prolific and mainly have a short life cycle, quick growth rate and high metabolic activity.

When living conditions deteriorate, meiobenthos responds by reduction of some groups, dominance of two–three groups and increase in the quantity of organisms with minimal body size having, as a rule, short life cycles [25].

The example of Sukhoy Liman confirms that two–three meiobenthos groups are present in the deepwater parts for most of the year. When the dissolved oxygen content in the near-bottom water layers increases from 3.5 to 5.0 mg dm^{-3}, the number of meiobenthos representatives already increases to 5–7 groups. For an

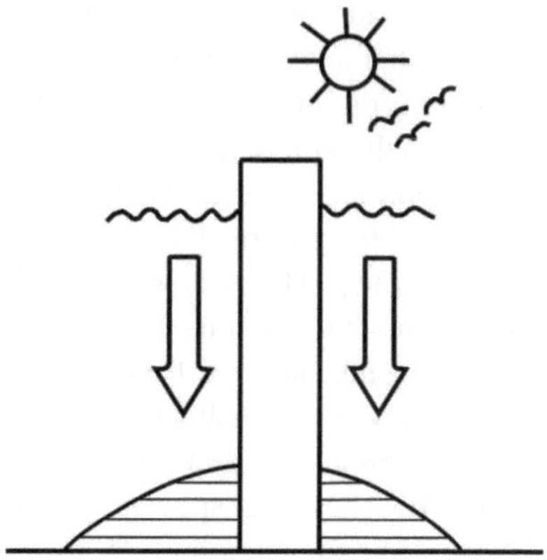

Fig. 4.9 Scheme of intensive accumulation of matter in marine ports and formation of ecotone zones

oxygen content up to 7.0–9.0 mg dm^{-3}, the meiobenthos composition includes all the known groups for the basin [25]. The vital processes of meiobenthos consume significant amounts of the bacterial and infusorial production besides detritus and diatoms [25, 54].

In MPs aquatories, meiobenthos also exists to a considerable degree due to vital activity of zooplankton, macrozoobenthos of HTS fouling and macrozoobenthos of the shallow waters and intermediate zone.

In MPs, the organisms of micro-, meio- and macrozoobenthos form complex trophic webs by feeding not only on dead OM and metabolic by-products of hydrobionts from other groups, but also on live OM. In all MP zones, meiobenthos communities continuously undergo complex reorganization processes because of the seasonal changes in the quantitative and qualitative characteristics of meiobenthos representatives, their food items, macrozoobenthos and ichthyofauna.

In MPs, the bottom sediments are the main habitat for the majority of meiobenthos organisms. Deterioration in the oxygen regime and sediment silting usually lead to significant reduction in the abundance of most meiobenthos groups accompanied by an increase in one or two of them. In MPs, such negative tendencies increase moving from the coastal shallow waters and sandbars around HTS through the intermediate zone and reach their maximum in the deepwater zones characterized by the presence of the black, semi-liquid, hydrosulphuric silt. Most groups of meiobenthos prefer depths at which the bottom sediments already do not experience significant mechanical influence from waves, but oxygen conditions still remain acceptable.

Because of its mechanical characteristics and hydrochemical peculiarities, the meiobenthos of silty sediments uses food resources that, for various reasons, can not be used by macrofauna. Where the hydrochemical conditions are still sufficiently dynamic, the use of these nutritional resources by macrofauna also becomes energetically unfavourable. Meiobenthos itself becomes food for many macrozoobenthos organisms and fish [25, 29, 132, 143, 146].

Groups such as nematodes, gastrotriches, kinorhynchs, harpacticoids, marine mites, foraminifera, tardigrades and others are constant components of the Black Sea meiofauna [26]. For example, the meiobenthos of Grigorievsky Liman has a marine character and is composed by 12 groups of big taxonomic units: nematodes, foraminifera, harpacticoids, ostracodes, halacarides, turbellarians, nemertines, kinorhynchs, oligochaetes polychaetes, bivalves and gastropods larvae and juvenile and cypris larvae of barnacles [25]. The maximum quantitative development of all meiobenthos groups was registered on the silty sediment with an admixture of coquina [56]. The species diversity of meiobenthos is most marked in mussels' biocenoses and extremely poor in the biocenosis of *Nereis* [25]. In MPs, mussels constitute the basis of fouling biocenoses and of some ecotone zones adjacent to HTS. They also inhabit the shallow-watered and intermediate zones. On soft sediments, nematodes and harpacticoids dominate in terms of density. In the NWBS sandy beaches, 52 species of interstitial meiofauna are usually present in the coarse-grained sands and 21 species in the fine-grained sands [25].

Most harpacticoids and halacarides are phytophagous, while nematodes include deposit feeders, detritophagous and predators. Predators prevail on sands and coquina, while detritivores and deposit feeders dominate on silts.

While moving from the shore towards the deepwater zones, the thickness of the sediment layer inhabited by micro-, meio- and macrozoobenthos organisms reduces and, in stagnation zones, benthic organisms inhabit mainly the sediment's surface.

Compared with the adjacent sea zones, the main zones of benthal population in MPs are shifted towards the shore because of the MPs protected character, artificially increased depths and weakened hydrodynamics.

Taking into account the important role of meiobenthos in all zones of MPs benthal zone, the biological peculiarities and the importance of its biggest systematic groups will be discussed in more detail.

Foraminifera. Foraminiferans belong to protozoans and usually have sizes in the range 0.1–1.0 mm. They have an exoskeleton in the form of shells (calcareous, sometimes chitinized and composed of agglutinated sand particles and others) that participate in the formation of bottom sediments.

The majority of foraminifera shows an alternance between haploid and diploid generations. Gametes are equipped with two flagella and swim actively, which contributes to the widespread presence of foraminifera in MPs benthal zone. They feed on unicellular algae and small-sized detritus particles. The smaller food particles are digested inside the shell and bigger ones outside it, in which case the food particle is covered by cytoplasm and enters the organism after digestion.

Together with nematodes and harpacticoids, foraminifera form the permanent meiobenthos fauna of the deepwater zone of Grigorievsky Liman, preferring silty sediments [51]. The density of foraminifera settlements sharply rises in proportion to the decrease of oxygen concentrations in the near-bottom water layer, but their species diversity reduces at the same time [92]. Foraminifera serve as food for many meio- and macrozoobenthos organisms.

Turbellaria. Turbellarians, or free-living flat worms, slide on the substratum surface using ciliar locomotion, but can also swim. During sliding they excrete mucus that serves as lubricant. Turbellarians length is usually less than 1.0 mm, but individuals from the order Rhabdocoela have lengths up to 2.0–3.0 mm. Although some turbellarians feed on diatoms, they are mainly predators and feed on nematodes and oligochaetes, rotifers, small-sized crustaceans and turbellarians of other species. Some species are interstitial. Turbellarians inhabiting silty sediment have a flattened body, which facilitates them in remaining at the surface of the semi-liquid substratum. Some turbellarians can inhabit environments containing hydrogen sulphide. Turbellarians have no pelagic stage of development. Turbellarians feed on juvenile polychaetes, belonging to families Spionidae, Capitellidae and Nereidae [152] whose representatives are usual in the benthal of the Black Sea MPs.

In the aquatories of Odessa MP and Grigorievsky and Sukhoy Limans, turbellarians are found in the peripheral zone adjoining the deepwater black silts [9, 112, 137]. In the majority of cases, turbellarians are present on sandy sediments and rocky bottoms [25]. They are typical inhabitants of the MPs fouling cenosis.

Gastrotricha. Gastrotriches, commonly referred to as hairybacks, are free-living, microscopic (0.05–1.50 mm), pseudocoelomate animals. Their body is covered by a cuticle and has a flattened form with cilia on the ventral side. The ciliated epithelium on the ventral side allows their creeping motion. They have hypogenesis, without a pelagic larva. They inhabit the surface of sand, silted sand and silts up to depths of 15 m. They can be found in macrophytes' bushes. Many species prefer middle- and coarse-grained sands and lead an interstitial way of life. They feed on detritus, bacteria, protozoans, diatoms and other unicellular algae. Some species can live into sand at a depth of several dozens of centimetres. Various species occupy their horizon of the littoral zone and inhabit certain sand layers.

Despite the absence of pelagic settling stages, gastrotriches have a rather wide distribution and take part in the transformation of OM in MPs benthal zone.

Nematoda. Nematodes are present in all MPs biotopes and include deposit feeders, detritophages and predators. The nematodes body is covered by a thick cuticle, which has a very different structure in the various groups. A chitinoid formation, derivated from the cuticle, protects the nematodes head during their movement through the interstitial spaces. The head capsule more often occurs in representatives of the order Enoplida and it also hampers their consumption by predators. Nematodes have no respiratory and blood circulatory organs. As a rule, they are dioecious and their eggs are enclosed in a solid capsule protecting them from physical damage and chemical impact of the environment. Their development takes place in the same habitats as those of adult forms. Larvae are also covered by a strong cuticle that they shed four times while increasing in size. Marine nematodes from the order Enoplida have predatory forms, armed with mobile teeth acting like strong jaws. The digestive apparatus of many nematodes has specific adaptations for fragmentation of the food pellets [41].

Detritophagous species prevail among nematodes. Remains of harpacticoids nauplii are found in the food pellets of some nematodes. The free-living nematodes secrete sticky, slimy strands in which even juvenile macrobenthos can be trapped. The free-living nematodes occupy the second and third trophic levels in MPs trophic chains.

Nematodes inhabit various sediments, from oxygen-rich ones to those of the reducing zone having high concentrations of hydrogen sulphide and lacking free oxygen. The saprobiotic nematodes of the group *Rhabdutida* have a size up to 1.0 mm. Development of eggs into adult specimen occurs very quickly, sometimes over a couple of hours. *Rhabdutida* moves and consumes food nearly continuously. Living in MP zones with excessive content of OM in the form of small-sized detritus (usually zones with black silts), they actively contribute to OM mineralization owing to the saprobiotic bacteria and primitive fungi which excrete strong enzymes. In MPs benthal zone, their main ecological functions are the processing of OM and the release of BM. Nematodes reach high density.

On silty MPs sediments, nematodes have the highest density of settlements compared with the other meiobenthos groups. They very rarely serve as a source of food for macrozoobenthos and fish [25]. Nematodes, first of all, destroy OM. Predatory nematodes perform OM transformation and regulate meiobenthos quantity. They usually comprise about 20.0–25.0% of the total quantity of this group of worms [26].

Most nematodes species concentrate in the uppermost 1.0 cm of surface sediments. For some populations, the maximum density is located in the subsurface sediment layer. The Thiobios group dwells deeper at 4–6 cm, in anoxic conditions and at high concentrations of hydrogen sulphide [90]. In the coastal zone, the nematodes quantity increases with the decrease in the sediment grain size and with increasing depth [56].

In 2001, 30 species of free-living nematodes were found in Odessa MP aquatory; 27 in silts and 8 in the fouling composition [9].

In Odessa sea-region, the most varied nematodes fauna is present in silty sediments at depths of 17–24 m. With regard to feeding habits, non-selective detritophagous organisms comprise, on average, 31.0–43.0% of the total in silty sediments, but only 8.8% in sand and coquina biotopes [63].

In Sukhoy Liman, nematodes are the dominating group in the composition of meiobenthos, although their density is relatively low and varies from 15,000 to 77,500 ind. m^{-2} [63].

In the lower and central parts of Grigorievsky Liman, where silty sediments prevail, meiobenthos composition included nematodes, foraminifera, harpacticoids and kinorhynchs. Nematodes and foraminifera comprise 99.1% of the total quantity of organisms. The total quantity of meiobenthos in Grigorievsky Liman varies widely ranging from 117,000 to 6.15 millions ind. m^{-2}. On average, nematodes had a density of 1.785 million ind. m^{-2}, representing 68.9% of the total meiobenthonic organisms [137].

In Grigorievsky Liman, the maximal density and biomass of nematodes are registered on silty sediments at 13–17 m depth. Out of 35 nematodes species found in this basin, 28 species occur in silts. More than 80.0% of nematodes in silts have size up to 2.0 mm. In fine-grained sediments, the share of nematodes in total meiobenthos composition varies from 60.0 to 90.0%. In the coastal zone, 17 nematodes species were found in sand with admixture of coquina [137].

Kinorhyncha. The spiny-crown worms kinorhynchs are aschelminthes. These are microscopic animals usually measuring less than 1.0 mm and less than 0.3 mm in most species. They are free-living organisms. Their body is covered by a well-developed cuticle, which plays the role of exoskeleton. Kinorhynchs larvae and adults occupy the same habitats. Kinorhynchs are common in shallow waters. They mainly inhabit in the interstitial spaces of sandy sediments, but can also be found in silts and on macrophytes. They cannot swim, but move by pushing their heads forward into the mud, extending their scalids, and then retracting their head into the body. They feed on unicellular algae and detritus.

The highest known indices of kinorhynchs quantity and biomass in the Black Sea basin were registered for the silty sediments of Grigorievsky Liman. An average quantity of 48,000 ind. m^{-2} was registered in June 1997 [137].

Nemertini. Nemertini or ribbon worms are visually similar to turbellarian worms. Their body is covered by a ciliated epithelium. Only benthic forms were registered in the Black Sea. The majority inhabit the shallow-water zone. All free-living nemertines are predators and catch other small-sized worms with the help of a proboscis. Pelagic stages of development are not known. Nemertines can actively move on substrata and swim in the near-bottom water layer. The smallest forms have lengths of 3.0–10.0 mm. Nemertines are mainly near-bottom animals settling on sandy and rocky sediments. They are registered in the aquatory of Odessa MP and in Grigorievsky and Sukhoy Limans on bottom zones adjoining to black silts [9, 112, 137].

Oligochaeta. Oligochaetes play an important role in aquatic ecosystems. Oligochaetes favour the processes of water and inorganic particles exchange between the superficial and deeper layers of silt by digging into it. Oligochaetes affect the physical and chemical characteristics of bottom sediments by passing detritus and rests of vegetal biomass through their intestines. Oligochaetes serve as food for bottom fish. They are considered a codominant meiobenthos representative in the coastal and intermediate zones of Odessa MP, Grigorievsky and Sukhoy Limans.

Harpacticoida. Harpacticoids are representatives of entomostracans and belong to Copepoda. They are dioecious without pelagic larvae. They inhabit the bottom and move mainly by crawling. However, many species can swim over short distances if needed, which allows them to move from fouling biocenoses into the sediments and across different depths. Some species have an interstitial way of life and, in MPs, are confined to the coastal, sandy, shallow waters or to the sandbanks adjacent to HTS. Harpacticoids breathe via the whole body surface and avoid bottom zones contaminated by hydrogen sulphide. They feed on detritus, unicellular algae and others. They play an extremely important role as food for larvae and fries of the bottom and near-bottom fish in whose diet they substitute for pelagic copepods and cladocerans. It is known that harpacticoids form accumulations on faecal pellets of macrozoobenthos organisms [56].

When the oxygen content lowers to less than 4.0 mg dm^{-3}, the harpacticoid density sharply reduces [25]. On average, the total harpacticoid density on silts is 1.8-fold higher than on silted sands [30].

In Grigorievsky Liman, the quantity of harpacticoids in the whole aquatory increases during the spring months. In particular, a more marked increase in their share in the total amount of meiobenthos organisms is observed in the shallow-water zone, while their quantity reduces by almost twofold at depths of 4–17 m. At the end of summer, cases of reduced quantities of the benthic copepods were registered in the coastal as well as in the deep part of the liman. At the beginning of autumn, harpacticoids of the deepwater zones are constantly depressed (6.6% of the total meiobenthos abundance), while nematodes and foraminifera dominate in terms of quantity. At the opposite, in the coastal zone, September is characterized by the highest relative abundances of harpacticoids in terms of total quantity (41.5%) and of the total biomass (63.7%) of meiobenthos [30].

Harpacticoids inhabit interstitial spaces of the splash zone, algal substrata and soft sediments [25].

Ostracoda. Ostracodes, also known as seed shrimps, are a class of Crustacea. Their body is protected by a bivalve shell. They are benthic animals inhabiting the surface of bottom sediments. There are species that crawl on aquatic vegetation and burrow into oxidized sediments. Many species can swim in the near-bottom water layer. Species inhabiting silty sediment have thin and smooth shells. Larvae hatch from brooded eggs with already hard shells. Ostracods nauplius larvae can both crawl and swim. There are detritivorous and predatory species among ostracodes. Generally, they feed on unicellular algae, dead plants and animals' residues. Ostracodes play an important role for benthos-feeding fish. In MPs, they mainly inhabit the shallow waters and algal substrata and belong to the co-dominant component of meiobenthos.

Halacaridae. Halacarides, or marine mites, have sizes of 0.2–1.7 mm. They have widely spread legs that allow them to grip tightly to substrata and are not adapted for swimming. Mites can easily attach to underwater objects and macrophytes moving along them. Eggs are laid on underwater substrata or are carried on posterior legs. They feed mainly by sucking algal juices, decomposing organisms, small-sized invertebrates and their eggs. Predator species are also known.

Marine mites prevail in the shallow-water zones where there is an algal substratum [25]. In Odessa Gulf, in the 1970s, five species of marine mites were registered on sandy sediments at a depth of more than 10.0 m and 14 species in the fouling composition of HTS (mussels and macrophytes) in the coastal zone (0–5.0 m). In 1994–2001, seven species were reported for the Odessa Gulf. Five of them were registered on macrophytes and six in mussel druses [9].

Polychaeta. Various species of bristle-bearing polychaete worms are present in all MPs biotopes and include detritophages, deposit feeders and predators. Many of them can switch from one feeding type to another. Some polychaetes inhabit the substratum surface, while others burrow into sand or silt. Some species build corneous, sand or lime tubes. Some polychaetes can inhabit both in the benthal zone and fouling, but the variety of polychaetes forms living in bottom biotopes is generally much higher [74].

In MPs aquatories, polychaetes favour the decomposition and transformation of OM and are the main component of fish forage. Polychaetes also play a significant role in the formation of bottom sediments structure by forming and displacing faecal pellets and by bioturbation of the upper sediment layer. In MPs, they play an important role in matter and energy turnover. Owing to their high caloric content, polychaetes are completely assimilated by fish [70]. Species inhabiting silty sediments predominantly feed on detritus deposits, while those living on hard substrata feed on detritus suspended in the water column or are predators. Most polychaete species inhabiting the benthal zone have pelagic stages of development [69].

For polychaetes inhabiting saprobic zones, sediment is not only a substratum and habitat, but also a source of food [70]. Polychaetes inhabiting MPs include small-sized forms of about a couple of millimetres such as *Syllidae* and others, such as *Nereidae*, that have sizes of more than 10 cm. The respiration of polychaetes usually takes place through cutaneous covering.

The small-sized *Sillidae*, which are part of meiobenthos, form high-density settlements. The quantity of these worms reaches 6430 ind. m^{-2} within mussel's community on silty sediments [54].

As a result of the connection of Sukhoy Liman with the sea and of the construction of Ilyichevsk MP, the fauna of this basin was enriched by 16 new species of polychaetes during 15 years [68, 71, 112].

In Grigorievsky Liman, 11 polychaetes species previously not registered there were observed after construction of the permanent channel connecting the liman with the sea [137].

In MPs benthal, the most widespread polychaetes are *Capitella capitata, Melinna palmata, Polydora cornuta (P. limicola), Spio filicornis, Nephthys hombergii, Heteromastus filiformis, Harmothoe imbricata, Neanthes succinea* and *Hediste diversicolor* [52, 53, 70, 83, 113, 116, 117].

The free-living polychaetes of the orders Phyllodocemorpha, Nereimorpha and Eunicemorpha are mainly predators that find their food on the black silts adjacent to MPs HTS [54].

The silt-inhabiting *H. diversicolor* and *N. succinea* can withstand the absence of oxygen for 7 days and the presence of hydrogen sulphide for 3–6 days. *H. diversicolor* and *N. succinea* can bury into the black silt at depths up to 30 cm and 10 cm, respectively. In laboratory conditions, polychaetes leave their shelters and climb up the walls of culture dishes when the water has been stagnating for a period ranging from a couple of hours to a few days [19]. In natural conditions, they have the same possibility on berths' walls, piers and breakwaters of MPs.

Losovskaya [70] indicates that *N. succinea* is often present in sediments close to HTS. Besides detritus, the rests of polychaetes and amphipods were found in the food composition of *N. succinea*. Foraminifera, nematodes, ostracodes, halacarides and gastropod mud snails (Hydrobiidae) were found in the intestines of *H. diversicolor* [70].

Harmothoe imbricata lives on stones, among macrophytes and on silts. It feeds on small crustaceans, gastropods and other bottom invertebrates. Instances of cannibalism are also known [146]. Mineral particles, detritus, diatoms' shells, foraminifers, gastropods and other items were found in the intestines of *H. filiformis* [70].

In MPs, *N. hombergii* inhabits all types of silty sediments at all depths. It feeds both within the sediment and on its surface. The rests of its food included algae at various stages of decomposition, diatoms, detritus, foraminifera and nematodes [70].

It is known that the polychaete *C. capitata* is capable of living in the most polluted harbours, where oil and fuel float on the water surface and cover the whole bottom and all underwater objects [41]. *C. capitata* feeds on detritus and diatoms. No meiobenthos organisms were found in its food [70]. *C. capitata* larvae are registered in plankton nearly all year round, i.e. there is always the possibility of forming new settlements and renewing old ones on the bottom after their death.

Melinna palmata is connected with zones rich in organic detritus. It inhabits loose silty sediments surmounted by a near-bottom water layer having oxygen concentrations of less than 3 mg dm^{-3} [70]. *Melinna palmata* forms big settlements associated with semi-liquid black silt and with tubes located close to each other. Polychaetes feed on detritus, collecting it around tubes.

The small-sized polychaetes *P. cornuta* and *S. filicornis* are mass, detritivorous forms. In Odessa MP aquatory, 19 species of polychaetes were registered, including 17 species in benthos and 9 species in fouling. Average indexes of polychaetes quantity and biomass are higher on shells with silt than on silts. In zones with relatively free water exchange, the average quantity of polychaetes was fourfold, and their biomass fivefold, higher than in areas with limited water exchange. In 2001, the polychaetes species composition in Odessa MP benthos was richer (17 species) than in Odessa Gulf and in the whole sea region of Odessa between Sukhoy and Grigorievsky Limans [74].

In 2001, the polychaetes most often registered in the macrozoobenthos composition of Odessa MP deepwater zone were *N. succinea*, *H. filiformis*, *C. capitata*, *P. cornuta* and *S. filicornis*.

In HTS fouling, *N. succinea* accounted for 81.5% of the average quantity and 97.7% of the average biomass of all polychaetes. In benthos, this species was dominant only in terms of biomass (63.9%), while *P. cornuta* and *H. filiformis* dominated in terms of quantity (37.8 and 27.9% of the average polychaetes quantity, respectively) [74].

The main taxa of polychaetes were *H. imbricata*, *N. succinea* and *P. cornuta* in Odessa MP and *N. hombergii* and *H. filiformis* in Grigorievsky Liman. These species are mainly detritophages. In Grigorievsky Liman, *N. succinea*, *P. cornuta* and *M. palmata* were found as codominant bottom inhabitants [117, 137].

In the macrozoobenthos of the deepwater zone of Sukhoy Liman, polychaetes represented 86.6% of quantity and 53.7% of biomass; with *H. diversicolor*, *N. succinea*, *P. cornuta* and *S. filicornis* being quantitatively dominant [112].

Studies of the MPs meiobenthos showed that the representatives of eumeiobenthos dominated the meiobenthos of Grigorievsky Liman (Yuzhny MP) during all seasons. Their share comprised 77.6–84.8% of the total quantity of organisms in the coastal zone and 93.0–99.8% at depths of 2–13 m. Nematodes and harpacticoids were the main component of eumeiobenthos in the liman [137].

In Grigorievsky Liman, the share of pseudomeiobenthos varied from 4.12 to 36.2% of the total quantity of organisms in the deepwater zones and from 30.6 to 89.0% in the nearshore zone [137]. In the deepwater zone of Odessa MP and in Sukhoy Liman, the appearance of pseudomeiobenthos species had a sporadic character.

In the benthal zone of Grigorievsky Liman, representatives of 76 macrozoobenthos taxa were registered, including 3 coelenterates, 22 worms, 16 molluscs, 32 crustaceans, 2 chironomids larvae and 1 tunicate. The highest taxa number (59) was found in the coastal zone; the corresponding figure in the deepwater zone being 44. Crustaceans were represented by 28 species in the coastal zone and by 15 species in the deepwater zone. In the coastal zone, the maximal number of taxa reached 27 at one station, but the average number for the liman was 8.8 [137]. In 1992–2006, *N. succinea*, *P. cornuta*, *H. filiformis*, *Hydrobia acuta* were among the permanent macrozoobenthos taxa of Grigorievsky Liman, while *N. hombergii*, *M. palmata*, Oligochaeta g. sp., *M. lineatus*, *C. glaucum*, *A. ovata*, *Mya arenaria*,

Sphaeroma pulchellum, *Idotea baltica basteri*, *Ampelisca diadema* and *Chironomus salinarius* belonged to codominant taxa [137].

In MPs ecosystems, the main role belongs to mussel's settlements. In these water basins, mussels inhabit on soft sediments as well as on HS of natural origin (pebbles, stones). A significant part of macrozoobenthos in the mentioned basins is however connected with the coastal belt of macrophytes, which usually forms at depths between 0.5 and 1.5–2.0 m.

Forty-three macrozoobenthos taxa were found in the deepwater zone of Odessa MP and 44 in Grigorievsky Liman. For comparison, only 10 taxa were reported in the meio- and macrozoobenthos composition of Constanza MP at the end of the 1970s–beginning of 1980s [149]. The benthos of Tuapse MP comprised 13 species [84]. In 1997, 48 and 43 taxa of macrozoobenthos were found in Sevastopol Bay and in Kamishovaya Bay, respectively [4].

4.2.2 Biotic Peculiarities of the Shallow Waters of the Benthal Zone

In the coastal shallow waters and sandbanks of MPs, the biota (quantity, species composition, size, morphological peculiarities, feeding, reproduction) is first of all regulated by wave action and by sediments stability and composition.

In MPs aquatory, the construction of berths, moles and breakwaters can, along with dredging activity, totally destroy the coastal shallow waters and the biocenoses connected with them; as it happened in Constanza, Odessa, Yalta, Sochi and Batumi MPs. The coastal biocenoses typical for MPs aquatories are partly preserved, for the moment, where the shores are not totally covered by HTS; as in the case of Sukhoy and Grigorievsky Limans, Varna and Beloslav Lakes and Sevastopol and Kamishovaya Bays. In MPs aquatories, routine dredging works result in narrowing of the coastal shallow-water zone, reduction of its area and quick increase in water depth. As it was mentioned above, in the Black–Azov Sea basin, the littoral zone occupying the coastal part of the shallow waters in which light penetrates down to bottom is divided into supra-, pseudo- and sublittoral [44, 45, 61]. Hydrobionts inhabiting MPs shallow waters are represented mainly by oxyphilous and eurioxybiont species.

In MPs, the surface layer of coastal and shallow-water sediments is populated by bacteria, autotrophic and heterotrophic flagellates, infusoria and other protozoans and organisms of meio- and macrozoobenthos.

In the 1960s–1970s, during construction of Ilyichevsk and Yuzhny MPs, the main biocenoses in the supralittoral zone of Odessa marine region (and of the NWBS as a whole) included: biocenosis of *Talorchestia* on sandy beaches with admixture of vegetative casts, biocenosis of *Orchestia* on algal storm casts, biocenosis of seagrass (*Zostera*) on sandy beaches and biocenosis of amphipoda (*Gammarus*, *Orchestia*) on sandy-pebble beaches. During the same period, the composition of pseudolittoral biocenoses comprised: biocenosis of *Donacilla*,

Ophelia and *Pontogammarus* on sandy beaches, biocenosis of *Gammarus* and *Eurydice* on beaches with pebbles and stones, biocenosis of polychaete (*Microspio*) on pseudolittoral sands and biocenosis of nemertines (*Lineus*) on silty-sand coastal sediments [12]. Biocenoses of *Lentidium* and *Mya* were described later [112]. The quantity of organisms in these biocenoses reached thousands and tens of thousands of individuals per square meter [12, 112]. In Odessa sea-region, fouling of the brown alga *Cystoseira barbata* with a specific complex of invertebrates and fish formed on all big stones, rocky beds and rocks up to depths of 4–5 m.

In the middle of the 1970s, the negative tendencies in the development of coastal biocenoses in the NWBS started to become clear. As a consequence, various reorganizations of biocenoses took place and part of the species completely disappeared from the biota composition. All these large-scale ecological processes, chiefly connected with the anthropogenic activity, also appeared in MPs aquatories. *Cystoseira* community has practically disappeared in the NWBS. Species composition and ratio among macrophytes species also has significantly changed [112].

Overall, the presence of a high number of micro-, meiobenthos and macrozoobenthos organisms in the littoral zone is determined by the abundance of food resulting from the massive development of attached vegetation as well as from the entry of OM of vegetal and animal origin in the form of storm casts. Part of OM also originates from land.

In MPs, previously existing or newly formed shallow waters play an important role in the ecosystems functioning and frequently represent specific reserves facilitating the preservation of biodiversity. Shallow waters are the most dynamic zones of MPs where, in spite of significant fluctuations in temperature, salinity, illumination and wave action, the content of dissolved oxygen does not usually drop below the critical level. Various sediment types are found in the coastal zone: rocky bottoms, sands of different grain sizes, sands with coquina, silted sands and silts. Such substrata diversity increases the number and spectrum of ecological niches. The macrophytes bushes are connected with this zone.

In coastal bottom communities, the primary OM production is connected with unicellular algae, especially diatoms belonging to microphytobenthos, and with macrophytes.

In the benthal of MPs, unicellular algae, macrophytic algae and high aquatic plants form the autotrophic level. In this zone, the development of aquatic photosynthetic plants is controlled by available area, illumination, hydrodynamics and hydrochemical conditions.

The production of microphytobenthos on hard substrata of the coastal zone exceeds that on soft sediments by about tenfold [127]. Increase in HS areas via coastal protection by riprap and concrete tetrapods lead to a significant increase in the primary production of microphytobenthos. Diatom algae are the main component of the microphytobenthos of MPs coastal shallow waters. They play an important role in the biological productivity of aquatic ecosystems and feeding of bottom organisms. The pennate diatoms comprise more than 80.0%, and centric less than 20.0%, in the macrozoobenthos food ration [108]. Depending on season and water temperature, microphytobenthos dwelling on rocky sediments of the

Black Sea coastal zone synthesizes 0.45–28.2 g of carbon per month per 1 m^2 [81]. The number of bottom diatoms on sublittoral silts is rather high and their biomass comparable with that of zoobenthos [102, 109].

In MPs aquatories, the significant artificial increase of depth worsens the living conditions of many microphytobenthos representatives on the bottom. At the same time, introduction of vast areas of artificial HS into MPs aquatories creates the possibility for mass development of attached microphytobenthos forms directly on HS as well as epiphytes on various big-sized representatives of the fouling community.

In the Black and Azov Seas coastal zone, significant interaction of microphytobenthos and phytoplankton communities is registered because of turbulent mixing, offshore–onshore phenomena and stirring up of bottom sediments [109].

During a year, diatoms with sizes of 20–60 μm are the main contributors to the size structure of microphytobenthos, while diatoms and dinoflagellates (with the size of 20–100 μm) mainly contribute to that of phytoplankton [109].

According to Ryabushko [117], 132 species of microphytobenthos and 86 species of phytoplankton (33 species being common with microphytobenthos and 50% of them being present in the water column) were found under the eutrophic conditions of Kazachya Bay (Sevastopol). The development of microalgae in the bay exhibited two peaks: the first one in spring (in terms of quantity) for microphytobenthos and the second one in summer (in terms of biomass) for the phytoplankton. The maximal values of the Shannon index were registered in spring for microphytobenthos ($H = 4.25$) and in summer for phytoplankton ($H = 4.26$). Quantity and biomass of microalgae were 2–2.5-fold higher in the bay than in the adjacent open sea zones [109].

Seventy-six species potentially toxic for biota and man were registered in the composition of the Black Sea microalgae; 50% of them living in the benthal zone [109].

The number of taxa found in microphytobenthos was 103 in the NWBS, 318 in Crimea and 156 in the Zernov phyllophora field in the central part of the NWBS. The seashores of Caucasus and Crimea had 42 species in common [109].

In the coastal shallow waters of Sukhoy and Grigorievsky Limans, the fennel pondweed *Potamogeton pectinatus* (sometimes called ribbon weed) and the seagrasses *Zostera marina* and *Zostera minor* have an important role together with other macrophytes.

The ribbon weed grows on silty sediments and forms tall, ramified stems (up to 100 cm and more) with a well-developed, ramified root system. It blooms in May–June. The vegetative reproduction prevails over seed propagation. Juvenile grey mullet, atherines and gobies can be found among the pondweed beds.

Z. marina and *Z. minor* are found in the same habitats and under similar conditions, but with varying relative abundances. The prevalence of one of the two *Zostera* species depends mainly on sediment characteristics. *Z. minor* predominates on clean sand or sand with coquina, while *Z. marina* prevails on silty sediments. *Zostera* biomass is minimal in November–December, starts to increase in February–March and reaches its maximum in June–July. *Zostera* leaf abscission takes place in August–September. Vegetative development completely stops in autumn at water temperature of 12.0–13.0 °C and anthesis takes place in May. Reproduction is mainly vegetative by sprouting repeatedly from the rhizome. The seasonal changes of *Zostera* biomass are not connected with temperature, but rather with seasonal fluctuations in light regime [93]. *Zostera* is a source of detritus in the coastal zone. *Zostera* and the associated diatom cover are a source of food for many invertebrates (amphipoda, gastropods, polychaetes and others).

Zostera minor is most often found on sandy sediment with a small proportion of silt at depths of 0.5–4.0 m, while *Z. marina* mainly occupies the sandy-silty sediments at the depths below 3.0 and 4.0 m.

The coastal part of macrophytes belt is regularly washed ashore by waves where it is utilized by various organisms. Amphibiotic freshwater shrimps of the genera *Orchesia* and *Talorchestia* are among its main consumers. *Orchesia* population accumulates the energy of primary OM in the form of protoplasm of its body, but also forms faeces. Upon availability of sufficient amount of algal storm casts, *Orchesia* can form 200–500 kg (wet weight) of faeces per square meter per annum. These faeces are washed out by waves and enrich coastal waters with organic suspension [123].

In MPs, macrophytes bottom beds are observed in some zones of the coast, both in shallow waters and in the benthal intermediate zone, as well as on silty-coquina bars. In the Black Sea, 162 species of zoobenthos were identified in the coastal macrophytes belt [79].

Two complexes of organisms are distinguished in the swash zone of the Black and Azov Seas: 1—active pseudolittoral (constantly displacing to remain in the zone of breaking waves and represented by filter feeders animals from amphipoda and bivalves); 2—passive pseudolittoral (does not displace along the littoral following the sea level and awaits water arrival to its dwelling habitat). The zone of the second complex is connected with the middle position of the water line. Both pseudolittoral complexes are absent in MPs shallow waters adjacent to HTS, because of the lack of the classical near-swash zone. In MPs, these complexes are found exclusively in the coastal shallow waters of the preserved natural coastal line.

Organisms of the supra-, pseudo- and sublittoral zones consume a significant part of the dead OM from the biologic turnover, immediately after its formation. More than 95.0% of the underwater vegetation of the Black Sea undergoes destruction in the coastal line communities [130]. In MPs, in which the coastal line is absent, the dead OM of plant origin mainly accumulates on the bottom.

At water oxygen concentration of 7.0–9.0 mg dm^{-3}, typical for the MPs coastal zones, the representatives of nearly all its big taxonomic groups are usually registered in the composition of meiobenthos [25]. The main part of meiobenthonic forms of MPs shallow waters can belong to epi- and infauna at the same time, because of its constant displacement between the sediment surface and its thickness during its life cycle [26]. Conditions for pseudomeiobenthos development are more favourable in MPs coastal shallow waters and sandbars than in the deepwater zones, especially compared with water-rich black silts. The distribution of meio- and macrozoobenthos in the coastal zone of MPs is closely connected with the physico-chemical characteristics of bottom sediments and macrophytes debris.

In rocky zones with active hydrodynamics, fine-dispersed bottom sediments do not accumulate appreciably and as in the periphytal, biocenoses are dominated by epifaunal filter feeders (comprising up to 64.0–84.0% of total biomass), infaunal organisms (2.0%), collectors (2.0%) and deposit feeders (4.0%) [42].

Interstitial meiofauna is especially developed in sandy zones with coarse- and middle-grained sands. The role of infaunal filter feeders, collectors and deposit feeders organisms increases in silty-sand and silty sediments.

Meiobenthos has an uneven distribution with depth. As a rule, the highest abundances are found in the pseudolittoral, upper sublittoral and, especially, phytal zone. This zone is also the richest in terms of faunal taxonomic composition which impoverishes as depth increases [27]. The main factors determining the depths of animal penetration into the sediment are its mechanical characteristics and its oxygen regime. At sandy beaches, the content of pore water in interstitial spaces also plays an important role [28].

In sandy sediments (middle- and coarse-grained sands), 85.0% of total meiobenthos settlements density is confined to the uppermost 5 cm, although organisms from various groups are present up to depths of 20 cm. In silty sediments with a high quantity of detritus, 94.0% of all meiobenthos organisms inhabit the upper first centimetre. Eumeiobenthos groups such as harpacticoids and turbellarians are registered only in the upper two centimetres layer; while oligochaetes penetrate up to depths of 4 cm and nematodes, small-sized polychaetes and gastropods are registered up to depths of 10 cm. However, 50.0–80.0% of these animals are concentrated in the upper two centimetres of bottom sediments.

In the coastal zone, hydrodynamics is less active at depths of 2–5 m. Here sediments formation dominates over their transportation so that the MP bottom is covered by silty-sand, silt and clay. Feeding on detritus is energetically justified at sufficient high speed of its sedimentation. This is why infaunal communities of filter feeders develop on the coarsest sediments (sands) where significant deposition of detritus does not take place. In such biocenoses, infaunal filter feeders comprise up to 74.0% of total density, epifaunal filter feeders 6.0%, collectors 14.0% and deposit feeders 4.0% [42, 62].

At further weakening of hydrodynamics between the coastal shallow waters and the deepwater zones of MPs, the sedimentation of detritus increases and organisms feeding on detritus, i.e. detritophages collectors and deposit feeders acquire a leading role. In biocenoses mainly formed by collectors, these groups comprise 64.0% of the biomass, while the corresponding contributions of epifaunal filter feeders, infaunal filter feeders and deposit feeders are 4.0, 9.0 and 16.0%, respectively [42, 62].

In the coastal zone of Grigorievsky Liman, meiobenthos is represented by 12 groups during spring, but some of them have very low occurrence in most cases; turbellarians—10.5%, halacarides—15.8%. At depths below 2 m, nematodes are quantitatively dominant and comprise, on average, nearly 70.2% of the total amount of organisms. The remainder is almost equally shared between harpacticoids and foraminifera, with the contribution of other groups to total quantity being less than

1.0%. A sharp quantitative and qualitative impoverishment of meiobenthos takes place at the approach of autumn and the number of groups reduces to eight [137].

In the coastal zone up to 1 m macrozoobenthos composition comprises taxa of worms (14), molluscs (13), crustaceans (28) and other 4 (coelenterates and chironomid larvae) [137]. The average quantity and biomass of macrofauna are 20,275 ind. m^{-2} and 334.1 g m^{-2}, respectively. The permanent taxa are *N. succinea, H. diversicolor, P. cornuta, Oligochaeta* g. sp., *H. acuta, M. lineatus, C. glaucum, A. ovata, M. arenaria, S. pulchellum, Idotea baltica basteri* and *C. salinarius* [137].

The early ontogenesic stages of some fish are connected with the NWBS coastal shallow waters, as in the case of the Black Sea flounder *Platichthys lucsus*, turbot *Psetta maeotica*, sand sole *Pedusa lascaris*, gobies *Gobiidae*, combtooth blennies *Blenniidae*, grey mullets *Mugilidae*, the Black Sea sand smelt *Atherina pontica*. The destruction of coastal shallow waters makes their normal development impossible. In MPs, the absence of the coastal shallow waters with their biocenoses, the creation of steep shore conditions with reduced wave action, the presence of vertical walls of quays, berths and moles prevent the destruction of dead OM, molluscs' shells, tests and skeletons of hydrobionts; all of which are accumulated in the bottom sediments of aquatories. In the coastal zone and in the shallow waters of MPs, bacterial biomass can reach 5.5 g per 1 m^2 in sandy bottom compared with 57.0 g in silts of the deepwater areas [45]. The lower bacterial biomass indicates the presence of relative good conditions in the shallow waters. When critical conditions appear in the deep waters and in the intermediate zones of MPs, some mobile hydrobionts survive by moving to shallow waters and into HTS fouling cenosis.

Ship waves have significant influence on macrozoobenthos composition in the passage zone into Grigorievsky Liman aquatory, inside and outside it. Waves are generated not only by big ships, but also by tugs, pilot and roadstead cutters. In shallow waters, owing to the ships' waves action (0.1–0.4 m), sediments adjacent to the shipping channel consist of well-washed, small-grained sands. Accumulation of the dead OM, predominantly macrophytes, bivalves and jellyfishes, takes place in the zones of interaction between ships' waves and the shore line. Nevertheless, an oxygen deficit occurs very rarely at the bottom of these shallow waters having depths up to 1 m.

In bottom zones adjacent to edges of the passage into Grigorievsky Liman aquatory, bottom macrofauna usually comprises small-sized vagile species with low quantitative indices, which is typical for such type of bottom sediments and for the open seashores of similar depth in the NWBS. Representatives of 16 taxa were found in the passage zone: worms and crustaceans—6 taxa each and 4 molluscs' taxa. In 2013, the quantitative indices were 1153 ind. m^{-2} for density and 6.072 g m^{-2} for biomass. In the coastal zones of the liman, a total of 37 taxa was found at the same depths, including worms (10 taxa), molluscs (9 taxa), crustaceans (17 taxa) and chironomid larvae (1 taxon). According to the results of five surveys carried out in 2013, the average quantity of macrozoobenthos comprised 56,448 ind. m^{-2} for a biomass of 205.410 g m^{-2}.

In May 2013, an unusual mass development (754,700 ind. m^{-2}) of the small-sized detritivorous polychaete *Pygospio elegans* (body mass less than 0.001 g) occurred in the coastal shallow-water zone located on the right bank of Grigorievsky Liman, slightly south of Odessa Port Plant. It could be related with accumulation of a high quantity of detritus in autumn 2012, transformed by winter storms.

The settlements of seaweeds, pondgrass and especially green algae are connected with the coastal shallow waters which, in MPs aquatories, are regularly impacted by ship waves. As the ship wave's influence weakens, a higher number of epiphytes appear on macrophytes and the number of gastropods and crustaceans feeding on periphyton increases.

On the small-grained sand found in the shallow-water part of Grigorievsky Liman, the bivalves *Lentidium mediterraneum* (up to 6–7 mm in size) and the deeply borrowing *M. arenaria* (up to 100 mm in size) form the bulk of quantity and biomass of macrozoobenthos. The sand shrimp *Crangon crangon* is also characteristic among the large-sized forms of macrozoobenthos in this biotope.

Flounder juveniles, various species of gobies and mullets gain weight along the water edge of the small-grained sand zones of Grigorievsky Liman exposed to the action of ships' waves.

On well-washed, small-sized sandy sediments of MPs, infaunal organisms prevail over epifaunal ones both in quantity (92.1%) and biomass (88.2%) and vagile hydrobionts (99.7% of quantity and 99.6% of biomass) over sessile ones.

In the coastal zone of Grigorievsky Liman, 26–29 taxa are present in various biotopes. In 2013, the small-sized and compact "silted sand biotope" had the highest average indices of density (138,003 ind. m^{-2}) and biomass (327.345 g m^{-2}), while the biotope of sand with shell had the lowest ones (1588 ind. m^{-2} and 11.502 g m^{-2}).

4.2.3 Biotic Peculiarities of the Intermediate Depth Zones of the Benthal

In MPs, biota of the intermediate depth zone having reduced hydrodynamics is regulated by the input of detritus and longshore drifts flow. In this zone, most of SOM settling to the sediment surface is transformed by organisms inhabiting bottom sediments, while only a small part of it is buried into sediment thickness. In the intermediate zone, the dissolved oxygen content reduces sharply (below 3.5–4.0 mg dm^{-3}) at depths ranging from 1–2 to 4–6 m. Representatives of 5–7 meiobenthos taxonomic groups are usually present under such conditions [25]. The majority of meio- and macrobenthos, with representatives typical for each specific port location, usually inhabits the intermediate zone of MPs.

Natural sandy-coquina bars and sandbanks remain in the aquatories of some MPs (Sukhoy and Grigorievsky Limans, Varna and Beloslav Lakes). In some ports of the Black–Azov Sea basin, protective constructions consist of slightly changed natural geomorphologic formations. During dredging, part of the extracted sediment is sometimes deposited near the fairway. Dombovy Island in Sukhoy Liman and sand islands in Dniester Liman originated in such way. More or less stable sandbanks, which serve as biotopes for coastal hydrobionts, form around such islands. Owing to their content of natural materials (clay, sand, coquina and stones), they are quickly colonized by organisms complexes typical for the coastal zone. Such areas represent shallow-water zones created into aquatories at depths typical for the intermediate zone.

Detritus is usually less energetically valuable than food of vegetable and animal origin, but it is present in significant amounts in MPs ecosystems. For adequate nutrition, hydrobionts inhabiting the intermediate zone also use other food (bacteria, microphytobenthos, meiobenthos and small-sized macrozoobenthos).

The silts of polluted harbours, to which MPs aquatories belong, are inhabited by a rich saprophitic microflora. The quantity of alive saprophitic microflora decreases by several-fold within the sediment thickness, but it persists even at depths up to 60 cm [120].

In MPs, the sediment character influences bottom microalgae photosynthesis. The intensity of photosynthesis is 1.5–2.0-fold lower on silts than on silty-sands, both being the main sediment types in the intermediate zone. Respiration processes prevail on silts. For an increase in depth from 1 to 2 m, the concentration of chlorophyll "*a*" decreases both on sandy sediments and on silts in the majority of cases [126].

On silty sediments, the microphytobenthos production is low and destructive processes prevail. On sands, the microalgae production is three- to fourfold higher than on silts [127]. Thus, in the intermediate zone, the role of bottom microalgae photosynthesis is significantly lower in communities of silty-sands sediments than in those of sands. The qualitative composition of microphytobenthos changes significantly depending on sediment character, with 40.0–50.0% of diatom species being common for silts and sands. On soft silts, the wet biomass of microphytobenthos comprises on average 89.0 g m^{-2}, although variations are possible depending on season, depths and sediment character [127].

In the coastal and intermediate zones of Grigorievsky Liman, meiobenthos composition includes big taxonomic groups such as foraminifera, nematodes, harpacticoids, ostracodes, halacarides, turbellarians, kinorhynchs, oligochaetes, polychaetes, larvae and spats of bivalves and gastropods and cypris larva of barnacles. In spring–summer period, the occurrence of halacarides, turbellarians, gastropods and barnacles comprised 27.0, 16.0, 44.0 and 33.0%, respectively. In Grigorievsky Liman, the meiobenthos spatial distribution is uneven and depends on sediment type, hydrodynamics intensity, depth, presence of algal substrata, salinity and level of anthropogenic pressure [137].

In the intermediate zone of Grigorievsky Liman, foraminifera occurred, on average, in 91.5% of samples, harpacticoids in 86.4%, larvae of bivalves in 57.6%, polychaetes in 47.4%, oligochaetes in 39.0% and ostracodes in 35.6%. Nematodes were commonly present. In the pseudomeiobenthos composition of the intermediate benthal zone, bivalves, polychaetes and oligochaetes had the highest densities, while marine mites, turbellarians and gastropods could be considered as relatively rare [137].

On soft and loose sediments of the coastal shallow waters and intermediate zones, macrobenthos has an important ecological role consisting in bioturbation and structuring of the surface layer of bottom sediments [102]. Macrozoobenthos can transfer bottom sediments both horizontally and vertically. At the bottom of the intermediate zone, traces of animals crawling on silt and black hillocks are observed on the surface of grey oxidized sediment. Hydrobionts passages dig through the oxidized layer into the reduced zone.

During their vital activity, benthos and near-bottom organisms change the characteristics of individual detritus particles by aggregating them into faecal pellets and redepositing them onto sediments. While feeding, benthic organisms change the nutritional value of sediment particles. Polychaetes of the Capitellidae family pass the deeply laying layer of black silt through their intestines and transport it to the surface. As a result of this process, previously buried OM returns to the oxidizing zone where it is partly used by detritus feeders inhabiting the sediment surface and partly mineralized [55].

In MPs, the ecological role of bottom-feeding and borrowing animals (along with their faecal pellets, passages and tubes) consists in maintaining the structure of the sediments surface layer. This structure and the constant bioturbation change the sediment diffusion properties. As a result, the transfer of oxygen and nutrients to the surface layers of bottom sediments is not governed solely by simple physical diffusion and oxygen exchange and nutrients recirculation are more intensive [102]. In the Black Sea, the thickness of the processing (bioturbation) layer of sediments by infauna is usually 10–15 cm [1].

The semi-liquid black silts (up to 50.0–55.0% of water content) are not affected by bioturbation. In MPs, sediment mechanical characteristics not allowing animals to hold on its surface and to create holes in it are the actual border between the transitional and the deepwater zones.

The so-called "silty" mussels' settlements form on sandy and silty-sandy sediments and in the dense silts of the intermediate zone. A number of macrophytes, barnacles, polychaetes, bryozoans, amphipods, gastropods and other organisms are connected with these settlements. On the silty-sand sediment, the byssus of the mussels' druzes can be covered by silty-sand. Unlike in fouling biocenosis, the juveniles found themselves in unfavourable conditions for settling down. The majority of surviving young-of-the-year are registered on shells in the upper part of the druzes, where larvae of the barnacle *B. improvisus* also settle. The significant mussel settlements at the bottom attract the sea snail *Rapana venosa* which feeds on them.

In various MPs, *H. acuta* and other gastropods inhabit on compact silt at a distance of 20–50 m from HTS. In Crimean ports and in the NWBS, predatory gastropods are also registered in this biotope.

In the intermediate zone (3.0–4.5 m depths) of Odessa MP, 33 taxa (worms—11, molluscs—9, crustaceans—11, coelenterates and chironomids larvae—1 each) were registered in the silted shell biotope. The quantitative indices of benthos development were much lower in the silts' biotope than in the silted coquina one

(Table 4.2). Sixteen to twenty-four taxa were found in individual samples; for an average density of 10,608.3 ± 1791.0 ind. m^{-2} and an average biomass of 4175.7 ± 928.7 g m^{-2}.

The following species were found in the silted coquina biotope: *Actinothoe clavata*, *Turbellaria* g. sp., *Nemertini* g. sp., *H. imbricata*, *H. reticulata*, *Exogone gemmifera*, *N. succinea*, *Platynereis dumerilii*, *P. cornuta*, *Prionospio cirrifera*, *H. filiformis*, *Oligochaeta* g. sp., *Mohrensternia lineolata*, *Setia valvatoides*, *H. acuta*, *Doridella obscura*, *M. lineatus*, *Mytilus galloprovincialis*, *C. glaucum*, *A. ovata*, *M. arenaria*, *B. improvisus*, *Palaemon elegans*, *Rhithropanopeus harrisi tridentata*, *Iphinoe tenella*, *S. pulchellum*, *Idotea baltica basteri*, *Stenothoe monoculoides*, *Marinogammarus olivii*, *Melita palmata*, *Dexamine spinosa*, *Microdeutopus gryllotalpa* and *C. salinarius* [116].

On the silted coquina in Odessa MP, *B. improvisus*, *M. lineatus*, *M. galloprovincialis* and *M. gryllotalpa* were the mass species and accounted for 76.2% of the total quantity. In the intermediate zone, mussels species with lengths up to 60 mm contribute to most of the benthos biomass. Molluscs prevail in the silted coquina biotope (48.7% of average quantity, 98.5% biomass), while worms dominate in the black silts of the deepwater zone.

The apical part of Grigorievsky Liman, with silty-sand and silty sediments at 2.5–6.7 m depths, belongs to the intermediate zone of the MP aquatory. In this part of the liman, the permanent taxa are *H. imbricata*, *N. succinea*, *P. cornuta*, *H. filiformis*, *H. acuta*, *M. lineatus*, *C. glaucum*, *A. ovata*, *M. arenaria*, *A. diadema* and *Ch. salinarius* [137].

The silty-sand and silty-coquina underwater bars with depths of 3.0–7.2 m are included in the MPs intermediate zone. In Grigorievsky Liman, such bars are the most highly productive cenoses comprising 37 taxa (worms and molluscs 9, crustaceans 15, others (coelenterates, tunicates and chironomids larvae)—4). In 1992–2003, the macrofauna average density was 15,875 ind. m^{-2} and the average biomass 5485.2 g m^{-2}. The bulk of benthos quantity (97.9%) and biomass (99.9%) consisted of 18 taxa from various systematic groups: *H. imbricata*, *N. succinea*, *P. cornuta*, *H. filiformis*, *M. lineolata*, *H. acuta*, *M. lineatus*, *M. galloprovincialis*, *C. glaucum*, *A. ovata*, *M. arenaria*, *B. improvisus*, *S. pulchellum*, *I. baltica basteri*, *Gammarus insensibilis*, *D. spinosa*, *M. gryllotalpa* and *Ch. salinarius* [137].

Mya arenaria forms its pluriannual settlements on sandy and silty-sandy sediments in the intermediate zone which is also inhabited by shrimps (*Crangon crangon*) and by the burrowing flat-browed mud shrimp (*Upogebia pusilla*). In June–July, gobies from the genus *Pomatoschistus* spawn in the same biotope under various objects such as *Mya* valves, mussels' shells and small flat stones. The red mullet *Mullus ponticus* and juveniles of the Black Sea flounder *P. lucsus* and turbot *P. maeotica* also feed at the same depth.

In Odessa MP, results of studies performed in 1976–1977 identified 16 macrofauna taxa in sand and coquina with mussels biocenosis forming near the lateral surfaces of Reydovy mole, but genus and species were not reported [40].

During surveys carried out in 2001, 43 taxa (worms 19, molluscs 8, crustaceans 14, sponges and tunicates 1 of each) were found in the black and grey silts biotope (Table 4.1) [116]. Their average density was 912.0 ± 149.4 ind. m^{-2} and the average biomass 11.515 ± 2.280 g m^{-2}.

Table 4.1 Comparative characteristics of macrozoobenthos species composition and quantitative development in silt and silty coquina biotopes in Odessa MP aquatory in 2001

Taxon	Biotope					
	Silt			Silted coquina		
	N	*B*	*P* (%)	*N*	*B*	*P* (%)
Spongia						
Spongia g. sp.	0.1	0.008	0.9	–	–	–
Coelenterata						
Actinothoe clavata (Ilmoni)	–	–	–	50.0	0.177	66.7
Vermes						
Turbellaria g. sp.	3.0	0.028	5.1	205.0	3.002	100.0
Nemertini g. sp.	0.3	0.001	2.6	31.7	0.312	83.3
Phyllodoce mucosa Oersted	0.1	0.001	0.9	–	–	–
Eteone picta Quatrefages	0.1	0.001	0.9	–	–	–
Harmothoe imbricata (L.)	36.5	0.203	53.8	270.0	3.085	100.0
H. reticulata Claparede	0.3	0.001	2.6	3.3	0.007	33.3
Exogone gemmifera Pagenstecheri	–	–	–	8.3	0.003	33.3
Hediste diversicolor (Muller)	7.8	0.408	8.5	–	–	–
Neantes succinea (Frey et Leuch)	149.6	3.445	82.1	643.3	8.808	100.0
Platynereis dumerilii (Audouin & M.-Edwards)	0.3	0.001	1.7	86.7	1.757	66.7
Nephthys hombergii Savigny	0.3	0.046	1.7	–	–	–
Nephthys cirrosa Ehlers	0.9	0.092	5.1	–	–	–
Spio filicornis (O.F. Muller)	17.4	0.013	45.3	–	–	–
Polydora cornuta Bosc	266.7	0.121	80.3	638.3	0.380	100.0
Prionospio cirrifera Wiren	0.5	0.001	4.3	15.0	0.015	33.3
Heteromastus filiformis Claparede	221.4	1.031	29.9	3.3	0.012	33.3
Capitella capitata (Fabricius)	4.8	0.008	18.8	–	–	–
Capitomastus minimus (Langerhans)	0.1	0.001	0.9	–	–	–
Melinna palmata Grube	0.3	0.003	2.6	–	–	–
Oligochaeta g. sp.	24.5	0.027	41.0	6.7	0.005	16.7
Mollusca						
Mohrensternia lineolata (Michaud)	0.3	0.001	2.8	1.7	0.007	16.7
Setia valvatoides Milachevitch	–	–	–	8.3	0.007	33.3
Hydrobia acuta (Draparnaud)	34.5	0.089	5.1	80.0	0.155	100.0
Doridella obscura Verrill	1.5	0.003	7.7	8.3	0.045	50.0
Mytilaster lineatus (Gmelin)	2.9	0.011	4.3	4108.3	195.000	100.0

(continued)

Table 4.1 (continued)

Taxon	Biotope					
	Silt			Silted coquina		
	N	*B*	*P* (%)	*N*	*B*	*P* (%)
Mytilus galloprovincialis Lamarck	1.8	0.260	1.7	925.0	3903.283	100.0
Cerastoderma glaucum Poiret	12.6	2.338	15.4	6.7	9.585	50.0
Abra ovata (Philippi)	15.3	1.210	15.4	25.0	2.653	50.0
Mya arenaria L.	0.8	0.002	5.1	3.3	0.015	33.3
Crustacea						
Balanus improvisus Darwin	1.7	0.134	3.4	978.3	39.800	100.0
Palaemon elegans Rathke	0.1	0.041	0.9	1.7	0.600	16.7
Crangon crangon (L.)	0.1	0.504	0.9	–	–	–
Upogebia pusilla (Petagna)	0.1	0.006	0.9	–	–	–
Rhithropanopeus harrisi tridentata (Maitland)	2.1	0.507	14.5	28.3	2.675	100.0
Iphinoe tenella G.O. Sars	4.1	0.005	17.9	28.3	0.033	50.0
Sphaeroma pulchellum (Colosi)	1.7	0.050	2.6	71.8	1.356	100.0
Idotea baltica basteri Audouin	0.7	0.005	2.6	30.0	0.172	66.7
Ampelisca diadema A. Costa	1.2	0.002	4.3	–	–	–
Stenothoe monoculoides (Montagu)	–	–	–	128.3	0.038	50.0
Gammarus insensibilis Stock	0.1	0.001	1.7	–	–	–
Marinogammarus olivii M.-Edwards	–	–	–	5.0	0.048	33.3
Melita palmata (Montagu)	20.3	0.054	21.4	91.7	0.382	83.3
Dexamine spinosa (Montagu)	2.3	0.003	3.4	36.7	0.100	83.3
Microdeutopus gryllotalpa A. Costa	72.5	0.080	33.3	2076.7	2.217	100.0
Corophium bonelli (M.-Edwards)	0.1	0.001	0.9	–	–	–
Tunicata						
Molgula euprocta Drasche	0.3	0.768	2.6	–	–	–
Insecta, larvae						
Chironomus salinarius (Kieffer)	–	–	–	3.3	0.003	16.7
Total	912.0	11.515	–	10,608.3	4175.737	–

N—average density, ind. m^{-2}; *B*—average biomass, g m^{-2}; *P*—occurrence, %

Based on occurrence frequency, three polychaetes species (*H. imbricata*, *N. succinea* and *P. cornuta*) were dominant and comprised 49.6% of the total quantity and 32.7% of biomass. Other four taxa (*S. filicornis*, *H filiformis*, oligochaetes and *M. gryllotalpa*) accounted 36.8% of quantity and 10.0% of biomass, while others were occasional. Six out of seven of the main and minor taxa belonged

Table 4.2 Comparative characteristics of macrozoobenthos quantitative indices in biotopes of silt and silted coquina of Odessa MP in 2001

Index	Biotope	
	Silt	Silted coquina
Sample number	117	6
Sampling depth (m)	3–16.5	3–4.5
Sampling depth, average (m)	11.1 ± 0.3	3.8 ± 0.3
Number of taxa in sample	0–17	16–24
Number of taxa in sample, average	5.5 ± 0.3	21.0 ± 1.1
Number of taxa, total	43	33
Benthos density in samples (ind. m^{-2})	0–9630	5290–16,300
Average benthos density (ind. m^{-2})	912.0 ± 149.4	10,608.3 ± 1731.0
Benthos biomass in samples (g m^{-2})	0–145.07	864.23–6844.96
Average benthos biomass (g m^{-2})	11.52 ± 2.28	4175.74 ± 928.71

to the taxonomic group of worms, which indicates the presence of unfavourable abiotic conditions for benthic organisms at the aquatory bottom. Many invertebrates, especially crustaceans, rarely occurred in the fauna of silts and had low quantitative indices. Their presence in silts biotope is probably explained by the fact that most of the 72 samples were collected directly at berths, where these species occur in the periphyton (Table 4.2).

The macrofauna is characterized by a relative constancy in its quantitative composition. Based on occurrence frequency, two-thirds of the taxa of various taxonomic groups comprised 99.40% of quantity and 99.98% of biomass.

Four organisms of periphyton epifauna (the barnacle *B. improvisus*, the mussels *M. lineatus* and *M. galloprovincialis* and the freshwater shrimp *M. gryllotalpa*) were dominant and comprised 76.2% of the total quantity. Mussels represented by individuals reaching 60 mm in size accounted for 93.5% of the benthos biomass, while molluscs with sizes of 25–45 mm dominated in terms of quantity (55.5%).

The water exchange between the port aquatory and the adjacent part of the sea activates during onshore–offshore phenomena. For the rest of time, water exchange is weak, especially near the berths in the far side of aquatories and distant from the passages into the port. Two zones with a different type of water exchange were identified in silts' biotopes. In zones with relatively free water exchange (at a distance of 50 m from berths, near breakwaters and AC), fauna composition was diverse (39 taxa) and its indices (density 1750.0 ± 341.1 ind. m^{-2} and biomass 22.82 ± 5.14 g m^{-2}) were significantly higher than zones with limited water exchange (directly near berths in the far side of harbours) where 29 taxa were found for a density of 388.2 ± 64.3 ind. m^{-2} and a biomass of 4.45 ± 1.32 g m^{-2} (Table 4.3).

This trend is confirmed by the peculiarities of benthos spatial distribution in the various port zones. The highest density (3418 ind. m^{-2}) and biomass (34.94 g m^{-2}) of macrozoobenthos were registered in the AC and near breakwaters (2976 ind. m^{-2}

Table 4.3 Comparative characteristics of the quantitative indices of macrozoobenthos living in the silt biotope of Odessa MP aquatory in 2001 in zones with different regimes of water exchange

Index	Water exchange character	
	Limited	Relatively free
Number of samples	72	45
Sampling depth (m)	3–16.5	3–15
Sampling depth, average (m)	11.3 ± 0.3	10.9 ± 0.5
Taxa number in sample	0–12	2–17
Taxa number in sample, average	4.4 ± 0.4	7.2 ± 0.5
Taxa number, total	29	39
Benthos quantity in sample (ind. m^{-2})	0–1990	40–9630
Average benthos quantity (ind. m^{-2})	388.2 ± 64.3	1750.0 ± 341.1
Benthos biomass in sample (g m^{-2})	0–78.60	0.03–145.07
Average benthos biomass (g m^{-2})	4.45 ± 1.32	22.82 ± 5.14

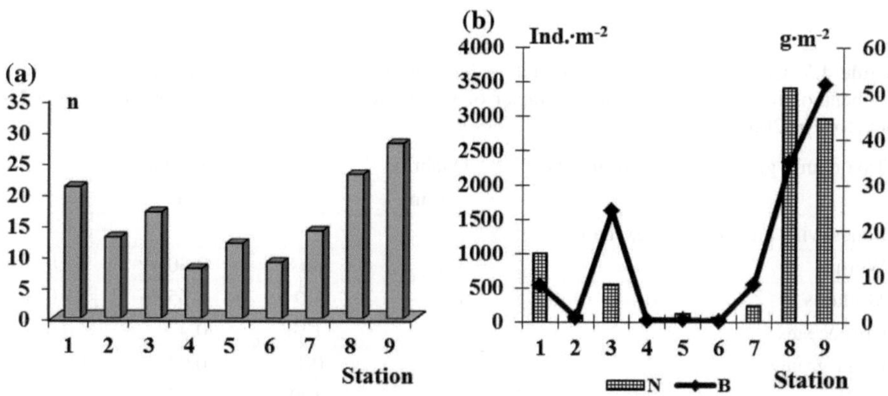

Fig. 4.10 Comparative characteristic of macrozoobenthos quantitative indices in the silt biotope of Odessa MP aquatory in 2001: **a** taxa number (n), **b** density (N, ind. m^{-2}) and biomass (B, g m^{-2}). Numbers on the horizontal axis corresponds to harbours: *1*—Oil, *2*—Khlebnaya (Grain), *3*—Zavodskaya, *4*—Prakticheskaya, *5*—Kabotazhnaya, *6*—Novaya, *7*—Karantinnaya; *8*—access channel, *9*—breakwaters

and 52.05 g m^{-2}), as well as in Neftyanaya (Oil) (1009 ind. m^{-2} and 8.08 g m^{-2}) and Karantinnaya harbours (238 ind. m^{-2} and 8.26 g m^{-2}) located near the AC (Fig. 4.10). The other harbours located farther inside the port have limited water exchange (especially the semiclosed Voennaya harbour) and, consequently, are characterized by a relatively poor composition of bottom macrofauna along with low density and biomass. In Zavodskaya harbour, the relatively high benthos biomass is explained by the occasional presence of relatively big-sized (in comparison with other taxa) specimens of the bivalve *C. glaucum*. Compared with these zones, the water exchange in zones occupied by silted coquina should be considered as relatively dynamic.

Table 4.4 Comparison of the quantitative indexes of the main taxonomic groups of macrozoobenthos in biotopes of silted coquina and silt in Odessa MP aquatory in 2001

Taxonomic group	Taxa number	Quantity		Biomass	
		Ind. m^{-2}	%	g m^{-2}	%
Silted coquina biotope					
Worms	11	1911	18.0	17.39	0.4
Molluscs	9	5167	48.7	4110.75	98.5
Crustaceans	11	3477	32.8	47.42	1.1
Other groups	2	53	0.5	0.18	–
Total	33	10,608	100.0	4175.74	100.0
Silt biotope					
Worms	19	735	80.6	5.43	47.2
Molluscs	8	70	7.7	3.91	34.0
Crustaceans	14	107	11.7	1.39	12.1
Other groups	2	–	–	0.78	6.7
Total	43	912	100.0	11.51	100.0

Table 4.5 Comparison of the quantitative indexes of the main taxonomic groups of macrozoobenthos in areas with various character of water exchange in the silt biotope of Odessa MP aquatory in 2001

Taxonomic group	Taxa number	Quantity		Biomass	
		ind. m^{-2}	%	g m^{-2}	%
Zones with limited water exchange					
Worms	14	337	86.8	2.30	51.8
Molluscs	6	4	1.0	1.65	37.0
Crustaceans	7	47	12.1	0.47	10.6
Other groups	2	<1	0.1	0.03	0.6
Total	29	388	100.0	4.45	100.0
Zones with relatively free water exchange					
Worms	18	1371	78.3	10.44	45.8
Molluscs	7	176	10.1	7.54	33.0
Crustaceans	13	203	11.6	2.86	12.5
Other groups	1	<1	–	1.98	8.7
Total	39	1750	100.0	22.82	100.0

In the biotope of silted coquina, molluscs dominated in terms of quantity (48.7%) and biomass (98.5%), while the number of taxa of molluscs, worms and crustaceans was practically the same (Table 4.4). In silts biotope, worms predominated in terms of taxa number (19), quantity (80.6%) and biomass (47.2%). The number of worms taxa (14) was less and their relative quantity (86.8%) and biomass (51.8%) a bit higher in the zone of silts biotope having a limited water exchange than in the zone with relatively free water exchange (18 taxa, 78.3% of quantity and 45.8% of biomass, respectively) (Table 4.5).

Table 4.6 Comparison of the quantitative indices of macrozoobenthos trophic groups in biotopes of silted coquina and silt in Odessa MP aquatory in 2001

Main trophic groups	Taxa number	Quantity		Biomass	
		ind. m^{-2}	%	g m^{-2}	%
Silted coquina biotope					
Detritophagous	11	3516	33.1	12.79	0.3
Seston feeders	5	6022	56.8	4147.68	99.3
Phytophagous–detritophagous	4	198	1.9	1.96	0.1
Phytophagous	4	265	2.5	3.40	0.1
Carnivorous	8	605	5.7	9.31	0.2
Polyphagous	1	2	–	0.60	–
Total	33	10,608	100.0	4175.74	100.0
Silts biotope					
Detritophagous	17	821	90.0	6.45	56.0
Seston feeders	7	20	2.2	3.52	30.6
Phytophagous–detritophagous	4	23	2.5	0.11	1.0
Phytophagous	3	3	0.3	0.01	–
Carnivorous	10	45	5.0	0.88	7.7
Polyphagous	2	<1	–	0.55	4.7
Total	43	912	100.0	11.52	100.0

Six main trophic groups were recorded in benthos composition. Detritophages (90.0% of quantity and 56.0% of biomass) were dominant in the silts biotope, while seston feeders dominated in the silty-coquina biotope (56.8 and 99.3%, respectively). However, detritophages predominated in both biotopes in terms of taxa number (Table 4.6). The index of uniformity of food structure in the silts biotope and in the biotope of silted coquina was 0.30 and 0.95, respectively.

In silt biotopes, the ratio of quantitative indices of macrobenthos trophic groups (Table 4.7) to the index of uniformity of food structure was relatively similar (0.24–0.31) across zones with a different regime of water exchange.

In silts biotope, the number of epi- and infaunal taxa was practically the same (20 and 23, respectively). The infaunal invertebrates dominated in terms of quantity (80.0%) and biomass (80.5%). The vagile hydrobionts dominated in number of taxa (34), quantity (99.2%) and biomass (89.7%), in agreement with the sediments type.

Epifaunal organisms (20 taxa, 86.9% of quantity and 99.4% of biomass) were dominant in the biotope of silted coquina. Representatives of three sessile organisms (*B. improvisus*, *M. lineatus* and *M. galloprovincialis*) dominated on quantity (56.7%) and biomass (99.1%), while the remaining 30 taxa belonged to vagile forms.

Overall, 49 macrozoobenthos taxa were registered in 2001 in Odessa MP aquatory at depths of 3.0–16.5 m: worms—20, molluscs—9, crustaceans—16 and others—4. In the biotope of silts, covering most of the aquatory bottom, 43 taxa were registered with an average density of 912 ind. m^{-2} and biomass of 11.52 g m^{-2}. Two areas with different types of water exchange were studied and

Table 4.7 Quantitative indices of the main trophic groups of macrozoobenthos in zones with different regimes of water exchange in silts biotope of Odessa MP aquatory in 2001

Main trophic groups	Taxa number	Quantity		Biomass	
		ind. m^{-2}	%	g m^{-2}	%
Zone with limited water exchange					
Detritophagous	13	332	85.5	2.21	49.6
Seston feeders	6	2	0.4	1.60	36.0
Phytophagous–detritophagous	2	15	4.0	0.04	0.8
Phytophagous	2	1	0.2	<0.01	0.1
Carnivorous	6	38	9.9	0.60	13.5
Total	29	388	100.0	4.45	100.0
Zone with relatively free water exchange					
Detritophagous	15	1603	91.6	13.23	58.0
Seston feeders	6	49	2.8	6.60	28.9
Phytophagous–detritophagous	4	35	2.0	0.22	1.0
Phytophagous	2	6	0.4	0.01	–
Carnivorous	10	56	3.2	1.34	5.9
Polyphagous	2	1	–	1.42	6.2
Total	39	1750	100.0	22.82	100.0

compared. Benthos indices (39 taxa, 1750.0 ind. m^{-2}, 22.82 g m^{-2}) were much higher in the zone with a relatively free water exchange (i.e. at a distance of 50 m from berths, near breakwaters and AC) than in the zone with a limited water exchange (i.e. near berth in harbours apical part)—29 taxa, 388.2 ind. m^{-2}, 4.45 g m^{-2}. Among taxonomic groups, worms dominated in terms of quantity (80.6%) and biomass (47.2%), while detritophages were the dominant trophic group (90.0% quantity and 57.0% biomass). Vagile infaunal invertebrates dominated depending on sediments characteristics.

Thirty-three taxa for average quantity and biomass equal to 10,608.3 ind. m^{-2} and 4175.74 g m^{-2}, respectively, were registered in the silted coquina biotope, which forms local aggregations. *B. improvisus*, *M. lineatus*, *M. galloprovincialis* and *M. gryllotalpa* were the mass species accounting for 76.2% of the total quantity. Mussels comprised 93.5% of the average biomass. Molluscs predominated among systematic groups (48.7% of quantity and 98.5% of biomass) and seston feeders dominated among trophic groups (56.8 and 99.3%, respectively). The sessile hydrobionts of epifauna dominated the epifauna with regard to both quantity and biomass.

4.2.4 Biotic Peculiarities of the Deepwater Areas of the Benthal Zone

Rocky bottoms, sandy sediments, oxidized and non-oxidized silts are present in the deepwater areas of MPs aquatories of the Black–Azov Sea basin. In Odessa MP, native clays were uncovered after dredging. The distribution of water-rich silts near berths, moles and breakwaters follows a general pattern. The more compact silts are found at some distance from HTS. In MPs, the deepwater aquatories have depths above 8–10 m and up to a maximum of 20–21 m. In the deepwater zones, sedimentation is a continuous process because of the weakened hydrodynamics and limited water exchange. Up to 90.0–95.0% of the bottom area is covered by sediments in many MPs. In these zones, it can be considered that biota is mainly governed by the process of sediment formation.

The surface of the semi-liquid silts (with up to 50.0–55.0% of water content) formed in MPs is hardly a suitable substratum for the movement and settlement of big-sized benthic forms, which would sink down into the sediment. The excessively water-rich, fluid silt also does not allow hydrobionts to burrow and construct passages inside it. The easily stirred suspension, which sediments to the bottom of the deepwater zones of MPs, hampers macrophytes' photosynthesis and hypoxia and anoxia zones form under it.

In the deepwater areas of MPs, the decomposition of accumulated OM occurs via fermentation carried out by bacteria and fungi. The sediments of these areas mainly consist of fine particulate material. The content of alevropelites in sediments is an important parameter responsible for the physico-chemical conditions, geochemical peculiarities and ecological conditions in which animals live [37]. In MPs, the most acidic sediments ($pH < 7$) are characterized by a high proportion of argillous-silt, and an enrichement in OM. Sediment acidity is determined by the presence of hydrogen sulphide and high concentrations of humic and other organic acids.

Owing to the high OM content, bacterial biomass in silted sediments can reach 57.0 g m^{-2}, which is twice than in the coastal, well-washed sandy sediments [45].

In the deepwater zones of the Black Sea, growth is slower, and secondary production lower, than in the euphotic zone and above the thermocline [35]. This feature is also observed in the MPs benthal zone where protozoans and meiobenthos form the basis of silts fauna. Among meiobenthos, nematodes, foraminifera and harpacticoids are the most common organisms during most of the year. Only in winter, during complete mixing of the water column, representatives of other meiobenthos groups and small-sized forms of macrozoobenthos appear on areas covered by the black silts; mainly at their periphery. Detritophagous and deposit feeders, mostly from different taxonomic groups of worms with lengths of 0.5–5.0 mm, prevail among them. These organisms do not require high oxygen concentrations, are able to remain at the surface and in the upper layers of semi-liquid silts and are resistant to the presence of hydrogen sulphide.

In conditions of sufficient oxygenation, the dead and living OM in the MPs deepwater zones of MPs used by vagile organisms belonging to the HTS fouling composition, as well as by invertebrates and fish from the intermediate zone. Fish larvae and juvenile participate in OM elimination from the deepwater zones.

The energetic exchange in bottom communities of silty sediments is dominated by bacteria, microalgae, protozoans and meiobenthos. In the deepwater zones, a significant part of the untransformed OM can accumulate on the bottom as a "fresh" silt layer and get buried because of the excessive input of detritus.

Organisms inhabiting the bottom zones highly enriched in OM can usually reproduce rapidly all year round or in winter, when the processes of decomposition and fermentation are slowed down and the ventilation of the near-bottom layer intensifies. Several benthic species have pelagic settling stages or resting and non-active forms to survive during critical conditions and quickly reestablish themselves as soon as conditions improve.

Bottom-dwelling organisms pass detritus and fine sediment through their digestive tracts and aggregate the undigested rests, enriched in bacteria and oxygen, into their faecal pellets. The form and size of faecal pellets vary depending on species, but their size is many times bigger than that of the original detritus particles and silt. MPs bottom sediments therefore become more resistant to stirring owing to the activity of detritophages and deposit feeders.

Microphytobenthos and microzoobenthos, along with bacterial detritus, are an important nutrient source for many groups of pseudo- and eumeiobenthos. In the coastal marine ecosystems, the contribution of bottom and periphyton microalgae to primary OM production is not taken into account and there are reasons to surmise that it is comparable to that of the phytoplankton [43, 81]. Indeed, the significant role of diatoms present in microphytobenthos composition at various depths and on all MPs sediments types was already mentioned.

In the same way as bacteria, microphytobenthos algae actively migrate following changes in light conditions and hydrochemical parameters in their living environment. The character of microphytobenthos vertical displacement influences, in turn, the meiofauna vertical migrations [25, 29, 132].

In MPs water basins, the most abundant diatoms accomplish a defined cycle. Like many unicellular algae and the general majority of hydrobionts, diatoms are heterotopic, i.e. the various stages of their life cycle take place in different ecotopes. After intensive blooms or during stress situations in the water column, the microphytes sink to the surface of the sediment, including silts, in the form of resting spores, single vegetative cells and colonies. Spores and vegetative cells return into the water column in winter following the improvement of abiotic conditions. Benthic diatoms can pass into the pelagial zone and from the water column into the benthal zone all year round [65]. The presence of such mechanisms ensures the preservation of diatom populations in MPs aquatories despite sharp changes in environmental conditions.

In MPs, the widespread cyanobacteria of the genus *Microcystis* hibernates in the silt sediments surface layer in the form of slimy colonies, whose living cells use mineral and organic compounds contained in the silt. At the onset of favourable conditions, they migrate into the water column and, staying in the plankton, switch to a predominantly autotrophic type of feeding [60].

As it was already mentioned, the body structure of pennate diatoms allows them to live in different ecological zones of basins. These diatoms are able to move actively on hard and soft substrata as well as in the water column [106].

In the benthal zone, the diversity of microalgal species during a given season differs significantly across the various years. The highest species diversity occurs in spring and autumn, while a significantly lower species number is registered in summer [81]. Even in extremely polluted zones of Sevastopol Bay, silts contain a high number of living diatoms [43]. In general, in Sevastopol Bay, the state of microphytobenthos on loose sediments is generally characterized as suppressed, but it could be estimated as steady on substrata without toxic properties [81].

In the MPs sediment layer, the sulphate-reducing bacteria use the easily oxidizable compounds such as lactate, ethanol and pyruvate as substratum. They use more complex or more oxidized OM only in mixed cultures with saprophytic bacteria. Sulphate-reducing bacteria are unable to utilize complex organic compounds without the help of concomitant saprophytic microflora and cannot live in sediments in the absence of the latter [120].

In the deepwater areas of MPs, these two groups of bacteria inhabit silts and have complementary functions in accomplishing OM mineralization and BM recycling.

There are all reasons to suppose that various bacteria groups could decompose any substance of natural origin whose structure is formed by carbon. The accumulation of OM in the deepwater zones of aquatories ensures the existence of a diversified microbial population. The higher is OM concentration, the quicker its decomposition [3]. Bacteria and primitive fungi destroy detritus in aerobic and anaerobic conditions. In the benthal zone, bacteria are an important food source for protozoans, worms, molluscs, crustaceans and other organisms.

Infusoria are numerous in the black silts. Many of them are resistant to a wide range of conditions in the aquatic environment and are less demanding than the more complex hydrobionts as to water oxygen saturation. This tolerance allows them to inhabit the anaerobic zones of MPs, where predators (consumers of protozoans) rarely penetrate, but food resources are present. Protozoa can inhabit the bottom of water basins characterized by high levels of carbon dioxide and methane formation. The presence in the protozoans' cytoplasm of methane consuming bacteria, neutralizing the toxic effect of methane, allows them to populate hypoxic and anoxic zones [106].

The benthic forms of protozoans can actively move to search more favourable conditions. In MPs, infusoria can reach high densities [85]. However, an important ecological and physiological characteristic of protozoans is that accumulation of their own metabolites in water restrains or completely stops their reproduction. Such mechanism is triggered by an excessive increase in organisms' quantity and a decrease in food amount. Some species can enter an anabiosis state or form resistant cysts [106]. These mechanisms specifically sustain the existence of protozoa on the semi-liquid, black silts in MPs.

Studies of the infusoria *Epichlintes ambiguus*, *Holosticha gibba*, *Orthodon hamatus* and *Trichotaxis crassa* showed that they feed on unicellular algae and bacteria. In general, they choose small-sized forms of diatom and peridinian algae [99], but protozoans can also be part of their ration. Infusoria play an important role in maintaining favourable conditions at the surface of bottom sediments by consuming detritus and bacteria. In turn, they serve as food for various invertebrates, larvae and juvenile fish.

Protozoans inhabit the sediment surface and upper (1–2 cm) layer and constitute a food source for detritophagous and predatory forms of meio- and macrozoobenthos. Even sediments polluted by oil contain significant amounts of infusoria although their quantity and taxonomic composition vary widely. The highest number of infusoria is observed in the upper sediment layer where the genera *Frontona*, *Diophris*, *Peretricha* and *Euplutos* dominate [87].

Inside MPs, zones covered by black silts are inhabited by detritus consumers (collectors and deposit feeders) and are mainly confined to bottom areas located at some distance from passages into aquatories and from berths. In these biocenoses, deposit feeders comprise up to 64.0% of the total biomass, epifaunal filter feeders 0.4%, infaunal filter feeders 6.0% and collectors 10.0% [42, 62].

In most areas of MPs bottom, benthos constantly lives in a non-steady state and periodically experiences stress situations leading to mass mortality phenomena of various intensities. During inter-bloom periods, micro- and meiobenthos recolonize zones with sufficient oxygen content and some forms of macrozoobenthos can appear at the periphery of such zones. Part of the pseudomeiobenthos representatives can pass into the macrozoobenthos size group. In periods of critically deteriorated conditions, the sessile forms of eumeiobenthos, pseudomeiobenthos and macrozoobenthos can survive in the narrow ecotone zones adjacent to HTS and raised above the bottom or in the fouling community of HTS.

In the black silts, the meiobenthos temporary components (pseudomeiobenthos) are represented by sunken larvae of crustaceans, bivalves, gastropods and big-sized polychaetes. They usually have a low possibility to survive until the adult stage in this biotope, but they replenish its food resources for some time with their live OM. Part of the OM and BM is removed from the benthal zone following grazing by near-bottom fish.

Depending on the actual species of heterotrophic organisms and on various circumstances, the annual production of a population may exceed its average annual biomass by 1.8–70 and, in some cases, even 100-fold [35]. During the inter-bloom periods, the presence of food resources at the black silts periphery makes it an attractive feeding ground for vagile bottom animals. Many organisms manage to recover their populations in a short time owing to individuals surviving in the ecotone zones and fouling of the coastal shallow waters and in the intermediate zones not subject to prolonged hypoxia and anoxia.

As it was mentioned, meiobenthos organisms inhabit various types of sediments, but reach the highest quantitative development on silted sediment with admixture of coquina. These sediments contain high levels of detritus that serves as a source of food for collectors and deposit feeders. Such combination of differently oriented hard surfaces and soft sediment creates an abundance of microniches for epifaunal, onfaunal and infaunal animals [56]. In MPs, silt with coquina constantly accumulates near the HTS lateral surfaces and constitutes the most extended biotope.

Predatory species are present among the representatives of different meiobenthos groups and small-sized forms of macrozoobenthos inhabiting the black silts. They include nematodes, harpacticoids, ostracodes, halacarides, nemertines, turbellarians and polychaetes. For them the ecotone zones serve as a "hunting ground".

In MPs silty sediments, micro- and meiobenthos have several ecological functions: 1—maximum possible decomposition of OM during the feeding process; 2—conservation of part of OM in their biomass; 3—transformation of part of OM into faecal pellets, i.e. aggregation and enrichment by bacteria of detritus particles; 4—participation in bioturbation (when allowed by the mechanical characteristics of sediment). The detritus particles aggregated into faecal pellets, together with bacteria, become much bigger and energetically more advantageous food items.

Owing to micro- and meiobenthos activity, part of the reduced OM returns from the black silts zone into the heterotrophic food chains of the pelagial and periphytal zones without undergoing a complete mineralization. In the benthal zone, small-sized meiobenthos organisms serve as food objects for heterotrophs of the second, third and higher orders inhabiting the benthal, pelagial and periphytal zones.

As a consumer of dead and alive OM in the benthal zone, meiobenthos represents a food resource for many meiobenthos species, as well as for big-sized bottom invertebrates (polychaetes, molluscs, crustaceans) and fish on various phases of their ontogenesis [26, 51].

In the deepwater zones of Sukhoy Liman, meiobenthos is mainly represented by nematodes and foraminifera, with a low number of harpacticoids or polychaetes, during most of the year [112]. In this aquatory, at oxygen concentrations in the near-bottom water layer lower than 2.0 mg dm^{-3}, organisms of four meiobenthos groups (foraminifera, nematodes, oligochaetes and polychaetes) were recorded [25]. Such conditions of hypoxia are typical for MPs deepwater zones and mass mortality phenomena are quite often observed. In the Black Sea, nematodes quite often comprise 85.0–98.0% of the total quantity of the bottom meiobenthos fauna; and even 100.0% at mass mortality [25].

In zones of Grigorievsky Liman with depths of 12.5–16.5 m, only nematodes and foraminiferans typically occur in 100% of the cases. The black silts of Grigorievsky Liman are the kingdom of nematodes and some detritophagous polychaetes.

In the black silts covering the deepwater zone of Grigorievsky Liman, the abundance of nematodes varies between 33,330 and 1.984 million ind. m^{-2} with average density and biomass of 529,771 ind. m^{-2} and 1.366 g m^{-2}, respectively and a size range of 0.8–3.6 mm. Forty-three percentage of the nematodes are non-selective detritus feeders, while 25.0% are selective detritus feeders [137]. They reach their peak of quantity in winter, which may be directly connected with the seasonal amelioration of oxygen conditions at the bottom.

In the deepwater zone (more than 10 m) of Grigorievsky Liman, the macrozoobenthos composition comprised 44 taxa: 15 taxa each of worms and crustaceans, 12 taxa of molluscs and 1 taxon each of coelenterates and chironomid larvae. During the study period (1992–2006), the average abundance and biomass were 505.5 ± 47.8 ind. m^{-2} and 8.67 ± 1.18 g m^{-2}, respectively. Based on the frequency of occurrence, only two polychaetes species, *N. hombergii* and *H. filiformis*, belonged to permanent ones, while the polychaetes *N. succinea*, *P. cornuta*, *M. palmata* and gastropods *H. acuta* belonged to codominant species. The remaining 38 taxa (25.0% of quantity and 46.8% of biomass) were occasional species. Four species, *N. hombergii*, *N. succinea*, *M. palmata* and *C. glaucum*, accounted for 56.7% of the average biomass [137].

In the 1970s, it was shown that black slits mainly inhabited by nematodes predominate on the Odessa MP bottom. The macrozoobenthos representatives were absent even in the compact black silts and on different substrata of anthropogenic origin (metal wreckage and wooden packing, rope scraps) which visibly projected from the bottom [40].

In 2001, 43 taxa of bottom macrofauna (worms—19, molluscs—8, crustaceans—14, sponges and tunicates 1 of each) were found in biotopes of the black and grey silts in the deepwater zone of Odessa MP. Their average quantity and biomass comprised 912.0 ± 149.4 ind. m^{-2} and 11.52 ± 2.28 g m^{-2}, respectively. Based on occurrence, the three polychaetes species *H. imbricata*, *N. succinea* and *P. cornuta* were the dominant species, comprising 49.6% of macrofauna quantity and 32.7% of its biomass. Other four taxa (the polychaetes *S. filicornis* and *H. filiformis*, oligochaetes and the freshwater shrimp *M. gryllotalpa*) were ascribed to codominant, accounting for 36.8% of quantity and 10.0% of biomass of the macrofauna. The rest of taxa were registered occasionally. From the seven dominant and codominant taxa, six belonged to taxonomic groups of worms [116].

In Odessa MP, polychaetes were one of the main groups of meiobenthos in terms of species number (16), density (707.1 ind. m^{-2}) and biomass (5.38 g m^{-2}).

Owing to their high resistance towards unfavourable factors such as pollution and natural disturbance, many polychaetes species can serve as indicators of marine bottom communities' condition. It is recommended to use parameters of polychaetes taxocen such as biomass, quantity and species number as well as various indices (diversity, Pielou's measure of species evenness, species abundance, dominance) for estimation of environmental quality [139].

In the Black Sea MPs, the following mass detritivores species of infaunal polychaetes are the most resistant towards the OM pollution of aquatories: *H. diversicolor*, *N. succinea*, *Staurocephalus rudolphi*, *P. cornuta*, *H. filiformis*, *C. capitata*, *Pectinaria neapolitana* and *M. palmata*. The epifauna representative *H. imbricata*, a vagile polyphagous polychaete, is able to survive up to 7 days in anoxic conditions and must also be included in the above list [72].

In Odessa MP, the highest average quantity for the polychaete *H. imbricata* was registered in the aquatories of open harbours and the densest patches in the zones with relatively free water exchange. In semiclosed harbours, characterized by stagnation phenomena, the average quantity of *H. imbricata* was a couple of times lower and the density of patches one order of magnitude lower than in other harbours. In semiclosed harbours, quantity does not exceeded 30 ind. m^{-2} compared with 290 ind. m^{-2} in open ones. At all harbours, the average quantity was higher immediately near the berths than at a distance of 50 m from them.

The highest average quantity and patch density of *Neanthes succinea* were observed in Odessa MP aquatories with relatively free water exchange. In semiclosed aquatories, their quantity reached 60 ind. m^{-2} compared with up to

980 ind. m^{-2} in the open harbours. In zones with free water exchange, the quantity of *N. succinea* near the berths was higher than at a distance of 50 m from them.

In a zone with relatively free water exchange, the maximal quantity of *P. cornuta* was 900 ind. m^{-2} and, on average, its quantity was higher near the berths than at a distance of 50 m distance from them.

In Odessa MP aquatory, *H. filiformis* has extremely high average quantity and high aggregation levels are registered in zones with free water exchange, reaching 4600 ind. m^{-2}. In open harbours aquatories, the quantity of this species was lower near the berths than at a distance of 50 m from them (88 vs. 146 ind. m^{-2}, respectively). This species avoids semiclosed harbours [72].

The contribution of *N. succinea* to the total quantity of polychaetes comprised 41% in the coastal zone of Odessa region versus a value of 25–30% in Odessa port harbours; which testifies about the high tolerance level of this species. In Odessa region, *P. cornuta* accounts for less than 10% of the average quantity of polychaetes, but its contribution reaches 52% in Odessa port aquatories. In Odessa MP aquatories, *P. cornuta* is an ecological analogue of the saprobic polychaete *C. capitata* whose density reaches 4000 ind. m^{-2} in the most polluted zones of MPs of the north-eastern coast of the Black Sea in Odessa MP, *C. capitata* was registered near the berths at an average density of 11 ind. m^{-2} [72].

In the aquatories of many MPs, *C. capitata* can be replaced by the spionid species *Streblospio gynobranchiata* and *P. cornuta*, which have been introduced into the Black Sea [72].

For polychaetes taxonomic group, the Simpson index of species diversity (based on density) was 0.72 and the dominance Berger–Parker index 0.37. The indicators of total and organic pollution, the α-mesosaprobic *C. capitata* and the species *N. hombergii* (tolerant to oxygen deficit), do not play a significant role in Odessa MP benthos. The former species was exceptionally found near the berths in quantities of 10–90 ind. m^{-2} and isolated individuals of the latter were very rarely registered in the port aquatory (1.6% of occurrence frequency). On the other hand, as already mentioned, these species belong to the most abundant life forms in the bottom biocenoses of polluted aquatories near the coasts of Caucasus [82]. In Odessa MP, *C. capitata* is quantitatively replaced by *P. cornuta* whose maximal densities are 6500 ind. m^{-2} in benthos and 9600 ind. m^{-2} in fouling. This species is resistant to reduction of oxygen content and to pollution, in particular from oil [70, 91]. Although *P. cornuta* belongs to the β-mesosaprobic species, it can also inhabit the α-mesosaprobic zone [91]. *P. cornuta* (indicated as *P. ciliata*) was the dominant species of macrozoobenthos in the strongly polluted aquatory of Constanza MP. Here, its density reached 1095 ind. m^{-2} whereas *C. capitata* was found only at one station with a density of 40 ind. m^{-2} [149].

In Odessa MP, the mass development of the detritophagous polychaetes *P. cornuta* and *H. filiformis* and of the mainly detritophagous *N. succinea* is an index of the eutrophication of this aquatory and of the high level of OM in water [9]. Composition and quantitative indices of polychaetes in the silt biotopes vary widely in the different zones of Odessa MP.

The polychaetes fauna, which comprised the bulk of quantity and biomass of the whole macrozoobenthos, was impoverished in the most isolated by HTS harbours of Odessa MP. These aquatories are characterized by the presence of stagnant waters and oxygen deficiency. Furthermore, their bottom sediments are mainly constituted by black semi-liquid silts smelling of hydrogen sulphide. In these zones, the total average density of polychaetes (in total 7 species) was 107 ind. m^{-2}. In some of the harbours (Novaya, Voennaya, Yacht club aquatory) [9], the detritophagous *P. cornuta* and *C. capitata* were present in low numbers (no more than 10 individuals per 0.1 m^2 sample) and *H. filiformis* was absent in spite of the very high OM content of the bottom sediments. Only at one station located at the entrance into Karantinnaya harbour, the polychaetes density was relatively high (580 ind. m^{-2}) due to the presence of *P. cornuta*.

Eleven polychaetes species were found in the aquatories of the less isolated harbours with an overall average density of 598 ind. m^{-2}. The average densities of *P. cornuta* and *N. succinea* were nearly the same (206 and 203 ind. m^{-2}), while that of *H. filiformis* was significantly lower (50 ind. m^{-2}). The highest number of *P. cornuta* (132) and *N. succinea* (98) individuals in a single sample was recorded near the berths, while that of *H. filiformis* (52) at 50 m distance from them.

Sixteen polychaetes species, for an average density of 1650 ind. m^{-2}, were found in the open part of the aquatory. The highest number of individuals in samples of 0.1 m^2 was registered at a distance of 50 m from the HTS of the Neftyanaya harbour mole (992 ind.) and in the AC aquatory (514 ind.). The average quantity of *P. cornuta* and *N. succinea* was nearly the same (346 and 344 ind. m^{-2}), while that of *H. filiformis* was significantly higher (830 ind. m^{-2}). At the same time, the density of *H. filiformis* exceeded that of *P. cornuta* only at three stations out of nine. Two of these stations were located in the AC beyond the borders of breakwaters, i.e. in the zone of the port with the best water exchange. The quantitative dominance of polychaetes species belonging to the family of Capitellidae over those of Spionidae in the less polluted aquatories is not characteristic for other regions of the Black Sea. In Odessa MP, this phenomenon can apparently be explained by the presence of OM in the relatively coarse fractions of the bottom sediments of the above-mentioned stations during the study period [9]. In such situation, this source of food was more available to the unselective deposit feeder *H. filiformis* than to the detritus collector *P. cornuta*.

The characteristics of zoobenthos of Odessa MP and, in particular, the parameters of polychaetes' taxocen provide more information about the eutrophication of the MP aquatory than about its pollution. The presence of sponges and diversified fauna of crustaceans also testifies about eutrophication. The latter are present not only in HTS fouling, but also in the benthos of loose sediments. In Odessa MP, the biocenosis of *Tritia reticulate* (typical of polluted regions) is not developed, while the pollution indicator species *C. capitata* is present in small amounts only near the berths. At the same time, *P. cornuta* and *H. filiformis*, indicators of eutrophic waters, show mass development along with the highly eurybiont and r-strategist species *N. succinea*.

Specific representatives of the fouling community were found in the benthos of the semi-liquid black silt at a distance of 1–3 m from the berths' walls: the crustaceans *S. pulchellum, Idotea baltica basteri, Gammarus insensibilis, M. palmata* and *D. spinosa* and the molluscs *M. lineolata* and *D. obscura.* Apparently, they move to the black silts where they encounter favourable feeding conditions [9].

Comparison of the macrozoobenthos qualitative composition of silts biotopes in the deepwater zones of Odessa (2001) and Yuzhny (1992–2006) MPs showed the presence of 56 taxa in their aquatories (worms—20, molluscs—13, crustaceans—16, sponges, coelenterates, ascidians and chironomids larvae 1 each), including 43 taxa in Odessa MP and 44 in Yuzhny MP [117] (Table 4.8).

Table 4.8 Comparative characteristics of species composition and quantitative indexes of macrozoobenthos in silt biotope of Odessa and Yuzhny MPs aquatories

Taxon	Odessa MP			Yuzhny MP		
	N	*B*	*P*	*N*	*B*	*P*
Spongia						
Spongia g. sp.	0.1	0.008	0.9	–	–	–
Coelenterata						
Obelia longissima (Pallas)	–	–	–	< 0.1	0.051	2.2
Vermes						
Turbellaria g. sp.	3.0	0.028	5.1	–	–	–
Nemertini g. sp.	0.3	0.001	2.6	–	–	–
Phyllodoce mucosa Oersted	0.1	0.001	0.9	0.8	0.009	2.7
Eteone picta Quatrefages	0.1	0.001	0.9	0.1	0.005	0.5
Harmothoe imbricata (L.)	36.5	0.203	53.8	3.6	0.043	10.8
Harmothoe reticulata Claparede	0.3	0.001	2.6	1.0	0.010	3.2
Neanthes succinea (Frey et Leuch)	149.6	3.445	82.1	46.6	1.777	47.6
Hediste diversicolor (Muller)	7.8	0.408	8.5	2.1	0.424	5.4
Platynereis dumerilii (Aud. et M.-Edwards)	0.3	0.001	1.7	–	–	–
Nephthys hombergii Savigny	0.3	0.046	1.7	32.3	1.048	53.0
Nephthys cirrosa Ehlers	0.9	0.092	5.1	1.7	0.007	4.3
Spio filicornis (O.F. Muller)	17.4	0.013	45.3	18.9	0.022	13.0
Polydora cornuta Bosc	266.7	0.121	80.3	60.1	0.053	43.8
Prionospio cirrifera Wiren	0.5	0.001	4.3	–	–	–
Heteromastus filiformis (Claparede)	221.4	1.031	29.9	135.1	0.536	74.6
Capitella capitata (Fabricius)	4.8	0.008	18.8	1.4	0.002	1.6
Capitomastus minimus (Langerhans)	0.1	0.001	0.9	–	–	–
Pectinaria koreni Malmgren	–	–	–	0.1	0.001	0.5
Melinna palmata Grube	0.3	0.003	2.6	59.0	1.052	44.3
Oligochaeta g. sp.	24.5	0.027	41.0	14.0	0.009	15.1
Mollusca						
Rissoa membranacea Adams	–	–	–	0.7	0.001	0.5

(continued)

Table 4.8 (continued)

Taxon	Odessa MP			Yuzhny MP		
	N	*B*	*P*	*N*	*B*	*P*
Mohrensternia lineolata (Michaud)	0.2	0.001	1.7	3.8	0.023	7.6
Setia valvatoides (Milachevitch)	–	–	–	4.4	0.010	2.2
Hydrobia acuta (Draparnaud)	34.5	0.089	5.1	47.4	0.149	29.2
Bittium reticulatum (Costa)	–	–	–	0.1	0.001	0.5
Retusa truncatella (Locard)	–	–	–	0.1	0.002	0.5
Doridella obscura Verrill	1.5	0.003	7.7	–	–	–
Mytilaster lineatus (Gmelin)	2.9	0.011	4.3	2.2	0.238	6.5
Mytilus galloprovincialis Lamarck	1.8	0.260	1.7	1.1	0.004	3.2
Cerastoderma glaucum Poiret	12.6	2.338	15.4	5.4	1.038	14.6
Parvicardium exiguum (Gmelin)	–	–	–	0.1	0.001	0.5
Abra ovata (Philippi)	15.3	1.210	15.4	1.4	0.036	5.6
Mya arenaria L.	0.8	0.002	5.1	11.7	1.770	21.1
Crustacea						
Balanus improvisus Darwin	1.7	0.134	3.4	1.3	0.068	3.2
Palaemon elegans Rathke	0.1	0.041	0.9	–	–	–
Crangon crangon (L.)	0.1	0.504	0.9	1.0	0.076	1.1
Upogebia pusilla (Petagna)	0.1	0.006	0.9	–	–	–
Rhithropanopeus harrisi tridentata (Maitland)	2.1	0.507	14.5	–	–	–
Iphinoe maeotica (Sowinskyi)	–	–	–	0.7	0.001	3.2
Iphinoe tenella G.O. Sars	4.1	0.005	17.9	0.1	0.000	0.5
Sphaeroma pulchellum (Colosi)	1.7	0.050	2.6	1.1	0.005	2.2
Idotea baltica basteri Audouin	0.7	0.005	2.6	3.2	0.035	3.2
Ampelisca diadema A. Costa	1.2	0.002	4.3	16.6	0.107	23.2
Stenothoe monoculoides (Montagu)	–	–	–	0.7	0.001	0.5
Perioculoides longimanus (Bate et Westwood)	–	–	–	0.9	0.001	1.1
Gammarus insensibilis Stock	0.1	0.001	1.7	2.5	0.012	2.7
Melita palmata (Montagu)	20.3	0.054	21.4	0.3	0.001	1.1
Dexamine spinosa (Montagu)	2.3	0.003	3.4	–	–	–
Microdeutopus gryllotalpa A. Costa	72.5	0.080	33.3	10.8	0.014	10.8
Erichthonius difformis M.-Edwards	–	–	–	1.4	0.001	2.2
Corophium bonelli (M.-Edwards)	0.1	0.001	0.9	1.3	0.001	2.7
Phtisica marina Slabber	–	–	–	0.1	0.001	0.5
Insecta, larvae						
Chironomus salinaris (Kieffer)	–	–	–	10.3	0.026	21.6
Tunicata						
Molgula euprocta Drasche	0.3	0.768	2.6	–	–	–
Total	912.0	11.515	–	505.5	8.687	–

N—average quantity, ind. m^{-2}; *B*—average biomass, g m^{-2}; *P*—occurrence, %

Table 4.9 Comparison of the quantitative indexes of macrozoobenthos in the silt biotope of Odessa and Yuzhny MPs aquatories

Index	Odessa MP	Yuzhny MP
Date of sampling	2001	1992–2006
Number of samples	117	185
Sampling depth (m)	3.0–16.5	10.0–17.5
Average sampling depth (m)	11.1 ± 0.3	14.6 ± 0.1
Taxa number, total	43	44
Including: worms	19	15
• Molluscs	8	12
• Crustaceans	14	15
• Other	2	2
Taxa number per sample	0–17	0–18
Average taxa number per sample	5.5 ± 0.3	5.1 ± 0.2
Benthos quantity per sample (ind. m^{-2})	0–9630	0–6370
Benthos average quantity (ind. m^{-2})	912.0 ± 149.4	505.5 ± 47.8
Benthos biomass per sample (g m^{-2})	0–145.07	0–343.70
Benthos average biomass (g m^{-2})	11.52 ± 2.28	8.69 ± 1.18
Biomass of feeding (for fish) benthos (g m^{-2})	10.74	8.62
Biomass of feeding (for fish) benthos (%)	93.30	99.40
Index of uniformity of food structure	0.30	0.26
Alien species, taxa number	5	3
• Quantity (ind. m^{-2})	272.8	73.1
• Biomass (g m^{-2})	0.77	1.89

In both ports, the macrozoobenthos of silts was exclusively formed by euryhaline marine complexes. The coefficient of taxa community similarity between aquatories (according to Jaccard–Alekhin) was 57.1%. The corresponding coefficients for worms, molluscs and crustaceans were 75.0, 53.8 and 52.6%, respectively. In both aquatories, the macrozoobenthos quantitative indices varied widely across stations (Table 4.9). The character of bottom sediments was the main factor determining the composition and quantitative development of bottom macrofauna in both ports. In Odessa MP, where the bottom sediments are more diversified, the average density (912.0 ± 149.4 ind. m^{-2}) and biomass (11.52 ± 2.28 g m^{-2}) of benthos were slightly higher than in Yuzhny MP (505.5 ± 47.8 ind. m^{-2} and 8.69 ± 1.18 g m^{-2}, respectively). In both MPs, worms quantitatively dominated among taxonomic groups (Fig. 4.11).

Despite the significant number of taxa found in both MPs, the number of dominant and codominant taxa in terms of occurrence frequency was 3 and 4 in Odessa MP and 2 and 4 in Yuzhni MP. Furthermore, all main taxa belonged to polychaetes: *H. imbricata*, *N. succinea* and *P. cornuta* in Odessa MP; *N. hombergii* and *H. filiformis* in Yuzhny MP.

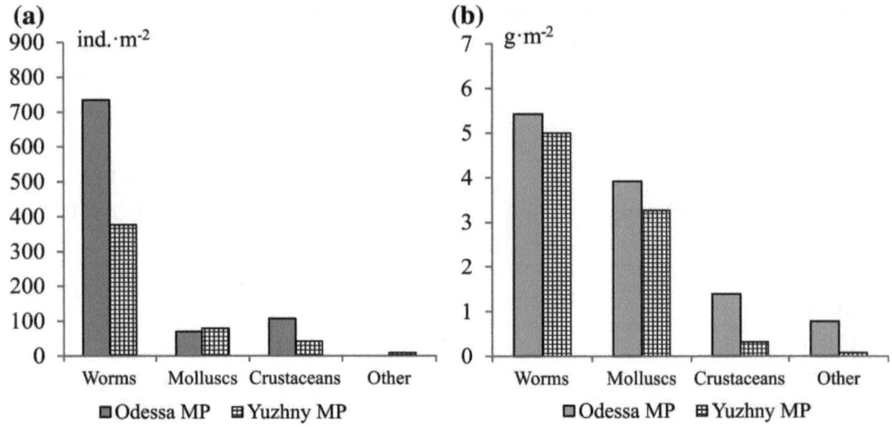

Fig. 4.11 Comparison of the quantitative indexes of the macrozoobenthos main taxonomic groups in the silt biotope of Odessa and Yuzhny MPs aquatories (**a** quantity, ind. m^{-2}; **b** biomass, g m^{-2})

The fact that 8 out of 10 dominant and codominant taxa belonged to taxonomic group of worms testifies about the unfavourable abiotic conditions for development of the macrofauna at the bottom of these aquatories. In Odessa MP, dominant and codominant taxa comprised 86.5 and 42.7% of the total quantity and biomass, respectively. The corresponding figures in Yuzhny MP were 75.0 and 53.2%.

In both MPs, detritophages were the dominant trophic group in terms of number of taxa (17–20), relative abundance (90.0–84.2%) and biomass (56.0–48.7%) (Table 4.10). In MPs aquatories, the index of macrozoobenthos uniformity of food structure was nearly the same (0.26–0.30).

In both aquatories, the number of infaunal and epifaunal taxa was similar (24 and 19 in Odessa MP, 23 and 21 in Yuzhny MP). However, infaunal organisms dominated on quantity (80.3% in Odessa MP, 82.7% in Yuzhny MP) and biomass (80.7 and 92.3%, respectively).

In agreement with the characteristics of bottom sediments, representatives of vagile complexes dominated on all quantitative indexes (37 taxa, 99.2% of quantity and 89.7% of biomass in Odessa MP; 38 taxa, 87.4% of quantity and 83.7% of biomass in Yuzhny MP).

In Odessa and Yuzhny ports, five (Odessa MP) and three (Yuzhny MP) alien species were recorded among macrozoobenthos: *P. cornuta*, *D. obscura*, *M. arenaria*, *B. improvisus* and *Rhithropanopeus harrisi tridentata*. Their total quantity and biomass comprised 29.9 and 6.7% of the total, respectively, in Odessa MP and 14.5 and 21.8% in Yuzhny MP.

Long-term studies of macrozoobenthos in the deepwater part of Grigorievsky Liman showed that, after some stabilization in the last years, there is a tendency towards growth in the quantity and biomass of bottom inhabitants.

Table 4.10 Comparison of the quantitative indices of macrozoobenthos main trophic groups in silt biotope of Odessa and Yuzhny MPs aquatories

Main trophic groups	Taxa number	Quantity		Biomass	
		ind. m^{-2}	%	g m^{-2}	%
Odessa MP					
Detritophages	17	821	90.0	6.45	56.0
Seston feeders	7	20	2.2	3.52	30.5
Phytophagous–detritophagous	5	23	2.5	0.11	1.0
Phytophagous	2	3	0.3	0.01	0.1
Carnivorous	10	45	5.0	0.88	7.6
Polyphagous	2	–	–	0.55	4.8
Total	43	912	100.0	11.52	100.0
Yuzhny MP					
Detritophages	20	426	84.2	4.23	48.6
Seston feeders	6	22	4.3	3.12	36.0
Phytophagous–detritophagous	4	7	1.4	0.05	0.6
Phytophagous	4	9	1.8	0.03	0.3
Carnivorous	9	41	8.1	1.18	13.6
Polyphagous	1	1	0.2	0.08	0.9
Total	44	506	100.0	8.69	100.0

In the deepwater part of Grigorievsky Liman, the number of surveys and sampling stations was threefold less in 2007–2011 than in 1992–2006. The number of macrozoobenthos taxa was practically the same in both periods (44 and 42, respectively). On the other hand, between the two periods, the average number of macrofauna taxa registered during one survey increased by 1.5-fold (from 14.6 to 22.0), the average taxa number per station by 1.6-fold (from 5.1 to 8.3), the average quantity by 3.1-fold (from 506 to 1591 ind. m^{-2}) and the average biomass by 3.2-fold (from 8.69 to 28.05 g m^{-2}). In particular, the average biomass of worms increased by threefold (from 5.02 to 14.99 g m^{-2}) and that of molluscs by almost fourfold (from 3.27 to 12.82 g m^{-2}).

Despite the general increase in the quantitative indices of alien species (taxa number from 3 to 5, density from 73 to 151 ind. m^{-2}, biomass from 1.89 to 7.16 g m^{-2}), their share in the macrozoobenthos composition (percentage abundance 14.4–9.5%, relative biomass 21.7–25.5%) remained nearly at the same level as in earlier studies (Table 4.11).

Over the whole monitoring period of the liman deepwater zone (29 surveys and 252 stations between 1992 and 2011), analysis of the macrozoobenthos quantitative indices dynamics showed that the average indices of quantity and biomass were somewhat higher starting from 2007 than in the previous period and had an increasing trend. The increase in biomass indices anticipated that of quantity indices. This is mainly explained by the increase in the number of bivalves, whose average body mass is significantly higher than that of worms, in the composition of

Table 4.11 Comparison of the macrozoobenthos quantitative indexes of the deepwater zone of Grigorievsky Liman in 1992–2006 and 2007–2011

Indexes	Period	
	1992–2006	2007–2011
Surveys number	22	7
Samples number	185	67
Sampling depth (m)	10.0–17.5	8.0–17.5
Average sampling depth (m)	14.7	14.7
Taxa number, total	44	42
Including worms	15	16
• Molluscs	12	12
• Crustaceans	15	11
• Other groups	2	3
Taxa number at one survey	6–25	15–27
Average taxa number at one survey	14.6	22.0
Taxa number at per sample	0–18	1–17
Average taxa number per sample	5.1	8.3
Average benthos quantity (ind. m^{-2})	506	1591
Including worms, ind. m^{-2} (%)	375 (74.2%)	1324 (83.2%)
• Molluscs, ind. m^{-2} (%)	79 (15.5%)	215 (13.5%)
• Crustaceans, ind. m^{-2} (%)	42 (8.3%)	19 (1.2%)
• Other, ind. m^{-2} (%)	10 (2.0%)	33 (2.1%)
Benthos average biomass (g m^{-2})	8.69	28.05
Including worms, g m^{-2} (%)	5.02 (57.7%)	14.99 (53.4%)
• Molluscs, g m^{-2} (%)	3.27 (37.7%)	12.82 (45.7%)
• Crustaceans, g m^{-2} (%)	0.32 (3.7%)	0.10 (0.4%)
• Other, g m^{-2} (%)	0.08 (0.9%)	0.14 (0.5%)
Invasive alien species, taxa number	3	5
• Quantity, ind. m^{-2} (%)	73 (14.4%)	151 (9.5%)
• Biomass, g m^{-2} (%)	1.89 (21.7%)	7.16 (25.5%)

bottom macrofauna. This change in the structure of bottom biocenoses is probably determined by the relative stabilization of the environment for hydrobionts in some zones of the bottom.

To better understand the macrozoobenthos peculiarities in the MPs aquatories ecosystems, a comparative analysis of macrofauna was carried out directly in the harbours of Odessa MP and at reference stations with the similar depths in the Odessa region.

In Odessa MP aquatory, 43 taxa (worms—19, molluscs—8, crustaceans—14 and 2 representatives of other groups) were found in the macrozoobenthos composition. In Odessa Gulf, at reference stations outside the port aquatory, the number of worm taxa (20) was nearly the same as in the MP aquatory, while the numbers of species belonging to molluscs (23), crustaceans (26) and other groups (5) were 2.9,

1.9 and 2.5-fold higher, respectively. In general, the total number of taxa at reference stations was 73 and it was 1.7-fold higher than in the MP aquatory. The number of taxa at reference stations ranged between 1 and 24. In MP aquatories, the macrozoobenthos was absent at 11 stations (9.4% of the total) and included 1–17 taxa at the other stations. The average number of taxa at the reference station (10.2) was 1.9-fold higher than in MP aquatory (5.5).

At both MP and reference stations, the bottom fauna was exclusively represented by euryhaline marine forms. The coefficient of taxonomic similarity (according to Jaccard–Alekhin) between the port and reference stations was only 39.8%; more specifically 65.2% for worms, 29.2% for molluscs and 37.9% for crustaceans. The macrozoobenthos quantitative indices of density and biomass varied markedly in both Odessa MP and Odessa Gulf, but the range of their fluctuations was significantly higher at reference stations. The average indices of density (2525 ind. m^{-2}) and biomass (752 g m^{-2}) at reference stations were 2.8 and 65.4-fold higher than in the port aquatory (912.0 ind. m^{-2} and 11.515 g m^{-2}, respectively). In both the port and the gulf, worms were dominant among the main taxonomic groups (80.6% in port aquatory, 57.3% at reference stations). In the MP, the worms also dominated in terms of biomass (47.2%), while this role belonged to molluscs at reference stations (96.9%).

In the MP, the indices of quantity and biomass of crustaceans were relatively high in spite of the relatively low indices of the macrozoobenthos. In MPs aquatories, samples were collected directly near the berths walls and at a distance of 50 m from them. Twenty-eight crustacean species were found in the periphyton composition on the berths. Part of the vagile crustaceans has the possibility to migrate from the berths' walls into the adjacent zones of the bottom. Ten species of crustaceans were found directly near the berths, with an average density and biomass of 157.7 ind. m^{-2} and 1.011 g m^{-2}, respectively. Thirteen crustacean species were found at a distance of 50 m from berths. Their average density (17.1 ind. m^{-2}) was nearly tenfold lower than immediately near the berths, while their biomass (2.072 g m^{-2}) was twofold higher, mainly owing to the presence of few (0.2 ind. m^{-2}), big-sized individuals of the sand shrimp *Crangon crangon* (average biomass 1.045 g m^{-2}).

In MPs aquatories, representatives of the polychaetes *P. cornuta* (266.7 ind. m^{-2}), *H. filiformis* (221.4 ind. m^{-2}) and *Neanthes succinea* (149.6 ind. m^{-2}) were the most abundant organisms. Their share in the average total quantity of the macrozoobenthos was 69.9%. At the reference stations in Odessa region, the highest average density was registered for *H. filiformis* (682.9 ind. m^{-2}), *M. galloprovincialis* (578.1 ind. m^{-2}) and *N. succinea* (330.9 ind. m^{-2}), which on the whole comprised 63.0% of the average quantity. It is worth noting that the average densities of the two dominant species of polychaetes (*H. filiformis* and *N. succinea*) at reference stations were, respectively, 3.1 and 2.2-fold higher than in the MP benthos.

In MPs aquatory, the highest contribution to the macrozoobenthos average biomass was registered for *N. succinea* (3.45 g m^{-2} or 29.9% of the total) and for the bivalve *M. galloprovincialis* (668 g m^{-2} or 88.8%) at reference stations. The average quantity of *M. galloprovincialis* in MP benthos was only 1.8 ind. m^{-2} for an average biomass of 0.260 g m^{-2}.

With regard to feeding guilds, detritivores dominated in terms of number of taxa (17—in port aquatory, 33—at reference stations) and quantity (90.0 and 61.3%, respectively) both inside and outside the MP. Detritivores species also dominated in terms of biomass in MPs aquatory (56.0%), while seston feeders (97.0%) dominated at reference stations. In MPs aquatory, the bottom macrofauna index of food structure uniformity (0.30) was 2.9-fold lower than at reference stations (0.86).

Both in the MP and in the gulf, the number of infaunal and epifaunal taxa was relatively close (23 and 20 in the port aquatory vs. 42 and 32 at reference stations), with infauna representatives dominating on quantity (80.0% in the port aquatory and 63.7% at reference stations). In Odessa MP, representatives of infauna also dominated the biomass (80.5%), while epifauna biomass dominated at reference stations (90.0%).

Vagile organisms had the highest number of taxa both inside and outside the MP (37 or 86% in the port aquatory, 65 or 87.8% at reference stations) and the highest quantity (99.2% and 66.0%, respectively). In Odessa MP aquatory, vagile species also comprised most of the biomass (89.7%), while sessile organisms accounted for 90.1% of the biomass at reference stations.

Eight alien species from various taxonomic groups were recorded: five in the MP aquatory and six at reference stations. Their average density was nearly the same in the port aquatory and at reference stations (272.8 ind. m^{-2} vs. 237.9 ind. m^{-2}), but their share in the macrozoobenthos composition was nearly threefold higher (29.9%) in the MP than at reference stations (10.6%). In Odessa MP, the seasonal changes in the composition of bottom macrofauna do not have a well-defined character.

In Varna and Beloslav Lakes, intensive formation of black silts has started after their connection with the sea and the subsequent salinization of their waters. In these lakes, bivalves of the genus *Dreissena* and *Monodacna* have disappeared and have been replaced by mass development of *M. galloprovincialis*, showing the highest accumulations on settlements at 5–8 m depth. *Modiolus lineatus*, *Cardium edule* and *A. ovata* also widely spread. Two gastropod species, *H. acuta* and *Rissoa venusta*, started to occur often and, with regard to crustaceans, the Dutch crab *Rh. harrisi tridentata* and the shrimps *Palaemon adspersus* and *P. elegans* became the mass species. Some polychaete species such as *H. diversicolor*, *Neanthes succinea*, *P. cornuta* and *Mercierella enigmatica* as well as chironomids larvae became relatively widespread [128].

On the soft silts of Sevastopol Bay, 44 macrozoobenthos species were recorded in 2001–2002; 14 out of which were mass species occurring at no less than 50.0% [131].

The maximal number of species (10) and the maximal quantity of gastropods (up to 800 ind. m^{-2}) were registered at the entrance into Sevastopol Bay. In total, 19 gastropods species, 7 of which at the shoreline were found in the bay in 2002–2003. It turned out that the more polluted was the bay zone, the lower were the number of species and the quantity of molluscs [77].

In the upper part of Sevastopol Bay, i.e. at the longest distance from the passage into it, is located oil harbour which is protected from the rest of the aquatory by a breakwater existing since 1903. In oil harbour, 41 species of macrozoobenthos were found in the piers zone: 7 bivalve, 11 gastropod, 9 malacostracan and 11 polychaete species. In the bottom zones adjacent to piers, gastropods dominated on quantity (49.0%), while bivalves dominated on biomass. On the other hand, gastropods dominated both on quantity and on biomass at a distance of 20 m from the piers [69, 70]. In summary, *Mytilidae* (filter feeders) dominated the biomass near the pier, while gastropods (predators) dominated at a distance of 20 m from it. At the bottom, 38 species of bottom macrofauna were found in the breakwater region: bivalves—7, gastropods—10, malacostracans—7 and polychaetes—13. At a distance of 10 m from the breakwater, the gastropods dominated on quantity, while *Mytilidae* had the highest biomass both close to and far from the breakwater [69, 70].

In Sevastopol Bay, in 1997, the macrozoobenthos composition on various sediments type included 48 taxa. Twenty-five of them were found in the bay's apex, 19 in its middle part and 41 in the zone of passage into the harbour. The highest average values of bottom macrofauna biomass were found at the bay's mouth (354.97 g m^{-2}), while the highest densities occurred in the apex and middle parts (3597 and 3920 ind. m^{-2}, respectively). In the apex and middle parts of the bay, the bottom sediments were mainly inhabited by the community of *C. glaucum–A. ovata–Tritia reticulata–H. acuta*. The importance of *Chamelea gallina*, *Modiolus adriaticus*, *Diogenes pugilator* and other species increased in the passage region. Lately, a tendency towards an increased role of *T. reticulata* and *Bittium reticulatum* is observed for adult bivalves in Sevastopol Bay [4]. For comparison purposes, it should be mentioned that, in Odessa MP, the mollusc *T. reticulata* (known for its resistance to pollution) was registered only one time (in 2001) in a qualitative sample [73].

In Kamishovaya Bay, bottom macrofauna composition in 1997 comprised 43 species, 26 out of which were found in the apex part, 18 in the middle part and 17 at the bay's mouth. The most widespread species were the polychaetes *H. filiformis* (occurrence 88.9%) and *C. capitata* (66.7%), the bivalves *C. glaucum* and *Ch. gallina* (66.7% each) and the gastropods *T. reticulata* and *B. reticulatum* (55.6% each). Four species were found only in the apex part of the bay and 16 only at its mouth. The maximal values of biomass were registered at the bay's mouth and those of density in its apex part [4].

Studies of the macrozoobenthos composition in Sevastopol and Kamishovaya Bays [23] showed that biomass comprised 10–50 g m^{-2} at most stations located at a distance of 30–90 m from moles. In general, the average biomass of macrozoobenthos was somewhat higher at a distance of 90 m than in zones adjacent to piers where a comparative reduction of density and biomass of the bottom macrofauna was observed. Detritivores usually dominated on quantity (53.0–90.0%) at a distance of 30–40 m from the piers, while the share of seston feeders increased (50–74%) at a distance of 90 m.

Table 4.12 Quantitative characteristics of the main macrozoobenthos species of Sevastopol Bay. Modified from [107]

Taxon	Occurrence (%)	Biomass (g m^{-2})		Quantity (ind. m^{-2})	
		Harbour waterway	Berths zone	Harbour waterway	Berths zone
Polydora cornuta Bosc	80.0	0.217	0.178	140	94
Capitella capitata (Fabricius)	80.0	0.087	0.360	88	327
Heteromastus filiformis (Claparede)	70.0	0.388	0.024	408	30
Hydrobia acuta (Draparnaud)	70.0	0.164	1.456	83	739
Mytilaster lineatus (Gmelin)	70.0	0.040	6.555	4	242
Rissoa parva (Costa)	60.0	0.057	0.176	13	54
Cerastoderma glaucum Poiret	60.0	–	0.640	–	41
Neanthes succinea (Frey et Leuch)	50.0	0.068	0.103	4	17
Nephthys hombergii Savigny	40.0	0.977	0.215	18	4
Tritia reticulata (L.)	40.0	3.241	4.823	4	9
Iphinoe elisae Bacescu	40.0	0.004	0.007	4	6
Parvicardium exignum (Gmelin)	40.0	–	0.240	–	47
Harmothoe reticulata Claparede	30.0	–	0.014	–	24
Oligochaeta g. sp.	30.0	0.005	0.001	35	6

After cleaning works on the bottom area adjacent to the berths of Sevastopol Bay, the macrozoobenthos composition in this zone comprised 32 taxa, including: Polychaeta—11, Malacostraca—3, Mollusca—15 (Gastropoda—5, Bivalvia—10) and also Nematoda, Nemertini and Oligochaeta. The number of macrozoobenthos taxa varied from 5 to 18 at the different stations. Fourteen taxa had a total occurrence of more than 30.0% and belonged to the number of dominant ones. About 80.0% of quantity and more than 90.0% of biomass was represented by this group of organisms [107]. Table 4.12 lists the main macrozoobenthos species of Sevastopol Bay benthal zone.

Along the fairway into Sevastopol Bay, *T. reticulata*, *N. hombergii* and *H. filiformis* were the dominant species and the total benthos biomass and density were 0.89–13.51 g m^{-2} and 446–1079 ind. m^{-2}, respectively. In the berths zone, the total density varied between 144 and 6000 ind. m^{-2} for a biomass range of 0.32–55.22 g m^{-2} [107].

In Gollandiya Bay, located in Sevastopol Bay, samples were taken at distances of 2 and 20–30 m from the berths. The depth varied from 3 m near the berths up to 4–9 m. Sediment consisted of sand and sand with an admixture of coquina. In the zone of the regional power station, located in the upper part of Sevastopol Bay, samples were taken at a distance of 2 and 30–35 m from the berths. Sediment was composed of grey sand with silt admixture. Fifty macrozoobenthos species were found in samples taken in Gollandiya Bay: 20 polychaetes, 9 crustaceans, 10 gastropods, 8 bivalves as well as bryozoans, chitons and oligochaetes.

In the zone of the regional power station, the number of bottom invertebrate species was 26 and included 10 polychaetes, 4 crustaceans, 9 gastropods, 4 bivalves, as well as bryozoans, oligochaetes and nemertines. Fifteen species occurred at more than 50.0% of the stations. Organisms resistant to the impact of oil pollution were found at all sampling locations in the zone of the regional power station. In general, species abundance was higher at a distance of 30–35 m from the berths. The increase in species abundance at stations more distant from the berths was mainly caused by the increase of polychaetes and bivalves share in the benthos diversity. In the regions of Gollandiya Bay berths and in the zone of the power station, the macrozoobenthos densities were 3775–43,930 ind. m^{-2} and 864–4202 ind. m^{-2}, respectively. The corresponding ranges for biomass were 27.2–202.7 g m^{-2} in Gollandiya Bay and 1.4–39.3 g m^{-2} in the zone of the power station [24].

In the latter zone, detritophages dominated (79.4% of the total quantity) near the berth, but their share decreased to 58.5% at a distance of 30–35 m from the berths as a consequence of the increased quantity of filter feeders *Mytilidae* mussels and barnacles growing on their shells. At the opposite, in Gollandiya Bay berths region, sestonophages dominated near the berths (81.2–96.5% of the total quantity) and their contribution decreased to 54.3–62.7% of the macrozoobenthos quantity at a distance of 30–35 m from berths [24].

In general, a reduction of species abundance, Shannon indices of diversity and Pielou's measure of species evenness is observed near berths, along with a simplification in the trophic structure of bottom communities [24].

In our opinion, the differences in macrozoobenthos species composition, quantity and biomass highlighted by Viter [24] between Gollandiya Bay and the zone of the regional power station are, first of all, connected with their different distance from the passage into the aquatory of Sevastopol Bay and with the different characteristics of their bottom sediments. In addition, it is necessary to consider that sediments usually become denser as the distance from berths increases, which creates more favourable conditions for populations of seston feeders. Such situation is observed especially in the zone of the berth of the regional power station where silted sands occur.

In Novorossiysk Bay, the composition of bottom fauna changed from its entrance towards its apex, gradually impoverishing both quantitatively and qualitatively. These changes can, first of all, be explained by the negative influence of Novorossiysk MP. The benthos biomass decreases from 45.00 to 104.00 g m^{-2} at the mouth to 7.95 g m^{-2} in the apex part [32].

Tuapse MP is similar to Novorossiysk MP with regard to composition and biomass of the zoobenthos [32, 82, 83].

In the immediate proximity of Novorossiysk MP, the bottom sediments were formed by toxic, nearly lifeless black silts with a hydrogen sulphide odour. In the various harbours of Novorossiysk MP, the sulphide content of the upper layer of bottom sediments varied from 80 to 1980 mgS dm^{-3} of wet silt, depending on the proximity to pollution sources. The critical concentration (>600 mgS dm^{-3} of wet silt) was registered in the zone of berths and near the discharge of the urban sewer. Such level of sulphides results in a practically total die-off of the macrozoobenthos at the bottom. High concentrations of free sulphides near the berths are explained by OM inputs from the settlements of HS foulers [114].

The main feature of Novorossiysk MP bottom biocenoses is their quantitative poorness. In 2006 and 2007, the density of macrofauna settlements varied from 2.5 to 12,500 ind. m^{-2} and biomass ranged from 0.1 to 40.0–60.0 g m^{-2}; with average values of 6400 ind. m^{-2} and 9.0 g m^{-2}, respectively) [114].

In 2006–2007, the bottom fauna of Novorossiysk MP included 38 taxonomic forms (excluding hydroids, ostracodes, nemertines and gammarids). In the port aquatory, nematodes, oligochaetes, polychaetes and nemertines dominated (75.0–95.0%) at most stations.

The species *C. capitata*, an indicator of bottom sediments eutrophication, dominated the polychaetes' composition. This species is able to survive under practically anaerobic conditions in liquid black silt with hydrogen sulphide odour. In 2006–2007, biocenosis of *C. capitata* occupied more than half of the bottom area of Novorossiysk port. The polychaetes *N. hombergii*, *H. filiformis* and *Neanthes succinea* were also present in the biocenose composition [114].

In 2001, the invasive species *S. gynobranchiata* (Spionidae) was registered for the first time in the estuarial biotope of the Tsemes River, with a density of 980 ind. m^{-2} [96]. In 2007, the density of this species in the estuarial zone reached 9000 ind. m^{-2} and it was already present at the majority of port stations [115].

Streblospio gynobranchiata is a small-sized worm (up to 10 mm) which inhabits the upper layer of silty substrata with hydrogen sulphide odour. Polychaetes of this genus are indicators of organic pollution [114].

The sediments of the estuarial zone of the Tsemes River are highly toxic with a methane content of 11.6 cm^3 kg^{-1} [114].

A gradual replacement of the polychaetes' community by the community of the bivalve *Plagiocardium papillosum* towards the exit from Novorossiysk port is observed. The biocenosis of *P. papillosum* is located in the region of the port entrance and outside its perimeter. The gastropods *Bittium reticulatum*, *T. reticulata* and *R. venosa*, the bivalves *M. lineatus* and *Pitar rudis*, polychaetes and other taxa were recorded in the biocenosis composition [114].

In the 1970s, the whole Tuapse MP was populated by one biocenosis of *Cerastoderma edule–N. hombergii.* The highest indices of species diversity and biomass of this biocenosis were registered in winter. The minimal biomass and maximum quantity was registered at the same time in summer owing to the high abundance of juveniles of *C. edule.* In total, 13 species were found in the macrozoobenthos composition of Tuapse MP and their average density was 69 ind. m^{-2}. The small-sized polychaetes were scarce. The annual average biomass of the bottom macrofauna comprised only 5.0 g m^{-2} [84].

In August 2010, the biomass of macrozoobenthos in Tuapse MP varied from 0.82 to 25.0 g m^{-2}. The highest densities were registered for *C. capitata*, *N. hombergii*, *H. filiformis*, *Nereis sp.* and *M. palmata.* All these polychaetes species gravitated around aquatories polluted by OM. In the region of the coal port complex and in oil harbour, the macrozoobenthos was extremely poor with densities and biomass equal to 200 ind. m^{-2} and 2.0 g m^{-2} in the former and 450 ind. m^{-2} and 1.9 g m^{-2} in the latter [114].

The characteristic feature of the modern bottom biocenoses of Tuapse port is a remarkable poorness (annual average density 1900 ind. m^{-2} and 6.5 g m^{-2} biomass). The polychaete worms *C. capitata*, *H. filiformis*, *N. hombergii* and *M. palmata*, species of the family Nereidae, nematodes and oligochaetes are the main component of the bottom fauna [114].

In Tuapse port, the bottom fauna comprised 32 taxonomic forms (excluding nematodes, oligochaetes, nemertines and ostracodes). Bottom biocenoses of Tuapse MP are poorer than those of Novorossiysk MP. Polychaetes (13 taxonomic forms), nematodes and oligochaetes were the main components of the fauna in the port silt biotope. Molluscs, crustaceans and other zoobenthos animals were not found in most parts of the bottom area, except for single specimen of molluscs, barnacles, and hermit crabs *Diogenes pugilator* in the port centre. All the central part of the port was occupied by the biocenosis of *H. filiformis* where polychaetes reached a density of 14,400 ind. m^{-2} and a biomass up to 21.1 g m^{-2} during some periods [114].

Despite being less polluted than Novorossiysk and Tuapse MPs, Sochi MP, resembles them in terms of species composition and zoobenthos biomass. In Sochi MP, the biocenosis of *N. hombergii* is found on the black silt at depths of 8.0–9.0. In summer, it is composed only of polychaetes and flat-browed mud shrimps *Upogebia pusilla.* In winter and spring, six molluscs species enter in its composition (*T. reticulata*, *Fabulina fabula (Angulus fabulus)*, *Abra nitida milachewichi*, *Chamelea gallina*, *Cerastoderma edule* and *Parvicardium exiguum*), but the quantity of each of them does not exceed 5 ind. m^{-2}. Owing to the seasonal presence of these molluscs, the zoobenthos biomass in Sochi MP is higher in winter and spring, than in summer and autumn. The zoobenthos quantity reaches its maximum in autumn owing to polychaetes. The average annual quantity of the bottom macrofauna comprises 235 ind. m^{-2} for a biomass of 5.55 g m^{-2} [84].

In relatively clean bays of the NWBS, the zoobenthos density usually increases in summer and autumn and decreases in winter and early spring. An opposite situation is observed in the aquatories of Novorossiysk, Tuapse and Sochi MPs [84]. This circumstance can apparently be connected with the specific particularities of MPs.

In Kerch MP, the bottom macrofauna develops best near the berths' walls. Here, species diversity is several times higher than in the middle part of the port and at the exit from it. A similar situation is observed for quantity and biomass which are threefold and 10–30-fold higher, respectively, near the berths' walls [21].

In zones with favourable conditions, transitional ecotone zones with width from a couple of metres up to dozens of metres exist between the fouling biocenosis on MPs HTS and the black silt biocenosis. In relatively shallow-watered aquatories (3.0–5.0 m), as in the zone of oil harbour of Sevastopol MP and Kerch MP, macrozoobenthos is better developed close to berths' walls. However, in aquatories with dense, silty and silty-sand bottom sediments, the representatives of meio- and macrozoobenthos also develop significantly in the central parts of the harbours.

Owing to the presence of protective structures, lowered hydrodynamics and increased production of pelagial and periphytal communities, the bottom of MPs presents conditions that, in natural ecosystems, are typical of much deeper waters. In the benthos composition of the Black–Azov Sea basin MPs, different and characteristic complexes of organisms can be distinguished for shallow-watered (0.1–2.0 m), transitional (2.0–6.0 m) and deepwater aquatories (8.0–21.0 m). Accelerated accumulation of OM on the bottom of the deepwater zones occurs because of gravitational sedimentation and sedimentative activity of hydrobionts from plankton and periphyton. At high water temperatures, hypoxia and anoxia zones form in such deep waters and only organisms with special adaptations can survive.

In the ecosystems of MPs aquatories, preservation of existing biodiversity requires application of existing environmental legislation and the preservation of the biotopes' diversity; especially the coastal shallow waters at shore zones with lengths of about 300–500 m. The main problem in sustaining a stable, normal functioning of all the components of MPs aquatories ecosystems, including the bottom biocenoses of silts, is to ensure the presence of a removal mechanism for the excess amounts of BM and OM and to create active ventilation conditions for the near-bottom water layer.

It is necessary to outline that, as a rule, a sharp boundary between the benthal, periphytal and pelagial subsystems is absent in MPs aquatories ecosystems.

A continuous temporal shift of biocenoses is observed in MPs aquatories, i.e. successions take place in all the subsystems of MPs ecosystems. Successions can take place as a result of natural causes as well as under the influence of anthropogenic factors. Numerous succession forms are distinguished; all of them originating as a consequence of rules, principles and laws already described in ecology. In MPs aquatories, primary successions are defined as those starting on new substrata not previously populated by hydrobionts, while secondary succession indicate those observed in aquatories on substrata already existing before MP construction or populated since a long time. In MPs, successions could have a reversible or irreversible character. Thus, for example, the anthropogenic eutrophication of MPs aquatories inevitably leads to the disappearance of many species of brown and red algae.

In the coastal sea zones unaffected by hydrotechnical constructions activity, various life forms persist on the bottom. The sediments uncovered during dredging of MPs aquatories and AC are usually reduced and sterile. After dredging, they undergo full oxidation for some time and simultaneously become populated by microbes, protozoans, diatoms, fungi and various invertebrates and regular successions take place within them. In MPs aquatories, invertebrates enter the bottom zones with newly exposed sediments from the water column (after termination of pelagic development stage), from the coastal shallow waters (if they are preserved) and from the HTS fouling community. The primary successions also take place on new HTS surfaces. In MPs, the ballast waters with their population and the periphyton of the ships' hulls can be a source of new species. Because all the living are born from the living (Redi's principle), new species can enter MPs aquatories in the form of resting stages, eggs, larvae and adult individuals. The entry of new species into communities is always accompanied by specific reorganization processes.

According to the Odum–Pinkerton principle, successions are always accompanied by fundamental shifts of the energy flow towards an increase of the energy amount required for on ecosystem sustention. During this process, the stability of ecosystems reduces for some time. Thus, in MPs aquatories, successions caused by anthropogenic activity always increase the energy expenses needed to sustain their ecosystems. Ecosystems lacking a sufficient amount of energy simplify and partially deteriorate because additional energy from outside would be necessary to sustain them.

In various MPs aquatories ecosystems, an impoverishment of biota can already be observed during their construction as well as during their functioning. Changes in this sense include the disappearance of some species and even of whole biocenoses. On the other hand, enrichment of biota owing to the introduction of new species and formation of new biocenoses can also occur. The removal of a species from the composition of MPs biota can be connected with the formation of unfavourable conditions for at least one of its life cycle stages. For introduction and proliferation of new species, favourable conditions must occur for all of their ontogenetic stages.

In all subsystems of MPs aquatories, organisms form large agglomerations and patches in the coastal shallow waters zone and continuous and dotted belts on HTS vertical surfaces. As a rule, following W. Allee's principle of aggregation of individuals, biota accumulations strengthen competition for food and space, but simultaneously increase the ability of groups of individuals to survive. Single specimens do not usually survive for long under new conditions.

According to the principle of system complementarity [38, 104] within a natural ecosystem, the development and functioning of each subsystem provide the preconditions for the successful development and self-regulation of other subsystems. In the case of MPs aquatories, this principle applies to the subsystems of pelagial, periphytal and benthal. In ecosystems, species use all the ecological niches provided by the environment, minimizing competition among themselves and maximizing productivity under the actual, local conditions [103, 104]. In such situations, space (water column and contact surfaces) is filled up to the maximum possible extent, in agreement with R. MacArthur "density packing" principle [104].

The diversity principle formulated by A. Thienemann states that the higher is the diversity of conditions within a biotope, the larger is the number of species present in the corresponding biocenosis. Consequently, the higher is the number of biotopes, the more species occur in the ecosystem as a whole. As a rule, diversity and total number of biotopes reduce in MPs, but new biotopes, which were not previously characteristic of the ecosystems of the region, can appear at the same time. The quicker occurs the transformation of natural biotopes, the harder is for species to adapt to such changes [104] and, therefore, the more destructive are the consequences for biocenoses. This is why in MPs, biocenoses could be combined for some period of time, or forever, and could acquire new features and evolve during the recovery processes.

In MPs aquatories, edificator species are usually registered at the bottom and on the HTS surfaces. These are species of plants and animals with a strong ability to, directly or indirectly, make the environment easier to colonize for other species. In MPs, examples of such species include the mussels *Mytilus* and *Mytilaster*, barnacles, hydroids, bryozoans, macrophytic algae and flowering plants (angiosperms). On soft sediments, edificators can be represented by the polychaete *M. palmata*, the bivalve *M. arenaria*, the amphipod *A. diadema* and the flat-browed mud shrimp *Upogebia pusilla*.

Some edificator organisms, their druses or small settlements not only impact the environment, but also create conditions for the formation of consortiums, i.e. associations of autotrophic and heterotrophic organisms connected among themselves by their habitat and by trophic links. In MPs benthal and periphytal zones, many agglomerations of living organisms are confined to consortiums, where dozens of epiphytes, symbionts and parasites species can live in close connection.

In MPs aquatories, consortiums are observed on the surface of navigational and mooring buoys as well as on lines, chains and loads holding them. In MPs, consorts can form small (up to couple of dozens square metres) patchy settlements, confined to soft sediments, e.g. seagrass *Zostera*, pondweeds *Potamogeton* and, on small, isolated hard substrata, *Cystoseira* and some other macrophytes. Consortiums connected with mussels' druses are frequent on the silty bottom of MPs. In the first part of the twentieth century, settlements of the sea oyster *Ostrea edulis* were the species consort in many MPs of the Black Sea. After the disappearance of *Cystoseira* and oysters from the MPs biota, a big group of hydrobionts connected with them has disappeared. This is an example of the "no one dies alone" ecological principle [38, 104]. With the disappearance of consort species, which constitutes and forms the specific consortium, many species consorbents also disappear from MPs biota because they do not enter in other consortiums of the same ecosystems.

Artificial aeration of the near-bottom layer with atmospheric oxygen could be used to oxidize hydrogen sulphide and prevent the rise of the redoxcline into the water column zones capable of sustaining "blooms". The oxidation of sulphides to thiosulphates follows a first-order kinetics and is very rapid. However, although lowering the toxic influence of H_2S, the formation of thiosulphates confers to water an odour of rotten eggs similar to that of hydrogen sulphide solutions, so that their presence in surface waters also is a negative phenomenon.

Mechanical surface aerators with a horizontal and a vertical spinning axis are used for aeration of shallow-watered aquatories. In deepwater basins, stratification is destroyed by blowing compressed air directly into the near-bottom layer. Aerators consuming small amounts of electrical energy and not requiring constant technical maintenance are specifically designed for this purpose. They are made of fibreglass and other non-toxic lightweight materials. For example, "Schönbrunn" type deep sea aerator is equipped with four pumps that convey an air–water mixture to the depth of 10 m and has an output capacity of 9200 m^3 per h [57].

The forced aeration should be used at the beginning of critical situations, generally during the hottest months of the year. Ideally, clean energy sources such as solar and wind power, wave action or heat pumps should be used to power aeration systems. Intensification of hydrodynamics in the water layer below the pycnocline could induce the formation of artificial internal waves.

Chapter 5
Main Structure and Peculiarities of the Functioning of Marine Ports Aquatories Ecosystems

Kommoner [54, 55] stated in a simple and understandable form the following "laws of ecology": 1—everything is connected to everything else, 2—everything must go somewhere, 3—there is no such thing as a free lunch, i.e. energy expenses are necessary, 4—nature "knows" better, i.e. natural ecosystems are very complex and consists of dynamic components so that it is impossible to foresee and predict all consequences of anthropogenic interference. These laws are also valid in MPs aquatories ecosystems.

According to the law of "cumulative action" of natural factors stated by E. Mitscherlich, A. Thienemann and B. Baule, productivity of biological systems (also including ecosystems of MPs aquatories) is determined by the totality of acting ecological factors, i.e. it is an integrated index.

The law of internal dynamic balance states that any change in a given environment (matter, energy, information, dynamical qualities of ecosystems) inevitably leads to the development of a natural chain of reactions that tend to neutralize the change itself or to form new natural systems; a process that can be irreversible following significant environmental changes [82]. The thesis that any local transformation in any part of a basin provokes regional and global response reactions in a whole complex ecosystem is one of the consequences of the above-mentioned law.

From the law of equivalency of all life conditions, it follows that all natural environmental conditions necessary for life are equally important. In MPs, the life conditions of hydrobionts are often severely violated.

V. Shelford's law of tolerance states that species well-being in an ecosystem may be limited by any factor outside minimum or maximum values of tolerance for that species. The interval between the values of tolerance is the optimum zone. In MPs, artificial disturbance of factors occurs which narrows the optimum zone for some species and broadens it for others.

In MPs, anthropogenic eutrophication can cause intensive development of phytoplankton and decrease water transparency to 0.1 m. The sharp reduction of

A.K. Vinogradov et al., *Ecology of Marine Ports of the Black and Azov Sea Basin*, DOI 10.1007/978-3-319-63062-5_5

light quantity entering the water column affects not only photosynthesis, but also causes a series of other negative consequences.

According to the one percent rule, a 1% increase or decrease in the natural ecosystem energy brings an ecosystem out of its homeostasis condition. MP protecting constructions reduce the hydrodynamics by several folds compared with the adjoining ecosystems. MPs are polluted by sewage waters containing OM, which is a potential source of energy.

Israel [52] noted that the ecological consequences of marine ecosystems pollution become apparent in the following processes and phenomena: 1—disturbance of ecosystems stability; 2—progressive eutrophication; 3—"red tides" (mass water "blooms") phenomenon; 4—accumulation of toxic chemicals in biota; 5—reduction of biological productivity; 6—appearance of mutagenesis and cancerogenesis in marine environment; 7—pollution of the near-coastal sea regions by pathogenic microflora. All these phenomena manifest in MPs aquatories ecosystems to various degrees.

Burkovsky [23] lists the following main components of natural ecosystems:

1. inorganic matter (natural resources);
2. organic matter (primary and secondary resources, as food);
3. environmental factors (climatic factors);
4. producers (creators of primary production);
5. macroconsumers (main consumers of primary production);
6. microconsumers (main decomposers and reducers);
7. energy (light, energy of chemical bonds, heat);
8. information;
9. space-time, determining scale and borders of ecosystem.

Some components belong to two–three categories at the same time; for example, light and heat are both types of energy and environmental factors.

Any structuring of an organism's community is the sum of the responses of individual species to the physical complexity of a biotope. Biotope acts as a special matrix on which communities form. Hydrodynamic processes and food are the main factors structuring marine communities [23].

According to contemporary visions about ecosystems and evolution of life on the planet, the environment is not a simple receptacle of life but one of its constituents. Life is a physical–chemical–biological system existing in the form of ecosystems. Matter, energy and information are literally "flowing" through organisms and this flow ensures their existence [23].

In the Earth's biosphere, biomass can only be self-renewed and redistributed between organisms (via death and birth) every time it reaches a physically determined (natural) limit. The total biomass of organisms on the planet remained stable over the last 70–100 million years, which indicates that its saturation limit has been attained under the existing conditions. In the biosphere, the living biomass is potentially limited by the Earth's space and resources [23]. The same principles apply to smaller ecosystems including those of MPs protected aquatories.

In MPs aquatories, the content of BM and OM is artificially increased and a series of other factors (oxygen content, general hydrodynamics) is decreased. This situation leads to a periodical increase in organisms' concentrations which is accompanied by a decrease in the average individual mass within the population. Because the intensity of major vital functions, including growth, is higher for smaller living organisms, the intensity of communities functioning eventually increases, but is realized by much smaller organisms.

In comparison with terrestrial ones, aquatic ecosystems are characterized by a higher number of links and interdependences because of the peculiarities of their habitat.

During the evolution of aquatic ecosystems, the morphological variety of biota contrasts the unity of living organisms' main functions: production, use, transformation and reduction of OM, reproduction of genetic information and maintenance of the general turnover of biogenic matter. Both primitive microorganisms and complex, multicellular ones are able to fulfil these functions.

Changes in the organisms performing the various functions (i.e. successions) took place during the whole period of the biosphere existence and, at shorter time scales (seasonal, interannual and others) occur right in front of us.

The theory formulated by Clements [130] stipulates that ecological successions are a process of ecosystem development and not only a shift in some separately functioning species. Succession is actually a way to preserve the ecosystems' functional elements and organisms' habitats.

Such well-known phenomena as "water blooms" and "red tides" are neither the final nor the main outcome of anthropogenic eutrophication, but only one of the evident aspects of the process of succession—in some way a side effect [71].

The transformation processes of matter and energy entering an aquatic ecosystem and accumulating in it constitute the nature of a succession. The species rotation is one of the indirect consequences of the increase in OM and BM amounts. This is a major adaptive mechanism of ecosystems and is directed towards the maximal and most rapid use of BM and its integration into the biotic turnover. Ricklefs [84] underlines that, in modern conditions, "a succession is the order of changes that take place in disturbed habitats". Thus, the main aim of preservation and management of natural ecosystems should be to sustain the environmental ecological integrity and the ecological processes in order to preserve biodiversity [139].

Intense economical activity, specific conditions and inputs of BM and allochthonous OM turn MPs aquatories into constantly "disturbed habitats", preventing them from reaching natural climax conditions [130] and transforming successions into destructive processes [43].

In the modern literature, the self-cleaning possibilities of ecosystems remain quite often substantially overestimated, especially when it concerns anthropogenic pollution whose harmful consequences are practically legalized and masked. In the vast majority of cases, ecosystems are not even able to manage small but constant pollution, as it becomes evident with time [72, 145].

Water basins do not always and not entirely manage even natural gradual pollution, which leads, albeit slowly, to their ageing, i.e. eutrophication. For aquatic ecosystems, anthropogenic pollution of any kind and quantity can become excessive and unacceptable if it gives a net acceleration to the destructive succession process. It is not possible to count on ecosystems' self-cleaning and to justify something based on it. It is important to consider water basins as storage tanks filled by wastewaters and toxicants and to make all the calculations of economic activity on these bases [71].

Taking the above considerations into account, it is possible to judge about the condition and development of a specific ecosystem only by evaluating and measuring what happens to matter and energy inside it. Many conceptions of ecosystems self-cleaning are merely terminological substitutes for the processes of abiotic and biotic transformation of matter and energy, as well as their diffusion and accumulation on the bottom of water basins and in the coastal zone. Such conceptions hardly comply with the notions about integrity of aquatic ecosystems and with fundamental laws of nature, in particular, with the Lomonosov-Lavoisier law of matter and energy conservation. They also become a theoretical ground for allowing anthropogenic pollution and leave loopholes for its practical implementation. Regardless of the pollution extent, the mechanisms of destructive successions are always launched and their consequences eventually become evident with time. In large ecosystems, such manifestations are recorded after years or decades.

5.1 Main Structure

The life in aquatic ecosystems is maintained by external and internal sources of energy that determine the ecosystems' general structure and productivity. As it is known, energy is the ability to execute work and, according to the first law of thermodynamics, can only be transformed from one form into another without being created or destroyed. The second law of thermodynamics states that any type of energy transforms into heat. Thermal energy disperses most easily and is less suitable for transformation into work. Hydrobionts more easily use the energy of light and food formed in an ecosystem or coming from outside. Any natural, relatively isolated or protected aquatic ecosystem, such as MPs aquatories, develops towards steady-state conditions.

Organisms produce positive entropy and, to stay alive, should constantly extract negative entropy from the environment. They produce energy using quanta of light for photosynthesis and chemical compounds for chemosynthesis, i.e. biochemical transformations, or receive energy with OM. In total, the entropy increases during destruction of matter and reduces at its creation [156].

According to the definition of Alimov [4, 7, 9], ecosystems represent aggregations of organisms living together that are localized in space and dynamical in time and environmental conditions and form a system of interdependent biotic and abiotic processes. The flows of matter, energy and information in an ecosystem

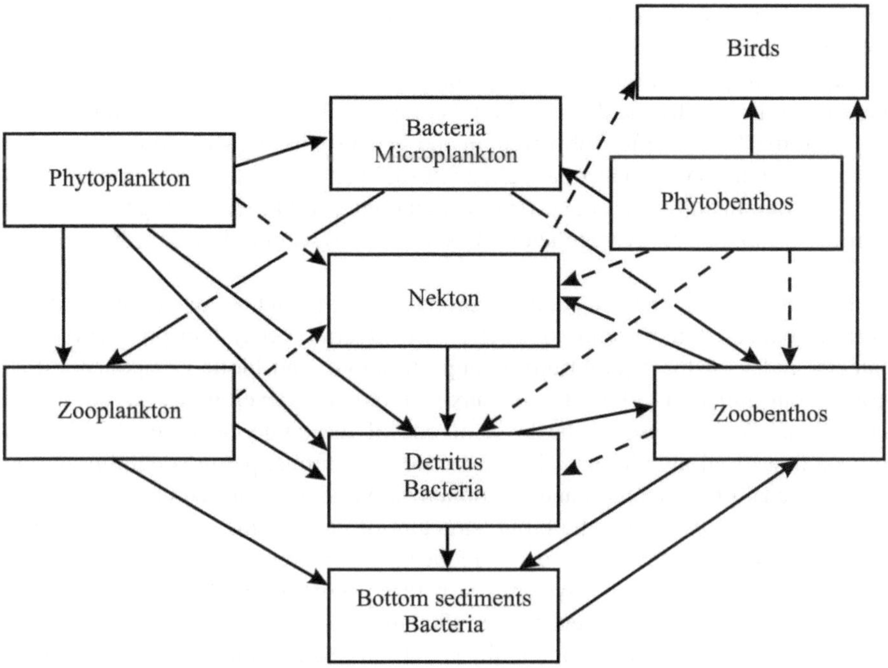

Fig. 5.1 Idealized scheme of matter turnover in MPs aquatories ecosystems (*dashed lines*—secondary links, *solid lines*—main links)

form as a result of the organisms' interactions between themselves and the environment. In ecosystems, the turnover of biological matter and the transformation of energy are realized by various interactions of the organisms between themselves; including food chains and trophic webs which can be complex and ramified. An ecosystem sustains its integrity owing to varied links between its components, realized via flows of energy, matter and information. In ecosystems, OM is produced, decays and is used during the maintenance process of biotic balance [7, 72, 73, 76, 81, 84, 85, 114] (Fig. 5.1).

The MPs ecosystems have the same characteristics as the integral, fully structured ecosystems discussed above. Production, transformation and reduction of the primary and secondary autochthonous OM and partial utilization of allochthonous OM take place at different levels in MPs ecosystems. As said above, ecosystems of MPs aquatories possess a series of peculiarities and have more or less well-defined natural and artificial borders along their perimeter while preserving at the same time some connection with the adjacent ecosystems via the access channels. With regard to many characteristics, they resemble ecosystems of estuaries, limans, lagoons, lakes and water storage basins [30].

Compared with the neighbouring ecosystems, the eutrophic ecosystems of MPs aquatories have a high productivity, i.e. can create higher amounts of OM. In general, ecosystems production is defined as the difference between primary production and the total metabolic expenses of the hydrobionts of all communities inhabiting there.

Ecosystem production occurs when only part of the energy captured in primary production is used by the different biotic components and dead OM accumulates on the bottom. This is often observed in the ecosystems of MPs aquatories where excessive, additional inputs of allochthonous BM and OM can quite often aggravate the situation. In eutrophic ecosystems, the amount of OM sedimenting together with detritus comprises on average about 37% of the production, but can reach 70% [1, 7, 9, 10, 124].

By comparison, the total annual production approaches, or slightly exceeds, zero, in balanced ecosystems of cyclic type. Overall, ecosystem production is an integral index of the equilibrium among all abiotic and biotic components. Each component functions close to its maximal possible level under given specific conditions. An increase in the level of one of the components, e.g. OM, while others, first of all abiotic ones, have reached their limits, perturbs equilibrium and causes a chain of responses and a reorganization of the other components.

As productivity grows, the biotic structure of ecosystems simplifies, their food chains shorten and species diversity reduces with increasing dominance by some species. As a result, the ecosystem's structure becomes simpler and energy dissipation decreases leading to energy accumulation.

The structure of aquatic communities and ecosystems' biota is preserved owing to constant energy expenses for maintaining order and renewing elements and structures. Ecosystems' state depends on the influence of specific environmental factors. When changing them, as it happens in MPs, their biotic structure is also destroyed. All energy of an ecosystem biotic components is connected with their structure or is spent on performing work (metabolism, respiration).

The whole structure of animal communities and ecosystems is mainly determined by resources' availability, competition and predation. In MPs aquatories there are conditions for an increase of primary production, but permanent predators (invertebrates, fish) disappear and competition slows down because of excessively high productivity. In ecosystems, the diversity of zooplankton and zoobenthos communities' progressively decreases proportionally to the increase of primary production and inputs of allochtonous OM.

The biotic structure of organisms' communities and ecosystems can change in time and space under the influence of both natural and anthropogenic factors. Superposition of the negative influence of anthropogenic factors on natural ones can lead to catastrophic consequences such as fauna mass mortality in MPs.

The species richness is connected with peculiarities of habitats (biotopes) and their diversity. The wider are the habitats' spectrum and the range of available food resources, the higher is the diversity of animal species. In MPs ecosystems, the coastal shallow waters with heterogeneous sediments are destroyed, the coastal line simplifies, the depths' spectrum changes and the bottom becomes even and silty. Conditions in MPs become more uniform because of the weakened wind influence and solar illumination. Only the presence of big areas of artificial HS favours the increase of species diversity in the periphytal zone.

Ecosystems offer a wide range of available resources and biotopes diversity and are inhabited by a great number of species specialized with respect to their needs. All other conditions being equal, the species richness inevitably decreases when part of OM, especially dead OM in the form of detritus, is underexploited. The most eurybiont species are advantaged and widen their habitat being able to quickly increase their quantity in proportion to the amount of released and forming resource. This general ecological feature manifests in the ecosystems of MPs aquatories that, although not rich in species, are rich in individuals. This is clearly visible, first of all, in epipelagial, epiperiphytal as well as in benthal zones of MPs ecosystems.

The epipelagial and the adjacent epiperiphytal zone of MPs aquatories represent a trophogenic ("nourishing") layer. In epipelagial and epiperiphytal zones, production of primary OM is performed by micro- and macrophytes and detritus from external sources also enters these zones. In bathypelagial, bathyperiphytal and benthal zones, forming tropholitic layer, the processes of OM decomposition prevail thanks to the metabolism of consumers belonging to different groups (bacteria, protozoa, fungi and others).

Anthropogenic influence on aquatic ecosystems, particularly in MPs, usually leads first to a quantitative decrease and then to disappearance of most specialized species with narrow ecological windows, especially oxyphile species. These processes are followed by reduction of trophic links, changes in trophic nets and release of resources. All this creates the preconditions for the introduction and mass development of eurybiont invasive alien species (for example, the comb jellies *Mnemiopsis* and *Beroe* in the pelagial zone and the small-sized polychaetes species in benthal zone, haarder, etc.).

MPs aquatories ecosystems typically have high levels of BOD_5, which serves as an index of the OM amount readily available to hydrobionts. It is well known that the diversity of zooplankton and zoobenthos communities reduces as BOD_5 value increases. In the pelagial zone, the increase of primary production causes water alkalinization and the diversity of phyto- and zooplankton communities reduces progressively when pH increases beyond values of 7.8–8.4 units for which the species maximum is registered [6, 8].

Biomass increases progressively with simplification of the ecosystem's structure (biodiversity). At any given time, the amount of hydrobionts biomass represents the resultant of OM production and utilization processes. In simplified ecosystems, mass species can reach very high quantity and biomass compared with previously existing conditions.

In communities and ecosystems, there are tight quantitative links between the structural and functional characteristics. Ecosystems' structure is preserved owing to constant expenses of energy on the regulation, reproduction and maintenance of ecosystem's components (organisms, populations, communities). The loss of certain species or changes in trophic links cause changes in the ecosystems' functional characteristics. As a result, ecosystems reach a new state determined by their new structural and functional characteristics [3, 6, 10].

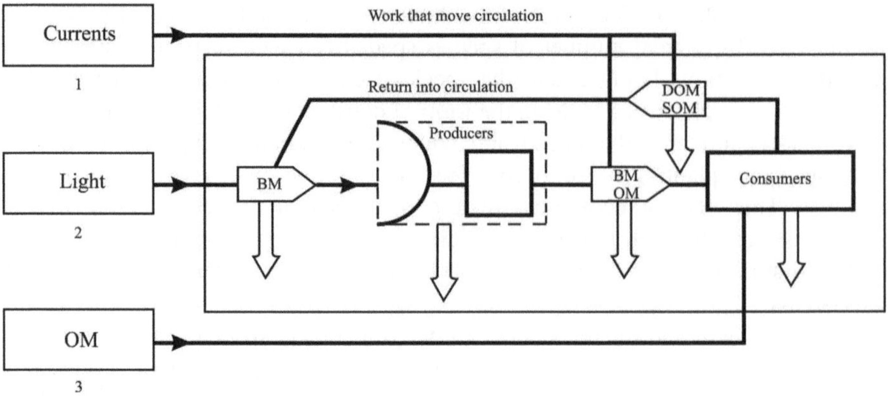

Fig. 5.2 Scheme of the three main flows of potential energy, entering into the coastal marine ecosystem (acc. to [72] with changes). *1–3*—types of sources of energy. *Bold lines*—links to the main components of ecosystem. *White arrows*—energy diffusing and becoming unavailable for use in a food chain. *Pentagonal indicators*—energy flows, assisting other flows to pass through energetic barrier. *BM*—biogenic matter, *OM*—organic matter, *DOM*—dissolved organic matter, *SOM*—suspended organic matter

Illumination, hydrodynamics, autochthonous and allochthonous inputs of dead OM, anthropogenic pollution by temporary and persistent toxicants are the major abiotic factors determining structural and functional characteristics of MPs aquatories ecosystems.

Figure 5.2 shows the general scheme of water movement, illumination and influence of allochthonous OM on MPs ecosystems (acc. to [72] with changes).

To properly understand the peculiarities of MPs aquatories ecosystems functioning it is important to understand the abiotic and biotic processes of transfer and transformation of matter, energy and information flowing through them.

It is known that organisms of the water column and bottom sediments posses resistance mechanisms against heavy metals. They are able to produce hydrogen sulphide, extracellular exopolymers and a series of other substances which interact with heavy metals and transform them into less toxic forms [85, 110, 132, 133, 136].

Light (solar radiation) is a primary source of energy without which functioning of natural ecosystems is not possible or seriously hampered. Light is not only a power resource for photosynthesis, but also a major integrative ecological factor.

Only about 1% of solar energy is converted into food and other biomass, while about 70% of it is transformed into heat or spent on evaporation, change of atmospheric pressure, wind, precipitation and others. This energy sustains the conditions that are necessary for the existence of aquatic ecosystems [73]. In marine ecosystems, thermal, haline and density water stratification forms along with the corresponding transition zones under the influence of heating, evaporation, wind-induced mixing and advection. In such zones, the redistribution of matter and energy always proceeds with much more intensity than outside them.

Winds and water foaming in the sea, in combination with Langmuir convection, lead to concentration of organic particles (and of the nutritive substances adsorbed on them) in the surface layer [88, 160].

Numerous studies showed that DOM can be transformed into structured OM due to abiotic processes in the water column. Thin semi-transparent organic films form by adsorption onto air bubbles. The dissolved OM is transformed into food suitable for crustaceans in smooth Langmuir cells covered by ripples [69].

5.1.1 Peculiarities of the Abiotic Structure

Figure 5.3 illustrates the general scheme of the main abiotic components of MPs aquatories ecosystems.

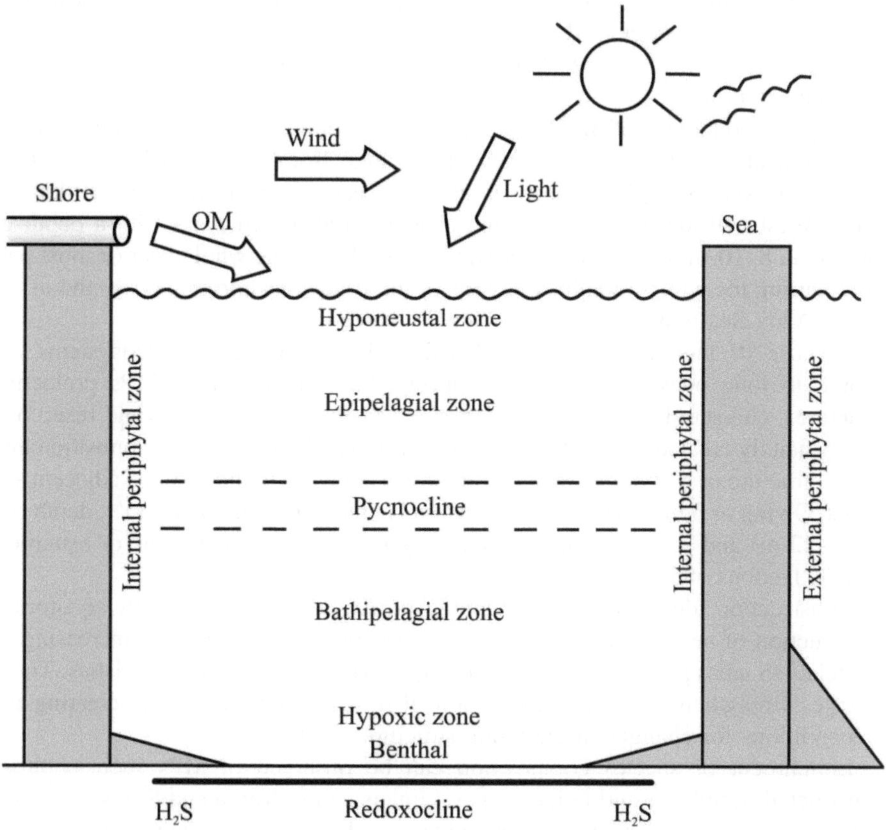

Fig. 5.3 Generalized scheme of the main abiotic components of marine ports aquatories ecosystems

Ecosystems of all MPs aquatories comprise abiotic and biotic components of extra- and intra-ecosystem origin that functions in tight interconnection and interdependence. The ensemble of morphometric peculiarities and abiotic characteristics of MPs aquatories ecosystems can be considered as the analogous of exoskeleton and endoskeleton, owing to which the biotic structure forms.

Frequently it is impossible to separate the abiotic and biotic components of ecosystems. Thus, dead OM, which enters into MPs aquatories ecosystems from outside in suspended and dissolved form, can be considered as an abiotic component from one side and as a biotic one from the other. In ecosystems, the biogenic elements are present in dissolved mineral form and then in the composition of living OM. In the same way as any mineral suspension, the live plankton organisms suspended in the water column increase turbidity and worsen the intensity of illumination of the underlying layers.

Benthic forms inhabiting on the surface and in the upper layer of sediments cause bioturbation, increase bottom roughness and favour the formation of microturbulent water movements even above silted sediments. In MPs periphytal, settlements of attached hydrobionts also favour the formation of turbulent flows and microvortexes. The filter feeders organisms favour accumulation of detritus on the bottom, thus decreasing the depths of MPs harbours. This process is termed biosedimentation.

The MPs aquatories of the Black–Azov Sea basin are located in various climatic zones and along all its shores. Their sizes range from dozens of hectares up to 26 km^2 in Constanza MP, located on an open seashore, and more than 20 km^2 in Varna-West MP located in Varna and Beloslav Lakes. All ports with a depth of more than 8–10 m have a two-layer water mass during the whole year or most part of it. During the winter months, the ports of the Black Sea northern coast and of the whole Azov Sea may be covered by ice for some time.

Because of their numerous peculiarities, the marine coastal ecosystems are similar to lotic ones such as rivers. During the construction of MPs protective structures, characteristics of lentic ecosystems such as lakes, limans and reservoirs are artificially created on the open marine coast. The biota species composition and biotic structure of MPs aquatories communities are close to those of the adjacent sea zones. Owing to protective structure, relatively small area (up to 26 km^2), depth (up to 21–22 m) and scales of phenomena, they resemble to ecosystems of estuaries, limans, lagoons and reservoirs.

Construction, deepening and widening of AC, deepening of MPs aquatories, construction of new berths and ports, reconstruction of old ports and increasing of artificial HS areas provide MPs protected aquatories with new characteristics. These changes strengthen or weaken the action of different abiotic factors, creating the preconditions for changes in the biotic structure of MPs.

Enrichment of species composition can be observed in MPs such as those constructed in Sukhoy and Grigorievsky Limans and in Varna and Beloslav Lakes. After construction of the permanent wide and deep access channels, their fauna became enriched by dozens of new species in the pelagial, benthal and periphytal zones. Change in dominant species took place and their ecosystems generally became more complex. At the same time, there are many examples of reduction of species diversity in old MPs.

Depth is one of the important characteristics of MPs aquatories. In shallow-water harbours with depths up to 3–5 m, the steady stratification of water masses does not usually take place and a pycnocline is not registered. More or less prolonged stratification of water masses can be observed in harbours with depths of 8–10 m and stratification may be constant in harbours with depths of 12–22 m. Figure 5.4 shows the annual course of stratification.

In shallow-water ecosystems, differences between the highest and the lowest values of biomass are minimal during the year. The variability of phyto- and zooplankton communities' biomass increases along with natural or artificial increases in depths [3, 4, 7, 9].

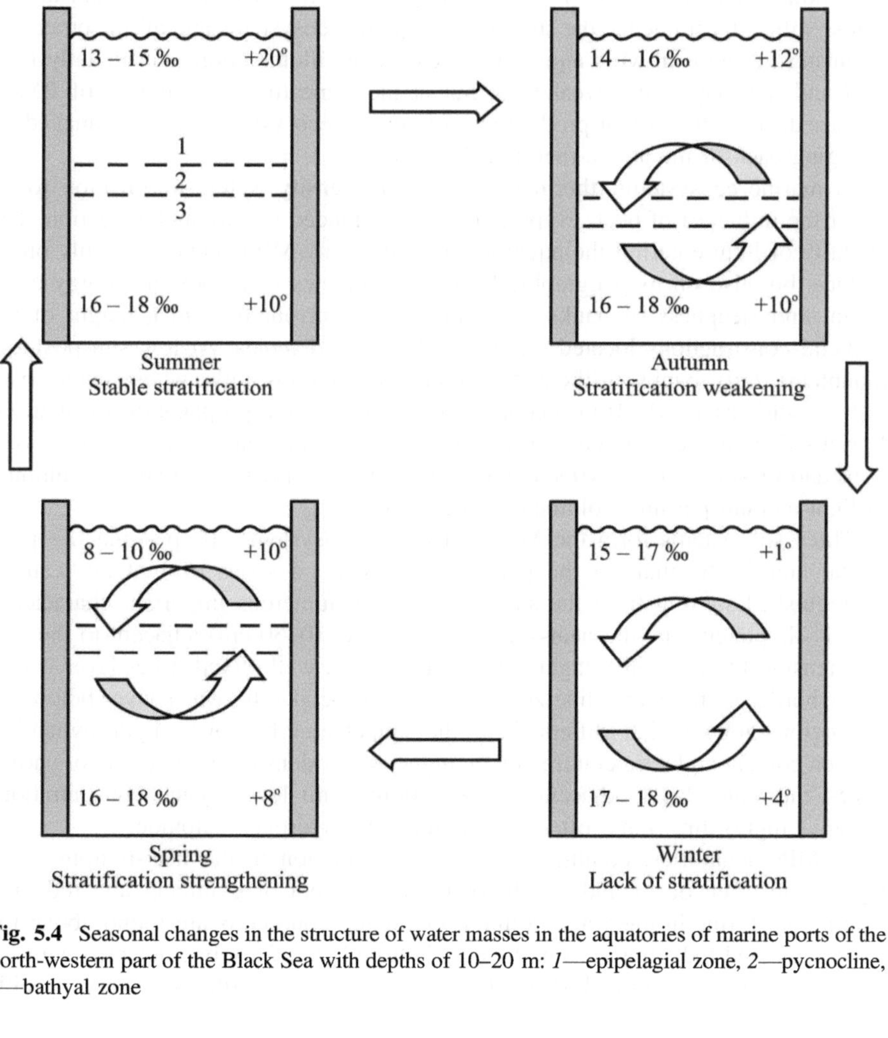

Fig. 5.4 Seasonal changes in the structure of water masses in the aquatories of marine ports of the north-western part of the Black Sea with depths of 10–20 m: *1*—epipelagial zone, *2*—pycnocline, *3*—bathyal zone

Ecological peculiarities of MPs constructed in regions with rocky deepwater shores and in regions with flat shores and loose sediments differ noticeably.

In MPs, the increase in depths causes: 1—reduction of light penetration; 2—decreased thickness of the photic layer, where the primary production generally takes place, in the epipelagial and epiperiphytal zones; 3—decrease in temperature and its seasonal fluctuations in the bathypelagic and benthal zones; 4—weakening of wind-driven circulation. The deeper becomes an aquatory, the lower is the possibility of bottom sediment's stirring up.

It is commonly believed that, in aquatic ecosystems, the species richness is connected with: 1—conditions stability (the more stable, the better); 2—organisms size (the bigger the organisms, the better); 3—energy flows (the more intense the better, up to a certain limit); 4—lifetime (the longer, the better).

The quantity of solar energy entering the aquatories surface, wave frequency and the amount of BM, dead OM and polluting toxic substances are the main factors that significantly influence the structure of aquatic ecosystems and all the processes occurring in them. In MPs aquatories ecosystems, illumination is artificially lowered and hydrodynamics weakened, but at the same time the amount of BM is increased. It is known that productivity of aquatic ecosystems can be managed by changing each of the above-mentioned factors.

In marine ecosystems, thermal, salinity and density water stratification forms under the influence of heat, evaporation, wind-induced mixing and advection. The quantity of light entering the aquatory surface of any MP depends not only on its surface, but also on its geographic location, frequency of cloudy and foggy days, height and steepness of banks, elevation of the surrounding area, height of the artificial constructions located on shores, berths and moles. At low sun position, mountains, hills, banks, artificial constructions and ships hulls, whose height over water could reach 50–60 m, become an obstacle to the propagation of sunrays. Various light shades originate in MPs because of large quantity of high constructions and mechanisms. In MPs, the artificial increase of depths reduces the amount of light amount per unit volume of water mass.

Three subsystems form the MPs aquatories ecosystems: 1—pelagial, 2—periphytal and 3—benthal. In the pelagial subsystem, a series of biotopes can be distinguished starting from the surface based on numerous important characteristics: 1—hyponeustal—the near-surface water layer (0–30 cm) adjacent to the surface tension film; 2—epipelagial—the near-surface, well-mixed water layer whose lower border is the pycnocline zone; 3—bathypelagial—the water layer below the pycnocline zone which differs from the epipelagial by lower hydrodynamics, oxygen content and temperature and by higher water density; 4—hypoxic or anoxic zone—the water layer adjacent to the bottom, with low oxygen concentrations (about 1 mgO_2 dm^{-3}) and often contaminated by hydrogen sulphide.

In MPs aquatories, conditions of oxygen depletion in the near-bottom water layer quite often occur and a redoxocline (a gradient of oxidizing and reducing conditions) forms in the water column. Oxidizing conditions are found above the pycnocline and reducing ones in the near-bottom layer. Closely connected with the pycnocline is the chemocline gradient zone which constitutes a very specific biotope. Frequently, the maximal quantity and biomass of hydrobionts groups (in particular, phototrophic organisms and phytoplankton) and chlorophyll maxima are specifically registered in the chemocline. In this zone, specific oxygenic and anoxygenic phototrophic species dominate [48]. Many of the predominating species are mixotrophic and adapted to lowered concentration, or total absence, of oxygen and the presence of organic and reduced inorganic substances.

Along with the phototrophic microorganisms, the pycnocline zone (chemocline) is populated by infusorians and rotifers. Three subzones are distinguished in the chemocline: the pre-redoxocline subzone with hypoxic water masses (<1 mgO_2 dm^{-3}) and high redox (oxidation–reduction) potential; the proper redoxocline and the post-redoxocline anoxic subzone which already has reducing conditions [24, 38, 48].

Many organisms inhabiting the chemocline, in particular protozoans, have an active mobility. Heterotrophic prokaryotes have gas vacuoles. The ability to change position in the water column allows chemocline organisms to avoid zones with unfavourable conditions (high concentrations of hydrogen sulphide for aerobic organisms and the presence of free oxygen for anaerobes). A significant part of the pycnocline zone (chemocline) organisms is mixotrophic, i.e. combine autotrophic and heterotrophic feeding. Biotic processes realized by the pycnocline (chemocline) zone communities control the entry of organic and reduced inorganic compounds and BM into the layer under the pycnocline [38].

The biotopes of the pelagial and benthal zones share common borders and, depending on aquatories depths, part of them is quite often not distinct. The properties of benthal biotopes are tightly connected with those of the overlying water layers. In MPs benthal zone, portions of coastal shallow waters with various sediment types (stones and rocks, native clays, broken shell, various grain-sized sand, silty-sand, silty-shell, silt) may be preserved. Sediments character is closely connected with hydrodynamics. In MPs benthal zones, the depths from 3–5 m up to 8–10 m form the intermediate zone between the coastal shallow waters and the deepwater bottom zones covered by black silt and containing hydrogen sulphide. Each of these three zones has its own bottom biocenoses depending on the sediment characteristics. The coastal macrophytes belt is connected with the coastal shallow waters.

Two zones are distinguished in the MPs periphytal formed by the HTS lateral walls: epiperiphythal zones in contact with the epipelagic horizon and bathypelagic zones washed by the bathypelagic horizon. On the HTS basement, ecotone zones in which periphytal and benthal fauna hydrobionts are registered can be formed in conditions of lowered hydrodynamics. The external surfaces of protecting HTS are subject to increased hydrodynamical influence and, consequently, the zones with the highest biomass of fouling organisms are located somewhat deeper. In the MP epiperiphytal zone, the near-surface macrophytes belt forms on HTS external and internal surfaces.

The general cycle of matter passes through all the subsystems and communities of MPs aquatories ecosystems from hyponeustal to benthal zones. Furthermore, in warm months, a near-surface circulation periodically forms in the epipelagial layer owing to wind-drift convection, surface currents and Langmuir convection. Its upper and lower borders are the film of surface tension and the isopycnic surface of thermo- and halocline, respectively. In the bathypelagic layer, the near-bottom circulation (lower border of pycnocline—bottom surface) is formed owing to offshor—onshore wind-induced flows, internal waves and waves formed by deep-draft ships.

Ecosystems of MPs aquatories are characterized by heterogeneous conditions. In the water column of such ecosystems, OM occurs as live suspended organisms, dissolved OM and detritus. The spatial heterogeneity (spottiness, patchiness) of suspended live OM (i.e. plankton) distribution in MP ranges from dozens of centimetres up to dozens or hundreds of metres. The peculiarities of hydrobiological processes are determined not only by the characteristics of a given water volume, zone or water layer occupied by various species, but also depend on the characteristics of the adjacent zones and water layers, the near-wall water layers of HTS, the near-bottom water layers and pore waters.

Pelagic and bottom communities, as well as communities of the periphytal, are distinguished in MPs aquatories ecosystems. The pelagial zone of MPs aquatories ecosystems is characterized by a rapid biotic turnover. This rapid turnover is determined by both the small size of producers (phytoplankton) and consumers (zooplankton) and by lowered hydrodynamics (compared with the adjacent zones), which results in slow mixing and creates conditions for bacterial destruction, dissolution and dispersion of mineralization products. Nevertheless, a significant part of OM of dying phytoplankton is quite often not utilized by phytophagous organisms and becomes detritus.

Processes that influence the transformation of OM, BM and the general flows of matter are active inside the MPs aquatories ecosystems. They manifest most vividly via interaction and matter exchange at the water-bottom, water-air and water-HTS interfaces. Vertical and horizontal redistribution of matter in protected aquatories also belongs to these processes and takes place under the influence of local water movements and input of allochthonous OM from land.

Table 5.1 presents the time scales of some processes typical of aquatic ecosystems, particularly MPs aquatories ecosystems.

Water movement is a major factor in determining differences between habitats, communities and aquatic ecosystems [47, 50, 72, 73, 85]. The total energy of all water movements (hydrodynamics) determines the structure of hydrobionts' communities and their seasonal and interannual changes.

Depending on hydrodynamics, ecosystems can be divided into groups. In particular, transit ecosystems are energetically better provided than standing ecosystems.

In MPs aquatories, important phenomena for aquatic communities and ecosystems are related to water motion: upwelling (rising of deep waters rich in biogenic elements), outwelling (entrance of estuarine waters rich in dead OM and containing various polluting substances) and downwelling (sinking of surface waters containing dead OM, biogenic elements and oxygen to the pelagial deepwater layers).

Table 5.1 Some time scales of natural variability and the most relevant scales for hydrological processes. Modified from [155]

Physical variable and biological processes	Time scale (s)													
	10^{-2}	10^{-1}	10^{0}	10^{1}	10^{2}	10^{3}	10^{4}	10^{5}	10^{6}	10^{7}	10^{8}	10^{9}	10^{10}	10^{11}
Microturbulent pulsations	■	■	■											
Changeability of superficial illumination		■	■	■	■	■	■	■						
Duration of wind mixing in shallow-watered ecosystems					■	■	■							
Seasonal changes of water basin thermal regime								■	■	■				
Interannual changeability of hydrophysical processes										■	■	■		
Climate variability											■	■	■	■
Cycles of biogenic elements						■	■	■	■	■				
Changes in proportion between limiting elements								■	■	■	■	■		
Processes of eutrophication and systems transition into a new state										■	■	■	■	
Processes of fluorescence	■	■	■	■	■	■								
Photosynthetic rate				■	■	■	■	■						
Breathing rate							■	■	■	■				
Duration of reproduction									■	■	■			
Process of matter consumption		■	■	■	■									
Assimilation process				■	■	■	■							
Formation of community species structure														

In MPs ecosystems, water movements slow down the loss of biogenic elements via their deposition to bottom sediments.

In MPs ecosystems, hydrobionts' communities and flows of live and dead matter always form as a consequence of various water motions, i.e. energy flows. Various communities can function in energetic conditions favourable for them.

Artificial changes in the natural energetic supply inevitably causes changes in the structure of aquatic ecosystems' biological components and various types of reorganization at different scales.

Odum [73] notes that a lower energy seashore, subject to weaker wave action, is more populated and more diversely populated than a high-energy shore under the influence of strong waves. In MPs aquatories, the biota composition in the highly energetic epipelagial and epiperiphytal zones significantly differs from that of the lower energy bathyal, benthal and bathyperiphytal zones.

Unlike many MPs located in various regions of the World Ocean, the ports of the Black–Azov Sea basin are practically not subject to the regular influence of tidal currents. In MPs, their contribution to the total energy of the water motions is negligible. In MPs epipelagial zone, superficial wind-drift currents, Langmuir convections, rivers discharge currents and ship waves formed by ships' hulls and propellers have primary importance. The role of internal waves, seiches and flows, created by the propellers and hulls of the deep-draft ships, intensifies in the pycnocline layer and below.

In MPs, the periodical offshore–onshore wind phenomena, and the compensational currents connected with them, affect the whole, or most of the, water column. During periods of prolonged wind calm, the water movements in the hyponeustal layer (0–30 cm) are driven by the surface thermal (gravitational) microconvection. In the near-bottom layer (0–30 cm), water movements are created by the near-bottom thermal microconvection caused by the Earth's heat.

In MPs, HTS lateral surfaces and ships' hulls dampen the energy of waves caused by surface wind, rivers' discharge, seiches and ships. Following waves breaking, the flows become turbulent. The presence of turbulent flows creates conditions for periphyton development and concentration of high biomasses at the HTS lateral surfaces.

In the pelagial, periphytal and benthal zones, the energy of water motion helps to transfer BM from consumers to producers and speeds up the delivery of suspended food to consumers. Turbulent water movements not only carry BM, food and oxygen, but also remove metabolites away from the mass settlements of hydrobionts.

It is known that weakening of the water movements frequently reduces the yield of aquatic ecosystems in the same way as a reduction of available solar light [72]. At the same time, the total and, in particular, the plankton primary production could increase up to a certain level at the beginning of the hydrodynamics' reduction before decreasing [73, 85, 114].

According to the flow or nutrients supply, aquatic ecosystems are divided into continuous and discontinuous or cyclic [72]. Owing to various water movements, river run-off and discharges of wastewaters, the MPs aquatories quite often become habitats with a continuous flow of BM and the adjacent open waters with an intermittent one. As a rule, limitation of MPs ecosystems by absence or low BM content is not observed [30, 91].

In MPs aquatories, invertebrates constituting the epi- and infauna and fish are the organisms most resistant to water movement. Turbulent water movements with speeds in the range of 5–15 cm s^{-1} are optimal for the majority of hydrobionts occupying various MP habitats. More specifically water speeds of 1–10 cm s^{-1} are favourable for plankton organisms, 5–25 cm s^{-1} for attached macrophytes and filter feeders animals of the periphyton, 5–50 cm s^{-1} for the bottom, near-bottom and pelagic fish and 1–25 cm s^{-1} for infaunal invertebrates.

Table 5.2 reports velocities linked to various hydrodynamic processes (with the exception of occasionally catastrophic water movements leading to death of organisms and biotopes destruction) that play various roles in hydrobionts' life.

The settlements and the entire communities of benthos and periphyton destroyed by catastrophic water movements can recover relatively quickly because of the practically constant presence in the water column of pelagic larvae of many bottom invertebrates, coming from the adjacent sea zones.

Table 5.2 Major hydrodynamic processes in marine ports aquatories and their average velocities

Water movement type	Velocity, (cm s^{-1})
Circular movements at sea scale	20–40
Coastal circulations	10–50
Wind-drift flows, downwelling	5–100
Langmuir circulations	5–10
Seiches	5–10
Internal waves	5–10
Ship waves	10–15
Fading of turbulent energy dissipation velocity	1–5
Surface thermal microconvection	1–5
Near-bottom thermal microconvection	1–5
Compensational flows	10–15
Wind-induced flows	10–30
Flows causing bottom erosion	25–50
Flows causing longshore drift flows	25–50
Water motion, providing normal survival and feeding of plankton organisms	1–10
Water motion, providing normal survival and feeding of periphyton	5–25
Water motion, providing normal survival and feeding of benthic organisms	1–25
Water motion, creating conditions for upwelling	15–30
Water motion, creating conditions for mineralization of organic matter on bottom	1–5
Water motion, creating conditions for hydrobionts reproduction, eggs fertilization in external environment and survival of organisms at early stages of ontogenesis	1–5
Biological water mixing during hydrobionts vital functions	1–5
Rip currents	50–100
Surface flows in passages to marine ports aquatories	25–100

More or less autonomous biological cycles could form in MPs owing to water movements, the absence of biogenic matter deficit, ubiquitous presence of bacteria and protozoans in all subsystems (pelagial, periphytal and benthal zones) and in all main communities (hyponeustal, epipelagic, bathypelagic, hypoxic, periphytal and benthal zones) and diversity of food chains. Depending on the size of biotic components and the level of their organization, turbulent water motions of different speed create conditions favourable to a variety of communities.

Autotrophs, chemotrophs and heterotrophs with sizes from 2–5 to 200–500 μm, macrophytes with epiphyte periphyton and big-sized invertebrates with fish can form closed biotic cycles. However, the essential condition for these processes is the presence of bacteria and protozoans accelerating OM mineralization. Biotic turnovers (OM production, transformation and destruction) on pico-, nano- and microplankton level can take place within dozens of hours, but require months to years at the level of macroforms.

In sterile seawater, sufficient quantity and proper proportions of dissolved mineral matters and appropriate values of pH, temperature and oxygen content are not enough for the normal and prolonged survival of hydrobionts. The presence of various metabolites of hydrobionts in seawater, first of all antibiotics, ferments and vitamins, is also necessary. Only when this condition is fulfilled, the artificial seawater acquires the properties of a "live" one. Bacteria, moulds, yeasts, microphytes and others are the main suppliers of the biologically active dissolved substances.

MPs ecosystems are characterized by increased detritus content of autochthonous origin or entering from land and atmosphere. Detritus has various influences on both abiotic and biotic processes.

Detritus, consisting in a suspension and solution of OM, influences water transparency and light transmission. The so-called yellow substance, dissolved in water, changes its optical properties. Detritus, concentrating under the film of surface tension, disturbs the exchange between the aquatic and atmospheric environments and influences the near-surface layer hydrodynamics. By accumulating at the isopycnic surfaces, detritus can change the light quantity reaching the benthal zone. High concentrations of detritus can mechanically hamper the feeding of zooplanktonic filter feeders. Sinking to the bottom and forming a layer of dead OM, detritus affects the living conditions of bottom animals and macrophytes. A great amount of the oxygen dissolved in water is consumed on the oxidation of detritus, which is present in the form of suspension, OM solution or deposited on bottom sediment.

From a biological point of view, detritus represents one of the forms of readily assimilable food, source and stock of biogenic matter in aquatic ecosystems. A high stock of potential energy is concentrated in detritus.

Figure 5.5 shows a scheme of the use of dissolved, suspended and sedimentary dead OM, as well as of the turnover of oxygen and of the mineral forms of nitrogen and phosphorus.

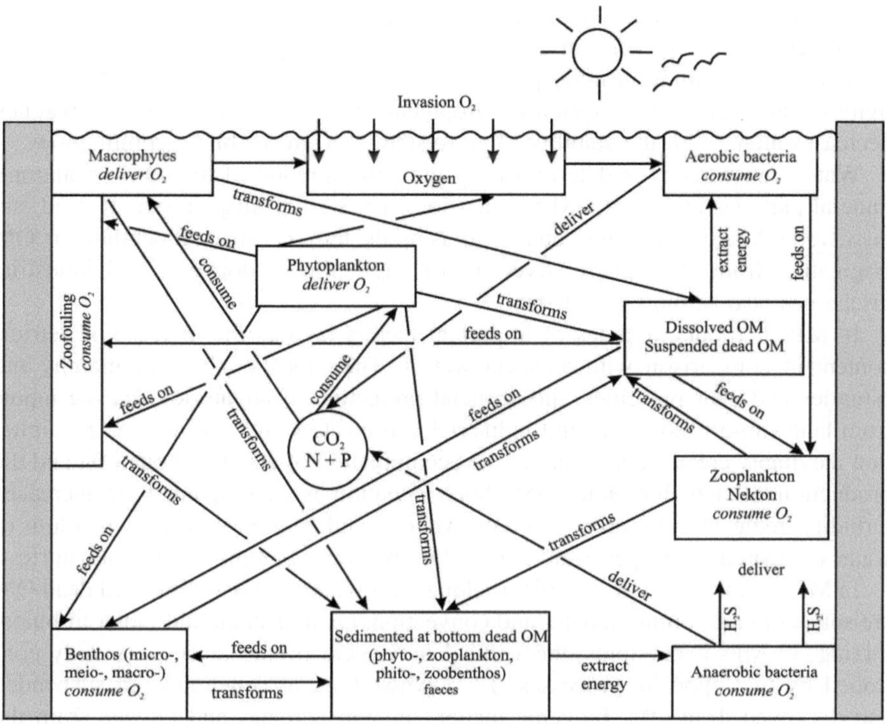

Fig. 5.5 Generalized chart of oxygen, nitrogen and phosphorus mineral forms turnover and use of dissolved, suspended and sedimentary dead organic matter in MPs aquatories ecosystems

In detritus, the content of phosphorus, which is an important biogenic element for aquatic ecosystems functioning, sometimes ranges from 43 to 97% of the total stock [4, 7].

When there is a big amount of organisms feeding on live phytoplankton not forming excessively dense concentrations, interrelations between autotrophs (phytoplankton) and heterotrophic zooplankton (phytophagous) occur via grazing. In MPs, the main mass of primary OM, formed by intensively developing phytoplankton, actually becomes available for the feeding of pelagic heterotrophic organisms (zooplankton) only after its death and bacterial decomposition, i.e. already in the form of detritus. In rather complex ecosystems such as MPs aquatories, the food relationships of hydrobionts belonging to various levels form complex food webs. In them, most of the energy flows through the detrital pathway, i.e. detritus consumption by heterotrophic organisms contributes additional production to the primary one of autotrophs and chemotrophs [72, 91, 98]. In particular, detritus of autochthonous and allochthonous origin is a main component in the biotic turnover of matter and energy in marine coastal ecosystems [99, 100].

Stability and peculiarities of ecosystems functioning directly depend on water transparency that represent an integral index of the degree of eutrophication and with which planktonic primary production is inversely correlated [3]. As productivity or the degree of ecosystems eutrophication increase, the water transparency decreases along with the stability of ecosystems and major biotic communities.

Water transparency is determined by the concentration of live OM (plankton), mineral particles and dead OM (suspension), i.e. seston. Suspensions of dead and dissolved OM are commonly referred to using the term detritus. Allochthonous OM originating from atmosphere, river run-off, agricultural, domestic and industrial discharges also belongs to detritus.

In MPs harbours, a series of important factors favours the increase of detritus content in comparison with adjacent waters. Their location in river mouths and estuaries and near big cities, the artificial protection of aquatories and the inputs from land run-off, domestic and industrial wastewaters and atmospheric precipitation inevitably create conditions for the accumulation of allochthonous OM and the production of autochthonous OM. Such situation is accompanied by increased turbidity, reduced transparency, a more yellow, yellow-green and brown colour of water compared with open waters and often by water blooms in MPs aquatories.

In MPs, planktonic live organisms along with suspended and dissolved dead OM are subject to advection, transfer and convection, i.e. molecular and molar turbulent mixing. In MPs ecosystems, the vertical movement of substances is mainly controlled by two opposite processes: 1—downwelling, or penetration of suspended and dissolved dead OM, biogenic matter, mineral particles and oxygen from the surface layer into the water column; 2—upwelling, or rise up of biogenic matter, suspended and dissolved dead OM and gases forming at OM decomposition from the near-bottom layer into the water column and to the surface.

The coastal line, surface tension film, isopycnic surfaces in the water column, HTS lateral surfaces and aquatories bottom act as a specific "screens" that concentrate matter and energy by interrupting their various flows. According to the vivid expression of V.I. Vernadsky, life "concentrations" originate in MPs ecosystems because of these processes. From the list of mentioned "screens", only HTS lateral surfaces are completely artificial, while all others are exposed to various anthropogenic modifications.

To understand the peculiarities of MPs aquatories ecosystems' functioning, it is necessary to consider the distribution of living and dead OM as well as the direction of its main fluxes under the influence of major natural abiotic and biotic factors and of anthropogenic activity.

Transfer of gases and other substances occurs through the aquatic surface and their transport velocity is connected with changes in the evenness of the water surface. Compared with a calm water surface, the appearance of even small ripples increases the velocity of molecular diffusion transport by up to fourfold [70]. In protected MPs aquatories, periods of calm surface waters increase, especially in summer months, causing drops in the transportation speed.

Gas bubbles and light-weighted mineral and organic particles of different origin, including dead organisms, rise to the film of surface tension from below. Dust particles, suspensions and OM in the form of pollen, spores, seeds and dead insects also settle down onto this surface from above. The presence of such particles increases water evaporation [70]. In protected MPs aquatories, accumulation of the small-sized particles on the smooth water surface is more intensive than in the open sea.

In the water column, suspended live OM (plankton) usually forms structures resembling clouds. For a variety of reasons, the larvae of many bottom organisms (meroplankton) are unevenly distributed in the water column and are carried like bigger- or smaller sized clouds in the same way as phyto- and zooplankton.

In the benthal, living matter usually has a belt-like distribution near the shore and, depending on the biotopes variety, a patchy or mosaic one in the deepwater part of MPs aquatory. On HTS lateral surfaces, the living matter has a belt-like distribution in the same way as near the shore. The belt distribution of benthal hydrobionts in the coastal biocenoses and their stratified structure in the MPs water column in MPs are, so to speak, projected onto the HTS lateral surfaces.

Dead OM, suspended in the water column, settles to the bottom as detritus and is permanently concentrated on a flat surface. Dead OM accumulation can be particularly high in some bottom zones, especially small cavities and shipping channels under the influence of flows. The phenomenon is especially evident at mass dying-off of phytoplankton that falls to the bottom and forms a layer of easily stirred-up sediment with a thickness up to 10–20 cm and more.

The input of allochthonous OM in MPs ecosystems always increases the total detritus content and the number of detritophages in water column, on the bottom and in the fouling. The surplus of dead OM, unused by zooplankton and periphyton, accumulates as detritus in the water column and on the bottom, thus creating a risk of secondary organic pollution.

At the beginning, the external inputs of dead OM into MPs ecosystems could favour the increase of detritophages and filter feeders in the water column, in the periphytal and in the benthal. However, further increase in the quantity of detritus is frequently harmful for these organisms.

The reduction of allochthonous OM entrance into MPs ecosystems is the main factor reducing the primary production and a method for their bottom-up management via the control of biogenic matter and its consumers.

In MPs aquatories ecosystems, estimating the ratio between formation and transformation speed of detritus permits to predict their future.

Hydrodynamical conditions influence detritus spatial distribution in MPs ecosystems, the speed of its sedimentation and its accumulation on various bottom areas.

In modern publications, the term detritus indicates all forms of OM removed from a given trophic level as a result of death of the organisms inhabiting the aquatory, secretion, excretion and others. Allochtonous OM entering a specific aquatic ecosystem is also part of detritus. Detritus is composed of suspended particles of dead OM together with bacteria, fungi and protozoans that colonize them. As a matter of fact, many detritus particles constitute peculiar microecosystems. Detritus is a complex dynamic system in which chemical composition, energy value, ratio of dead and alive OM are constantly and continuously changing. For a certain period of time, detritus accumulates energy which, later on, is diversely involved into biotic turnover. The feeding value of detritus depends on its origin, residence time in the basin ecosystem and degree of OM transformation [3, 4, 7, 9, 10].

Being a specific trophic reserve of ecosystems, detritus accumulates and, for some time, preserves substantial stocks of energy within zones of life "concentration". It thus becomes an important stabilizing element of ecosystems and it is considered that one of its main functions is to weaken oscillating process originating in aquatic ecosystems. Detritus also controls, to a large extent, the general processes of destruction and mineralization of OM [7, 9].

On the surface of detritus particles take place different physical and chemical reactions, biochemical and biological processes, in particular hydrolysis and chemical and microbial synthesis. As a result, the dissolved OM is further included into biotic turnover. Chemotaxis of bacterioplankton and active search and capture of detritus particles by many zooplanktonic organisms are biological mechanisms accelerating the detritus biological decomposition [7, 20, 40, 59].

The speed of biochemical destruction of detritus components is controlled by various abiotic and biotic factors. The former include temperature, water pH, mineralization level and solar radiation in the ultraviolet spectral range. The latter are grazing, viruses influence, competition between filter feeders and detritophages along with their ecological peculiarities. The predator zooplanktonic organisms inhabiting the water column and viruses could effectively control the quantity and biomass of detritophagous organisms at the bottom [23, 33, 81, 85]. The presence of zooplankton and viruses usually speeds up the process of detritus decomposition. Intra- and interspecies competition between detritophages in various ecosystem communities could slow down detritus decomposition [23, 37, 40, 59].

In any case, detritus participates in the majority of processes which take place in MPs aquatories ecosystems. All biomolecular compounds can be a source of detritus. In aquatic ecosystems, the highest amount of detrital carbon is present in the dissolved form. About 50% of the dissolved detritus consists of the same biomolecular compounds composing the living matter. The average ratio of dissolved to suspended carbon content is 10:1 [163, 164].

As it was already underlined, in excessively eutrophicated ecosystems a significant part of detritus is neither utilized in the water column nor by benthic organisms and thus accumulates into the sediments. Here, its decomposition by bacteria quite often leads to anoxia and hypoxia phenomena. In MPs, such phenomena could not only directly affect the near-bottom water layer, but also spread up to the pycnocline lower border and, sometimes, up to the water surface.

The removal of BM from MPs aquatories ecosystems occurs through the relatively narrow passages of the access channels and through the film of surface tension (via gas bubbles and water drops) and the zone of coastal shallow waters and shoreline (via foam formation and consumption of dead OM by larvae of terrestrial insects and other animals). It also occurs via long-term incorporation into the biomass of long-lived (years, decades) hydrobionts, migration of fish and other hydrobionts outside MPs and deposition into bottom sediments (Fig. 5.6).

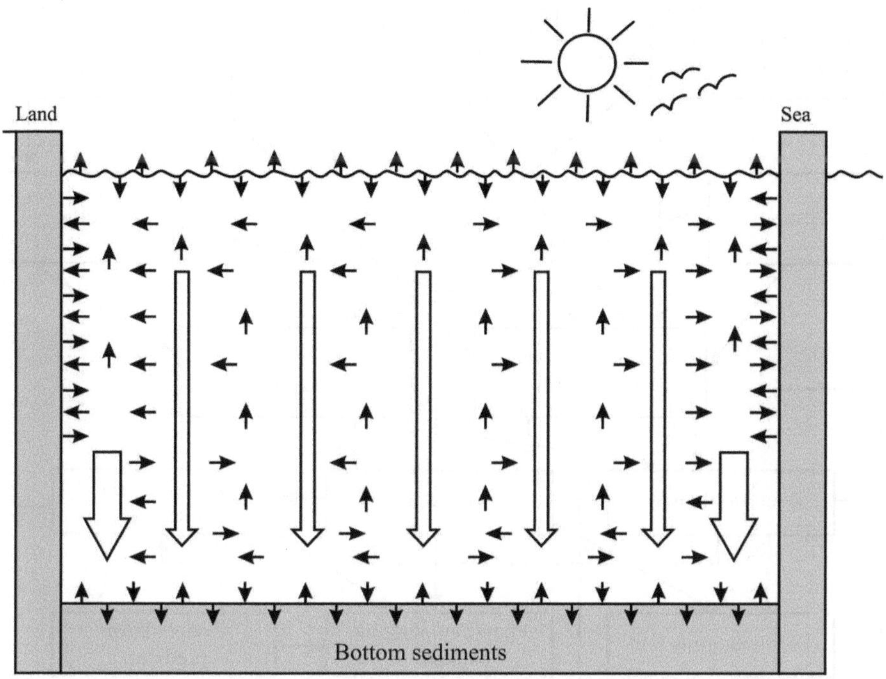

Fig. 5.6 Representation of the main flows of dead organic matter in marine ports' ecosystems

The internal surfaces of the HTS protecting MPs are impenetrable for dead OM that is accumulated at the foundations of berths and moles in the form of a loose sediment layer.

Various persistent toxic substances enter MPs aquatories ecosystems from the atmosphere, with surface and river run-off, with industrial and municipal wastewaters, as a consequence of the MPs infrastructure functioning, from the ships' paint coatings and from other possible sources. Among toxic substances, inputs of heavy metals (mercury, cadmium, lead, copper, zinc and others) are particularly important.

Figure 5.7 shows the pathways of persistent toxicants in aquatic ecosystems. In MPs aquatories, the heavy metals content increases as a result of slowed water exchange, hydrobionts ability to incorporate metals into food chains and detritus ability to adsorb ions [29, 34, 39].

It is known that microorganisms of water column and bottom sediments have defensive mechanisms against heavy metals. They are able to produce hydrogen sulphide, extracellular exopolymers and other substances which interact with heavy metals and transform them into less toxic forms.

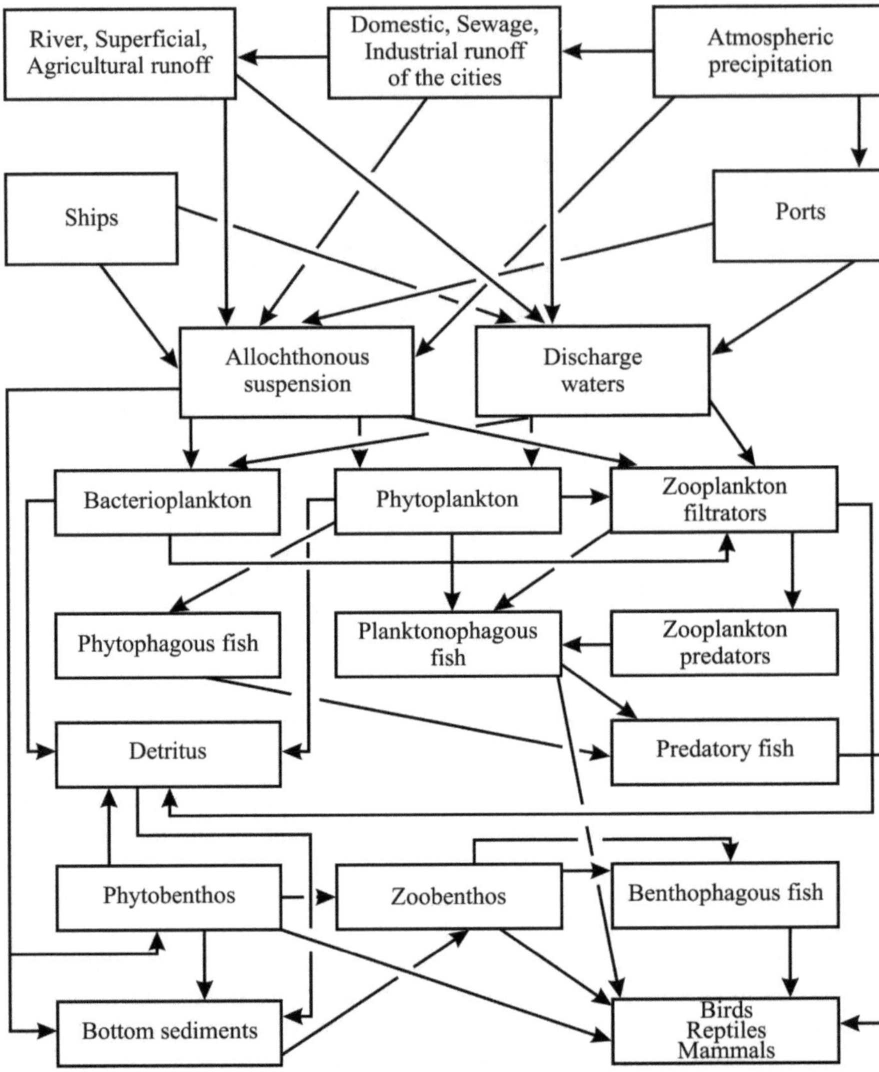

Fig. 5.7 Pathways of persistent toxicants in marine ports aquatories ecosystems

Above certain concentrations, persistent and non-persistent toxic substances (oil products, pesticides, heavy metals, synthetic surface active substances) originating from various sources have a negative impact on all biota components and accumulate in harbours. However, excessive OM decomposition and, first of all, labile sulphides that form directly in MPs aquatories are the biggest danger for the majority of hydrobionts [30, 91, 102, 103].

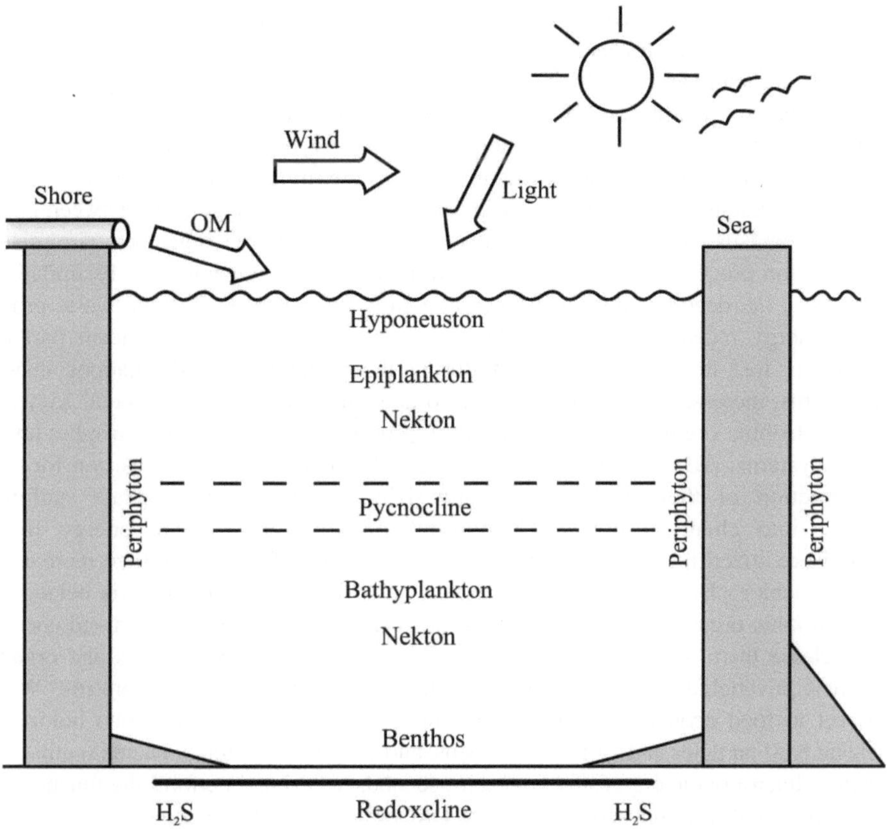

Fig. 5.8 Generalized overview of the main biotic components of MPs aquatories ecosystems

5.1.2 Peculiarities of the Biotic Structure

Figure 5.8 presents a generalized overview of the main biotic components of MPs aquatories ecosystems. Aquatories ecosystems have a complex biotic structure influenced by seasonal and interannual changes.

The general biotic structure of MPs comprises hyponeuston, plankton, periphyton and benthos communities. Depending on the number of biotopes, various hydrobionts complexes (biocenoses) contribute to the composition of each of these communities. The water column above the pycnocline (epipelagic zone) is inhabited by a complex of planktonic organisms distinct from that of the MPs bathipelagic zone. The epipelagic and the adjacent epiperiphytal zone are characterized by better conditions for phototrophic and oxyphilic hydrobionts. The bottom biocenoses are tightly connected with the sediments' characteristics and with hydrodynamics.

In the structure of each big community, producers, consumers and decomposers belonging to various levels of organization are present.

Biotic structure and peculiarities of hydrobionts' distribution in MPs aquatories are determined by quantity and quality of food resources and by feeding habits of the various species along with the biotopes diversity. Hydrobionts forming the major biotic components of MPs aquatories ecosystems are divided into main groups based on feeding habits. A constant exchange of matter and energy goes on between each living organism and the environment. Metabolism inside the organisms comprises assimilation processes, i.e. synthesis of cell molecular components, and dissimilation processes, i.e. decomposition and excretion from organism of the final waste products. Energy redistribution takes place via these processes. A significant part of energy is lost in the form of heat during the metabolic transformations inside organisms, increasing the entropy of the "organisms—aquatic environment" system.

Autotrophic organisms are OM primary producers and are the first trophic level in ecosystems. All of them use CO_2 as the only or main source of carbon for the construction of their body. Photosynthesizing autotrophs incorporate sunlight energy into chemical bonds. Chemosynthetic autotrophs obtain energy from slightly oxidized inorganic compounds. Some groups of bacteria utilize methane.

The achlorophyllous algae, fungi, animals and the majority of bacteria belong to heterotrophic organisms, which use exogenous OM as a source of carbon and energy and release them at their decomposition. Organotrophic microorganisms and organotrophic invertebrates feeding on DOM belong to heterotrophic organisms. With respect to food acquisition, heterotrophic organisms are distinguished into holozoic, seizing hard particles and various organisms, and osmotrophic (fungi, bacteria) utilizing DOM. Heterotrophic organisms synthesize secondary OM and perform the function of consumers and decomposers, closing the OM cycle by the formation of CO_2.

Mixotrophic microorganisms combine different feeding strategies, i.e. use of SOM and/or DOM and oxidation of inorganic substances. Organotrophic microorganisms use DOM as an oxidizable substratum. As mentioned above, two types of producers (photo- and chemotrophs) are present in aquatic ecosystems, including MPs aquatories ecosystems. Photosynthetic organisms, i.e. macrophytes and microphytes, usually have advantages. Macrophytes of MPs aquatories include not only algae, but also higher plants; while microphytes are exclusively algae. They inhabit the water column and attach to various substrata and are mainly represented by diatoms in the benthos.

In the coastal marine ecosystems, the total production of microphytobenthos and macrophytobenthos is comparable. The production of epiphytic microalgae fouling macrophytes thalli could be of the same magnitude as the production of macrophytes themselves. Different benthic forms, among which small-sized gastropod mass species play an important role, feed on macrophytes fouling. It is considered that fouling of macrophytes and the detritus forming during their bacterial decomposition have a higher nutritional value than the macrophytes themselves. In the near-surface belt of HTS, i.e. in epiperiphytal, macrophytes fouling by epiphytes is hampered owing to high hydrodynamical activity. In the benthal zone, the presence of a high number of epiphytes frequently testifies about macrophytes ill-being.

In relatively clean MPs aquatories, phytoplankton is the main OM producer. Heterotrophic organisms and bacteria exist at its expense and, with the help of different enzymes, transform the detritus that forms at phytoplankton's death and recycle its simple components into the biotic turnover. This picture slightly changes in the presence of external sources of detritus.

The concentration of suspended detritus and the OM content in the upper layer of bottom sediments determine the dominance of various trophic groups in the different parts of MPs aquatories.

Heterotrophic hydrobionts use two main feeding modes on detritus: 1—water filtration; 2—consumption of the OM contained in bottom sediments. The inhabitants of the water column comprise filter-feeding organisms and organisms selectively feeding on detritus particles and live hydrobionts. Detritus accumulates in the fine-grained sediment fractions (pelite, aleurite) in amounts sufficient for the feeding of mass bottom animals. Coarse-grained sands do not provide adequate living conditions for the majority of detritophagous species.

Table 5.3 presents different classifications of plankton organisms [51]. In aquatic ecosystems, the majority of matter, energy and information flows originates at the organization level of live and dead phytoplankton.

Table 5.3 Classification of plankton organisms (acc. to [51])

Classification criterion		Group of organism
Size	Up to 5 μm	Picoplankton (bacteria, cyanobacteria)
	From 5 μm up to 50 μm	Nanoplankton (phytoplankton)
	From 50 μm up to 1 mm	Microplankton (phyto- and zooplankton: protozoans, rotifers, etc.)
	From 1 mm up to 5 mm	Mesoplankton (zooplankton: cladocerans, copepods)
	From 5 mm up to 15 mm	Macroplankton (crustaceans, fish larvae, insects larvae)
Functional role in ecosystems	Producers (autotrophs)	Phytoplankton
	Producers (chemotrophs)	Bacterioplankton
	Reducers	Bacterioplankton
	Consumers	Zooplankton
Habitat	Littoral zone	Littoral plankton
	Pelagial zone	Pelagic plankton
Duration of stay in plankton	Whole life cycle	Holoplankton
	Part of life cycle	Meroplankton (larvae of fish, insects, benthos organisms, numerous parasites)
Systematic classification position	Prokaryotes	Bacterioplankton (bacteria), phytoplankton (cyanobacteria)
	Eukaryotes	Algae, animals

The control mechanisms of aquatic ecosystems, comprising those of MPs aquatories, include microbial communities regulating accumulation and release of biogenic elements, behavioural mechanisms, “predator–prey” interactions regulating the population densities of separate species and other biological processes [10, 40, 56, 73, 96].

Organisms inhabiting the benthal zone and macrophytes beds are commonly termed benthos. Benthos is divided into three size groups: 1—microbenthos (protozoans, unicellular algae) up to 0.1 mm; 2—meiobenthos (different types of protozoans and invertebrates) up to 2 mm; 3—macrobenthos (various types of invertebrates) above 2 mm.

In MPs, hard substrata of natural and artificial origin are populated by vagile and sessile seston feeders, macro- and microphytes, representatives of micro- and meiobenthos and macrozoobenthos.

Sediments formed mainly by molluscs’ shells and fragments thereof are called coquina. There are other sediment types that contain big amounts of barnacles’ shells fragments [47]. In the Black and Azov Seas, these sediments are formed on small, relatively shallow-watered bottom zones because of ice friction against the HTS surfaces and of the activity of strong storms.

Broken shells are usually less populated by benthic forms, although various meiobenthos organisms could settle there in the presence of silt admixture [32, 53].

Hydrobionts form various settlements depending on sediment compactness, its mineral and granulometric compositions and the amount of detritus contained in it. Particular animal communities form in coastal and surface macrophytes belts.

In MPs, the periphyton could be considered as an analogue (impoverished in species) of the benthal zone population of adjacent regions. Among the invertebrates inhabiting benthal and periphytal zones, filter feeders and detritophages are distinguished based on their feeding strategies. The filter feeders are divided into two main groups: 1—epifauna of hard substrata, 2—epifauna of soft sediments. There are two main types of detritophages: 1—collectors, 2—sediment feeders. The former consume detritus on the surface of bottom sediments, while the latter feeds within its upper layers [47, 81, 85].

Predatory animals also enter in the composition of benthos and periphyton. They feed on other live animals and on dead ones which have not yet begun to decompose.

Epifaunal filtrators dominate in MPs zones with active hydrodynamics, such as the epiperiphytal. In zones with lowered hydrodynamics, such as HTS foundations, the leading role passes to infaunal filter-feeding organisms. At further attenuation of hydrodynamics, the leading role passes to detritophagous organisms.

The planktonic organisms (bacterioplankton, phytoplankton and zooplankton) inhabiting the water column are usually connected with certain water masses where cell division, reproduction, release of reproductive products, fertilization and initial phases of development take place. Temporary inhabitants of the pelagial zone, such as mero- and ichthyoplankton (larvae of bottom invertebrates, eggs and larvae of fish), are connected with water masses in a somewhat different way. The water column is subject to different disturbing influences. Water masses transform and mix up over time under the influence of various water movements. During these processes, the primary gathering of passively and slowly moving planktonic organisms, resembling clouds, disintegrates first at the periphery and then in the very nucleus of such clouds and the concentration of planktonic organisms reduces. New accumulations of passively moving planktonic organisms are possible when the organisms get caught into vortex-type flows of various origin and scale. Thus, dispersion as well as concentration of planktonic organisms can be observed in different zones of the water column. Gravity and water buoyancy also play a certain role. Because of MPs aquatories protection, the possibilities for dispersion of planktonic organisms are restricted. The presence of HTS vertical surfaces from the bottom to the surface of the water column creates conditions for the formation of organisms' accumulations. During these dispersion and accumulation phenomena, planktonic organisms pass various stages of ontogenesis, satisfy their trophic needs and excrete metabolites.

In the MPs pelagial zone, the greatest quantity of planktonic organisms and fish schools form more or less sparsely populated accumulations. In the benthal zone, hydrobionts typically have a patchy distribution on area of flat surfaces of various sizes or inside a thin layer of sediments. Zonal or patchy distribution is typical for organisms in the coastal shallow-watered zone and in the periphytal zone.

The structure of aquatic ecosystems can change in space and time owing to natural and anthropogenically disturbed external and Intra-system factors. Ecosystems stability can be short-term (successional) and long-term (evolutional). Stability is considered as a system's ability to preserve a relatively unchanged state under the influence of non-catastrophic phenomena and to withstand changes of abiotic and biotic factors while preserving a dynamic equilibrium (homeostasis) [44]. Preservation of the average level of species composition and diversity usually reflects stability of MPs ecosystems over time [7, 22, 23, 89, 92, 93, 105, 111, 113, 116, 123, 131, 151].

High quantities of migrating fish come into MPs aquatories (and similar ecosystems) to uptake, use and redistribute part of the excessive and underused production present in them. Thus, on the one hand, they accomplish a control "from above" and, on the other, link, by matter and energy flows, separate, relatively small, but comparatively autonomous ecosystems to much bigger ones.

At changing of external conditions, the structural changes in ecosystems which most distinctly manifests following anthropogenic impacts, start with disappearance of species with narrow ecologic spectra, i.e. stenobiotic and specialized species. These modifications cause changes in the structure of trophic links, usually towards a simplification of system complexity. These events liberate ecological niches which are filled by short-lived species with wide ecological spectra, high productivity and high quantity of biomass (the so-called r-strategists). This new production is not always useful for humans and frequently is even harmful.

The ecosystems' structure and functioning depend on the turnover of matter and energy within the environment as well as on the links between the processes which increase or decrease entropy inside the system.

Usually, the MPs ecosystems evolve towards structure simplification and decrease in the ratio of dissipating energy to energy enclosed in production. This happens mainly because of gradual OM accumulation in the system, included OM deposited in bottom sediments. The speed of matter turnover increases with the organic input to the ecosystem, but the time of OM destruction slows down leading to OM accumulation. Accumulation of OM in bottom sediments is always higher in more productive water bodies.

Oxidation of excess dead OM uses significant amounts of oxygen, inevitably leading to its deficit. A decrease in oxygen concentration causes the disappearance of oxyphile species and their substitution by species with lower oxygen requirements. Usually, these are the short-cycled species (r-strategists) with high speeds of biomass turnover and characterized by significant fluctuations of their quantity and biomass. During anthropogenic eutrophication (waters pollution by non-toxic OM), the same processes occurring during ecosystems evolutional development due to internal factors are observed, but with higher velocities. Such systems age quicker.

Three structural levels can be recognized in the organization of the biota of MPs aquatories ecosystems. Bacteria, autotrophic and heterotrophic pico- and nanoplankton, protozoan plankton and benthos form the first level. The second one is formed by bacteria, fungi, net phytoplankton, metazoan zooplankton, micro- and meiobenthos. The third level includes bacteria, fungi, macrophytes, big-sized zooplankton forms, macrozoobenthos, pelagic and bottom fish, reptiles, aquatic and semi-aquatic birds and dolphins. Organisms belonging to all the three levels form, via a multitude of connections, the three-dimensional trophic webs and function as a whole ecosystem. At the same time, organisms within each level can form closed cycles of matter turnover. The presence of bacteria, fungi and protozoa closes the trophic chains. The three levels of MPs aquatories ecosystems structural organization differ, on average, in their speed of BM turnover (from many-folds to hundred-folds per annum) and thus productivity.

Figure 5.9 shows the general biological structure of MPs aquatories ecosystems along with levels of structural organization.

The widely used and informative Shannon index considers the abundance as well as the degree of dominance of individual species, thus providing a good estimate of the organizational complexity of the community structure in MPs aquatories ecosystems. A higher value of this index indicates an increase in structure uniformity, while its decrease is associated with a less uniform structure and increased dominance of separate components. An increase of the index points to a more complexly organized structure. Both increases and decreases of the index are possible depending on the life span of MPs aquatories ecosystems.

As the structure of communities becomes more complex, the share of energy diffused in the form of thermal energy increases with respect to the energy incorporated into the production of the examined biological systems. More complexly organized systems are characterized by lower productivity and a larger share of dissipated to produced energy. The complexity of communities' structure is "paid" by energy dissipation [7]. Thus, the ratio between production and dissipation speed

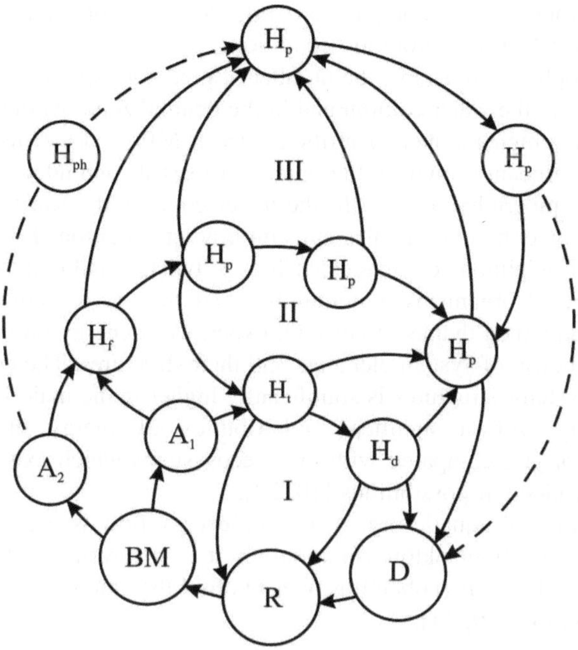

Fig. 5.9 General scheme of the ecological structure of marine port aquatory ecosystems: *I*—level formed by bacteria, unicellular producers of primary organic matter, mixotrophic, unicellular and small-sized multicellular invertebrates and fungi. *II*—level formed by the colonial unicellular and multicellular producers of primary organic matter, unicellular animals and invertebrates. *III*—level formed by the large-sized invertebrates, fish and other vertebrates. A_1—microphytes, A_2—macrophytes, H_{ph}—phytophagous invertebrates and vertebrates animals, H_f—pelagic and sessile invertebrates filter feeders, H_p—predatory invertebrates and vertebrates of various order, H_T—necrophages, H_d—detritophagous, mud-eaters, *D*—detritus, *R*—reducers, *BM*—CO_2, phosphorus, nitrogen and other biogenic matter

decreases as the structure of a community gets more complex, while it increases when the community structure becomes simpler.

In ecosystems, the growth of primary production causes structural reorganizations of zooplankton and zoobenthos communities, which traduces in a decrease in the complexity of the communities' structures. A high productivity of the heterotrophic components of ecosystems is possible only following structure simplification. On the other hand, complexly organized systems are more stable [23, 45, 71, 107, 108], but cannot reach a high productivity [10].

Informational interconnections, regulating intra- and interspecific interactions and interrelations, play an important role in forming ecosystems structure [3, 7, 68, 73].

In MPs aquatories subject to pollution and eutrophication, the share of production towards the diffusing energy increases, the communities' structure simplifies and their species composition shifts towards dominance of more eurybiont species resistant to saprobiotic conditions.

In MPs eutrophic ecosystems, the number of predatory species and filter feeders animals reduces in the water column and in the benthal zone. At the same time, the construction of a high number of artificial HS in MPs ecosystems provides conditions for a significant growth in the biomass of sessile and attached filter feeders animals in the epiperiphytal zone. In the water column, the same structures offer protection and lower hydrodynamics, thus improving conditions favourable to mass development of gelatinous organisms such as jellyfishes and comb-jellyfishes.

The structure of organisms' communities and MPs ecosystems, i.e. their biodiversity, is maintained thanks to constant expenses of energy on the maintenance of order and renewal of system elements and their structures. The energy expenses to sustain ecosystems structures is significantly higher in the majority of MPs (that function mainly with a significant contribution of energy of allochthonous anthropogenic origin) compared with those ecosystems which exist mainly owing to their own production possibilities [10, 23].

In communities of aquatic organisms, the energy flow is proportional to their production. In bacterioplankton communities it is on average 214-fold, in zooplankton eightfold and in zoobenthos over twofold the amount of energy enclosed in their biomass [4, 7, 9, 11].

5.2 Trophodynamic Processes

Various interspecific biotic interactions exist in natural ecosystems: 1—neutralism, 2—amensalism, 3—commensalism, 4—competition (interference), 5—predator–prey (exploiter), 6—mutualism [73, 85, 118, 119]. These interrelations do not always directly correspond to trophic connections. However, the hydrobionts participating in these interactions are all components of each specific ecosystem and they all take part in biotic turnovers and in the formation of matter and energy flows.

According to V.I. Vernadsky law of constancy, any change in the amount of living matter in one of the ecosystems' subsystems (including MPs) inevitably leads to a change of the same magnitude but of opposite sign in the other subsystems. Thus, in eutrophicated ecosystems such as MPs, the highly developed species and communities are substituted by other standing at an evolutionary and successionally lower level and the big-sized organisms by smaller ones [82, 83]. As a result, the ways of obtaining and processing food, the speed of its use and the growth of production change.

In all cases, a decrease in the number of active species (i.e. diversity) leads to a reduction of the total level of matter biotic turnover in ecosystems [10, 23, 72, 119].

In ecosystems similar to those of MPs aquatories, the presence of internal sources of easily accessible energy is combined with external inputs of BM and dead OM in amounts exceeding those required by the functioning organisms of biocenoses. For these ecosystems, low species diversity is more effective. As it was already mentioned, maintenance of high level of diversity requires additional energy expenses, for example for saturation of the water column by oxygen. As long as production (*P*) and OM entering from outside are higher than respiration (*R*), OM will accumulate in the system which will remain eutrophicated.

It was determined that functional complexity of ecosystems increases their stability to a greater extent than structural complexity [4, 10, 73, 161].

As in marine ecosystems, ecosystems of MPs aquatories experience regular diurnal and seasonal changes in the distribution and flows of OM and energy. Even significant seasonal changes in species composition belong to the general characteristics of a specific ecosystem, because they regularly repeat each year. The integrity, functional stability and essential characteristics of a specific ecosystem are preserved in the course of diurnal and seasonal changes. In an ecosystem, disturbance of the character and scales of diurnal and seasonal changes is an evidence of ill-being.

It is known that species abundance and diversity of biocenoses in ecosystems result from adaptation to the quantity and qualitative composition of the corresponding energetic flow and to the associated input of feeding resources [23, 72, 73, 114, 119]. The development "strategy" of natural ecosystems consists in increasing diversity, but only to a point that does not lead to reduction in the efficiency of resource use by all biotic components. There is a diversity optimum that is determined by the efficiency of the "energetic machine" of a particular ecosystem [23, 114].

Based on H. Odum and E. Odum law of maximal energy (which is also valid for information), the most stable and competitive ecosystems are those which favour the flow of energy and information and use their maximal amounts in the most effective way.

MPs aquatories ecosystems are frequently simpler, but also more stable and competitive, than the ecosystems of the adjacent sea zones with which they can establish trophic links and interrelations. Such ecosystems reach their maximum efficiency by switching on primary OM production and transformation and reduction of autochthonous and allochthonous OM at the level of microorganisms (bacteria, nanoflagellates and others).

The term "feeding" indicates several processes, including uptake by organism, disintegration, hydrolysis, digestion, absorption and fixation of nutrient materials in the biomass. Feeding is a process of metabolism and is essential for all life forms [17]. Via feeding, organisms obtain the different chemical compounds that they need for vital activity, growth and reproduction. As mentioned above, three main nutrition types are distinguished: intracellular, extracellular (or distant) and membranous. Mineral compounds as well as organic substances of biogenic and abiogenic origin can be used as food. It is possible to speak about feeding even when only one species and the forage resource it needs are present in an ecosystem. However, it is possible to speak about trophic links only in the presence of at least two species, one of which uses the other one as a food source. Trophic chains are the most important component of communities and ecosystems. They play a leading role in organizing the relations between the populations of various species belonging to the same, or to different, functional groups [3, 9, 16]. As it was already underlined, material, energetic and informational links are the basis of an ecosystem internal structure [23].

Sorokin [94, 95] has introduced the notion of “first-food”, which comprises the primary production and dead OM (suspended and dissolved). The extent of “first-food” utilization by bacteria and protozoans (which are responsible for the heterotrophic part of the metabolism of the planktonic community constituting the first level of ecosystems structural organization) can approach 90%. Bacteria and protozoans transfer the “first-food” to the higher trophic levels. In the pelagial zone of the Black–Azov Sea basin MPs, links of food chains are formed by micro-heterotrophs (bacteria, protozoans) and metazoans feeding on detritus (sea sparkles and rotifers of the genus *Synchaeta*, cladocerans, barnacles larvae and copepods of the genus *Acartia* with a wide food spectrum) develop [30, 91, 122].

In accordance with the trophodynamic approach, ecosystems accumulate the energy consumed by biological components in the process of feeding. Hydrobionts transform matter by using organic and mineral compounds as well as suspended and dissolved components of the external environment, excreting metabolic products, substituting tissues during moulting, growing and dying. The increase of organisms’ biomass occurs at the same time as matter transformation and nutrients turnover in a given ecosystem.

The relationships between organisms during transfer of food energy via predator–prey interactions or utilization of dead OM are called food or trophic chains. Two types of trophic chains are distinguished: the grazing food chain and the detritus food chain [44]. Autotrophic organisms and dead OM (suspended or dissolved) are the first link of grazing and detritus food chains, respectively. Within an ecosystem, the totality of the organisms occupying a defined position in food chains forms the so-called trophic levels (producers and consumers). In an ecosystem, there are usually no more than four–five trophic levels. According to the theory of organization and functioning of ecosystems, the food energy stocked in one ecosystem trophic level is transmitted to the next one. However, this process has a cascade character rather than a continuous one. Each trophic level assimilates only 10–20% of the food it consumes and spends the rest of the energy contained in the food on its own metabolism. Thus, the net production of each level is usually one order of magnitude lower than that of the previous one. The longer are the trophic chains, the less is the production of the top level compared with the bottom ones [23, 73, 85, 114, 119].

Ecosystems’ energetic pyramids can be built by considering all the nutritive energy sources and ordering them on successive levels. The pyramids of energy indicate the direction and amount of energy flowing through an ecosystem and (or) the productivity of successive trophic levels. They reflect the speed of the food passage through the food chain. The changes in size and metabolic intensity of an individual do not influence the form of these pyramids.

Unlike energy, which flows through ecosystems, matter circulates in the closed cycles and passes through turnover loops. The biogenic elements forming the organisms' biomass can enter in the composition of different molecules during the metabolic processes and can be repeatedly used. Grazing food chains include photosynthesizing plants that are consumed by non-predatory animals which, in turn, are consumed by carnivorous animals. Detritus chains start from dead OM which is transformed by microorganisms into forms accessible to detritophages, which are then eaten by predators [23, 72, 81, 85, 119]. Because of microorganisms' activity, there is a turnover of BM which becomes available again to autotrophs.

BM may also be released from tissues of dead plants and animals and from animal faeces via abiotic physico-chemical reactions; i.e. without the participation of microorganisms. However, such decomposition reactions are very slow. In MPs aquatories, the microorganisms' concentrations quite often reach extreme values for marine coastal ecosystems and the contribution of purely physical and chemical processes to BM replenishment can be neglected.

The organisms that decompose OM are the final link in grazing and detritus food chains. These are mainly bacteria, yeasts and saprophytic fungi that settle on and gradually destroy dead plants, dead animal bodies and animal faeces. Microorganisms are always present, in various amounts, in all marine ecosystems and their quantity can quickly increase when additional food resources become available. The decomposers release various substances into ecosystems, including antibiotics (growth inhibitors), vitamins and growth stimulators.

The concepts of food chains, trophic levels and pyramids of biomass and energy were developed in the first half of the twentieth century. These concepts provided a logical and harmonious framework to model the trophic relationships in aquatic ecosystems and to understand the main functioning principles of ecosystems themselves [4]. However, in the second half of the twentieth century, it was noticed that linear food chains could exist only in some ideal or very simple ecosystems. The existence of discrete trophic levels was also put in doubt. In the last decades, the literature usually agrees that transformation of matter and energy in natural aquatic ecosystems always takes place along more or less complex trophic webs. Many trophic webs include species with a wide spectrum of food objects and species capable of being autotrophic, mixotrophic or heterotrophic depending on environmental conditions. For this reason, the attribution of some species to a precise trophic level quite often becomes impossible.

Notions questioning the linear conception of food chains first emerged when the microbial community was still considered as a separate trophic level. Later, the idea of a microbial trophic web gained importance, because microorganisms occur everywhere. This more modern conception postulates an active participation of microorganisms (bacteria and protozoans) in matter and energy flows [20]. Such participation is referred to as "microbial loop", "microbial community" and "microbial food web". In aquatic ecosystems, the "microbial loop" is one of the major factors in feedback mechanisms.

In MPs, appropriate conditions for the formation of "microbial loop" effect in trophic webs originate in the epipelagial and pycnocline zones, in the near-wall water layer adjoining to HTS, in the bathypelagic zone, in the near-bottom water layer and in the pore waters of bottom sediments. In the epipelagial zone, Langmuir convection provides the main manifestation of the "microbial loop" effect.

In MPs, the water movement contributes, to a variable extent, to energy dissipation and to mixing and reduction of matter concentration, according to the second law of thermodynamics. At the opposite, the filtration activity of organisms feeding on SOM and DOM and predation favour its accumulation in the form of biomass and faeces.

The development of the "microbial food loop" [126, 149] concept allowed a detailed study of trophic webs and trophic links in aquatic ecosystems. Bacteriophagous flagellates with sizes of less than 20 μm are one of the main components of the "microbial trophic web". They represent the tiniest of the consumers and are collectively referred to as heterotrophic nanoflagellates [23, 59, 61, 158].

According to the classical trophodynamic theory of food webs, the total efficiency of organic carbon transfer from one level to another decreases with each additional level and, on average, only 10% of organic carbon is transmitted across levels. Studies of microbial communities' structure and functioning showed that the presence of microorganisms increased the total trophic efficiency of planktonic food webs [3, 20, 23, 59, 154].

In the pelagial zone of MPs aquatories, there are trophic chains and webs formed by bacteria, zooflagellates and infusoria, i.e. the already mentioned "microbial loops" in which OM is produced and reduced [61, 91].

In MPs water column, high concentrations of detritus and microbial biomass are formed in situ or enter from land. Microbial food chains comprising bacteria and protozoans have an important role in OM destruction. In the aquatories of Novorossiysk and Tuapse MPs, bacteria account for 70–85% of the total respiration [91]. The microorganisms are the main users of OM inputs (unused heterotrophic production, fauna faeces and allochthonous OM) into MPs bays and aquatories. As observed by Sorokin [100], all sources of dead OM contribute to the total detritus pool of aquatic ecosystems, i.e. the stock of labile suspended and dissolved OM available to bacteria which should be considered during ecological evaluations.

Detritus can enter MPs aquatories from various sources, but it is transformed in two main ways: destruction by bacteria and fungi, or burial in bottom sediments. Dead OM in the detritus pool participates in the detrital chain and is subject to destruction via the bacterial link. Zooflagellates, infusoria and others participate in this process along with bacteria and fungi and, in MPs, their share in OM reduction is in the range 5–24% [91].

It was shown that, particularly in Novorossiysk and Tuapse MPs, most of the energy passes through the detritus food chain with the participation of bacteria and protozoa. In MPs, more than 90% of the energy is provided via dead OM and the corresponding value is 80% outside MPs [91].

It is commonly believed [64] that the leading role in OM destruction belongs to zooplankton in clean (oligotrophic) waters and to bacterioplankton in waters with high organic pollution typical of MPs.

The role of bacterioplankton and microzooplankton increases in the more eutrophic port waters. These organisms are able to thrive at a constant influx of allochthonous OM and the heterotrophic destructive activity is directed to the assimilation and mineralization of biomass having a mainly bacterial origin.

BM returns into the grazing food chain via the process of excretion by heterotrophic organisms and into the detrital food chain following the decomposition of suspended and dissolved dead OM by microorganisms. In MPs, the first process dominates in the epipelagial zone during autumn and winter and the second one in the bathypelagic, periphytal and benthal zones nearly all year round. In MPs aquatories ecosystems, the main flow of energy generally passes through the detrital food chain.

Structural and functional resistance of MPs aquatories ecosystems is provided by various regulatory processes based on trophic interrelations and, especially, on positive and negative feedbacks. It is considered that positive inverse links play a stabilizing role and negative ones realize self-regulation and self-correction [7, 23, 92, 111, 117].

In aquatic ecosystems, recycling of biogenic elements constitutes an example of positive inverse link realized in the "microbial loops" which function as specific "traps" and temporary BM deposits. Microorganisms have a relatively large surface to mass ratio and, hence, a relatively low sinking speed. Due to these characteristics, they create conditions for quick recycling and turnover of biogenic elements directly in MPs epipelagial zones. Wind-drift convection, near-surface thermal (gravitational) microconvection, Ekman's discontinuity and Langmuir convection are all factors contributing to the retention of microorganisms above the pycnocline. In the epipelagial zone, biogenic elements are therefore repeatedly used by planktonic autotrophs thanks to the high activity of microorganisms. Additional reserves and BM turnover can thus be created in the "microbial loops" of the MPs pelagial zone. The main flows of matter and energy into the pelagial zone can pass through such loops. Indeed, it was determined that, in some aquatic ecosystems, up to 50–60% of the plankton net production goes through "microbial loops" [16, 20, 23].

In marine ecosystems, pico- and nanoplankton represent the most important photosynthesizing organisms. The role of minuscule green flagellates with sizes of 2–25 μm is especially important. Tiny flagellates (about 5 μm) are present in the entire water column of marine coastal waters. Part of them contains chlorophyll, but their majority is colourless and feeds heterotrophically on dissolved OM during most of the time. Nanoplankton (small-sized phytoplankton organisms, bacteria, protozoans) plays a key role in the metabolism of coastal waters [61, 72, 98], particularly in MPs aquatories [91].

Mixotrophism, i.e. mixed feeding, is characteristic for many species of planktonic algae. Some iron- and thiobacteria present in eutrophic and polluted waters, such as those of MPs, are also mixotrophic. As mentioned above, in marine pelagial ecosystems, 60–90% of the total energy flow realized by the heterotrophic part of planktonic communities passes through bacterioplankton [98]. Coastal marine ecosystems are inhabited by representatives of all systematic and physiological groups of bacteria which enter them from the atmosphere, soil (via river run-off) and with industrial and domestic wastewaters. According to the hypothesis of "catabolic faultlessness of microorganisms", bacteria are able to decompose any natural substance whose structure is formed by carbon compounds [85].

In seas, the heterotrophic nanoplankton and microzooplankton are the main consumers of phytoplankton. These microscopic microorganisms, with sizes of 2–20 μm and 20–200 μm, respectively, include protozoans, mixotrophic and heterotrophic flagellates and can graze up to 90% of primary production [129, 146, 148].

In the water column and in the periphytal zone of MPs, heterotrophic bacteria and their consumers favour the cycling of organic carbon back to the "classical" part of the food chain. This organic carbon is composed of algal metabolites and faecal material excreted during zooplankton feeding on slime. The hardly digestible OM has no time to be decomposed by bacteria and mainly accumulates into sediments.

Bacteria (auto- and heterotrophic ones) and microplankton are able to grow and reproduce as fast as microautotrophs. This ability and the high efficiency of food consumption and assimilation are important factors in increasing the amount of OM that passes through the microbial community while moving towards the bigger consumers of a food web [59, 60, 97, 126, 132–136, 149, 150, 157].

Unlike other heterotrophic components of nanoplankton, heterotrophic flagellates are constantly present in all types of aquatic ecosystems and their quantitative development is linked to the water trophic state. In coastal marine waters, a significant amount of colourless flagellates is associated with detritus particles and big-sized phytoplankton cells. The ratio of bacterioplankton quantity to the quantity of heterotrophic nanoflagellates varies from 700 to 20,000, increasing from oligotrophic to eutrophic waters and also, with some exception, in waters experiencing significant and diversified anthropogenic pollution [59, 61].

The parasitic flagellates are also an important component of trophic webs. For example, they are able to stop water "blooms" of peridinian algae over short periods of time, which favours the transfer of organic carbon from the grazing to the detritus chain and the following strong development of heterotrophic bacteria and heterotrophic nanoflagellates [60, 96].

In natural conditions, diurnal, seasonal and annual cycles of heterotrophic flagellates are linked with those of bacteria. The maximal development of heterotrophic flagellates coincides with periods of maximal bacterial density or, more commonly, follows a significant increase in bacterioplankton concentration. The increase of dead OM content in the form of detritus precedes the periods of maximal bacterial quantity.

In aquatic ecosystems, heterotrophic flagellates use planktonic organisms of various trophic levels (heterotrophic bacteria, picoalgae, picocyanobacteria, phototrophic sulphur bacteria) and DOM as food. They are able to completely utilize bacterial production and up to 68% of the phytoplankton primary production [10, 23, 59].

It was determined that micro- and mesoplankton organisms actively consume the production of heterotrophic nanoflagellates.

Studies of planktonic microbial communities in different aquatic ecosystems revealed that the heterotrophic nanoflagellates not only play an important role as consumers of bacterial production and as a source of food for micro- and mesozooplankton, but also have a stimulating effect on bacterial metabolism. Phagotrophic flagellates are the main consumers of organisms smaller than 5 μm: heterotrophic bacteria, cyanobacteria, small-sized species of eukaryotic algae and phototrophic sulphur bacteria.

In MPs aquatories ecosystems, the heterotrophic nanoflagellates represent an important trophic link between primary producers and the higher trophic levels. The tiniest photosynthesizing microorganisms often form a significant part of primary production, while biomass and growth speed of protozoans in natural waters are high [7, 20, 23, 40, 41].

Studies of heterotrophic nanoflagellates highlighted that consumers' biomass in marine ecosystems is notably higher than it was believed on the sole consideration of micro- and macrozooplankton [8, 10, 20, 40, 59, 94–96, 99, 100, 1115].

It is commonly believed that yeast-like anaerobes were the first living organisms on Earth, inhabiting shallow-water basins and obtaining the energy required for respiration via fermentation in reducing conditions [23, 72, 81].

From an ecological perspective, the evolution of life and biosphere on Earth is comparable to a heterotrophic succession. It is well known that, in aquatic ecosystems, the major decomposers are not only bacteria, but also fungi. Fungi are present in all marine and continental water basins ecosystems in the same way as bacteria. Saprophytic fungi inhabit the whole water column, the upper layer of bottom sediments and all kinds of substrata [14, 67].

A recently elaborated notion proposes that, in seas, destruction and decomposition of OM is accomplished by the whole biota (and also by abiotic processes) along with bacteria and fungi; i.e. the organisms classically considered as decomposers [23, 73, 81, 85, 141, 153].

5.2.1 Trophic Chains

Organisms spend energy on basic metabolism, i.e. on their maintenance, and these expenses have an energetic and morphogenetic character at the same time. The tissues of an organism's body constantly renew during its whole life. Organisms spend energy to move, grow, reproduce and to form carbohydrate (plants) or fat (animals) deposits [23, 44, 72, 73, 81, 85].

Energetic processes in living organisms are based on redox reactions. Only a few microorganisms groups are able to perform these processes without oxygen by using glycolysis and fermentation. Aerobic oxidation of OM is energetically more profitable. Water motion is connected with the afflux of O_2 and the removal of CO_2 formed as a final product of OM oxidation.

In MPs, planktonic organisms such as bacteria, algae and animals are the main components of the hyponeustal, epipelagial, bathyal and hypoxial communities.

Planktonic communities function in the presence of nutrients and of suspended and dissolved dead OM of autochthonous and allochthonous origin. In MPs, all plankton components are linked by the processes of photosynthesis, chemosynthesis, consumption and mineralization of OM. These processes allow the transformation of energy and the transmission of matter from one trophic level to another.

In MPs, phytoplankton synthesizes OM using light energy and nutrients in the same way as in other aquatic ecosystems. Zooplankton decomposes and mineralizes the OM contained in food and excretes biogenic elements into water during its consumption of algae, bacteria and dead OM (detritus). Dead phyto- and zooplankton, faeces, and mucoid secretions pass into detritus and are consumed partly in the water column and partly in the bottom sediments. Bacteria, fungi and protozoans play an important role in the destruction of detritus, either suspended or deposited in sediments.

The quantity of BM could be a limiting factor for the primary production in oligotrophic and mesotrophic aquatic ecosystems, but such situation is unlikely in MPs eutrophic waters.

The typical ratio of P, N, and C in protoplasm is 1 P : 7 N : 40 C, i.e. 2 g of phosphorus, 14 g of nitrogen and 80 g of carbon are incorporated in 1 kg of organisms wet biomass. If we consider nitrogen and phosphorus in the photosynthesis–respiration equation, the result is:

$$106\,CO_2 + 16\,NO_3^{\,-} + HPO_4^{\,2-} + 122\,H_2O + 18\,H^+ \\ \leftrightarrow C_{106}H_{263}O_{110}N_{16}P + 138\,O_2 \uparrow$$

Because carbon dioxide occurs anywhere in seawater, nitrogen and phosphorus may limit the intensity of primary production. In a system containing only 1 g of phosphorus and 7 g of nitrogen, OM mass cannot exceed 500 g. The same situation will occur with 10 g of phosphorus and 7 g of nitrogen, as well as with 50 g of nitrogen and 1 g of phosphorus. The element whose proportion in the environment is less than its stoichiometric ratio in the protoplasm usually becomes the limiting factor [51].

In aquatic ecosystem where the ratio between the number of atoms of dissolved nitrogen and of dissolved phosphorus exceeds 16:1 (7:1 on a weight basis), phosphorus is the limiting factor. In such ecosystems, additional inputs of mineral or organic phosphorus promote algal development. When the ratio of nitrogen to phosphorus is less than 16:1 (7:1 on a weight basis), the water basin is limited by nitrogen and its addition causes further algal development.

It is known that some algae are able to accumulate phosphorus in their cells [143]. Accumulation takes places at the beginning of the vegetative period and P is then consumed during growth. Planktonic algae could accumulate 100 times more phosphorus than they need for photosynthesis. At the end of the vegetative period, algae do not react to increases in aqueous phosphorus concentrations [86, 87].

As already mentioned, planktonic animals participate in the cycling of biogenic elements and of phosphorus in particular. The relative importance of animals in this process is high in oligotrophic deep basins, where most of OM is mineralized in the water column. With increasing productivity, the relative contribution of zooplankton in providing phosphorus to algae gradually reduces and becomes insignificant in highly productive and shallow ecosystems. In such ecosystem, the processes taking place in the benthal zone are the most important ones. In MPs, animals of the periphytal community play an important role along with those of the benthal zone.

It was determined that mineral phosphorus can also limit zooplankton development. The amount of phosphorus in organisms most often increases along with the animals' weight. Among planktonic crustaceans, phosphorus content represents 0.55–1.5% of body mass on a dry-weight basis, with values of 1.0–1.2% being most common [86, 87].

The amount of phosphorus in the body of planktonic animals depends on their physiological state and on the presence of egg pouches and embryos. Females have a higher P content. The increase in animals' mass is accompanied by a regular increase in the speed of phosphorus excretion. If the weight of a crustacean changes by 1000-fold, then the intensity of phosphorus excretion becomes nearly fourfold higher. At that the same time, the rate of oxygen consumption increases fivefold. The trophometabolism of zooplankton unites all the plankton into a single community. Zooplankton activity decomposes and mineralizes the ingested food and excretes mineral compounds into the environment [86, 87].

Sometimes bacteria do not release phosphorus into the environment, but compete with phytoplankton for its consumption. An absence or deficit of biogenic elements in the environment can limit bacterial development and their ability to decompose OM. OM decomposition is significantly quicker in environments rich in mineral nitrogen and in presence of protozoans which excrete nitrogen and phosphorus during metabolism, thus providing bacteria with essential biogenic elements [109–112].

Besides phosphorus compounds, aquatic animals excrete other elements, particularly nitrogen, into the environment. The intensity of nitrogen excretion by crustaceans and fish is determined by body weight (within systematic groups), ambient temperature and motion activity. On average, 80% of the excreted nitrogen is represented by ammonia and urea. Planktonic copepods lose part of their nitrogen pool in the form of excrements (about 30% of the consumed nitrogen) and liquid excretions. The phytophagous zooplankton excretes nitrogen slower than the predatory one. In planktonic crustaceans, microelements concentrate in faeces, where their content exceeds that in the body by manyfold [137, 138]. Small-sized, plankton-eating animals excrete nutrients faster than the bigger ones. The excretion of mineral phosphorus by zooplankton varies according to the water trophic state [86, 87].

The quantity of biogenic matter excreted into water by phytophagous and predatory organisms of zooplankton during their vital processes exceeds by several folds that contained in dead bodies undergoing decomposition by microorganisms and fungi.

The excretions of zooplankton organisms contain dissolved inorganic and organic compounds of nitrogen and phosphorus along with carbon dioxide; all of which may be directly used by producers without preliminary decomposition by bacteria [40, 59, 72]. Free vitamins are present in seawater.

The coastal aquatic ecosystems, particularly those of MPs aquatories, are characterized by a high speed of biotic turnover with the participation of bacteria, microalgae and zooplankton, the last being usually the main, or one of the main, consumer of algal OM. Fungal spores can pass through the digestive system of planktonic crustaceans without damage. Planktonic animals feed on OM and mineralize it, excreting into the environment metabolites which are later used by bacteria and algae, thus closing the first level of biotic turnover.

Thus, in MPs ecosystems, the role of zooplankton consists in consuming microalgae, bacteria, dead suspended and dissolved OM and in transferring to higher trophic levels the energy contained in them. Planktonic animals are in turn food items for larvae, juvenile and adult fish, as well as for the big-sized multicellular animals of the periphyton and benthos.

In MPs, the main mass of zooplankton is formed by filter feeders [41]. This is the so-called "peaceful" zooplankton which can account for up to 80% of the total mass of marine planktonic communities [75, 98]. "Peaceful" zooplankton plays a leading role in the transformation of matter and transmission of energy from autotrophs to heterotrophs (i.e. from producers to consumers). Planktonic organisms that selectively use forage are collectively termed "predatory" zooplankton and, in turn, are usually selectively consumed by bigger predatory invertebrates and fish [77].

Owing to their small size, zooplanktonic animals have a quick metabolism, high fertility and high filtration activity [36, 40, 49, 86, 87]. However, some phytophagous planktonic organisms do not assimilate up to 90% of the total ingested food and excrete it as faeces. Faecal pellets are eaten by zooplankton as well as colonized by bacteria. Primary production is partly returned into the environment as liquid excretions which are quickly decomposed into the basic biogenic elements by bacteria [7, 23].

The planktonic, filter feeders crustaceans are operationally divided into "fine" and "coarse" filter feeders, depending on their ability to filtrate individual bacterial cells. The "fine" filtrators actively graze single bacterial cells, while "coarse" filtrators feed on aggregated bacterial cells and on detritus particles with sizes larger than individual bacterial cells [41, 77].

Most Cladocera belong to "fine" filtrators, while many Copepoda are "coarse" filtrators. Planktonic crustaceans mainly feed on aggregated bacteria or on bacteria inhabiting detritus particles.

The feeding of filter feeders is mildly selective, but big-sized particles, big cells and phytoplankton colonies are practically not used as food items [95]. At high concentrations of big-sized algae, mechanical impediments to filtration appear and the speed of consumption of small-sized edible algae is reduced [86, 87]. It was established that mass development of big-sized colonial algae reduces the fertility and quantity of many crustaceans despite the simultaneous increase in the quantity of small-sized, edible algae [40, 87].

Depending on aquatic ecosystems trophicity, protozoans, rotifers and crustaceans dominate planktonic communities. Cladocerans and copepods develop in relatively clean marine ecosystems, but the relative share of cladocerans increases with trophicity. Rotifers prevail in polytrophic waters [65, 66, 77]. The same observations apply to MPs aquatories ecosystems.

Zooplankton feeds mainly on cells of small- and middle-sized algae (mainly up to 30–50 μm). The large-sized, colonial phytoplankton ("net") is generally not used as food by filter feeders.

In the MPs pelagial zone, food links can become apparent in the form of a grazing or detritus food chain (or in the simultaneous presence of both) in ecosystem of different trophicity and during the various seasons or years. In highly productive ecosystems, to which the majority of MPs aquatories belongs, the detritus-bacterial food chains have a high importance [61, 90, 91, 96, 99, 100].

Grazing food chains, in which the energy of primary producers is directly transferred to primary consumers, prevail in ecosystems where small-sized ("edible") algae are most abundant. In ecosystems dominated by big-sized planktonic algae ("non-edible" by zooplankton), OM is utilized by consumers only after its transformation into detritus and bacterial mass. In such situations, a detritus, or detritus-bacterial, food chain originates. Detritus becomes available to filter feeders only after preliminary processing. The energy value of pico- and nanoplankton is usually higher than that of large-celled and colonial one [73, 75, 97].

As a rule, the majority of planktonic crustaceans mechanically and non-selectively filtrates the material suspended in water. However, large-sized copepods can perform selective filtration and actively seize relatively big food particles [75, 77].

The size of particles consumed by planktonic crustaceans is within 1 and 100 μm and depends on the crustaceans' body length and on the organization of the filtration apparatus. For crustaceans with a body length of about 1 mm, the optimal size of trapped particles is about 20 μm. The filtration apparatus of "peaceful" zooplankton is adapted for filtering small-sized suspensions. Owing to the high-speed growth of small-sized phytoplankton, the food reserve for "peaceful" zooplankton grows very quickly. Small-sized cells have a higher productivity per unit of algal biomass than big-sized ones, independently of the species. Small-sized coccoid green algae and cyanobacteria have higher values of the P (productivity)/ B (biomass) ratio than large-sized diatoms and dinophytes [77].

The algal species with a mucous membrane (cyanobacteria and green algae) remain viable when passing through the guts of planktonic crustaceans. The same happens with colonial algae which, upon passing through the animals' guts, are torn to small parts and increase their reproductive speed. The presence of animals always favours such growth in phytoplankton quantity. The quantity of some unicellular green algae (Clorophyta) and flagellates, not belonging to slimy colonial algae, also increases in the presence of zooplankton. With respect to zooplankton, planktonic algae are divided into species which increase their quantity in the presence of crustaceans and species that are indifferent to them [60, 77, 152].

Interrelationships between phyto- and zooplankton become significantly more complex when invertebrate predators, fish larvae and juvenile fish are present in the animal composition. At high fish densities, changes in the dominating species of planktonic crustaceans can occur. A reduction of the average mass of zooplankton also takes place, i.e. fish preferentially graze big-sized crustaceans and food items [7, 23, 35, 128].

The filtration speed of crustaceans with sizes of 1–2 mm decreases as the concentration of big-sized particles increases, which could even cause starvation and reduction of cladocerans' fertility [162].

Thus, food availability is determined not only by the morphological peculiarities of animals themselves, but also by the presence of big-sized, uneatable algal forms interfering with normal filtration in the water column.

As a rule, a zooplankton succession from calanoids towards cladocerans takes place during ecosystems' eutrophication. Copepods have a less dense filtration apparatus which allows them to effectively filtrate food particles at relatively low concentrations and with less energy expenses. Compared with cladocerans, they also have higher filtration activity and feeding rate. The increased share of copepods in eutrophicated conditions results in shorter developmental time to reach fertility and hence a higher number of generations and ability to consume algae of various sizes.

As the trophicity of a basin increases, the big-sized forms of zooplankton, including copepods, are usually replaced by the small-sized ones [77].

In eutrophic and highly eutrophic waters, many cladocera have distinct advantages over copepods, in particular the ability to feed at high concentrations of suspended matter in estuaries and in the desalted waters of the Azov Sea, the NWBS and the Dnieper-Bug and Dnestrovsky Limans.

Some big-sized species of crustaceans and decapod larvae start to disappear when the quantities of various cyanobacteria in plankton composition increase.

As a rule, the ecological and functional indices of zooplankton change proportionally to eutrophication as follows: 1—increases in the share of cladocera and rotifers and decreases in the copepods' share; 2—decreases in the average individual mass of animals; 3—higher volume to biomass ratios in zooplankton; 4—reduction in the ratio of zooplankton biomass towards that of phytoplankton [13].

The size of planktonic algal colonies, and therefore their palatability, mainly depends on hydrodynamics. Zooplankton consumes cyanobacteria when their colonies have small sizes and their concentrations are low, which occurs at relatively high hydrodynamics.

Small-sized green algae, diatoms, small-sized dinoflagellates and cryptomonads are the most actively consumed algae whose quantity is often regulated by zooplankton. Zooplankton does not feed on big-sized desmids and dinoflagellates and on big colonies of cyanobacteria and diatoms. At the same time, some algal species increase their quantity in the presence of zooplankton. Such algae include forms with strong walls or gelatinous capsules which can pass intact through the guts of zooplanktonic organisms [13, 75, 77, 86, 87].

As it was already mentioned, the colonial algae with mucous membranes are torn into small fragments during passage through the guts; which accelerates their reproduction [151, 152].

In natural conditions, grazing by zooplankton can regulate the quantity of planktonic algae only in cases when their concentration is not high.

Strengthening of fish pressure on zooplankton communities intensifies biological turnover and leads to an increased production of the small-sized, edible phytoplankton owing to changes in the zooplankton size structure during the process of grazing by larvae, fries and adult fish [7, 23, 35].

The ratio of algae, bacteria and detritus in the food of planktonic animals varies in different aquatic ecosystems. With regard to animals, the quantity of algae in the food ration increases proportionally with waters trophicity, while that of detritus decreases. These opposite changes reflect the variable content and ratio of these components in seston. The share of bacteria in zooplankton feeding is especially significant in highly trophic ecosystems, such as in MPs, where decomposing OM of autochthonous and allochthonous origin is plentiful.

The filtration speed of planktonic animals mainly depends on food concentration and it is the lowest in eutrophic waters.

In eutrophic ecosystems, phytoplankton development is determined by allochthonous inputs of biogenic elements as well as by inputs from bottom sediments. During the summer period, zooplankton consumes no more than 10–20% of the primary production per day and, in so doing, regenerates the mineral phosphorus necessary for photosynthesis [87]. Given their relatively limited role in the biotic turnover of eutrophicated ecosystems, zooplanktonic organisms are unable to limit algal development by grazing, which leads to water "blooms" and secondary contamination of waters. Such situation is observed, for example, in Grigorievsky and Sukhoy Limans, in Varna and Beloslav Lakes and in many ports located on the open seashore, in gulfs and bays [30, 91].

As already mentioned, two main conceptions are presently used for describing the typical features of aquatic ecosystems functioning: 1—linear food chains and trophic cascade and 2—microbial trophic chain.

According to the first conception, aquatic ecosystems are characterized by a series of ascending trophic levels in which organic and biogenic matter flows move up along the food chain and the food pyramid narrows from the primary producers towards zooplankton and fish. Flows are accompanied by losses of excretable substances and regeneration of biogenic elements. Copepods, cladocerans and rotifers, collectively termed "metazoan plankton" have the leading role in transferring energy from producers towards the upper trophic levels. It is considered that the primary producers determine the state of the higher trophic levels; which is called "bottom-up" control. In turn, predatory fish create a cascade of biotic impacts onto the lower levels and regulate the general ecosystem state; the so-called "top-down" control [9, 10, 20, 23, 40, 59].

According to the organization and functioning theory of ecosystems, predation by fish and zooplankton regulates the number of preys and determines the structure of the corresponding lower trophic layer. As noticed by Bulyon [20], the concentration of biogenic elements determines only the potential productivity of a water basin. The actual production is determined by the community structure and by the predator–prey interrelationships in each water basin.

In eutrophic ecosystems (such as MPs aquatories), changes in ichthyofauna composition usually affect zooplankton only slightly, i.e. changes in the grazing rate of planktonic animals is usually compensated by their high fertility. Under eutrophic conditions, changes in high trophic levels (e.g. fish) practically do not influence phytoplankton whose development is constantly stimulated by the high concentrations of biogenic elements [12].

The fundamental principles of the "trophic cascade effect" concept follow the classical notion of linear food chains in which flows are in the order biogenic elements, phytoplankton, zooplankton, fish as well as birds, reptiles and mammals.

The activity of bacteria and protozoans (heterotrophic microbial organisms) or "protozoan plankton" is not considered in the linear food chain concept. However, as shown by numerous studies, these organisms are able to use a significant proportion of the matter and energy flows in ecosystems [59, 96, 99, 100].

Various studies have shown that a significant part of OM is produced by phytoplankton in the dissolved form and is very quickly transformed into "living" bacterial matter [9, 20, 86, 87]. Production of dissolved OM is connected with phytoplankton death and exudates from live cells. Such labile OM is used especially quickly by microorganisms. The turnover time of phytogenic substrata varies from a couple of minutes up to a couple of hours and these substrata are not accumulated in the environment. More refractory DOM and suspended OM, which have a turnover of days to weeks, usually dominate in aquatic environments.

"Peaceful" zooplankton has sizes from 2–5 up to 150–200 μm and is divided according to its food spectra. Protozoans (flagellates and ciliates), rotifers (peaceful and facultatively predatory), copepods and cladocerans feeding by filtration interact in a predator–prey community where they create a complex network of direct and inverse links.

The above-described situation required the elaboration of a new and more detailed conception of ecosystems' functioning. According to this new conception, bacteria and protozoans, as well as the part of phytoplankton used by the latter, form the so-called "microbial loop" or "microbial trophic web" [9, 20, 23, 40].

The microbial community is formed by bacteria and protozoans (heterotrophic flagellates and ciliates) which serve as an intermediate stage (level) in the transfer of matter from autotrophic organisms towards net zooplankton. The use of DOM by heterotrophic bacteria and bacteriophagous protozoans comprises a considerable part of the pelagic carbon cycle [59]. Organic carbon which enters the "microbial loop" is converted into DOM [20]. Transmission of organic carbon towards higher trophic levels via the "microbial loop" is most important in ecosystems with relatively low primary production.

The "microbial loop" represents a complexly organized community of prokaryotic and eukaryotic unicellular organisms, including planktonic algae, bacteria, microscopic fungi, protozoans and viruses. Besides transferring matter and energy from the fine-sized autotrophs to the metazoic plankton, the microbial planktonic community participates in the production of primary OM (by photosynthesis and chemosynthesis) and in recycling of biogenic elements [59]. Microzooplankton is considered as a connecting link between microbial organisms and the large-sized zooplankton. Thus, a microbial food web represents a complex, interacting community (including phytoplankton, bacterioplankton and protozoan plankton) collectively responsible for autotrophic synthesis, regeneration of biogenic elements and transmission of OM to metazoic plankton.

The autotrophic component of the microbial community can be represented by picoalgae, picocyanobacteria, photosynthetic bacteria as well as by symbiotic algae and chloroplasts of protozoans. In salty aquatic ecosystems with high concentration of hydrogen sulphide, such as the ecosystems of MPs aquatories, OM production via chemosynthesis acquires a particular importance [23, 59, 72, 73].

Protozoan plankton serves as food for micro- and mesozooplanktonic organisms. Heterotrophic flagellates and infusoria with the same sizes as phytoplankton partially return energy from the "microbial loop" into the "classical" food chains. Increase in the quantity of predatory or parasitic protozoans in the plankton leads to a decreased transmission and transformation of microbial organisms' carbon towards the higher trophic levels and to replenishment of the organic carbon resources within the "microbial loop" [21, 23, 59, 92, 110, 112].

The microbial trophic web is considered as the most active biotic part of an ecosystem, where regeneration and accumulation of biogenic elements take place. Microbial organisms transform 45–90% of the primary production [20]. Bacteria alone assimilate 40–60% of the primary production in the form of dissolved OM [7, 9, 10]. By accumulating biogenic elements connected with OM, bacteria can utilize their mineral forms contained in water at low concentrations. This process favours the preservation and accumulation of biogenic elements in the photic and epipelagial zones and reduces their losses by sedimentation. Biogenic matter contained in bacterial cells is quickly incorporated into protozoan plankton which, in turn, excretes dissolved forms of nitrogen and phosphorus necessary to the synthesis of OM by phytoplankton. By grazing bacteria, ciliates decrease their quantity and thereby prevent excessive bacterial exploitation of the trophic resources supplied by phytoplankton [57–60, 109–112].

The function of bacterial consists in the regeneration and accumulation of biogenic matter from the environment. The biogenic elements accumulated by bacterial cells are incorporated into the consumers' bodies and are only later excreted in mineral form. Thus, protozoans favour quick BM recycling by consuming bacteria and eventually excreting the biogenic elements accumulated by them into the environment.

Compared with large-sized plankton, microbial organisms have lower sedimentation speeds. For this reason, the biogenic elements fixed by bacteria are preserved in the photic zone of the MPs epipelagial over a long time. Owing to the "microbial loop", biogenic elements are retained within the planktonic community and hence repeatedly available to algae [10, 20, 40, 159].

Switching the substances flow to microbial organisms favours the accumulation of biogenic substances in the photic zone; which helps the planktonic community to survive in conditions of nutrients deficit and creates preconditions for the autonomous functioning of the first ecosystem level.

In some eutrophic ecosystems, including MPs aquatories, bacterial growth is not balanced via grazing by planktonic organisms and the growth of planktonic organisms is not balanced by fish. In these ecosystems, mortality of planktonic organisms is determined by the activity of viruses, which act as a "top-down" controlling factor in the same way as protozoans. As a result, a significant share of bacterial carbon is not transmitted to higher trophic levels, but it circulates within the bacteria—viruses—dissolved OM "loop". In such ecosystems and situations, the "microbial loop" does not fulfil its function of intermediate link in the transfer of organic carbon from autotrophic organisms to metazoan plankton [9, 10, 20, 40, 59].

In MPs aquatories ecosystems, primary production mainly comprises plankton, microphytes and macrophytes in the periphytal and benthal zones. The contribution of each group to primary production is different across ecosystems. Macrophytes production grows slowly in ecosystems where plankton production comprises more than 90% of the total [9, 10, 23].

In shallow-watered ecosystems, the contribution of macrophytes of benthal and periphytal zones to primary production increases. In the deep-watered MPs ecosystems (10–20 m), primary production is mainly created by the photosynthetic activity of phytoplankton. In various MPs, the importance of periphytal macrophytes is determined by the amount of hard substrata suitable for their growth and by water transparency. As depth increases, the production of benthic macrophytes gradually drops to zero. As the macrophytes' share in ecosystems' primary production decreases, the importance of zooplankton communities increases with respect to that of benthic communities [10, 23].

Only an insignificant part of the net primary production of macrophytes is directly consumed by animals such as invertebrates or fish. Rather, its "processing" is performed by microorganisms communities (bacteria, protozoans). For this reason, the production of the coastal and near-surface macrophytes' belts forming inside MPs is usually not included into the pelagic food chain. Some energy exchange between the pelagial zone, from one side, and the periphytal and benthal zones, from another side, can be realized by active and relatively big invertebrates, fish and birds. Predatory invertebrates and fish use organisms of meiobenthos and macrozoobenthos which feed on microbenthos and detritus.

Together with plankton, benthic organisms represent a second big subsystem of water basins ecosystems, particularly in MPs aquatories. The role of benthic organisms changes depending on the water basin productivity. In productive ecosystems such as artificially deepened and eutrophicated MPs aquatories, the importance of benthos reduces, while that of plankton increases. The partial reduction of zoobenthos role in the benthal zone is compensated by settlements of benthic organisms on artificial hard substrata (periphyton) [30, 91].

The macrozoobenthos species composition depends on autochthonous and allocthonous forage resources and on habitat conditions; i.e. substratum diversity, availability of shelters, depth, hydrodynamics peculiarities and hydrochemical regime. Zoobenthos species composition is mainly determined by environmental conditions [74, 120] and most reliably characterizes habitat quality and the whole state of an aquatic ecosystem. Species diversity is the main characteristic of MPs benthic communities and determines their peculiarities.

In all ecosystems of MPs aquatories, various changes take place because of anthropogenic impact. Such changes most frequently lead to a decrease in the species number in communities, due to the disappearance of stenobionts, and affect the evenness of their population density values [3, 6, 10, 106, 121]. However, the biota composition of MPs aquatories located in limans and estuaries initially becomes richer after their connection with the sea.

Based on the dominating trophic group, all benthos and periphyton communities are divided into autotrophic and heterotrophic communities, occupying most part of the bottom surface of MPs aquatories and HTS lateral surfaces. Macrophytes autotrophic communities are connected with the coastal shallow waters, where they are preserved, and with the near-surface belt on HTS.

The representatives of macrozoobenthos (as well as of nekton) have specific life spans up to couple of years and some mollusks up to a dozens of years. Macrozoobenthos usually has 1–2 generation per year. For comparison, meiobenthos and planktonic animals have up to 4–12 generation per year, each one lasting weeks to months. The life duration of meroplankton and microbenthos, in particular infusoria, comprises hours or dozens of hours. In MPs, the majority of zoobenthos species have a two-phase life cycle including a larval stage. Their pelagic larvae are most often part of the zooplankton composition and form the merozooplankton. Microbenthonic organisms usually have a single-phase life cycle and live hours or dozens of hours.

The diversity of benthic animals communities and the average mass of individuals living in them are functionally connected. Important characteristics of communities such as turnover rate, production and energy flows are also functionally connected with the animals' mass. Anthropogenic-induced changes in diversity and structure complexity of benthic communities modify the above-mentioned characteristics [3, 6, 7, 9, 10].

In ecosystems, the diversity of zoobenthos communities is inversely proportional to plankton primary production. As BOD_5 increases, the diversity of zooplankton communities drops, while the diversity of benthic animal communities remains relatively high as long as the share of predatory animals in benthos remains within 10–30%. A further decrease in the proportion of predators in the community leads to a sharp reduction of its biodiversity [7, 9, 10]. The same situation occurs at excessive eutrophication within MPs, with decapods (crabs, shrimps) being usually the first to disappear.

Preservation of the macrobenthos species composition indicates that the original qualities of the aquatic environment are being preserved and is a proxy of the state of the whole MP ecosystem.

The characteristics of communities and ecosystems are determined by the whole complexity of interspecific relationships (trophic, allelochemical, ethological and others). The disappearance of species, particularly fish, being registered occasionally or accidentally does not represent a serious danger. The basis of ecosystems' biota comprises mass species, or "nucleus", and species which are regularly present, albeit in small amounts, and play an important role in ecosystems. The disappearance of such species from the biota structure could cause catastrophic reorganizations in ecosystems [120, 121, 125, 142].

When one of the mass species of the "nucleus" community disappears from the macrozoobenthos species composition, it is possible to talk about substitution of one community by another. Similar situations do cause the introduction and stabilization of new mass species.

Numerous studies showed that exogenous alteration of communities' species composition are irreversible [26, 46, 144]. Changes in species composition of benthic communities go along with changes of other cenotic characteristics which are also irreversible. Such changes create preconditions for significant, sharp and less predictable changes in a whole ecosystem.

The stability of communities' species composition provides the possibility of reestablishing the ecosystem initial characteristics after weakening, or cessation, of anthropogenic impact [22, 39, 127].

The variability of biomass dynamics usually increases gradually together with the production of natural aquatic ecosystems and reduction of the individuals' size in hydrobionts' communities. A change in the biomass of the benthos community has no influence on the amount of dead OM suspended in water and on OM content in bottom sediments. The biomass ratio between zooplankton and zoobenthos increases along with basin production [4, 6, 9, 10].

In aquatic ecosystems not affected by strong pollution and eutrophication, long-term fluctuations of benthos biomass are, on average, about threefold. In ecosystems where significant changes in biotic structure and functioning of the biological community occur, interannual fluctuations of benthos biomass can be abrupt [4, 6, 9, 10].

The differences between the highest and the lowest values of benthos biomass during the year decreases proportionally to the increase in depth and the uniformity of the bottom relief, such as during MPs dredging, and therefore general conditions. The annual fluctuations of benthos biomass are less in aquatories with more uniform conditions for bottom-dwelling animals than in basins with more diversified conditions [3, 6, 9, 30, 91, 122].

In various regions of the Black–Azov Sea basin, the permanent ichthyofauna of MPs aquatories usually comprises a relatively low number of species; varying from 10–20 at the northern shores to 20–40 along the shores of Crimea, Caucasus, Turkey and Bulgaria. The species diversity usually increases in the spring–autumn period. Besides the dominating coastal near-bottom and bottom fish, planktophagous and predatory pelagic fish which perform distant migrations enter into the ports for various periods of time.

In contrast with the sea zones adjacent to MPs aquatories, conditions for the reproduction and dwelling of juvenile and adult individuals of benthic fish are absent in large areas of deep-watered ports. In MPs, the bottom and near-bottom ichthyofauna accumulate in the zone of coastal shallow waters (0–5 m) if they are preserved or created. They are also found in the intermediate zone (8–10 m) between the coastal shallow waters and the deepwater bottom areas (21–22 m) covered by black silt and containing hydrogen sulphide. Benthic fish is represented by species from the families Gobiidae, Blenniidae, Gobiesocidae, Singnathidae, Scorpaenidae, Callionimidae and others.

The near-wall layer of the HTS in the periphytal zone offers favourable conditions for a variety of coastal pelagic fish. The following species are registered in these zones: the Black Sea sand smelt *Atherina pontica*, the Bonapartii's sand smelt *A. bonapartii*, the striped mullet *Mugil cephalus*, the golden mullet *Liza aurata*, the leaping mullet *Liza saliens*, wrasses Labridae, seabreams Sparidae, the gilthead *Diplodus annularis*, the banded sea perch *Serranus scriba*, pipefish Singnathidae, the sea-horse *Hippocampus guttulatus*, the transparent goby *Aphia minuta* and others.

Feeding and spawning migrations of mass fish species through MPs aquatories are an example of temporary flows of matter, energy and information. The feeding and release of reproductive products by adult fish lead to grazing, utilization and transformation of forage organisms into dead OM and faeces on the one hand and to the appearance of new food resources (eggs, larvae and newly hatched fish) in the plankton composition on the other one.

Adult pelagic planktophagous fish predominantly feed on holo- and meroplankton as well as floating eggs, pelagic larvae and fries of fish.

During their whole life cycle, various species of fish consume different types of food. Pelagic larvae and fries of the pelagic, bottom and near-bottom fish feed on bacteria, protozoans and zooplankton. Fries and juvenile of bottom fish switch to feeding on meiobenthos and macrozoobenthos only after reaching a certain size. Larvae and fries of bottom fish without pelagic stages of development feed on microbenthos and, later, on meiobenthos.

The bulk of the bottom ichthyofauna of the coastal shallow waters and intermediate zones of MPs comprises fish that feeds on meio- and macrozoobenthos as well as periphyton. Fish that inhabit the HTS near-wall water layer feed on meiobenthos, active larvae and adults living in fouling biocenoses.

Some species of fish (e.g. Blenniidae, Labridae, gilthead *D. annularis* and others) inhabiting the coastal shallow waters and periphytal zone feed mainly on macrophytes and their fouling. The majority of adult bottom fish feeds predominantly on benthic animals. The following species can be assigned to the number of predominantly ichthyophagous organisms: the knout goby *Mesogobius batrachocephalus* and the ratan goby *Neogobius ratan*, the greater weever fish *Trachinus draco*, the stargazer *Uranoscopus scaber*, the juvenile of the Black Sea turbot *Psetta maeotica* and the European black scorpion fish *Scorpaena porcus*.

Schools of mass migrating fish species occasionally (or regularly) enter into MPs aquatories where they become part of the food chains. The planktonophagous species entering into MPs aquatories include the Black Sea kilka *Clupeonella cultriventris*, the sprat *Sprattus phalericus* and the European anchovy *Engraulis encrasicolus*. They are followed by predatory fish feeding on them such as the horse mackerel *Trachurus ponticus*, the garfish *Belone euxini*, the bluefish *Pomatomus saltatrix,* the spiny dogfish (which also eats other predatory fish) and by dolphins. During their feeding, all these animals incorporate part of the living matter produced in MPs aquatories ecosystems into their biomass, but also transform the living matter of MPs into detritus via faeces production. When leaving the MPs ecosystems, migrating fish export a certain amount of transformed OM outside them.

Aquatic and semi-aquatic birds together with dice snakes are present as high-level consumers in the ecosystems of some MPs. In Grigorievsky and Sukhoy Limans, dice snakes eat the males of gobies which, during their spawning period, protect their nests with eggs. In this situation, the main damage to the gobies' populations is not the direct consumption of adult individuals, but the rapid consumption by invertebrates and fish of the gobies' eggs left without protection.

In MPs aquatories, invertebrates and fish form new, temporary trophic links and flows of matter, energy and information by releasing their gametes. By grazing the fries, young and adult fish control, to a certain extent, the quantity and size composition of planktonic organisms' populations. At the same time, predatory zooplanktonic forms and big-sized filter feeders (comb-jellyfish, jellyfish) can control the size of fish populations at the egg, prelarval and larval stages. Obviously, this is one of the ecosystem's mechanisms of autoregulation.

Various parasites and fungi participate in fish control along with viruses.

In aquatic ecosystems, fish regulate the number of preys and control the structure of lower trophic levels to a considerable degree. In oligo- and mesotrophic systems with large fish, communities of forage organisms use more energy for production than in the ecosystems with prevalence of the small-sized fish. In the eutrophic ecosystems of MPs, small, short-lived, planktonophagous fish species have a competitive advantage. Changes in the ichthyofauna composition of MPs usually affect zooplankton only slightly, i.e. grazing of planktonic animals is anyway compensated by their high fertility and by the presence of favourable conditions for feeding.

In ecosystems, an increase in the fish average size leads to more complex structures in the communities of the corresponding forage objects. This phenomenon is explained by the fact that predatory fish consume in the first place all forage objects of bigger sizes, which are energetically more favourable. In this way, predatory fish reduce the dominance of large forage objects in pelagial, periphytal and benthal communities. In animal communities, fish contribute to maintain high species diversity and stabilize energy flows [147].

Variability of biomass dynamics in ichthyocenoses is connected with the increase of ecosystems production which, in turn, leads to a reduction in individuals' size. However, ichthyocenoses variability is low compared with the phyto- and zooplankton and zoobenthos communities [2–5, 7, 9, 10].

Reduction of fish quantity could increase the complexity of the structure of bottom animals' communities. Following such reductions, the ratio of predatory to non-predatory animals changes towards a strengthened role of predators whose species diversity increases. The decrease in the quantity of fish increases the importance of benthic animals [8, 12].

An increase in fish catches quite often leads to a gradual reduction of plankton primary production and to an even slower reduction of zooplankton and benthos production. Additionally, water transparency increases and the process of phosphorus transport from the deeper layers into the near-surface layer weakens [8, 12, 23].

5.2.2 Flows of Matter, Energy and Information

Organization of energy flows is a fundamental characteristic of natural ecosystems [7, 72, 73, 164].

During the turnover of biological matter, the energy flows remain unidirectional. The solar radiation energy (photosynthetically active radiation; PAR) accumulated in the tissues and protoplasm of primary producers is gradually lost in the form of heat as it passes through all the links of trophic chains and along trophic levels. At the same time, synthesis of OM occurs in each link and level, resulting in the accumulation of energy in chemical bonds. Living organisms therefore hamper the quick dissipation of energy to various degrees and thus contrast the second law of thermodynamics [23, 73, 85, 119].

All transformations of matter require energy expenses. Photosynthesis is the first stage of use and transformation of solar energy. It creates the substances for the construction of plants bodies and, following their ingestion, those of phytophagous and predator organisms.

The law of energetic conductivity states that, in fully structured ecosystems, the flows of matter, energy and information should pass through the whole system. This circumstance is a condition for the self-regulation of the ecosystems themselves. According to the law of the unidirectional energy flow, the energy received by an ecosystem and assimilated by its producers is lost during respiration or irreversibly passed to consumers belonging to various trophic levels and, finally, to decomposers. The energy flow decreases at each of the trophic levels because of the losses accompanying respiration. The inverse energy flow from reducers to producers corresponds to 0.25–0.50% of the original one, so that it is not possible to speak about a real energy turnover [44, 73, 82].

In ecosystems of MPs aquatories, actively migrating planktonic and some benthic organisms create diurnal vertical flows of matter, energy and information having a seasonal maximum in spring and autumn.

In general, in the ecosystems of MPs aquatories, the largest part of matter and energy is transferred along the detritus food chains during the whole year. At the same time, in the epipelagial zone of MPs, the role of big-sized copepods increases and the concentration of bacteria reduces during the coldest months, which indicates a somewhat increased importance of the pascual trophic chain in this period. In the summer months, detritus trophic chains form in both the bathypelagial and epipelagial zone.

The input of solar energy, the amount of OM and hydrodynamics appear to be the main energetic factors, which determine the type and the intensity of the processes occurring in aquatic ecosystems. The balance between illumination, OM content and water motion is an important requirement for any ecosystem.

The use of the term energy "flow" allows to describe with a certain precision the conversion of extra- and intra-system energy in water basins. The transfer of energy across the biotic components of an ecosystem takes place from the low trophic levels towards the higher ones, from organisms producing and accumulating energy in OM towards organisms consuming it, from one community towards another. Energy is transformed within complex trophic webs, but reaches each trophic level. For all these reasons, the presence of organized energy flows is one of the fundamental properties of ecosystems [44, 73, 82, 83].

The trophic links and energy flows in ecosystems are complex and ramified. The processes of energy transformation involve energy contained in biotic components and energy of external and internal abiotic components of ecosystems (light energy, wind mechanical energy, waves and flows, waters thermal energy, energy of chemical bounds of mineral substances, etc.).

In aquatic ecosystems, information flows, besides the flows of matter and energy, also start at the level of autotrophic and heterotrophic pico- and nanoplankton. All flows are inseparable from one another. Chemical regulatory communication has a significant importance in marine coastal ecosystems. DOM molecules can be a source of energy and information and many DOM components often regulate homeostasis in ecosystems [7, 9, 10, 23, 82, 83].

Water transports regulatory substances of natural origin, as well as substances discharged by humans, which are frequently toxicants. They influence the individuals' survival, reproduction and growth as well as the growth rates, development and sex ratios of populations and the species ratio and abundance of a whole ecosystem [4, 5, 9, 10].

Owing to the presence of various interfaces, the ecosystems of MPs aquatories function as if they were delimited or restricted by a system of screens via which the flows of live and dead OM and of energy take place. The general properties and the individual peculiarities of MPs ecosystems are mainly determined and characterized by the flow of dead OM. Dead OM, detritus and its components represent a potential energy source for living organisms.

Because of physical principles, matter and energy dissipates through the water column. However, in ecosystems of MPs aquatories, dead and living OM (the latter consuming the former) concentrate in contact zones such as the water–atmosphere interface, the isopycnic surfaces in thermo- and halocline layers, and the interfaces between water and the bottom, water and coastal shallow waters, and water and the surface of hydrotechnical structures. In semi-enclosed MPs ecosystems, the forces favouring the dispersion of matter and energy are constantly counterbalanced by forces accumulating them in certain zones.

During the last decades, the biological productivity of the brackish waters of the NWBS and Azov Sea has increased because of the excessive amounts of OM. Following this situation, a natural worsening of the oxygen regime during the warm period could be observed without any new anthropogenic pressure. In semi-enclosed MPs aquatories, the concentration of suspended and dissolved OM can increase by one to twofold owing to the well-known "mud trap" effect [34, 72].

Based on the examples of Novorossiysk and Tuapse MPs, Selifonova [91] showed that the highest content of digestible OM in the waters of MP aquatories is confined to the most remote, separate harbours and gradually decreases towards the exits from the ports. Similar results were obtained for Odessa, Ilyichevsk and Yuzhny MPs aquatories [15, 30, 91].

Coastal settlements of benthic animals exist in conditions of increased hydrodynamics and movement of suspended and bottom longshore drifts flow. Suspended loads contain a significant amount of OM and food resources become available in places of their accumulation. Flows from river mouths and estuaries, drains and other flows carry suspended live and dead OM of allochthonous and autochthonous origin. The sedimentation of suspension occurs where flows slow down. Settlements of mass bottom filter feeders organisms and, somewhat farther, of detritophages form in such zones.

In MPs ecosystems, a two-way transit of dead OM, mainly in the form of detritus, takes place through the pelagial zone. It can happen owing only to physical and chemical processes, but it can involve the participation of living organisms. Dead OM enters into the hyponeustal zone from the atmosphere and from the underlying water column. In the benthal zone, the accumulation of the dead OM takes place under the influence of gravity and biological sedimentation. In the water column, the flow of dead OM at lowered hydrodynamics is parallel to the periphytal from the surface to the bottom. Periphyton organisms use this flow to uptake and use part of the OM available to them in the near-wall water layer. In MPs, a horizontal transfer of dead OM is also present owing to horizontal flows, currents and vortexes. HTS surfaces located along the pathway of these flows slow them down and stop them. As a consequence, accumulation of dead OM, which serves as an additional forage base for fouling animals, takes place in the near-wall layer. After utilization of part of the dead organic matter and associated energy, a significant amount of OM is rapidly transported to the benthal zone near the HTS foundations in the form of faeces and pseudo faeces. A much higher quantity of matter and energy therefore passes via the near-wall water layer and periphyton community than through the water column in the central parts of harbours. The populations of HTS lateral surfaces take advantage of this situation to increase their biomass.

In fouling biocenoses with ongoing ecological successions, the processes of autoregulation lead to increased biotic diversity, elongation of food chains and formation of complex food webs. In MPs, the energy stored in periphyton community increases proportionally with the maturation of the community itself.

In the periphytal and benthal zones of MPs, living OM can be considered as distributed on the surfaces, while it is located in the volume of the water column in the pelagial zone.

Life is present in all natural waters. Its highest diversity is connected with waters whose main parameters are close to the modern average of the World Ocean. The more the major parameters of a natural water differ from average ones, the less its species diversity and the smaller the size of its hydrobionts'. The biota composition of some natural aquatic ecosystems includes only few to dozens of microorganism's species. In other ecosystem, the number of invertebrate and vertebrate species quantity reaches many thousands. Despite significant differences in their biota, all aquatic ecosystems have the ability to self-organize and auto-regulate themselves.

The cumulative activity of all living organisms (living OM) in aquatic ecosystems is directed to the maintenance of an inorganic environment suitable for the existence of life. Apparently, the first ecosystems originating in the aquatic environment already had groups of organisms providing the production of primary organic matter, its transformation and its destruction to simple compounds. In the course of evolution, an increase in the organisms' sizes and complexity took place, but the general functions of ecosystems have been preserved. Thus, in the course of time, the first and most ancient levels of ecosystems organization were completed by new levels of more organized and more specialized groups of organisms also realizing the turnover of matter.

To correctly understand and forecast the processes occurring in modern aquatic ecosystems in general, and in MPs aquatories ecosystems in particular, it is necessary to consider a series of important circumstances. After having originated in the aquatic environment, life has never ceased to exist there and living matter has constantly adapted to new, changing conditions and has, in turn, changed the surrounding environment. The first aquatic ecosystems were functioning under anaerobic reducing conditions. Oxygen originated in the aquatic environment and reached its modern concentration thanks to the activity of living organisms. In aquatic ecosystems, the continuous processes of synthesis and decomposition of living matter reached a certain balance among them owing to the presence of some amount of dead OM. During the evolution of the World Ocean, and of the Black–Azov Sea basin belonging to it, both living matter and seawater composition changed [23, 29, 81].

Over a couple of billion years, the functioning of living matter provided a total turnover of matter in marine ecosystems and more or less closed turnovers of simple substances (biogenic elements) necessary for the construction of living matter itself. Major natural disasters in the past and anthropogenic activity in the present time affected and continue to affect the ecosystems' steady states. Living matter responds to such influences by various processes of reorganization.

To correctly estimate the future changes of the processes which take place in MPs aquatories ecosystems with a depth of more than 8–10 m, it is necessary to examine the evolution of the World Ocean waters and of life within them.

As shown above, a significant part of dead OM is not used in the water column and in the periphytal zone of MPs and settles to the bottom. As a result of its decomposition, hypoxia and anoxia phenomena often originate and high levels of labile sulphides appear. Thus, reducing conditions form in the upper layer of bottom sediments, in the near-bottom water layer and, occasionally, in the whole water column.

Groups of hydrobionts species having originated in the World Ocean during different stages of life evolution are present in modern coastal marine ecosystems, including those of MPs aquatories. As a rule, the most ancient forms of life are preserved and actively function under reducing saprobiotic conditions.

In saprophytic conditions, anoxic respiration is the basis of the vital activity of bacteria, yeasts, moulds and protozoans inhabiting MPs bottom. A significant role belongs to methane-oxidizing bacteria, which form methane (CH_4) by reducing organic carbon or the carbon contained in carbonates. An acidic environment destroys carbonates.

It is considered that the main water masses were formed at an early stage of the Earth development. Initially, a reducing environment prevailed in ocean waters and the chemical composition of oceanic, continental and rain waters was identical. Classical theories of water origin on Earth assume that it evolved and continues to evolve via the process of crustal melting and mantle degasification [25, 27, 28, 31].

Part of the water masses has cosmic origins. Some hypotheses draw attention towards the role of biota in the formation of the hydrosphere. In nature, many ecological groups of microorganisms use inorganic substrates and produce water as a result of their metabolic activity. Microorganisms able to oxidize reduced gases in water (and to obtain energy from this process) are still widely distributed [17, 23, 47, 81, 85].

Initially the ocean was salty. Despite being aggressive, the primary ocean waters had lower mineralization than the present ones because of the low solubility of magmatic rocks. At later stages, the ocean's salinity mainly increased, although it fluctuated during the various stages of its history. It is believed that the concentrations of the major cations and anions in the ocean has changed hardly more than two to threefold compared with modern levels [18, 19, 23, 25, 31, 70, 80, 81].

The origin of life (about four billion years ago) and the oxygen production connected with it have changed the chemistry of the oceans and the atmosphere. Very significant changes in the composition of ocean water and the transformation of the oceanic environment from a reductive into an oxidative one are connected with the beginning of oxygen entrance into the atmosphere. According to most estimations, the amount of free oxygen in the Earth atmosphere reached 0.001 parts of its present value (point Urey) about 1.2 billion years ago. About 600 million years ago, the amount of free oxygen was already 0.01 parts of the present one (point Pasteur) [42].

The first prokaryotes originated in aquatic ecosystems during the Archean (3.5 billion—2.6 ± 0.1 billion years ago). It is considered that bacterio-algal communities already existed in that period. Archaebacteria are considered an ancient group of microorganism with a prokaryotic cell organization. They sharply differ from true bacteria (eubacteria) as to physio-biochemical characteristics and include aerobes and anaerobes as well as chemoheterotrophs and chemoautotrophs. To aerobic bacteria belong those whose feeding strategy is based, similarly to what happens during photosynthesis, on the assimilation of CO_2 following oxidation of other inorganic compounds. To anaerobic bacteria belong organisms that are able to reduce sulphates, CO_2 and other compounds [17, 81].

During the Proterozoic (2.6 billion years—650–680 $\pm$ 20 million years ago), corresponding to the second eon of Earth history, cyanobacteria underwent mass development and the first unicellular eukaryotes, later followed by multicellular ones, appeared. Organisms not possessing a nuclear membrane are classified as prokaryotes. All bacteria, including archaebacteria and cyanobacteria, belong to this type of organisms. Prokaryotes lived in reducing conditions, were heterotrophic and used organic molecules of abiogenic origin for their vital activity. Eukaryotes originated in the Proterozoic and already possessed a nucleus. All unicellular and multicellular algae, fungi, protozoans as well as all metazoans and plants belong to eukaryotes [17, 81].

In the Proterozoic, at the end of which the earliest representatives of multicellular organisms appeared, ocean waters had already become slightly alkaline. Marine skeletal fauna using calcium carbonate appeared at the boundary between the proterozoic and palaeozoic (570 ± 20 million years—230 ± 10 million years ago). Selective uptake of calcium from water affected the previously existing cation balance and the ratio between the monovalent ions Na^+ and K^+ and the divalent ions Ca^{2+} and Mg^{2+} gradually approached the modern one [18, 19, 25, 80].

In the Phanerozoic, quantitative and qualitative changes in continental run-off had reached such an extent that they started to have a significant influence on the hydrochemical conditions of marginal and inner seas [25, 31, 80].

In the Palaeozoic, the atmosphere and hydrosphere still had a reducing character, but the water reaction was changing from slightly acidic to weakly alkaline. Wind waves and flows realized the mixing of the surface waters. Because of differences in temperature and total dissolved solids content of surface and deep waters, a more or less marked pycnocline could form in marginal and intercontinental seas. Dissolved oxygen, necessary for the majority of modern hydrobionts, was absent from the whole water column. The influence of ultraviolet radiation on the surface water layer was more powerful than in the present time [23, 25, 31, 80, 81].

"Calcium chloride seas" existed in the period of life origin. Marine skeletal fauna was able to live both in ancient "calcium chloride" and in the following "magnesium chloride" seas; the latter being also widespread in the present time. According to many scientists, warm, shallow, acid seas were the first environment to support life with organisms inhabiting the bottom and the near-bottom water layers [25, 80].

The age of the oldest microscopic forms of life is estimated to be at least 3.8 billion years. Primitive eukaryotes apparently lived 3.2 billion years ago. The age of the first photosynthesizing organisms exceeds 3 billion years. The origin of aerobic respiration dates back to 1.75 billion years. The chord rudiments were found in organisms that appeared 0.6 billion years ago and the first vertebrates appeared 0.45 billion years ago [17, 81]. It is commonly believed that marine fauna has not undergone big and important changes from the Cambrian (570 ± 20 million years—490 ± 15 million years ago) until our time [81].

In modern aquatic ecosystems, conditions for the life of photosynthesizing bacteria in the absence of oxygen exist within silty sediments found in the transition zones between oxidative and reductive layers where light of low intensity still penetrates. In natural ecosystems, photosynthesizing sulphur-reducing bacteria (also known as thiobacteria) usually account for 3–6% of the total annual photosynthetic production. However, this figure increases to 25% of the annual photosynthesis in ecosystems rich in hydrogen sulphide (H_2S) [72, 91, 101].

Non-sulphur photosynthesizing bacteria are usually facultative anaerobes and can survive in both presence and absence of oxygen. In the absence of light, they can behave as heterotrophs in the same way as many algae. Bacterial photosynthesis becomes significant in the eutrophicated and polluted waters of MPs.

Chemosynthetic and photosynthetic bacteria are ancient forms of life and occupy an intermediate position between autotrophs and heterotrophs. Chemosynthetic bacteria do not obtain the energy required to fix CO_2 in cells' using photosynthesis, but via the oxidation of simple chemical compounds. In these processes, ammonium oxidizes to nitrite, nitrite to nitrate, sulphide to sulphur and iron (II) to iron (III). Chemosynthetic bacteria can live in darkness, but the majority of them needs some oxygen.

Accumulation of high amounts of OM in sediments, accompanied by microbial destruction, could even lead to an increase of silt temperature compared with the overlying water layer [72].

Modern inhabitants of sediments contaminated by H_2S and of the near-bottom waters are capable of performing respiration in the dark and under anoxic conditions. In this way, in MPs, they fulfil ecosystems functions connected with the partial preservation and turnover of matter and energy; in tandem with aerobic organisms.

In the functioning of ancient and modern aquatic ecosystems, a huge role belongs to protozoans forming a subkingdom including unicellular eukaryotic organisms belonging to five–nine types. These organisms do not have a common architecture and present great differences among them. Asexual reproduction via different ways of scission is common among them along with various forms of sexual reproduction. In relation to MPs saprobiotic conditions, it is important to note that many protozoans are able to encyst and outlive the most critical situations; quickly recovering at amelioration of the ecological situation [17, 72, 73, 81].

At different stages of the World Ocean evolution, specific biotic communities, including producers, consumers and reducers (from the small-sized microorganisms up to the large-sized invertebrates and fish) were forming depending on the peculiarities and complexity of organisms. It is important to consider that archaebacteria, photosynthesizing bacteria, chemobacteria, microalgae and protozoans have been preserved in the composition of natural ecosystems during nearly 4 billion years. In modern seas, we can thus observe functional elements of ancient ecosystems that have been preserved until today. It is possible to consider that marine ecosystems have evolved towards progressively more complex structures, including into biotic turnovers more highly organized and more specialized groups of hydrobionts.

Ecosystems' reaction to excessive anthropogenic impact and, first of all, to eutrophication points to the validity of this concept. In such ecosystems, the large-sized consumers of upper trophic levels are the first to disappear, while bacteria, algae, microorganisms, protozoans and the smallest-sized multicellular animals persist.

Excessive eutrophication and pollution of MPs aquatories can recreate conditions characteristic of the coastal marine ecosystems of 2–3 billions years ago. The evolution of these ecosystems apparently went towards a reduction of waters' trophicity via an increase of the diversity and specialization of heterotrophs. With the help of such organisms, the utilization efficiency of all food resources increased and biogenic substances were kept in the biomass of the big-sized, long-living heterotrophs for a long time. Many of them were migrating species, coming from ecosystems with a lower degree of trophicity.

Reduction of biological diversity in MPs, i.e. simplification of biota owing to the presence of much bigger and highly organized heterotrophs, is usually considered the result of anthropogenic activity. In such situations, ancient mechanisms of substances turnover realized by microheterotrophs in conditions of hypoxia, anoxia, deficit or lack of light start to play a greater role in ecosystems' functioning. Such organisms form the first level (step) of the turnover loop and can function in oxidizing as well as in reducing conditions.

It is obvious that biotic components cannot completely disappear in the eutrophicated ecosystems of MPs aquatories. However, in such ecosystems, the major ecological functions will be performed by the most ancient and smallest organisms, represented by a small number of the most universal species.

In natural aquatic ecosystems free from periodical stress impacts, the populations' abundances and the structure of biotic components are regulated by either resource limitation or predatory pressure. Two types of control mechanisms can be distinguished: from "below" (bottom-up) and from "above" (top-down) [7, 9, 10].

In the ecosystems of many MPs aquatories, the regulation from "below", i.e. from the lower trophic levels, is quite often realized by an excess of BM, primary production and dead OM, rather than by their deficit.

In aquatic ecosystems, phosphorus compounds are not only nutritional elements for plants, but are present in all living OM and participate in the synthesis and transformation of OM as well as in cells' and tissues' metabolism. Phosphatides (phospholipids) and phosphoproteins are the most widespread phosphorus compounds. Phosphatides enter into the composition of cellular and subcellular structures of plants, animals and microorganisms. Phosphoproteins participate in the regulation of nucleus activity, mitochondrial oxidative processes and ion transport in cells. The process of ADP phospholyration leading to the formation of ATP has the most important role in the energetic balance of living cells. This process can take place in both anaerobic and aerobic conditions, suggesting its extremely ancient origin [17].

Inside biological systems, the energy flow is transferred with the help of macroergic phosphates. Binding and removal of phosphate-ions also serves for the transmission and accumulation of information so that, in nature, phosphorus carries both energy and information [23].

As already mentioned, MPs aquatories ecosystems function on the basis of the important general principles of self-organization and self-regulation, in the same way as other aquatic ecosystems. Processes occurring in MPs ecosystems have different temporal and spatial scales. The main principles of ecology concerning the close relations and interactions between separate functional components of ecosystems fully apply to all ecosystems of MPs aquatories [23, 73, 81, 84, 85, 118].

The basis for the functioning of ecosystems, both natural and anthropogenically modified ones such as MPs, are the flows of matter, energy and information. These flows strengthen or weaken owing to the activity of living organisms [23, 78, 79, 108].

Regulation of matter, energy and information flows belong to the main intra-system mechanisms of control. These mechanisms are the concentration of BM and partial grazing for autotrophs and grazing by relatively large-sized hydrobionts and fish for the small-sized heterotrophs. Viruses and parasites play an important role in the processes of ecosystems' control at all levels of organization of living matter.

Viruses are widespread in nature with some of them affecting animals and some others affecting plants. Viruses are non-cellular forms of life which are able to selectively penetrate into living cells and which can reproduce only inside living cells. They are intracellular genetic parasites and are able to infect all groups of living organisms. In natural conditions, viruses spread with the help of different vectors or with water movement. Viruses exist in resting or extracellular forms (viral particles or virions) and endocellular or reproductive forms. Virion penetration into the cell starts with its adsorption onto the surface of the cell membrane. All active processes of viruses occur inside the host cells. Some viruses are able to reproduce inside the nucleus and other cell structures, others exclusively in the cytoplasm and a third group both inside the nucleus and in the cytoplasm. Some viruses quit cells simultaneously, thus causing their destruction, while others quit cells gradually without destroying the infected cells. Viruses can cause mass mortality of unicellular and multicellular organisms, affecting whole populations and disrupting flows of matter and energy [63, 81, 104].

In aquatic ecosystems, internal flows of matter and energy at the primary autotrophic level are mainly regulated by water motion, illumination, concentration of CO_2 and biogenic matter in the photic zone, O_2 concentration and pH value. The large-sized mass predatory zooplankton forms and fish control ecosystems by grazing autotrophic organisms and small-sized heterotrophs. In MPs ecosystems, the settlements of barnacles and bivalves on HTS lateral surfaces play a similar but much wider function.

Besides viruses, true endo- and ectoparasites also exert a regulating influence on intra-ecosystem biotic processes. In MPs, a special regulating role belongs to moulds and fungi [122]. In natural conditions, various species of hydrobionts inhabiting MPs and belonging to various systematic groups are also prone to various illnesses. In semiclosed environments affected by human activity, the appearance of various illnesses becomes more likely one of the possible causes of disruption in the flows of matter, energy and information.

Odum [72] underlines that, in relatively big ecosystems, the interaction of matter and energy turnover flows creates conditions of homeostasis that do not require external inputs except for solar radiation and wind energy. However, such genuine, reliable homeostatic control can be formed only after relatively long periods of adaptation. MPs aquatories ecosystems constantly experience various anthropogenic impacts, quite often having contrasting effects. As a consequence, homeostatic controls, i.e. specifically ordered and balanced flows of matter and energy, do not usually form within them.

In protected aquatories such as those of MPs, the possibilities for a wide distribution of viruses' increase because of the high density of transmission vectors and hosts. Discharge of various polluted waters from land also favours the proliferation of viruses in MPs aquatories. In eutrophic MPs ecosystems, viruses create a "short circuit effect" when striking populations of various organisms. In such cases, a big mass of specifically organized living matter, taking active part in the flows of matter and energy, quickly passes into the pool of dead OM. The accumulation of dead OM opens, in turn, the possibility for mass development of bacteria, moulds and fungi. Viruses reduce populations' density, lead to temporary imbalance of ecosystems and kick-off mechanism that disrupts homeostasis. MPs ecosystems thus undergo stages of "rejuvenation".

Within some limits, natural ecosystems are able to withstand natural and anthropogenic influences (i.e. atypical flows of matter, energy and information over a certain time) with only weak, albeit measurable, disturbances. However, when the above-mentioned influences exceed a certain threshold, ecosystems quickly react by changing their biotic structure and passing into a qualitatively new state. In the scientific literature, there are numerous examples of such qualitative reorganizations becoming irreversible. Weakening or complete cessation of an impact does not therefore necessarily lead to the restoration of the preexisting state of complex ecosystems. The original parameters can only be approximately reapproached during processes of ecological succession occurring over dozens or hundreds of years. Over such time scales, ecosystems are very likely to experience new, additional disturbances.

The resilience of aquatic ecosystems usually indicates their ability to withstand changes caused by external factors and return to their initial state (i.e. preserve their structure and functional peculiarities) after cessation of an impact [44].

After natural or anthropogenic impacts, many ecosystems cannot return to their original structure because part of the species, or their biotopes, have been directly destroyed by such impacts. At sharp changes of living versus dead OM ratio, part of the latter can be buried in bottom sediments. Disappearance of forage species can be a cause of mortality and cessation of mass fish migrations.

Upon a change in their structure linked to any reason, ecosystems usually continue to function as one unit via trophic webs and reach a new steady state, i.e. they restart homeostasis.

It is supposed that a couple of equilibrium states are possible for ecosystems so that, after a stress, they do not return to the original equilibrium state, but move a new one [140].

Communities and aquatic ecosystems with higher diversity and dominated by stenobiontic species show the highest stability [3, 4, 6, 7, 9, 10, 23, 81, 85].

In eutrophic ecosystems, to which MPs aquatories also belong, a natural reduction of biodiversity takes place during certain stages. In general, the structure of communities and ecosystems simplifies, while the amplitude of seasonal fluctuations of their functional characteristics, including variability of interannual biomass dynamics, increases.

In MPs aquatories ecosystems, the conservation of matter, energy and information flows at some average level ensures their constant and stable functioning. Changing in even one of the main energy flows must be balanced by a weakening or strengthening of the other flows. For limiting the increase in primary production, it is necessary to limit (or to eliminate) BM inputs, reduce illumination and intensify hydrodynamics. The entrance of dead and live OM from outside requires strengthening of hydrodynamics and increasing water oxygen saturation by mechanical means. In the end, each MP ecosystem requires its own management model.

Conclusion

The construction of hydrotechnical structures, connected with transformation and reinforcement of shores, is widespread along the coasts of Black and Azov Seas. During the 1960–1980s, a stretch of more than 12 km of shore was anthropogenically changed by the construction of groynes, traverses, breakwaters and sand wash in the Odessa region. About 600 groynes were built on the South Coast of Crimea, along a zone of more than 30 km, between the 1960s and the 1990s. Analogous works, though on a smaller scale, were carried in the region of Zhelezny port (NWBS) and at Berdyansk Spit (Azov Sea). In any case, the construction and functioning of HTS in MPs and their ACs lead to very strong influences on the coastal marine ecosystems. These consequences could be immediately visible, but various processes also take place on the bottom.

Thus, the bottom layer southward of Sochi MP was depleted of drifts layer over a zone up to 400 m width a couple of years after its construction (1937–1938) [1]. A similar situation developed eastward of Sukhoy Liman at the end of the 1960s, after construction of Ilyichevsk MP and its AC. At a depth of 5–6 m, the sand cover carried by drifts, disappeared and native clays became exposed. Construction of MPs HTS affects and redistributes the longshore drifts flow, leading to drifts accumulation at some zones and to shore destruction in others. Yalta MP, located in the central part of Yalta city beach zone, disturbed the drifts movement and their differentiation causing the disappearance of part of the beaches [2]. The construction of Yuzhny and Ilyichevsk MPs and of the Neftyannoy (Oil) mole and AC to Oil harbour in Odessa MP increased the destruction of the shores located eastwards and westwards of them. Such processes are registered on both flat and steep shores and have various influences on coastal biocenoses.

All MPs are located in the "land–sea" or "land–river–sea" contact zones and consist of aquatories, external and internal HTS and ACs with the same depths as the aquatories. Thus, MPs are specific ecosystems that comprise natural and artificial components in various combinations.

In fact, HTS and ACs to MPs become constant positive and negative elements of the underwater landscape over long times. In MPs, moles, breakwaters and other HTS constitute HS for hydrobionts and accomplish AR functions from a

A.K. Vinogradov et al., *Ecology of Marine Ports of the Black and Azov Sea Basin*, DOI 10.1007/978-3-319-63062-5 hydroecological point of view. The hydrobionts' living conditions are different on the internal and external surfaces of HTS.

In the MPs of the Black and Azov Seas northern coast, the relatively high depths of 10–20 m are directly brought in the "land–sea" contact zone. Such depths, usually encountered at distances of 1–3 km from the shore, are brought directly in the "land–sea" contact zone where, with the exception of Crimean shores, their gradual increase is observed. Their presence within semi-closed MPs aquatories creates the preconditions for thermal and density vertical stratification of the water masses. In MPs aquatories, conditions typical for the edge of rocky steep shore are artificially created, but with lowered hydrodynamics and water exchange. These conditions are characteristic of natural ecosystems at much greater depths and affect hydrobionts distribution.

Pelagial, periphytal and benthal subsystems can be distinguished in all MPs ecosystems. In port aquatories, these subsystems function in conditions that differ, to a different extent, from those found in the adjacent sea zones. In MPs, the presence of a stable pycnocline leads to the formation of epi- and bathyplankton complexes. Furthermore, the periodical rise up of the hydrogen sulphide gradient significantly decreases the water volume suitable for survival of the majority of planktonic organisms.

In MPs aquatories, the lowered hydrodynamics ameliorates the living conditions of some microphytes, but, at the same time, is not suitable for the development of the majority of macrophytes. Indeed, in MPs, the coastal shallow waters and their typical biocenoses are destroyed. In MPs, a near-surface macrophytes belt forms on the HTS, although with a very impoverished species composition, instead of the usual near coastal macrophytes belt.

In general, a zonal distribution of foulers is characteristic for MPs HTS. The highest fouling biomass (up to 30–35 kg m^{-2}) forms at a distance of 1–3 m from the water surface on the inner HTS surfaces and at the depth of 3–5 m on their external surfaces. The fouling biomass reduces towards the surface and bottom parts of HTS. The MPs aquatories being most distant from the passage have the worse life conditions for the majority of hydrobionts species.

In the deepwater zones of MPs, the accumulation of dead OM on the bottom is accelerated because of gravitational sedimentation and sedimentation linked to the activity of the plankton and periphyton. At elevated water temperatures, hypoxic and anoxic zones appear and only organisms with special adaptations survive. In MPs, semi-liquid, hydrosulphuric black silt is the most typical bottom biotope. In MPs, saprobiotic conditions often appear at their bottom and in the near-bottom layer. Three groups can be distinguished in the composition of MPs meio- and macrozoobenthos: the first one, connected with the shallow waters up to 2.0–3.0 m; the second one at the depth 3.0–8.0 m and the third inhabiting the depths between 10 and 20 m. Provided that enough oxygen is present, transitional ecotone zones (with width from a couple to dozens of metres) exist between the fouling biocenoses of the MP HTS fouling and the biocenoses of black silt at the bottom. In the deepwater aquatories (10–20 m), macrozoobenthos in the central parts of harbours is, as a rule, poorer than at the HTS foundations. In relatively shallow-watered harbours (3–6 m), with more dense (less watered) sediments, these differences disappear.

In MPs, the distinction between the inhabitants of the pelagial, periphytal and benthal zones is not clear-cut. Many sessile species and active benthos forms have pelagial stages during their development. Indeed, the larvae of benthos species play an important role in plankton composition of MPs. At the onset of critical situations, benthos representatives that are not able to swim, but have adaptations for slow crawling, could leave the benthal zone by moving along the lateral surfaces of HTS.

It is noteworthy that construction of permanent deepwater channels and dredging in such "dying" water basins of estuarine type, such as Sukhoy and Grigorievsky Limans, not only stabilized their hydrological and hydrochemical regimes, but also led to significant enrichment of their biota and "rejuvenation" of their ecosystems despite their transformation into MPs aquatories.

The main general peculiarities of MPs aquatories ecosystems are: 1—reduced hydrodynamics and water exchange; 2—absence of coastal shallow waters and of their biocenoses; 3—substitution of flat shores by steep ones and creation of vertical stratification in the water masses; 4—presence of vast areas of artificial HS; 5—increased production of primary and secondary OM; 6—accumulation of dead OM in the water column and on the bottom; 7—formation of near-bottom hypoxia and anoxia; 8—weakened mechanisms for transferring excess OM and BM outside MPs ecosystems. The presence of two or more passages into MPs favour a more normal functioning of their ecosystems. MPs are sources of larval material for neighbouring ecosystems because of their sheltered conditions and absence of fishery.

In MPs aquatories ecosystems, the mechanisms for the utilization, transformation, mineralization and elimination of the excessive amounts of OM and BM must be sustained by creating conditions ensuring active ventilation of the near-bottom water layer. To preserve biodiversity in MPs aquatories, it is important to leave shore zones of 300–500 m length unaffected by hydrotechnical construction.

Intensification of water exchange between the MPs aquatories and the adjacent sea zones can significantly improve the ecological situation of MPs.

In basins where water renewal occurs more than 15–20 times per annum, water "blooms" do not occur or are very weak and without significant consequences for the processes taking place inside the basin, biological production and water quality [3].

In the presence of one wide passage, the flow usually enters MPs at one shore and goes out at the other. When the aquatory and passage are narrow, the incoming and outgoing flows move one over the other in opposite directions. Wide passages provide better exchange as does the presence of two or more passages located in different zones of an aquatory.

The productivity of a basin depends on its depth and area. The more shallow-watered and bigger is the basin area, the more productive and intense is the process of OM formation and decomposition in the basin. During MPs construction, isolation of local ecosystems and water depth are usually artificially increased, while water exchange with the adjacent sea regions weakens. The increase in depths without a corresponding increase in the water surface area lowers MPs productivity, which is, however, increased by the simultaneous availability of vast areas of artificial HS. The artificially increased isolation and the creation, following dredging, of conditions for the onset of a stable pycnocline lead to the growth of entropy in MPs ecosystems and, consequently, to their reorganization.

Protection of aquatories, division into separate harbours and dredging impact the conditions of the pycnocline formation and Langmuir circulations scales in MPs.

In some harbours, the curving of interior angles, forming between the quays and piers with berths could contribute to decrease areas of stagnant zones in MPs.

In MPs aquatories, special ships equipped with deeply located propellers of big diameter and turning at low speed can be used to promote water mixing and destroy the pycnocline.

The use of air breakwaters is known in navigation practice. Air breakwaters are bottom pipes laid on the bottom and into which compressed air is pumped. The pipelines walls have small openings [4] and, while rising up, the released air bubbles disturb the normal, orbital movement of water particles in wind-driven waves, therefore preventing their formation. The use of similar systems in MPs of the Black–Azov Sea basin could improve their ecological situation. The air bubbles, rising up from the bottom, cause the development of turbulent mixing in the whole water column. In autumn and winter, when the water temperature at the surface is lower than in the near-bottom layer, the air-induced mixing raises the water temperature at the surface, thus preventing formation of an ice cover. On the basis of this principle, waterways are maintained constantly free from ice with the help of pneumatic units. In MPs aquatories, the same installations rise the colder, near-bottom water during the summer months. In this way, they create a turbulent mixing that, first of all, lowers the water temperature at the surface layer and, secondly, increases oxygen content in the near-bottom layer.

Because similar abiotic conditions are created in the MPs of various regions, identical biocenoses form inside MPs. Individual species transported from the biocenosis of one MP into the analogous biocenosis of another MP therefore encounter preconditions suitable for their acclimatization.

Literature

Literatures Introduction

1. Basic biological studies of Odessa marine port (August–December 2001). Final report. GloBallast Programme Monograph Series of Odessa Demonstration Centre, Odessa; 2004. Issue 7. 171 p (in Russian).
2. Vinogradov AK, Bogatova YI, Sinegub IA. Ecosystems of marine ports aquatories of the Black–Azov Sea basin (Introduction to the ecology of marine ports). Odessa: Astroprint; 2012. 528 p (in Russian).
3. Goryachkin YN. Anthropogenic impact on the Black Sea coasts of Crimea. In: Ecological safety of coastal and shelf zones and complex exploitation of shelf resources. Sevastopol; 2010. Issue 23. p. 193–7 (in Russian).
4. Leonov AK. Regional oceanography. Part I. Leningrad: Gydrometeoizdat; 1960. 765 p (in Russian).
5. The Black Sea sailing directions. Moscow: Publishing House of the Ministry of defence of USSR; 1976. 507 p (in Russian).
6. Reva YA. Interannual oscillations of the Black Sea level. Oceanology 1997;32(2):211–9 (in Russian).
7. Zaitsev YP, Aleksandrov BG, Minicheva GG, editors. Biology and ecology of the north-western part of the Black Sea. Kiev: Naukova dumka; 2006. 701 p (in Russian).
8. Slevich SB. Shelf. Development and use. Leningrad: Gydrometeoizdat; 1977. 238 p (in Russian).
9. Hanke H. People, ships, oceans (6000 years of navigation venture). Leningrad: Shipbuilding; 1976. 432 p (in Russian).
10. The ecosystem of Grigorievsky (Maly Adzhalyk) Liman. In: Vinogradov AK, editor. Odessa: Astroprint; 2008. 263 p (in Russian).
11. Shuisky VD. Impact of sea-port constructions on dynamics of connected natural shores within non-tidal seas. In: Wang V, Schafer CT, editors. Island Env. and Coast Development. Nanjing: Univer. Press; 1992. p. 393–404.

Literatures Chapter 1

1. Adobovskiy VV. The negative influence of dredging on marine ecosystems. In: IInd All-Union conference on shelf biology. Kiev: Naukova dumka; 1978. Part 1. p. 13–4 (in Russian).
2. Ambaryan OA, Bryum AN, Ivanov VG. Main sea ports of Ukraine. Odessa: Mayak; 1993. 335 p (in Russian).

A.K. Vinogradov et al., *Ecology of Marine Ports of the Black and Azov Sea Basin*, DOI 10.1007/978-3-319-63062-5

3. Ambroz AI. Fish of the Dnieper, Yuzhny Bug and Dnieper-Bug Liman. Kiev: Publishing house of the Academy of Sciences of Ukrainian SSR; 1956. 405 p (in Russian).
4. Astafurova SA, Kopylov SA, Krizhanovska IM, Svertilov OO, Cherednichenko OP. Interrelationships of ecosystems with the environment and impact on them of the engineering and economic activity. In: Ecological problems of the Black Sea. Odessa: IVAZ; 2008. p. 19–23 (in Ukrainian).
5. Basic biological studies of Odessa marine port (August–December 2001). Final report. GloBallast Programme Monograph Series of Odessa Demonstration Centre, Odessa; 2004. Issue 7. 171 p (in Russian).
6. Bashkirov GS. Problems of hydraulics of shallow stormy sea: manual. Moscow: Reklamburo MMF; 1971. 46 p (in Russian).
7. Vinogradov AK, Bogatova YI, Sinegub IA. Ecosystems of marine ports aquatories of the Black–Azov Sea basin (Introduction to the ecology of marine ports). Odessa: Astroprint; 2012. 528 p (in Russian).
8. All about Ukrainian sea ports. 1999–2000 (Ukrainian Ports): Hand-book. Odessa: Ukrainian Ports; 1999. 400 p (in Russian).
9. Geographical encyclopedic dictionary. In: Treshnikov AF, editor. Moscow: Soviet Encyclopedia; 1983. 528 p (in Russian).
10. Goryachkin YN. Anthropogenic impact on the Black Sea coasts of Crimea. Ecological safety of coastal and shelf zones and complex exploitation of shelf resources. Sevastopol; 2010. Issue 23. p. 193–7 (in Russian).
11. Goryachkin YN, Fomin VV. Wave current parameters in the Kalamita Gulf. In: Ecological safety of coastal and shelf zones and complex exploitation of shelf resources. Sevastopol; 2010. Issue 23. p. 159–65 (in Russian).
12. Yesin NV, Krylenko VV. Elaboration of environmental protection measures and combined nature protection evaluation of sea ports projects (on example of Gelendzhik sea port). In: Studies of the sea coastal zone. Scientific publication. Kiev: Carbon Ltd.;2001. p. 144–51 (in Russian).
13. Yesin NV, Peshkov MV, Ivanov AA, Krylenko VV. Evolution of the shores of Temryuk Bay and problems of their protection. In: Studies of the sea coastal zone. Scientific edition. Kiev: Carbon Ltd.;2001. p. 68–75 (in Russian).
14. Zakutskiy VP, Vinogradov KA. Macrozoobenthos. In: Biology of the north-western part of the Black Sea. Kiev: Naukova dumka; 1967. p. 146–57 (in Russian).
15. Zenkovich VP. Dynamics and morphology of sea coasts. Moscow; Leningrad: Marine transport; 1946. Part 1. 496 p (in Russian).
16. Ingle J. The movement of beach sand. Leningrad: Gydrometeoizdat; 1971. 225 p (in Russian).
17. Klenova MV. Geology of the Sea. Moscow: The State educational-pedagogical publishing house of the Ministry of Public Education of RSFSR; 1948. 495 p (in Russian).
18. The Black Sea sailing directions. Moscow: Publishing House of the Ministry of defence of USSR; 1976. 507 p (in Russian).
19. Minkovskaya RY. Local ecological catastrophes in the zone of Dnieper-Bug mouth. In: Ecological safety of coastal and shelf zones and complex exploitation of shelf resources. Sevastopol; 2010. Issue 23. p. 166–70 (in Russian).
20. Marine geomorphology. In: Zenkovich VP, Popov BA, editors. Terminological manual. Coastal zone: processes, notions, definitions. Moscow: Misl; 1980. 280 p (in Russian).
21. Marine fouling and its control. Moscow: Military publication of the Ministry of defence of USSR; 1957. 502 p (in Russian).
22. Moschenko AV. Role of microscale turbidity in the distribution and variability of benthic animals. Vladivostok: Dalnauka; 2006. 321 p (in Russian).
23. Safyanov GA. The coastal zone of the ocean in XXth century. Moscow: Misl; 1978. 263 p (in Russian).
24. Safyanov GA. Estuaries. Moscow: Misl; 1987. 190 p (in Russian).
25. Zaitsev YP, Aleksandrov BG, Minicheva GG, editors. Biology and ecology of the north-western part of the Black Sea. Kiev: Naukova dumka; 2006. 701 p (in Russian).

26. Selin NI. Sediment influence on settling sedimentation of *Crenomytilus grayanus* larvae. In: Ist All-Union conference on marine biology (Vladivostok, 26–29 September 1977). Vladivostok; 1977. p. 122–3 (in Russian).
27. Simov VG, Dyakov NN, Shevela LA. Inflow of river waters into the Azov Sea. In: Ecological safety of coastal and shelf zones and complex exploitation of shelf resources. Sevastopol; 2010. Issue 23. p. 145–52 (in Russian).
28. Cherednichenko AP, Kopyilov SA, Suchkov IA, Astafurova SA, Goncharov IO, Shapran SD. Interrelationships of hydrodynamic factors, drifts dynamics and formation of sediment fields of marine part of the coastal zone and shelf of the north-western part of the Black Sea. In: The Black Sea ecological problems. Odessa: CNTEPI; 2004. p. 531–35 (in Russian).
29. The Black Sea: Collection. Leningrad: Gydrometeoizdat; 1983. 405 p (in Russian).
30. Exogenous processes of development of depositional coasts in the north-western part of the Black Sea. In: Shuyskiy YD, Vykhovanets GV, editors. Ministry of high and secondary special education of USSR. Odessa State University. Moscow: Nedra; 1989. 198 p (in Russian).
31. The ecosystem of Grigorievsky (Maly Adzhalyk) Liman. In: Vinogradov AK, editor. Odessa: Astroprint; 2008. 263 p (in Russian).
32. Assaf G, Gerard R, Gordon H. Some mechanisms of oceanic mixing revealed in aerial photographs. J Geophys Res. 1971;76(27):6550–72.
33. Shuisky VD. Impact of sea-port constructions on dynamics of connected natural shores within non-tidal seas. In: Wang V, Schafer CT, editors. Island Env. and Coast Development. Nanjing: Univer Press; 1992. p. 393–404.

Literatures Chapter 2

1. Ayzatullin TA, Lebedev VL, Khaylov KM. Ocean. Active surfaces and life. Leningrad: Gidrometeoizdat; 1979. 194 p (in Russian).
2. Ayzatullin TA, Lebedev VL, Khaylov KM. Ocean. Fronts, dispersions, life. Leningrad: Gydrometeoizdat; 1984. 192 p (in Russian).
3. Sevastopol aquatory and coasts: ecosystem processes and services for the human society. In: Pavlova EV, Shadrin NV, editors. Sevastopol: Aquavita; 1999. 289 p (in Russian).
4. Aleev YG. Nekton. Kiev: Naukova Dumka; 1976. 391 p (in Russian).
5. Aleksandrov BG. Hydrobiological bases of management of the state of coastal ecosystems of the Black Sea. Kiev: Naukova dumka; 2008. 343 p (in Russian).
6. Aleksandrov BG, Zaitsev YP, Vorobyeva LV, Ryasintseva NI. Hydrobiological research. Chapter 2. In: Monitoring of biological diversity maintenance of the Ukrainian wetlands. Melitopol: Branta; 1995. p. 108–34 (in Russian).
7. Alimov AF. Seasonal and long-term changes in the biomass of zoobenthos in continental water bodies. Hydrobiol J. 1991;27(2):3–9 (in Russian).
8. Alimov AF. Elements of aquatic ecosystem functioning theory. St. Petersburg: Nauka; 2000. 147 p (in Russian).
9. Alimov AF. Studies on biodiversity in the plankton, benthos and fish communities in ecosystems of freshwater bodies with different productivity. Bull AS USSR. Ser Biol. 2001;1:87–95 (in Russian).
10. Alimov AF. On theory of aquatic ecosystems functioning. In: Actual problems in the rational use of biological resources of water reservoirs. Rybinsk: Rybinsk Publishing House; 2005. p. 18–5 (in Russian).
11. Alimov AF. Changes in the structure of animal communities accompanying eutrophication and pollution of aquatic ecosystems. Proc Acad Sci Gen Biol. 2010;433(2):269–72 (in Russian).

12. Altman EN, Bezborodov AA, Bogatova YI et al. Practical ecology of marine regions. In: Keondzhyana VP, Kudina AM, Terekhina YV, editors. The Black Sea. Kiev: Naukova Dumka; 1990. 252 p (in Russian).
13. Andrianov OR, Belevich RR, Shamraev YI, Burov AM. Statistical evaluation of upwelling near Odessa shores. In: Ecological problems of the Black Sea. Odessa: INVAZ; 2008. p. 8–13 (in Russian).
14. Anninsky BE, Datsyk NA. Biomass and predation of the jelly-fish *Aurelia aurita* L. in the Black Sea during October 2010. Mar Ecol J. 2013;13(1):27–33 (in Russian).
15. Artemchuk NY. Microflora of USSR's seas. Moscow: Nauka; 1981. 192 p (in Russian).
16. Basic biological studies of Odessa marine port (Aug–Dec 2001): final report. GloBallast programme Monograph Series of Odessa demonstration centre. Odessa; 2004. Issue 7. 171 p (in Russian).
17. Bascom W. The disposal of waste in the Ocean. In: Ocean science Moscow: Progress; 1981. p. 352–68 (in Russian).
18. Belyaev VI. Studies on the behaviour of marine communities and ecosystems by methods of mathematical simulation. In: Studies of biological resources and their preservation in southern seas. Kiev: Naukova Dumka; 1977. p. 31–8 (in Russian).
19. Berlinsky NA, Bogatova YI, Bolshakov VN, Garkavaya GP. The quality of aquatic environment and bottom sediments in the port of Odessa (Program «GloBallast»). Ecological problems of the Black Sea. Odessa: CNTEPI ONLA; 2003. p. 65–8 (in Russian).
20. Biology of the north-western part of the Black Sea. In: Vinogradov KA, editor. Kiev: Naukova dumka; 1967. 266 p (in Russian).
21. Khaylov KM, editor. Biochemical trophodynamics in marine coastal ecosystems. Kiev: Naukova Dumka; 1974. 175 p (in Russian).
22. Bogorov VG. Biogeocenoses of the ocean pelagial zone. Program and method of studies of aquatic environment biogeocenoses. In: Biogeocenoses of seas and oceans. Moscow: Nauka; 1970. p. 28–46 (in Russian).
23. Bolgova LV. Changes in the ichthyofauna of Novorossiysk Bay induced by the anthropogenic impact. In: Long-term dynamics of the structure of coastal ecosystems of the Black Sea. Krasnodar; 1984. p. 100–107 (in Russian).
24. Brayko VD. Role of sessile species of infusoria and their swarmers in fouling cenosis. In: Biology of shelf zones of the World Ocean. Abstracts of the 2nd All-Union conference on marine biology. Vladivostok, FESC AS USSR; 1982. Part 1. p. 6–7 (in Russian).
25. Brayko VD, Dolgopolskaya MA. Role of larvae of fouling organisms in Sevastopol Bay zooplankton. In: Questions of commercial fishing exploitation and sanitary and biological regime of the Ukrainian water bodies. Kiev: Naukova dumka; 1970. Part 1. p. 64–5 (in Russian).
26. Bradshaw P, Woods JD. Geophysical turbulence and buoyant flows. In: Turbulence. Moscow: Nauka; 1980. p. 178–99 (in Russian).
27. Bruevich SV. On the stability of ocean water chemical composition in the Proterozoic period. Oceanology 1968;8(6):939–48 (in Russian).
28. Bruevich SV. Salt composition of the World Ocean waters during the post-proterozoic period. Oceanology 1969;9(5):904–5 (in Russian).
29. Bulyon VV. Structure and function of the microbial loop in plankton of lake ecosystems. Inland Water Biol. 2002;2:5–14 (in Russian).
30. Burdak VD. Biology of the Black Sea whiting *Odontogadus merlangus euxinus* (Nordmann): PhD thesis synopsis. Leningrad; 1960. 17 p (in Russian).
31. Burkovsky IV. Ecology of free-living infusoria. Moscow: Publishing House of Moscow State University; 1984. 302 p (in Russian).
32. Burkovsky IV. Marine biogeocenology. Organization of communities and ecosystems. Moscow: Scientific Press KMK; 2006. 285 p (in Russian).

33. Burlakova ZP. Release of dissolved organic metabolites by marine algae (in a coastal zone): PhD thesis synopsis, Sevastopol; 1970. 29 p (in Russian).
34. Valyashko MG. Evolution of ocean water chemical composition. In: The history of the World Ocean. Moscow: Nauka; 1971. p. 97–104 (in Russian).
35. Vdodovich IV. Species diversity and feeding of summer-spawning fish species larvae in the coastal zone of the Black Sea: PhD thesis synopsis, Sevastopol; 2011. 23 p (in Russian).
36. Veil P. Popular oceanography. Leningrad: Gydrometeoizdat;1977:500 p (in Russian).
37. Vinberg GG. The primary production of water basins. Minsk: Publishing House of the Academy of Sciences of Belarusian SSR; 1960. 328 p (in Russian).
38. Vinogradov AK. Ichthyoneuston of the Black Sea (ecological-morphological studies): PhD thesis synopsis. Odessa; 1970. 19 p (in Russian).
39. Vinogradov AK, Khutornoy SA. Ichthyofauna of Odessa region of the north-western part of the Black Sea (biological, ecological, eco-morphological peculiarities). Odessa: Astroprint; 2013. 223 p (in Russian).
40. Vinogradov AK, Bogatova YI, Sinegub IA. Ecosystems of marine ports aquatories of the Black–Azov Sea basin (Introduction to the ecology of marine ports). Odessa: Astroprint; 2012. 528 p (in Russian).
41. Vorobyova LV. Meiobenthos of the Ukrainian shelf of the Black and Azov Seas. Kiev: Naukova dumka; 1999. 300 p (in Russian).
42. Gaevskaya AV. Parasites, diseases and pests of mussels (*Mytilus*, Mytilidae). IV. Viruses. Sevastopol: ECOSI-Hidrofisica; 2007. 96 p (in Russian).
43. Galkovskaya GA. Planktonic rotifers and their role in water basins production: PhD thesis synopsis. Minsk; 1965. 20 p (in Russian).
44. Gevorgiz NS, Kondratiev SI, Lyashenko SV, Ovsyany EI, Romanov AS. Monitoring results of the hydrochemical structure of Sevastopol Bay during the warm period of the year. Ecological safety of coastal and shelf zones and complex exploitation of shelf resources, Sevastopol, 2002. Issue 1 (6). p. 139–56 (in Russian).
45. Gerlach SA. Marine pollution. Diagnosis and therapy. Leningrad: Gydrometeoizdatl 1985. 263 p (in Russian).
46. Gershanovich DE, Elizarov AA, Sapozhnikov VV. Bioproductivity of the ocean. Moscow: Agropromizdat; 1990. 236 p (in Russian).
47. Gilyarov AM. Population ecology. Moscow: Publishing House of Moscow State University; 1990. 191 p (in Russian).
48. Gorbenko YA. Ecology of marine microorganisms of periphyton. Kiev: Naukova Dumka; 1977. 250 p (in Russian).
49. Gorbenko YA. Ecology and practical importance of marine microorganisms. Kiev, Naukova Dumka; 1990. 160 p (in Russian).
50. Goryainova LI. Zooplankton of the north-eastern part of the Black Sea during winter 1967. In: Proceedings of scientific conference for the 50th Anniversary of Novorossiysk biological station. Novorossiysk; 1971. p. 16–7 (in Russian).
51. Goryainova LI. Seasonal and long-term dynamics of zooplankton in the Tsemes river estuary. In: Long-term dynamics of the Black Sea coastal ecosystems structure. Krasnodar; 1984. p. 37–42 (in Russian).
52. Goryainova LI, Gromov VV, Kopycheva LN, Kostuchenko LP, Milovidova NY, Smolyar RI. Impact of pollution of the North Caucasian bays of the Black Sea on hydrobionts. In: Proceedings of scientific conference for the 50th Anniversary of Novorossiysk biological station. Novorossiysk; 1971. p. 119–121 (in Russian).
53. Greze VN. Biology of mass species of bottom invertebrates. In: Problems of marine biology. Kiev: Naukova Dumka; 1971. p. 77–83 (in Russian).
54. Grinbart SB. Zoobenthos of limans of the north-western Black Sea region and adjacent sea zones: synopsis of Doctor of Science thesis. Odessa; 1967. 52 p (in Russian).

55. Gutelmacher BL. Metabolism of plankton as a whole: trophometabolic interaction between zoo- and phytoplankton. In: Proceedings of ZIN AS USSR. Leningrad: Nauka; 1986, vol. 133, 156 p (in Russian).
56. Davitaya FF. Atmosphere and biosphere—past, present and future. Leningrad: Gydrometeoizdat; 1976. 36 p (in Russian).
57. Dediu II. Ecological encyclopedic dictionary. Kishinev: Chief editorial board of Moldavian Soviet Encyclopedia; 1990. 408 p (in Russian).
58. Denisova AI, Nakhshina EP, Novikov BI, Ryabov AK. Bottom sediments in reservoirs and their impact on water quality. Kiev: Naukova Dumka; 1987. 164 p (in Russian).
59. Dolotov VV, Kondratiev SI, Lyashenko SV. Intraannual (seasonal) changes in the total content of biogenic elements and oxygen in different regions of Sevastopol Bay. In: Ecological safety of coastal and shelf zones and complex exploitation of shelf resources, Sevastopol; 2005, Issue 12. p. 167–76 (in Russian).
60. Dreik C, Imbri J, Knaus J, Turekian K. Oceanography/Ocean by itself and for us. Moscow: Progress; 1982. 469 p (in Russian).
61. Duvanin AI, Lipunov VM, Shumilov AV. Processes in the ocean surface layer. In: Marine hydrophysical research, Sevastopol, 1974. № 3 (66). p. 141–53 (in Russian).
62. Eremenko TI. Mechanism of distribution of species composition and biomass of macrophytes of the north-western part of the Black Sea: PhD thesis synopsis. Odessa; 1968. 23 p (in Russian).
63. Jerlov NG. Optical oceanography. Moscow: Mir; 1970. 224 p (in Russian).
64. Erokhin VE. Some peculiarities of the utilization of seawater dissolved organic matter by the crustacean *Tigriopus brevicornis*. In: Questions of commercial fishing exploitation and sanitary and biological regime of the Ukraine water basins. Kiev: Naukova Dumka; 1970. Part 1. p. 112–14 (in Russian).
65. Erokhin VE. The uptake of external organic metabolites of algae by marine invertebrates: PhD thesis synopsis, Sevastopol; 1971. 26 p (in Russian).
66. Life of animals, vol. 1. In: Polyanskiy Yu I, editor. Moscow: Prosveshenie; 1987. 448 p (in Russian).
67. Life of animals, vol. 2. In: Pasternak RK, editor. Moscow: Prosveshenie; 1988. 447 p (in Russian).
68. Zhirkov IA, with contributions from Azovskiy AI, Maksimova OV. Life at the bottom. Bio-geography and bio-ecology. Moscow: Scientific press KMK; 2010. 453 p (in Russian).
69. Zavarzin GA. Lectures in environmental microbiology. Moscow: Nauka; 2003. 348 p (in Russian).
70. Zaika VE, Kiseleva MI, Mikhailova TV, Makkaveeva EB, Sergeeva NG, Povchun AS, Kolesnikova EA, Chukhchin VD. Long-term changes in the Black Sea zoobenthos. Kiev: Naukova Dumka; 1992. 248 p (in Russian).
71. Zaitsev YP. Marine neustonology. Kiev: Naukova Dumka; 1970. 264 p (in Russian).
72. Zaitsev YP. Contemporary state and expected changes in the biology of the north-western part of the Black Sea due to protection and exploitation of its resources. In: Investigations of biological resources and their protection in the southern seas. Kiev: Naukova Dumka; 1977. p. 48–51 (in Russian).
73. Zaitsev YP. The most blue in the world. Black Sea ecological series, vol. 6. New-York: UN publishing house; 1998. 142 p (in Russian).
74. Zaitsev YP. Introduction to the ecology of the Black Sea. Odessa: Even; 2006. 222 p (in Russian).
75. Zaitsev YP, Garkavaya GP, Makarov YN, et al. The ecosystem of the north-western Black Sea under conditions of anthropogenic impact. In: Anthropogenic eutrophication of natural waters: materials of All-Union symposium, Moscow, September 1983. Chernogolovka; 1985. p. 49–72 (in Russian).
76. Zenkevich LA. The Black Sea. Biology of the seas of the USSR. Moscow: Publishing House of the USSR Academy of Sciences; 1963. p. 98–360 (in Russian).

77. Zenkovich VP. Dynamics and morphology of sea coasts. Moscow; Leningrad: Marine transport; 1946. Part 1. 496 p (in Russian).
78. Zenkovich VP. The principles of seashore evolution theory. Moscow: Publishing House of the USSR Academy of Sciences; 1962. 710 p (in Russian).
79. Zernov SA. General hydrobiology. Moscow; Leningrad: Publishing House of the USSR Academy of Sciences; 1949. 587 p (in Russian).
80. Zubov NN. Selected works in oceanology. Moscow: Voennoe Publishing House of Ministry of Defence of USSR; 1955. 547 p (in Russian).
81. Zubov NN. The fundamentals of the World Ocean straits study. Moscow: Geografgiz; 1956. 236 p (in Russian).
82. Ivanov VA, Tuchkovenko YS. Applied mathematical water quality modelling of shelf marine ecosystems. Sevastopol: MGI (Marine Hydrophysical Institute) NAS Ukraine; 2006. 368 p (in Russian).
83. Kalugina-Gutnik AA. Changes in the composition of algal flora of Novorossiysk Bay during the last 40 years and its phytogeographic analysis. In: Hydrobiological studies of the north-western part of the Black Sea. Rostov-on-Don: Publishing House of Rostov-on-Don University. 1973. p. 15–9 (in Russian).
84. Klenova MV. Geology of the Sea. Moscow: The State educational-pedagogical publishing house of the Ministry of Public Education of RSFSR; 1948. 495 p (in Russian).
85. Knipovich NM. Hydrology of seas and brackish waters. Moscow; Leningrad: Pischepromizdat; 1938. 513 p (in Russian).
86. Kovalchuk AA. Free-living plankton infusoria of the Dnieper-Bug Liman and their role in production-destruction processes. Hydrobiol J. 1997; 33(4):44–7 (in Russian).
87. Kovardakov SA. Water movement as an ecological factor in aquatories of coastal cities. In: State, prospects of improvement and use of sea ecological system of the coastal part of Crimea: proceedings of the scientific and practical conference dedicated to the 200-years of Sevastopol city, Sevastopol; 1983. p. 58–60 (in Russian).
88. Kondratiev SI, Dolotov VV, Lyashenko SV. Intraannual (seasonal) changes in total concentration of biogenic elements and oxygen in the waters of Sevastopol Bay. In: Ecological safety of coastal and shelf zones and complex exploitation of shelf resources, Sevastopol; 2003, Issue 8. p. 84–9, (in Russian).
89. Konstantinov AS. General hydrobiology. Moscow: Vysshaya shkola; 1967. 427 p (in Russian).
90. Kopylov AI. Nutrition of aquatic infusoria. Inland Water Biol J Inf Bull. 1977;33:19–33 (in Russian).
91. Kopylov AI. Feeding and production of planktonic infusoria of the Rybinsk reservoir: PhD thesis synopsis, Sevastopol; 1983. 23 p (in Russian).
92. Kopylov AI. Role of heterotrophic nanoflagellates in the functioning of marine and freshwater ecosystems: Synopsis of Doctor of Science thesis. Moscow: Institute of Oceanology of Russian Academy of Science (IO RAS); 2003. 40 p (in Russian).
93. Kopylov AI, Tumantseva NI. Analysis of the contents of the tintinnids food vacuoles and evaluation of their contribution to the consumption of phytoplankton production off the Peru coast. Oceanology 1987;3:464–9 (in Russian).
94. Kopylov AI, Kosolapov DB, Zabotkina EA. Distribution of viruses and their impact on bacterioplankton in eutrophic and mesotrophic reservoirs. Inland Water Biol. 2008;1:49–57 (in Russian).
95. Korobkin VI, Peredelskiy LV. Ecology. Rostov-on-Don: Feniks; 2001. 576 p (in Russian).
96. Kosarev AN, Savenko VS. Exchange processes of gases and salts between ocean and atmosphere. In: Dunanin AI, editor. Ocean interaction with environment. Moscow: Publishing House of Moscow State University; 1983. p. 83–100 (in Russian).
97. Kostyuchenko LP. Ichthyoplankton of harbours of the north-eastern part of the Black Sea. In: Proceedings of scientific conference for the 50th Anniversary of Novorossiysk biological station. Novorossiysk; 1971. p. 90–2 (in Russian).

98. Kurilov AV. Planktonic infusoria of the coastal zone and Limans of the north-western Black Sea: PhD thesis synopsis. Odessa; 2005. 19 p (in Russian).
99. Kustenko NG. Effect of environmental stress factors on the reproduction of diatoms. Kiev: Naukova Dumka; 1991. 156 p (in Russian).
100. La Fond ES. Internal waves. In: Sea. Development of ideas and observations, connected with seas studies. Leningrad: Gydrometeoizdat; 1965. p. 346–88 (in Russian).
101. Lebedev VL. Processes in the upper layer of the ocean. In: Duvanin AI, editor. Interaction of the ocean with the environment. Moscow: Publishing House of Moscow State University; 1983. p. 37–53 (in Russian).
102. Lebedeva MN. Quantitative distribution of heterotrophic bacteria as an indirect index of hydrological phenomena in seas and oceans In: Questions of biooceanography. Kiev: Naukova Dumka; 1967. P. 101–11 (in Russian).
103. Levastu T, Larkins G. Marine commercial species ecosystem. In: Efimov YN, editor. Quantitative evaluation of parameters and regulation of fishing. Moscow: Agropromizdat; 1987. 165 p (in Russian).
104. Leonov AK. Regional oceanography. Part I. Leningrad: Gydrometeoizdat; 1960. 765 p (in Russian).
105. Lomakin PD, Popov MA. Structure and seasonal variability of thermohaline zone in Balaklava Bay and adjacent aquatory of the Black Sea according to data of standard hydrophysical and satellite observations. In: Ecological safety of coastal and shelf zones and complex exploitation of shelf resources, Sevastopol; 2010, Issue 23. p. 171–9 (in Russian).
106. Lomakin PD, Popov MA, Kuftarkova EA, Kovrigin NP. Upwelling manifestation in the hydrophysical and hydrochemical characteristics of the water area of Balaklava Bay. In: Ecological safety of coastal and shelf zones and complex exploitation of shelf resources, Sevastopol, 2010, Issue 23. p. 180–92 (in Russian).
107. Losovskaya GV, Sinegub IA. Some peculiarities of the macrozoobenthos of Odessa port. Hydrobiol J. 2007;43(4):43–8 (in Russian).
108. The Black Sea sailing directions. Moscow: Publishing House of the Ministry of defence of USSR; 1976. 507 p (in Russian).
109. Makarov YN. Distribution of decapods larvae in neuston and plankton of the Ponto-Azov in connection with habitat of adult individuals: PhD thesis synopsis, Odessa; 1976. 25 p (in Russian).
110. MacIntyre F. The top millimetre of the ocean. In: Science about the sea. Moscow: Progress; 1981. p. 195–218 (in Russian).
111. Maksimova AA. Natural cycles: causes of recurrence of natural processes. Leningrad: Nauka; 1989. 236 p (in Russian).
112. Mamaeva NV. Ciliates as a component of the planktonic community: synopsis of Doctor of Science thesis. Leningrad; 1986. 34 p (in Russian).
113. Mamaeva NV. Ciliates and ecology of the Black Sea in the Gelendzhik region. Biol Sci J. 1992;70:137–44 (in Russian).
114. Manuylov VA. Underwater landscapes of the Peter the Great Gulf. Vladivostok: Publishing House of Far Eastern State University; 1990. 168 p (in Russian).
115. Mashtakova GP. Dynamics of phytoplankton biomass and quantity in the north-western and eastern Black Sea: PhD thesis synopsis, Odessa; 1971. 14 p (in Russian).
116. Mileykovsky SA. Types of larval development in marine bottom invertebrates. Distribution, quantity and ecological role of their larvae in plankton: synopsis of Doctor of Science thesis. Moscow; 1973. 42 p (in Russian).
117. Moroz TG. Macrozoobenthos of limans and lower river reaches of the north-western Black Sea. Kiev: Naukova Dumka; 1993. 187 p (in Russian).
118. Moschenko AV. Role of microscale turbidity in the distribution and variability of benthic animals. Vladivostok: Dalnauka; 2006. 321 p (in Russian).

119. Narusevich TF. Influence of phytoplankton on water transparency in the coastal zone of the Black Sea during spring. In: Questions of marine biology: proceedings of the IInd All-Union symposium of young scientists. Sevastopol. Kiev: Naukova Dumka; 1969. p. 84–8 (in Russian).
120. Nesterova DA. The role of microphytes in the complex of neuston organisms. In: Biological problems of southern seas oceanography. Kiev: Naukova Dumka; 1969. p. 108–10 (in Russian).
121. Nesterova DA. Phytoneuston in the western Black Sea. Hydrobiol J. 1980;16(3), p. 26–31 (in Russian).
122. Nesterova DA, Vasilenko LS. Phytoplankton and phytoneuston in the shallow-water bays of the Black Sea. Ecol Sea. 1986;23:24–30 (in Russian).
123. Ovsyany EI, Kemp RB, Repetin LN, Romanov AS. Hydrological and hydrochemical regime of Sevastopol Bay in conditions of anthropogenic influence in 1998–1999. In: Ecological safety of the coastal and shelf zones and complex exploitation of shelf resources, Sevastopol; 2000. p. 79–103 (in Russian).
124. Odum Y. Ecology, vol. 2. Moscow: Mir; 1986. 376 p (in Russian).
125. Ozmidov RV. Energy distribution among movements of ocean water at various scales. In: Proceedings of the academy science of USSR, atmosphere and ocean physics. 1965, vol. 1, № 8. pp. 439–48 (in Russian).
126. Ozmidov RV. Horizontal turbulence and turbulent mixing in the ocean. Moscow: Nauka; 1968. 196 p (in Russian).
127. Orlova IG, Pavlenko NE, Popov YN, Ukrainsky VV, Komorin VN. Results of studies on the hydrological-hydrochemical regime of Odessa port in the framework of the international project «GloBallast». In: Ecological problems of the Black Sea. Odessa: OCNTEI; 2002. p. 156–60, (in Russian).
128. Pavlovskaya TV. Feeding and reproduction of infusoria mass species in the Black Sea: PhD thesis synopsis, Sevastopol; 1971. 20 p (in Russian).
129. Pavlyutina LP, Podpletnaya NF, Savin PT, Sekundyak LY, Skripnik IA, Kirsanova EV, Mihalechko EY. Characteristic of the ecological state of Sukhoy Liman. In: Ecological safety of coastal and shelf zones and complex exploitation of shelf resources, Sevastopol; 2005, Issue 12. p. 120–28 (in Russian).
130. Petipa TS. Feeding and energetic balance of some mass species of Black Sea planktonic copepods belonging to various ecological groups: PhD thesis synopsis, Moscow; 1965. 15 p (in Russian).
131. Petipa TS. Trophic dynamics of copepods in marine planktonic communities. Kiev: Naukova dumka, 1981. 245 p (in Russian).
132. Petipa TS, Pavlova EV, Mironov GN. Energy balance of plankton organisms from various ecosystems of the Black Sea. In: Questions of biogeography. Kiev: Naukova Dumka; 1967. p. 115–20 (in Russian).
133. Petrov KM. Underwater landscapes: theory, research, methods. Leningrad; 1989. 126 p (in Russian).
134. Pivkin MV. Secondary marine fungi of Japan and Okhotsk Seas: synopsis of Doctor of Science thesis, Moscow, 2010. 40 p (in Russian).
135. Piontkovskiy SA. The ecology of the behaviour of copepods. Kiev: Naukova Dumka; 1985. 116 p (in Russian).
136. Piontkovskiy SA, Seregin SA. Copepod behaviour. Sevastopol: ECOSI-Hydrofisica; 2006. 148 p (in Russian).
137. Plakhotnik AF. Ocean-atmosphere interactions. Moscow: Nauka; 1978. 201 p (in Russian).
138. Posokhov EV. Chemical evolution of the hydrosphere. Leningrad: Gydrometeoizdat; 1981. 285 p (in Russian).
139. Protasov AA. The life in the hydrosphere. Essays of general hydrobiology. Kiev: Academperiodica; 2011. 704 p (in Russian).

140. Rialkin AI. Colonization of hard bodies by benthic organisms. St. Petersburg: Publishing house of St. Petersburg State University; 2008. 427 p (in Russian).
141. Reymers NF. The usage of nature: a reference book. Moscow: Misl; 1990. 637 p (in Russian).
142. Ricklefs R. Fundamentals of general ecology. Moscow: Mir; 1979. 419 p (in Russian).
143. Rozhdestvenskiy AV. Role of abiotic factors dynamics in the ecology of the Black Sea. In: Ecology and chemistry of the Black Sea. Summary. 30.09–02.10.1991. IHS «Frederic Joliot-Curie», Druzhba. Varna, 1991. p. 19–20. (in Russian).
144. Rozhdestvenskiy AV. Influence of anthropogenic factors on the hydrology and hydrochemistry of Varna Lake. In: Works of the Institute of Oceanology. Varna: BAS; 1992, vol. 2, p. 48–57 (in Russian).
145. Romanenko VD. Fundamentals of hydroecology. Kiev: Geneza; 2004. 644 p (in Russian).
146. Rudyakov YA, Sapozhnikov VV, Agatova AI. Changes in the nutritional value of decomposing plankton and some peculiarities of nitrogen and phosphorus regeneration process. In: Frontal zones of the south-east part of the Pacific Ocean. Moscow: Nauka; 1984. p. 92–9 (in Russian).
147. Ryanzhin VS, Karlin LN, Kochkov NV. Mysterious circulations. Priroda; 2008, № 4. p. 1–8 (in Russian).
148. Ryasintseva NI, Savin PT, Sarkisova SA, Bogatova YI, Nikanorov VA, Dotsenko SA. Modern state of the ecosystem of Adzhalik Liman. In: Ecological safety of coastal and shelf zones and complex exploitation of shelf resources, Sevastopol; 2000. p. 114–26 (in Russian).
149. Savchuk MY. Fries of the haarder (Mugilidae) of the north-western Black Sea: PhD thesis synopsis, Kishinev; 1970. 27 p (in Russian).
150. Savchuk MY. Distribution and quantity of juvenile of haarder in the Azov Sea and possibility of its economic use. In: Modern state and perspectives of development of haarder breeding in the Azov-Black Sea basin: proceedings of the IIIrd scientific-and-technological conference on grey mullets (haarder) cultivation (23–26 November 1976), Belgorod-Dnestrovsky; 1976. p. 30–2 (in Russian).
151. Salehova LP. Sex inversion, reproduction and development of the annular sea bream *Diplodus annularis* L.: PhD thesis synopsis, Kaliningrad; 1966. 21 p (in Russian).
152. Safronov IN. Geomorphology of the Northern Caucasus. Rostov-on-Don: Publishing House of Rostov University; 1969. 217 p (in Russian).
153. Safyanov GA. The coastal zone of the ocean in XXth century. Moscow: Misl; 1978. 263 p (in Russian).
154. Zaitsev YP, Aleksandrov BG, Minicheva GG, editors. Biology and ecology of the north-western part of the Black Sea. Kiev: Naukova dumka; 2006. 701 p (in Russian).
155. Selifononva ZP. Structure and functioning of the anthropogenically transformed ecosystem of Tuapse Port (Black Sea). In: Proceedings of IVth International conference "Modern problems in hydrobiology" (11–15 October 2010). St. Petersburg; 2010. p. 155 (in Russian).
156. Selifonova ZP. Ecosystems of the aquatories of Novorossiysk and Tuapse ports (Black Sea). St. Petersburg: Nauka; 2012. 227 p (in Russian).
157. Selifonova ZP, Yasakova ON. Phytoplankton of port cities aquatories of the north-eastern shelf of the Black Sea. Mar Ecol J. 2012. 11(4):67–7 (in Russian).
158. Smirnov NN. Biology of Cladocera. In: Zoology of invertebrates. Moscow: VINITI; 1975, vol. 3, 116 p (in Russian).
159. Sorokin YI. Heterotrophic microplankton as a component of marine ecosystems. J Gen Biol. 1975;36(5):716–30 (in Russian).
160. Sorokin YI. Production of the microflora. In: Biology of the ocean, vol. 2. Moscow; 1977. p. 209–33 (in Russian).
161. Sorokin YI. The Black Sea. Moscow: Nauka; 1982. 216 p (in Russian).

162. Starushenko LI, Bushuev SG. Limans in the Odessa region of the Black Sea and their use for fishery. Odessa: Astroprint; 2001. 151 p (in Russian).
163. Stelmakh LV, Kuftarkova EA, Babich II. Phytoplankton growth rate and microzooplankton grazing by the autumn "bloom" of *Emiliania huxlevi* in the western Black Sea. Mar Ecol J. 2013;12(2). p. 51–62 (in Russian).
164. Stepanov VN, Svetlichny LS. Studies on the hydromechanical characteristics of planktonic copepods. Kiev: Naukova Dumka; 1981. 128 p (in Russian).
165. Sushchenya LM. Detritus and its role in the production process in water basins. Hydrobiol J. 1968;4(2). p. 41–7 (in Russian).
166. Tarasov NI. Biology of the sea and the navy. Stenograph of public lecture in Moscow. Moscow: Publishing House "Pravda"; 1949. 30 p (in Russian).
167. Terenko LM. Dinoflagellates trophic structure of the Black Sea coastal zone. In: Proceedings of international scientific practical conference "Dynamics problems of the Azov Sea and Kerch Strait ecosystems". Rostov-on-Don: Publishing House SSC RAS; 2004. p. 92–4 (in Russian).
168. Terenko LM. Dinoflagellates of the north-western Black Sea: species diversity and ecology: PhD thesis synopsis, Sevastopol; 2005. 23 p (in Russian).
169. Thomas-Bourneuf M, Mollo P. The plankton challenge. In: Eremeev VN, editor. NAS Ukraine, Institute of Biology of Southern Seas named after A.O. Kovalevsky. Sevastopol; 2011. 281 p (in Russian).
170. Trenkenshu RP, Avsiyan AL, Novikova TM. The simplest models of microalgal growth. Dark processes. Mar Ecol J. 2013;12(2):69–77 (in Russian).
171. Tuchkovenko YS, Ivanov VA, Sapko OY. Estimation of the influence of coastal anthropogenic sources on the water quality of Odessa region of the north-western part of the Black Sea. Sevastopol: ECOSI-Hidrofisica; 2011. 169 p (in Russian).
172. Tuchkovenko YS, Dotsenko SA, Nikanorov VA, Savin PT. Role of wind-driven coastal upwelling in the origin of hypoxia in Odessa region of the north-western Black Sea. In: Ecology of the Sea. 2003, Issue 63. p. 60–5 (in Russian).
173. Whittaker R. Communities and ecosystems. Moscow: Progress; 1980. 328 p (in Russian).
174. Ukrainskiy VV, Popov YN, Orlova IG. Summer-autumn regime of dissolved oxygen under conditions of eutrophication in the north-western shelf of the Black Sea waters. In: Ecological problems of the Black Sea. Odessa: OCNTEI; 2001. p. 335–6 (in Russian).
175. Umorin PP. Relationships between bacteria and flagellates in the destruction of organic matter. J Gen Biol. 1976;37(6):831–5 (in Russian).
176. Umorin PP. The role of protozoans in the destruction of dissolved organic matter. In: Proceedings of the Institute of internal waters biology, USSR Academy of Sciences. 1978, № 40. p. 137–58 (in Russian).
177. Umorin PP. Interrelationships among algae, bacteria and protozoans in the processes of self-cleaning and formation of water quality. In: Proceedings of the Institute of Internal Waters Biology, USSR Academy of Sciences. 1987. № 54/57. p. 132–48 (in Russian).
178. Fedorov VD, Gilmanov TG. Ecology. Moscow: Publishing House of Moscow State University; 1980. 464 p (in Russian).
179. Finenko ZZ. Primary production of the southern seas. In: Questions of bioceanography. Kiev: Naukova Dumka; 1967. p. 69–74 (in Russian).
180. Khaylov KM. Elements of ecological metabolism in the sea coastal zone: synopsis of Doctor of Science thesis, Moscow, 1969. 44 p (in Russian).
181. Khaylov KM. Ecological metabolism in the sea. Kiev: Naukova dumka; 1971. 252 p (in Russian).
182. Khalilova MR. Long-term changes of the hydrochemical regime of Novorossiysk Bay (1977–1980). In: Long-term dynamics of structure of the Black Sea coastal ecosystems, Krasnodar; 1984. p. 10–6 (in Russian).
183. Khlebovich TV. The role of planktonic infusoria in the biological transformation of matter and energy: PhD thesis synopsis, Leningrad; 1986. 19 p (in Russian).

184. Khutornoy SA. Rare species of the ichthyofauna of Odessa Gulf and adjacent sea aquatories of the Black Sea. In: Proceedings of the scientific conference of students, post graduate students and young scientists for the 180 years from the birth of L.S. Tsenkovsky (Odessa, 28 March 1 April 2003), Odessa; 2003. p. 184–94 (in Russian).
185. Tsikhon-Lukanina EA, Soldatova IN, Nikolaeva GG. Assimilation of organic and mineral substances by marine crustaceans. In: Questions of marine biology: Proceedings of the IInd All-Union symposium of young scientists, Sevastopol; 1969. Kiev: Naukova Dumka, 1969. p. 140, (in Russian).
186. Tsyban AV. Bacterioneuston and bacterioplankton of the shelf zone of the Black Sea. Kiev: Naukova Dumka; 1970. 272 p (in Russian).
187. Chepurnova EA. The characteristics of detritus at various stages of its mineralization. In: Questions of marine biology: proceedings of the IInd All-Union symposium of young scientists, Sevastopol; 1969. Kiev: Naukova Dumka, 1969. p. 142–43 (in Russian).
188. Cherednichenko AP, Kopyilov SA, Suchkov IA, Astafurova SA, Goncharov IO, Shapran SD. Interrelationships of hydrodynamic factors, drifts dynamics and formation of sediment fields of marine part of the coastal zone and shelf of the north-western part of the Black Sea. In: The Black Sea ecological problems. Odessa: CNTEPI; 2004. p. 531–35 (in Russian).
189. The Black Sea. Collected volume. Varna: Bakalov G. Publishing House; 1978. 635 p (in Bulgarian).
190. Chilikina NS. The biological activity of sea foam. Problems of the World Ocean. In: Proceedings of the conference of young scientists Moscow State University (MSU). Moscow: Publishing House of Moscow State University; 1970. p. 205–10 (in Russian).
191. Shubakov IV, Shubakov DI, Buskov IP, et al. Oceanography. Odessa: Abrikos; 2003. 332 p (in Russian).
192. Shuyskiy YD. Shore types of the World Ocean. Odessa: Astroprint; 2000. 480 p (in Ukrainian).
193. Shuleykin VV. Physics of the sea. Moscow: Publishing House of AN USSR; 1962. 470 p (in Russian).
194. Shtereva GP, Dimitrov AI. Въерху някой характеристики на органичното вещество във Варненско и Белославско езеро през есенний сезон. In: Тр. на Института по океанология. Varna: BAS; 1998, vol. 2, p. 44–8 (in Bulgarian).
195. Shtereva GP, Khristova OD, Nikolova TS, Dzhurova BS. Въерху състояннето на Белославского езеро през периода 1990–1998 г. In: Тр. на Института по океанология. Varna: BAS; 2001, vol. 3, p. 23–8 (in Bulgarian).
196. The ecosystem of Grigorievsky (Maly Adzhalyk) Liman. In: Vinogradov AK, editor. Odessa: Astroprint; 2008. 263 p (in Russian).
197. Yasakova ON. Seasonal dynamics of the phytoplankton of Novorossiysk Bay in 2007. Mar Ecol J. 2013;12(1):92–102 (in Russian).
198. Alimov AF. Towards a theory of functioning of aquatic ecosystems. Leiden: Backhuys Publ.; 2003. 130 p.
199. Analysis of Marine Ecosystems. Lonchurst AR, editor. London; New York; Toronto; Sydney; San Francisco: Academic Press, 1981; 741 p.
200. Assaf G, Gerard R, Gordon H. Some mechanisms of oceanic mixing revealed in aerial photographs. J Geophys Res. 1971;76(27):6550–72.
201. Brooks JL, Dodson SI. Predation, body size and composition of plankton. Science 1965;150:28–35.
202. Clements FE. Plant succession; an analysis of the development of vegetation. Carnegie Inst. of Washington Publ. 1916. № 242. 512 p.
203. Davis PG, Caron DA, Jonson PW, Sieburth LM. Phototrophic and apochlorotic components of picoplankton and nanoplankton in the North Atlantic: geographic, vertical, seasonal and diel distributions. Mar Ecol Prog Ser. 1985;21:15–26.

204. Ducklow HW, Carlson CA. Oceanic bacterial production. Adv. Microb. Ecol. 1992;12:113–81.
205. Ducklow HW, Purdie DA, Williams PI, Davies JM. Bacterioplankton: a sink for carbon in a coastal marine plankton community. Science 1986;232:865–8.
206. Fenchel TM. Suspension feeding in ciliated Protozoa: functional response and particle size selection. Microb Ecol. 1980;6(1):1–11.
207. Fowler SW, Small LF. Sinking rates of euphausiid faecal pellets. Limnol Oceanogr. 1972;17(2):293–6.
208. Gamenick JA, Jahn A, Vopel K, Giere O. Hypoxia and sulphide as structuring factors in a macrozoobenthic community on the Baltic Sea shore: colonization studies and tolerance experiments. Mar Ecol Prog Ser. 1996;144:73–85.
209. George DG, Edwards RW. *Daphnia* distribution within Langmuir circulation. Limnol Oceanogr. 1973;18(5):798–800.
210. Harrison WG. Nutrient regeneration and primary production in the Sea. In: Falkowski PG, editor. Environmental science research. New York, London: Plenum Press; 1980, vol. 19, 433–60.
211. Hutchinson GE. An introduction to population ecology. New Haven: Yale University Press; 1978. p. 256.
212. Langmuir J. Surface motion of water induced by wind. Science 1938;87(2250):119–23.
213. Lindeman RL. The trophic-dynamic aspect of ecology. Ecology 1942;23:399–417.
214. McIntyre AD, Johnston R. Effects of nutrient enrichment from sewage in the Sea In: Gameson A, editor. Discharge of sewage from Sea outfalls. Oxford: Pergamon Press; 1975. p. 131–41.
215. Moncheva S, Dontcheva V, Shtereva G, Kamburska L, Malej A, Gorinstein S. Application of eutrophication indices for assessment of the Bulgarian Black Sea coastal ecosystem ecological quality. Water Sci Technol. 2002;46(8):19–28.
216. Munn CB. Viruses as pathogens of marine organisms from bacteria to whales. Mar Biol Assoc UK 2006;86(3):453–67.
217. Odate T, Imai K. Seasonal variation in chlorophyll-specific growth and microzooplankton grazing of phytoplankton in Japanese coastal water. J Plankton Res. 2003;25:1497–505.
218. Putland JN, Iverson RL. Microzooplankton: major herbivores in an estuarine planktonic food web. Mar Ecol. 2007;345:63–73.
219. Rosenthal H, Hempel G. Experimental studies in feeding and food requirements of herring larvae (*Clupea harengus* L.). In: Steele JH, editor. Marine food chains. Berkeley CA: University of California Press; 1970. p. 344–64.
220. Rounsefell GA. Ecological effects of offshore construction. J Mar Sci. 1971;2:1–208.
221. Shanks AL. Surface slicks associated with tidally forced internal waves may transport pelagic larvae of benthic invertebrates and fishes shoreward. Mar Ecol Prog Ser. 1983;13:311–5.
222. Shelford VE. Ecological succession: pond fishes. Biol Bull. 1911;21:127–51.
223. Shelford VE. Animal communities in temperate America. Chicago: University of Chicago Press; 1913. p. 438.
224. Sherr EB, Sherr BH. Heterotrophic dinoflagellates: a significant component of microzooplankton biomass and major grazers of diatoms in the sea. Mar Ecol. 2007;352:187–97.
225. Shtereva G, Dzhurova B. Hydrochemistry of the Beloslav Lake-Varna Lake-Varna Bay system during spring. In: Proceedings of the 8th international conference "Black Sea 2006", September 2006 Varna; 2006. p. 249–55.
226. Shtereva G, Dzhurova B. Water quality of Varna Lake (Bulgaria). In: Proceedings of the international conference BALWOIS 2010 Ohrid, Republic of Macedonia, 25–29 May 2010. Ohrid; 2010. p. 1–6.
227. Sommer U, Stibor H, Katechakis A, Sommer F, Hansen T. Pelagic food web configurations at different levels of nutrient richness and their implications for the ratio fish production: primary production. Hydrobiologia 2002;484:11–20.

228. Stavn RH. The horizontal-vertical distribution hypothesis: Langmuir circulations. Limnol Oceanogr. 1971;16(2):453–66.
229. Stavn RH. The horizontal-vertical distribution hypothesis: Langmuir circulations. Limnol Oceanogr. 1971;16(2):453–66.
230. Suttle CA. Viruses in the sea. Nat London. 2005;437:359–61.
231. Thienemann A. Grundzuege einer allgemeinen Oekologie. Arch Hydrobiol. 1939;35:267–85.
232. Thode H. Comparative studies on the influence of oxygen deficiency and hydrogen sulphide on marine bottom invertebrates. Neth J Sea Res. 1973;7:244–52.
233. Urban JL, McKenzie CH, Deibel D. Nanoplankton found in faecal pellets of microzooplankton in coastal Newfounland waters. Bot Mar. 1993;36(4):267–81.
234. Weller RA, Price JF. Langmuir circulation within the oceanic mixed layer. Deep Sea Res. 1988;35(5A):711–47.
235. Wiebe PH. Small-scale spatial distribution in oceanic zooplankton. Limnol Oceanogr. 1970;15(2):205–17.

Literatures Chapter 3

1. Ayzatullin TA, Lebedev VL, Khaylov KM. Ocean. Active surfaces and life. Leningrad: Gidrometeoizdat; 1979. 194 p (in Russian).
2. Aleksandrov BG. Ecological consequences of anthropogenic transformation of the coastal zone of the Black Sea. Kiev: Carbon-LTD; 2001. p. 25–34 (in Russian).
3. Aleksandrov BG. Problem of aquatic organisms' transportation by ships and some approaches for risk assessment of new invasions. Mar Ecol J. 2004;3(1):5–17 (in Russian).
4. Aleksandrov BG. Hydrobiological bases of management of the state of coastal ecosystems of the Black Sea. Kiev: Naukova dumka; 2008. 343 p (in Russian).
5. Aleksandrov BG, Buzovkina TB, Egorov ES, Yurchenko AA. Influence of hydrophysical factors on initial stages of marine fouling and corrosion. In: Ships protection of from corrosion and fouling. Proceedings of the 3rd inter-sectoral scientific and technical conference. Kaliningrad; 1986. p. 46–7 (in Russian).
6. Alekseev RP, Zaitsev YP, Ivanova EV, Synegub IA. Formation of zoocenosis of various artificial reefs of the north-western part of the Black Sea. In: Artificial reefs for fisheries. Proceedings of the All-Union conference (Moscow, 2–4 December 1987). Moscow; 1987. p. 57–9 (in Russian).
7. Altman EN, Bezborodov AA, Bogatova YI, et al. Practical ecology of marine regions. In: Keondzhyana VP, Kudina AM, Terekhina YV, editors. The Black Sea. Kiev: Naukova dumka; 1990. 252 p (in Russian).
8. Andreyuk EI, Kozlova IA. Microbiological aspects of metal corrosion. Microbiol J. 1981;43(2):139–45 (in Russian).
9. Basic biological studies of Odessa marine port (Aug–Dec 2001): final report. GloBallast programme Monograph Series of Odessa demonstration centre. Odessa; 2004. Issue 7. 171 p (in Russian).
10. Bening AL. On the study of bottom life of the river Volga: monograph of the Volga biological stations. 1924;1:210 p (in Russian).
11. Biology of the north-western part of the Black Sea. In: Vinogradov KA, editor. Kiev: Naukova dumka; 1967. 266 p (in Russian).
12. Bocharov BV, Zevina GB, Ilyin IN, Kolbasov GA, Kruchkov GI, Negashev SE. Fouling in the southern part of Vietnam. In: Biodeterioration, fouling and protection against it. Moscow: Nauka; 1996. p. 37–9 (in Russian).
13. Brayko VD. The role of substratum in the sedimentation of bryozoan larvae. Biooceanography questions. Kiev: Naukova dumka; 1967. p. 75–83 (in Russian).

14. Brayko VD. Biological and faunistic studies of bryozoans of the Black Sea and their role in the fouling: PhD thesis synopsis. Sevastopol; 1973. 20 p (in Russian).
15. Brayko VD. Periphyton of the Black Sea. Kiev: Naukova dumka; 1985. 123 p (in Russian).
16. Vinogradov AK, Khutornoy SA. Ichthyofauna of Odessa region of the north-western part of the Black Sea (biological, ecological, eco-morphological peculiarities). Odessa: Astroprint; 2013. 223 p (in Russian).
17. Vinogradov AK, Bogatova YI, Sinegub IA. Ecosystems of marine ports aquatories of the Black–Azov Sea basin (Introduction to the ecology of marine ports). Odessa: Astroprint; 2012. 528 p (in Russian).
18. Vinogradov KA, Losovskaya GV. Class bristle worms—Polychaeta. In: Manual for the identification of the Black and Azov Seas fauna, vol. 1. Kiev: Naukova dumka; 1968. p. 215–359 (in Russian).
19. Vorobyova LV. Dynamics of colonization by meiofauna of substratum of artificial and natural origin in the marine environment. Hydrobiol J. 1996;32(6):15–20 (in Russian).
20. Galkina VN. Role of mussels metabolites in the enrichment of coastal waters by dissolved organic compounds. In: All-Union conference on shelf biology. Kiev: Naukova dumka; 1978. Part 1. p. 28–9 (in Russian).
21. Galkina VN. Role of animal mass species in turnover of organic substances in the coastal waters of the northern seas: PhD thesis synopsis. Leningrad; 1985. 23 p (in Russian).
22. Galkina VN, Kulakovsky EE, Kunin BL. The environmental influence of mussels aquaculture in the White Sea. Oceanology 1982;22(2):321–4 (in Russian).
23. Galtsova VV. Meiobenthos of marine ecosystems as exemplified by free-living nematodes. In: Proceedings of the Zoological Institute, vol. 224, AS USSR; 1991. 236 p (in Russian).
24. Gerlach SA. Marine pollution. Diagnosis and therapy. Leningrad: Gydrometeoizdatl 1985. 263 p (in Russian).
25. Gershanovich DE. Ecological role of artificial reefs in the sea. In: Artificial reefs for fisheries. Abstracts of the All-Union conference (Moscow, 2–4 December 1987). Moscow; 1987. p. 10–3 (in Russian).
26. Gorbenko YA. Ecology of marine microorganisms of periphyton. Kiev: Naukova dumka; 1977. 250 p (in Russian).
27. Gorbenko YA. Ecology and practical importance of marine microorganisms. Kiev: Naukova dumka; 1990. 160 p (in Russian).
28. Gorbenko YA, Benzhitskiy AG. On the production of vitamin B_{12} by periphyton microorganisms in the sea. In: Biology of the sea. Kiev; 1979. Issue 48. p. 5–8 (in Russian).
29. Gorbenko YA, Brayko VD, Bobkova AN. Fouling and its practical aspects. In: Artificial reefs for fisheries. Abstracts of the All-Union conference (Moscow, 2–4 December 1987). Moscow; 1987. p. 48–50 (in Russian).
30. Goromosova SA, Shapiro AZ. On the adaptation mechanisms of zoo-fouling organisms of the Black Sea towards hypoxia and poison effect. In: Proceedings of the All-Union symposium on state of knowledge of the Black and Mediterranean Seas on use and protection of their resources (Sevastopol, October 1973). Kiev: Naukova dumka; 1973. Part 4. p. 20–2 (in Russian).
31. Goryainova LI, Gromov VV, Kopycheva LN, Kostuchenko LP, Milovidova NY, Smolyar RI. Impact of pollution of the North Caucasian bays of the Black Sea on hydrobionts. In: Proceedings of scientific conference for the 50th Anniversary of Novorossiysk biological station. Novorossiysk; 1971. p. 119–121 (in Russian).
32. Greze II. On the biology of the freshwater shrimp *Ampelisca diadema* (A. Costa) in the Black Sea. In: Benthos. Kiev: Naukova dumka; 1965. p. 3–8 (in Russian).
33. Grinbart SB. Fouling of wooden and stone structures in Odessa Gulf. In: Proceedings of Odessa State University, vol. 3. Series Biology; 1938. p. 53–7 (in Ukrainian).
34. Grinbart SB. Zoobenthos of Odessa Gulf. In: Proceedings of Odessa State University, vol. 4; 1949. p. 51–73 (in Russian).

35. Gromov VV. Marine aquatic vegetation of Novorossiysk Bay in the upper sublittoral zone. In: Proceedings of the scientific conference for the 50th Anniversary of Novorossiysk biological station. Novorossiysk; 1971. p. 52–54 (in Russian).
36. Evstigneeva IK, Tankovskaya IN. Macroalgae of periphyton and benthos of Laspi Bay coastal zone (Crimea, the Black Sea). Ecol Sea. 2010;81:40–9 (in Russian).
37. Eremenko TI. Mechanism of distribution of species composition and biomass of macrophytes of the north-western part of the Black Sea: PhD thesis synopsis. Odessa; 1968. 23 p (in Russian).
38. Jerlov NG. Optical oceanography. Moscow: Mir; 1970. 224 p (in Russian).
39. Life of animals, vol. 1. In: Polyanskiy Yu I, editor. Moscow: Prosveshenie; 1987. 448 p (in Russian).
40. Zhirkov IA, with contributions from Azovskiy AI, Maksimova OV. Life at the bottom. Bio-geography and bio-ecology. Moscow: Scientific press KMK; 2010. 453 p (in Russian).
41. Zaika VE, Valovaya NA, Povchun AS, Revkov NK. Mytilidae mussel of the Black Sea. Kiev: Naukova dumka; 1990. 208 p (in Russian).
42. Zaitsev YP. Artificial reefs as an instrument of management of ecological processes in the sea coastal zone. In: Artificial reefs for fishery: abstracts of the All-Union conference (Moscow, 2–4 December 1987). Moscow; 1987. p. 3–5 (in Russian).
43. Zakutskiy VP. The shells of *Balanus improvisus* as shelters for other organisms. Zool J. 1965;44(7):1092 (in Russian).
44. Zvyagintsev AY. Marine fouling in the north-western part of the Pacific Ocean. Vladivostok: Dalnauka; 2005. 432 p (in Russian).
45. Zvyagintsev AY, Mikhaylov SP. Fouling communities of marine ships and hydrotechnical constructions in Southern Vietnam. In: Biology of Vietnam coastal waters. Hydrobiological studies of littoral and sublittoral zone of Southern Vietnam. Vladivostok: Far eastern division of the Academy of Science of USSR; 1988. p. 137–58 (in Russian).
46. Zevina GB. Fouling of hydrotechnical constructions of the Caspian Sea. In: *Proceedings* of the *Institute of Oceanology* of the *USSR Academy of Sciences, vol.* 49; 1961. p. 65–96 (in Russian).
47. Zevina GB. Fouling in the seas of the USSR. Moscow: Publishing House of Moscow State University; 1972. 213 p (in Russian).
48. Zevina GB. Biology of marine fouling. Moscow: Publishing House of Moscow State University; 1994. 134 p (in Russian).
49. Zenkevich LA. The Black Sea. Biology of the seas of the USSR. Moscow: Publishing house of the USSR Academy of Sciences; 1963. p. 98–360 (in Russian).
50. Zernov SA. Study of life in the Black Sea. St. Petersburg; 1913. 280 p (in Russian).
51. Ivanova OA. Estimation of the resistance to corrosion of deepwater drilling platform for the Black Sea conditions. In: Ecological safety of coastal and shelf zones and complex use of shelf resources. Sevastopol; 2010. Issue 23. p. 292–5 (in Russian).
52. Ilichev VD, Bocharov BV, Gorlenko MV. Ecological principles of defence against biodamage. Moscow: Nauka; 1985. 262 p (in Russian).
53. Kalugina-Gutnik AA. Changes in the composition of algal flora of Novorossiysk Bay during the last 40 years and its phytogeographic analysis. In: Hydrobiological studies of the north-western part of the Black Sea. Rostov-on-Don: Publishing House of Rostov-on-Don University; 1973. p. 15–9 (in Russian).
54. Kalugina-Gutnik AA. Phytobenthos of the Black Sea. Kiev: Naukova dumka; 1975. 246 p (in Russian).
55. Kaminskaya LD, Alekseev RP, Ivanova EV, Synegub IA. Bottom fauna of the Odessa Gulf coastal zone and of adjacent areas in the presence of hydrotechnical construction. In: Biology of the sea. Kiev; 1977. Issue 43. p. 54–64 (in Russian).
56. Karpov VA, Kovalchuk YL, Poltarukha OP, Ilyin IN. Complex approach towards protection from marine fouling and corrosion. Moscow: Scientific press KMK; 2007. 155 p (in Russian).

57. Kiryukhina LN, Kucherenko MI. Studies of sanitary and biological state of bottom sediments of Novorossiysk Bay. In: Proceedings of scientific conference for the 50th Anniversary of Novorossiysk biological station. Novorossiysk; 1971. p. 66–7 (in Russian).
58. Knipovich NM. Hydrology of seas and brackish waters. Moscow; Leningrad: Pischepromizdat; 1938. 513 p (in Russian).
59. Komulaynen SV. Ecological groups of algae in aquatic ecosystems and problems of classification. In: Algae: taxonomy, ecology and use in monitoring. Ekaterinburg: Ural Division of the Russian Academy of Sciences (URAN); 2011. p. 128–34 (in Russian).
60. Koryakova MD, Nikitin VM, Speshneva NV. Role of bacterial film under barnacles in the corrosion of high-alloyed steel in sea water. Prot Metals 1998;34(2):208–11 (in Russian).
61. Koryakova MD, Nikitin VM, Speshneva NV, Suponina AP. Fouling and corrosion of high-alloyed steel in waters of marine ports. Prot Metals 2001;37(3):279–83 (in Russian).
62. Kudelina EN, Zhuravleva SK. Feeding of copepods and barnacles larvae in the Azov Sea. In: Proceedings of Azov institute of fishery, vol. 6. 1963. p. 71–82 (in Russian).
63. Kuznetsov AP. Ecology of the World Ocean bottom communities (Trophic structure of marine bottom fauna). Moscow: Nauka; 1980. 244 p (in Russian).
64. Kuznetsova IA. Energy flow in barnacle population. In: Ist All-Union conference on marine biology (Vladivostok, 26–29 September 1977). Vladivostok; 1977. p. 88–9 (in Russian).
65. Lebedev EM. Fouling of the ships navigating in the Azov Sea and Kerch Strait. In: Proceedings of the Institute of Oceanology of AS USSR, vol. 49. 1961. p. 118–35 (in Russian).
66. Lebedev EM. Biological damage of materials in the aquatic environment. In: Problems of biological damage and fouling of materials, objects and constructions. Moscow: Nauka; 1972. p. 156–63 (in Russian).
67. Losovskaya GV. The ecology of Black Sea polychaetes. Kiev: Naukova dumka; 1977. 90 p (in Russian).
68. Losovskaya GV. Living forms of the Black Sea polychaetes. Hydrobiol J. 1998;34(3):44–50 (in Russian).
69. Losovskaya GV. *Macrozoobenthos* as an indicator for m*onitoring* the e*nvironmental quality* of the *Black Sea: a* review. Hydrobiol J. 2002:38(1):50–61 (in Russian).
70. Losovskaya GV. Changes in species and ecological diversity of Class Polychaeta in conditions of eutrophication of the north-western Black Sea. In: Ecological problems of the Black Sea. Odessa: *OCNTEI*; 2002. p. 124–6 (in Russian).
71. Losovskaya GV. Changes in species composition, ecological and morphological characteristics of the north-western Black Sea polychaetes over half a century. Ecol Sea 2003;63:41–5 (in Russian).
72. Losovskaya GV, Sinegub IA, Bondarenko AS. Diversity of polychaete fauna in fouling of hydrotechnical constructions and in benthos of Odessa port. In: Ecological safety of coastal and shelf zones and complex exploitation of shelf resources. Sevastopol; 2005. Issue 12. p. 617–21 (in Russian).
73. Losovskaya GV, Sinegub IA, Rybalko AA. Comparison of species composition and quantitative development of polychaetes in fouling and benthos based on the example of Odessa port. Marine Ecol J. 2004;3(1):51–8 (in Russian).
74. Lyublinsky EY, Yakubenko AP. Strategy and tactics of protection from marine fouling. In: Proceedings of All-Union scientific and technical conference "Protection of ships and technical facilities from fouling". Leningrad: Shipbuilding; 1990. p. 5–16 (in Russian).
75. Makkaveeva EB. Daily migrations of invertebrates in the biocenoses of vegetation beds. In: Proceedings of the scientific conference for the 50th Anniversary of Novorossiysk biological station. Novorossiysk; 1971. p. 69–70 (in Russian).
76. Makkaveeva EB. Invertebrates of macrophyte beds of the Black Sea. Kiev: Naukova dumka; 1979. 228 p (in Russian).

77. Malyutin OI. Influence of water flow on attached animals in controlled aquarium conditions. In: IIIrd All-Union conference on marine biology (Sevastopol, 18–20 October 1988). Kiev; 1988. Part 1. p. 227–8 (in Russian).
78. Microalgae of the Black Sea: problems of biodiversity, preservation and biotechnology usage. In: Tokarev YN, Finenko ZZ, Shadrin NV, editors. NAS of Ukraine, Institute of biology of the southern seas. Sevastopol: ECOSI-Hidrofisica; 2008. 454 p (in Russian).
79. Milovidova NY. Zoobenthos of harbours of the north-eastern Black Sea: PhD thesis synopsis. Rostov-on-Don; 1967. 18 p (in Russian).
80. Milovidova NY. Zoobenthos significance for sanitary assessment of the sea coastal zone. In: Theory and practice of biological auto-purification of polluted waters. Moscow: Nauka; 1972. p. 175–9 (in Russian).
81. Milovidova NY. Seasonal characteristics of zoobenthos in the harbours of north-eastern Black Sea. In: Hydrobiological studies of the north-eastern Black Sea. Rostov-on-Don: Publishing House of Rostov University; 1973. p. 73–7 (in Russian).
82. Milovidova NY, Kiryukhina LN. Sanitary and biological aspects of the Black Sea macrozoobenthos. Kiev: Naukova dumka; 1985. 102 p (in Russian).
83. Milovidova NY, Tsymbal IM. Importance of macrofouling of hydrotechnical structures in self-purification of port aquatories. In: State, perspectives of improvement and use of marine ecological system of the Crimean coast: Proceedings of a scientific-practical conference dedicated to the 200 years of Sevastopol. Sevastopol; 1983. p. 16–7 (in Russian).
84. Milovidova NY, Tsimbal IM. Phytofouling of hydrotechnical constructions in port aquatories of some Black Sea harbours. Ecol Sea. 1984. Issue 17. p. 76–9 (in Russian).
85. Minicheva GG. Morphofunctional fundamentals of marine phytobenthos formation: synopsis of Doctor of Science thesis. Sevastopol; 1998. 32 p (in Russian).
86. Minicheva GG, Bolshakov VN, Khomova ES, Shvets AV. Patterns of phytofouling formation as a function of the orientation of underwater surfaces. Marine Ecol J. 2011;4:56–66 (in Russian).
87. Mironov OG, Milovidova NY, Tsimbal IM. Formation of benthic communities on newly built pier constructions. Hydrobiol J. 1982;18(4):18–21 (in Russian).
88. Mikhaylov SR, Kashin IA. Fouling of piers and wrecked ships in Rassvet and Nagaev Bays (Taui Bay, Sea of Okhotsk). In: Fouling organisms of the Far East Seas. Vladivostok: FESC AS USSR; 1981. p. 3–15 (in Russian).
89. Morozova-Vodyanitskaya NV. Vegetative fouling in Tuapse port. In: Proceedings of the Novorossiysk biological station. Rostov-on-Don: Publishing House of Rostov University; 1961. p. 11–39 (in Russian).
90. Marine fouling and its control. Moscow: Military publication of the Ministry of defence of USSR; 1957. 502 p (in Russian).
91. Moschenko AV. Role of microscale turbidity in the distribution and variability of benthic animals. Vladivostok: Dalnauka; 2006. 321 p (in Russian).
92. Muravyova IP. Microphytofouling of the artificial underwater constructions in Sevastopol aquatory. In: Ist symposium of the Ukrainian hydroecological society (Kiev, 16–19 November 1993). Kiev; 1994. p. 36 (in Russian).
93. Nikitin VN. Biology of marine fouling on ships in the Black Sea. Proc AS USSR. 1947;58 (6):1183–5 (in Russian).
94. Odum Y. Fundamentals of ecology, vol. 1. Moscow: Mir; 1975. 743 p (in Russian).
95. Odum Y. Ecology, vol. 2. Moscow: Mir; 1986. 376 p (in Russian).
96. Oshurkov VV. Succession and structure of fouling communities in the Black Sea. In: Biology of shelf zones of the World ocean: Processing of IInd All-Union conference on marine biology. Vladivostok: FESC AS USSR; 1982. Part 1. p. 51–2 (in Russian).
97. Oshurkov VV. Dynamics and structure of some fouling and benthic communities in the White Sea. In: Ecology of fouling in the White Sea. Leningrad: Zoological institute of the Academy of Science of USSR; 1985. p. 44–59 (in Russian).

98. Oshurkov VV. Successions and dynamics of epibenthic communities from the boreal upper subtidal zone. Vladivostok: Dalnauka; 2000. 206 p (in Russian).
99. Partalyi EM. Seasonal changes in the formation of fouling biocenoses in the Azov Sea (Taganrog Bay). Mariupol; 1980. 293 p (in Russian).
100. Povchun AS, Revkov NK, Kozintsev AF, Valovaya NA. Aggregation of the Black Sea *Mutillidae*: various types of adaptation to the environment. In: IIIrd All-Union conference on marine biology (Sevastopol, 18–20 October 1988). Kiev; 1988. Part 1. p. 239–40 (in Russian).
101. Poltarukha OP, Karpov VA, Filichev NL, Tan NX. Studies of the systematic composition of marine fouling fauna. In: Climatic and biological materials resistance. Moscow; Hanoi: GEOS; 2003. p. 38–40 (in Russian).
102. Protasov AA. Periphyton: terminology and main definitions. Hydrobiol J. 1982;18(1):9–14 (in Russian).
103. Protasov AA. Some questions of general comparative and practical periphytology. In: Proceedings of the international scientific conference "Periphyton and fouling: theory and practice" (22–25 October 2008). St. Petersburg: Zoological Institute RAS; 2008. p. 68–70 (in Russian).
104. Protasov AA. The life in the hydrosphere. Essays of general hydrobiology. Kiev: Academperiodica; 2011. 704 p (in Russian).
105. Rialkin AI. Colonization processes and protection from fouling. St. Petersburg: Publishing house of St. Petersburg State University; 1998. 272 p (in Russian).
106. Rialkin AI. Colonization of hard bodies by benthic organisms. St. Petersburg: Publishing house of St. Petersburg State University; 2008. 427 p (in Russian).
107. Rachinskaya AV. Microphytobenthos of hard substrata in the biomonitoring of the coastal zone of the Black Sea. In: Ecological problems of the Black Sea. Odessa: CNTEPI ONLA; 2003. p. 275–81 (in Russian).
108. Rachinskaya AV. Diatoms of microphytobenthos of hard substrata as indicator of marine environment quality (Ilyichevsk region). In: Ecological problems of the Black Sea. Odesa: CNTEPI; 2004. p. 357–60 (in Russian).
109. Reznichenko OG, Soldatova IN. Experimental validation of the specificity of fouling biocenoses. In: Experimental ecology of marine invertebrates. Vladivostok; 1976. p. 150–3 (in Russian).
110. Romanenko VD. Fundamentals of hydroecology. Kiev: Geneza; 2004. 644 p (in Russian).
111. Romanenko VD, Zhukinskiy VM, Oksiyuk OP, et al. Method of ecological estimation of surface waters quality according to corresponding categories. Kiev: Simvol; 1998. 28 p (in Ukrainian).
112. Salskiy VO. Studies of shipwrecks fouling with the help of diving. In: Proceedings of the Odessa Biological Station. Kiev: Publishing House of the Academy of Science of Ukrainian SSR; 1962. Issue 4. p. 66–71 (in Ukrainian).
113. Sanitary norms and regulations for the protection of sea coastal waters against pollution in zones of water use by human populations (Sanitary Regulations and Standards/SanPiN). № 4631–88. Moscow: Ministry of Health of USSR; 1988. 14 p (in Russian).
114. Svirskaya MA, Melnichuk EP. Preliminary data on fouling composition in some ports of the Black Sea. In: Proceedings of the All-Union symposium on studies of the Black and Mediterranean Seas, use and protection of their resources (Sevastopol, October 1973). Kiev: Naukova dumka; 1973. Part 4. p. 107–8 (in Russian).
115. Zaitsev YP, Aleksandrov BG, Minicheva GG, editors. Biology and ecology of the north-western part of the Black Sea. Kiev: Naukova dumka; 2006. 701 p (in Russian).
116. Selin NI. Sediment influence on settling of *Crenomytilus grayanus* larvae. In: Ist All-Union conference on marine biology (Vladivostok, 26–29 September 1977). Vladivostok; 1977. p. 122–3 (in Russian).
117. Synegub IA. On broadening of the natural habitat of the nudibranchia *Doridella obscura* Verrill in the Black Sea. Hydrobiol J. 1994;30(3):107–9 (in Russian).

118. Synegub IA, Bondarenko AS, Rybalko AA, Kudrenko SA. Fouling macrofauna of the hydrotechnical structures of Odessa port. In: Periphyton and fouling: theory and practice: Proceedings of the international scientific and practical conference (22–25 October 2008). St. Petersburg: ZIN RAS; 2008. p. 82–4 (in Russian).
119. Slavina OY. Mussel growth in the *Bay of Sevastopol.* In: Benthos. Kiev: Naukova dumka; 1965. p. 24–9 (in Russian).
120. Soldatova IN, Lukasheva TA. Energy exchange of mass fouling molluscs and their benthic variants. In: Biology of shelf zones of the World Ocean: Proceedings of IInd All-Union conference of marine biology. Vladivostok: FESC AN USSR; 1982. Part 1. p. 168–9 (in Russian).
121. Solovyova OV. Vertical distribution of the molluscs *Mytilus galloprovincialis* Lam. and *Mytilaster lineatus* Gmel. on riprap of hydrotechnical construction. In: Proceedings of Ternopil pedagogical University, vol. 3(44). Series: Biology. Special issue: Hydroecology. 2010. p. 263–5 (in Russian).
122. Manual on the ecology of marine bivalves. Way of life of bivalves belonging to genera present in Cenozoic marine and brackish sediments of southern USSR. Moscow: Nauka; 1966. 350 p (in Russian).
123. Stepanyan OV. Fouling algae on port constructions of the Azov and Black Seas. In: Materials of the international scientific and practical conference 'Periphyton and fouling: theory and practice'. (22–25 October 2008). St. Petersburg: ZIN RAS; 2008. p. 205–7 (in Russian).
124. Tarasov NI. Biology of the sea and the navy. Moscow: Naval Pub. House of the Peoples Commissariate of the USSR Navy; 1943. 192 p (in Russian).
125. Tarasov NI. Biology of the sea and the navy. Stenograph of public lecture in Moscow. Moscow: Publishing House "Pravda"; 1949. 30 p (in Russian).
126. Titlyakov EA. Lability of the structure and functions of the photosynthetic apparatus of marine macroalgaes. In: Ist All-Union conference on marine biology (Vladivostok, 26–29 September 1977). Vladivostok; 1977. p. 134 (in Russian).
127. Thomas-Bourneuf M, Mollo P. The plankton challenge. In: Eremeev VN, editor. NAS Ukraine, Institute of Biology of Southern Seas named after A.O. Kovalevsky. Sevastopol; 2011. 281 p (in Russian).
128. Turpaeva EP. Fouling communities. In: Oceanology. Biology of the Ocean, vol. 2. Moscow: Nauka; 1977. Ocean biological productivity. p. 155–62 (in Russian).
129. Fedorov VD, Gilmanov TG. Ecology. Moscow: Publishing House of Moscow State University; 1980. 464 p (in Russian).
130. Fortunenko YA. On the question of engineering protection of coastal waters. In: Ecological problems of the Black Sea. Odessa: CNTEPI ONLA; 2003. p. 388–9 (in Russian).
131. Khaylov KM. Elements of ecological metabolism in the sea coastal zone: synopsis of Doctor of Science thesis. Moscow; 1969. 44 p (in Russian).
132. Khaylov KM. Ecological metabolism in the sea. Kiev: Naukova dumka; 1971. 252 p (in Russian).
133. Hanke H. People, ships, oceans (6000 years of navigation venture). Leningrad: Shipbuilding; 1976. 432 p (in Russian).
134. Cherbadzhi II. Influence of ecological factors on the intensity of microalgal photosynthesis in soft sediments. In: Ist All-Union conference on marine biology (Vladivostok, 26–29 September 1977). Vladivostok; 1977. p. 148–9 (in Russian).
135. Shadrin NV. Tendencies in the plankton changes after creation of artificial reefs. In: Artificial reefs for the fisheries. Moscow; 1990. p. 154–9 (in Russian).
136. Shurova NM. Structural-functional organization of population of the Black Sea mussels *Mytilus galloprovincialis*: synopsis of Doctor of Science thesis. Sevastopol; 2009. 32 p (in Ukrainian).
137. Shurova NM. Structural-functional organization of population of the Black Sea mussels *Mytilus galloprovincialis*. Kiev: Naukova dumka; 2013. 207 p (in Russian).

138. Shurova NM, Losovskaya GV, Zolotarev VN. Mussels zoocenosis as a biotope for introduced species. In: Ecological problems of the Black Sea. Odessa: OCNTEI; 2002. p. 298–301 (in Russian).
139. The ecosystem of Grigorievsky (Maly Adzhalyk) Liman. In: Vinogradov AK, editor. Odessa: Astroprint; 2008. 263 p (in Russian).
140. Gomoiu M-T. Potential role and ecological effects of artificial reefs constructed on the coastal sandy bottoms of the Black Sea (Romania). Trav Mus Hist Nat «Gregore Antipa» 1989;38:291–306.
141. Gomoiu M-T, Tiganus V. Structure qualitative et quantitative des salissures formées dans les eaux du large de la Mer Noire. Rapp Et Proc-Verb Reun Commis Int Explor Sci Mer Mediterr. 1981;27(2):183–4 (in French).
142. Gray JS. The measurement of marine species diversity, with an application to the benthic fauna of the Norwegian continental shelf. J Exp Mar Biol Ecol. 2000;250:23–49.
143. Grove RS, Sonu CJ. Review of Japanese fishing reef technology. In: Southern California Edison Company, P.O. Box 800, Rosemead, CA 91770, Tech. Rep. 83. RD–137. 1983. 112 p.
144. Jørgensen CB. Mortality, growth, and grazing impact of cohort of bivalve larvae *Mytilus edulis*. In: Droop M, Wood EF, editors. Advances in Microbiology of the sea, vol. 20, № 2. New York: Academic Press; 1981. p. 185–92.
145. Lucas A. La nutrition des larves de bivalves. Oceanis. 1982;8(5):363–88 (in French).
146. Manahan DT, Crisp DJ. The role of dissolved organic material in the nutrition of pelagic larvae: amino acid uptake by bivalve veligers. Amer Zool. 1982;22(3):635–46.
147. Ogawa Y. Basic theory. In: Vik SF, editor. Japanese artificial reef technology. Aquabio Inc., 2957 Sunset Blvd., Bellair Bluffs. 1982. FL. Tech. Rep. 604. p. 99–137.
148. Ohshima Y. Introduction: report from the consolidated reef study society. In: Vik SF, editor. Japanese artificial reef technology. Aquabio Inc., 2957 Sunset Blvd., Bellair Bluffs. 1982. FL. Tech. Rep. 604. p. 93–8.
149. Perron FE, Turner RD. Development, metamorphosis and natural history of the nudibranch *Doridella obscura* Verrill (Corambidae: Opisthobranchia). J Exp Mar Biol Ecol. 1977;27:171–85.
150. Rounsefell GA. Ecological effects of offshore construction. J Mar Sci. 1971;2:1–208.
151. Scheer BT. The development of marine fouling communities. Biol Bull. 1945;89(2):103–21.
152. Schoener A. Artificial substrates in marine environments. In: Cairns J, editor. Artificial substrates. USA: Ann Arbor Science Publishers; 1982. p. 1–22.

Literatures Chapter 4

1. Aybulatov NA, Marakuev VI. Lithodynamics and bioturbation of sediments in the north-eastern shelf of the Black Sea (based on underwater photographs). Problems of geomorphology, lithodynamics and lithology of shelf. Moscow: Nauka; 1982. p. 168–82 (in Russian).
2. Ayzatullin TA, Lebedev VL, Khaylov KM. Ocean. Active surfaces and life. Leningrad: Gidrometeoizdat; 1979. 194 p (in Russian).
3. Ayzatullin TA, Lebedev VL, Khaylov KM. Ocean. Active surfaces and life. Leningrad: Gidrometeoizdat; 1979. 194 p (in Russian).
4. Sevastopol aquatory and coasts: ecosystem processes and services for the human society. In: Pavlova EV, Shadrin NV, editors. Sevastopol: Aquavita; 1999. 289 p (in Russian).
5. Aleksandrov BG. Hydrobiological bases of management of the state of coastal ecosystems of the Black Sea. Kiev: Naukova dumka; 2008. 343 p (in Russian).
6. Aleksandrov BG. Hydrobiological bases of management of the state of coastal ecosystems of the Black Sea. Kiev: Naukova dumka; 2008. 343 p (in Russian).

7. Aliev AD. Benthos studies of Baku Bay. Hydrobiological and ichthyological studies in Southern Caspian and inland water reservoirs of Azerbaijan. Baku: Publishing House of the Academy of Sciences of Azerbaijan SSR; 1965. p. 42–7 (in Russian).
8. Alimov AF. Studies of biotic balances of freshwater basins ecosystems in the USSR. Hydrobiol J. 1987;23(6):3–9 (in Russian).
9. Basic biological studies of Odessa marine port (Aug–Dec 2001): final report. GloBallast programme Monograph Series of Odessa demonstration centre. Odessa; 2004. Issue 7. 171 p (in Russian).
10. Bashkirov GS. Problems of hydraulics of shallow stormy sea: Manual. Moscow: Reklamburo MMF; 1971. 46 p (in Russian).
11. Berlinsky NA, Bogatova YI, Bolshakov VN, Garkavaya GP. The quality of aquatic environment and bottom sediments in the port of Odessa (Program «GloBallast»). In: Ecological problems of the Black Sea. Odessa: CNTEPI ONLA; 2003. p. 65–8 (in Russian).
12. Biology of the north-western part of the Black Sea. In: Vinogradov KA, editor. Kiev: Naukova dumka; 1967. 266 p (in Russian).
13. Biological encyclopedia dictionary. In: Gilyarov SM, editor. Moscow: Soviet encyclopedia; 1986. 831 p (in Russian).
14. Bogorov VG. Biogeocenoses of the ocean pelagial zone. Program and method of studies of aquatic environment biogeocenoses. In: Biogeocenoses of seas and oceans. Moscow: Nauka; 1970. p. 28–46 (in Russian).
15. Burkovsky IV. Structural and functional organization and stability of marine bottom communities (an example from the sand littoral of the White Sea). Moscow: Publishing house of Moscow State University; 1992. 208 p (in Russian).
16. Burkovsky IV. Marine biogeocenology. Organization of communities and ecosystems. Moscow: Scientific press KMK; 2006. 285 p (in Russian).
17. Veil P. Popular oceanography. Leningrad: Gydrometeoizdat;1977:500 p (in Russian).
18. Vinogradov AK, Bogatova YI, Sinegub IA. Ecosystems of marine ports aquatories of the Black–Azov Sea basin (Introduction to the ecology of marine ports). Odessa: Astroprint; 2012. 528 p (in Russian).
19. Vinogradov KA, Losovskaya GV. Class bristle worms—Polychaeta. In: Manual for the identification of the Black and Azov Seas fauna, vol. 1. Kiev: Naukova dumka; 1968. p. 215–359 (in Russian).
20. Vinogradov ME, Sapozhnikov VV, Shushkina EA. The Black Sea ecosystem. Moscow: Nauka; 1992. 112 p (in Russian).
21. Viter TV. Macrozoobenthos of hydrotechnical constructions of Oil harbour (Sevastopol Bay, the Black Sea). Ecol Sea. 2009;18:28–33 (in Russian).
22. Viter TV. Benthic community in the region of hydrotechnical constructions of Oil harbour (Sevastopol Bay, the Black Sea). In: Proceedings of Ternopil pedagogical university. Series: Biology. Special issue: Hydroecology, vol. 3(44). 2010; p. 40–4 (in Russian).
23. Viter TV. Macrozoobenthos of hydrotechnical constructions of Sevastopol and Kamishevaya Bays. Ecological safety of coastal and shelf zones and complex exploitation of shelf resources. Sevastopol, 2011; Issue 25, vol. 1. p. 408–16 (in Russian).
24. Viter TV. Bottom communities in the region of berths located in Gollandiya Bay and in the region of GRES regional power plant (Sevastopol Bay). Ecological safety of coastal and shelf zones and complex exploitation of shelf resources. Sevastopol, 2013; Issue 27. p. 431–8 (in Russian).
25. Vorobyova LV. Meiobenthos of the Ukrainian shelf of the Black and Azov Seas. Kiev: Naukova dumka; 1999. 300 p (in Russian).
26. Vorobyova LV, Zaitsev YP, Kulakova II. Interstitial meiofauna of sandy beaches of the Black Sea. Kiev: Naukova dumka; 1992. 144 p (in Russian).
27. Galkina VN, Kulakovsky EE, Kunin BL. The environmental influence of mussels aquaculture in the White Sea. Oceanology 1982;22(2):321–4 (in Russian).

28. Galtsova VV. Some aspects of vertical distribution of the free-living nematodes. Evolution, systematics, morphology and ecology of free-living nematodes. Leningrad: Nauka; 1981. p. 27–31 (in Russian).
29. Galtsova VV. Meiobenthos of marine ecosystems as exemplified by free-living nematodes. In: Proceedings of the Zoological Institute, vol. 224, AS USSR; 1991. 236 p (in Russian).
30. Garlitskaya LA. Ecology of Harpacticoida (Crustacea, Copepoda) of the north-western part of the Black Sea: PhD thesis synopsis. Sevastopol; 2010. 20 p (in Russian).
31. Gerlach SA. Marine pollution. Diagnosis and therapy. Leningrad: Gydrometeoizdat; 1985. 263 p (in Russian).
32. Goryainova LI, Gromov VV, Kopycheva LN, Kostuchenko LP, Milovidova NY, Smolyar RI. Impact of pollution of the North Caucasian bays of the Black Sea on hydrobionts. In: Proceedings of scientific conference for the 50th Anniversary of Novorossiysk biological station. Novorossiysk; 1971. p. 119–21 (in Russian).
33. Goryachkin YN. Anthropogenic impact on the Black Sea coasts of Crimea. Ecological safety of coastal and shelf zones and complex exploitation of shelf resources. Sevastopol; 2010. Issue 23. p. 193–7 (in Russian).
34. Goryachkin YN, Fomin VV. Wave current parameters in the Kalamita Gulf. In: Ecological safety of coastal and shelf zones and complex exploitation of shelf resources. Sevastopol; 2010. Issue 23. p. 159–65 (in Russian).
35. Greze VN. The rate of production in populations of heterotrophic marine organisms. Questions of biooceanography. Kiev: Naukova dumka; 1967. p. 121–7 (in Russian).
36. Greze II. Amphipods of the Black Sea and their biology. Kiev: Naukova dumka; 1977. 155 p (in Russian).
37. Gurevich VI, Khasankaev VB. Lithological parameters of intertidal sandflat biogeocenoses on Dalniy beach. Ecological studies of sandy littoral. Apatity: Murmansk Marine Biological Institute; 1976. p. 3–25 (in Russian).
38. Dediu II. Ecological encyclopedic dictionary. Kishinev: Chief editorial board of Moldavian Soviet Encyclopedia; 1990. 408 p (in Russian).
39. Denga YM, Lisovsky RI, Mikhaylov VI. Oil pollution in the Black Sea ecosystem. Ecological problems of the Black Sea. Odessa: CNTEPI ONLA; 2003. p. 123–34 (in Ukrainian).
40. Dzhurtubaev MM, Chernolev MM. Bottom fauna of Odessa port. In: Proceedings of the IInd All-Union conference on shelf biology. Kiev: Naukova dumka; 1978. Part 2. p. 41–2 (in Russian).
41. Life of animals. In: Polyanskiy YI, editor. Moscow: Prosveshenie, vol. 1; 1987. 448 p (in Russian).
42. Zhirkov IA, with contributions from Azovskiy AI, Maksimova OV. Life at the bottom. Bio-geography and bio-ecology. Moscow: Scientific press KMK; 2010. 453 p (in Russian).
43. Zaika VE, Kiseleva MI, Mikhailova TV, Makkaveeva EB, Sergeeva NG, Povchun AS, Kolesnikova EA, Chukhchin VD. Long-term changes in the Black Sea zoobenthos. Kiev: Naukova dumka; 1992. 248 p (in Russian).
44. Zaitsev YP. The most blue in the world. Black Sea ecological series, vol. 6. New-York: UN publishing house; 1998. 142 p (in Russian).
45. Zaitsev YP. Introduction to the ecology of the Black Sea. Odessa: Even; 2006. 222 p (in Russian).
46. Zubov NN. The fundamentals of the World Ocean straits study. Moscow: Geografgiz; 1956. 236 p (in Russian).
47. Ivlieva OV. Influence of biotic factor on current sedimentation in the Azov Sea. In: Geology, geography and ecology of the ocean: proceedings of the international scientific conference dedicated to 100 years after birth of D.G. Panov (Rostov-on-Don, 8–11 June 2009). Rostov-on-Don; 2009. p. 135–7 (in Russian).

48. Kandyuk RP. Sterols in bottom sediments of the north-western part of the Black Sea. In: Biology of shelf zones of the World Ocean: proceedings of the IInd All-Union conference on marine biology. Vladivostok: FESC AS USSR; 1982. Part 2. p. 79–80 (in Russian).
49. Kiryukhina LN, Kucherenko MI. Studies of sanitary and biological state of bottom sediments of Novorossiysk Bay. In: Proceedings of scientific conference for the 50th Anniversary of Novorossiysk biological station. Novorossiysk; 1971. p. 66–7 (in Russian).
50. Kiryukhina LN, Kucherenko MI. Studies of sanitary and biological state of bottom sediments of Novorossiysk Bay. In: Proceedings of scientific conference for the 50th Anniversary of Novorossiysk biological station. Novorossiysk; 1971. p. 66–7 (in Russian).
51. Kiseleva MI. Qualitative structure and quantitative distribution of meiobenthos of the western coast of Crimea. In: Benthos. Kiev: Naukova dumka; 1965. p. 48–61 (in Russian).
52. Kiseleva MI. Benthos of soft sediments of the Black Sea. Kiev: Naukova dumka; 1972. 246 p (in Russian).
53. Kiseleva MI. Food spectra of some benthic invertebrates of the Black Sea. Zool J. 1975;54 (11):1595–601 (in Russian).
54. Kiseleva MI. Bristle worms (Polychaeta) of the Black and Azov Seas. Apatity: Publishing house Kola Science Centre Russian Academy of Science; 2004. 409 p (in Russian).
55. Kiseleva MI, Vityuk DM. Feeding habits of polichaetes of the Capitellidae family in the Black Sea. Ecological and morphological studies of benthic organisms. Kiev: Naukova dumka; 1970. p. 67–75 (in Russian).
56. Kiseleva MI, Slavina OY. Quantitative distribution of meiobenthos on the coast of Caucasus. Hydrobiological studies of the north-eastern part of the Black Sea. Rostov-on-Don: Publishing House of Rostov University; 1973. p. 82–91 (in Russian).
57. Klapper H. Experience on technologies for the improvement of lakes and storage water reservoirs conditions in GDR. In: Anthropogenic eutrophication of natural waters: proceedings of the All-Union symposium, Moscow, September 1983. Chernogolovka; 1985. p. 172–97 (in Russian).
58. Klenova MV. Geology of the Sea. Moscow: The State educational-pedagogical publishing house of the Ministry of Public Education of RSFSR; 1948. 495 p (in Russian).
59. Kondratiev SI, Lyashenko SV. Peculiarities of the distribution of biogenic elements in the water column and in pore waters of Sevastopol Bay. In: Ecological safety of coastal and shelf zones and complex exploitation of shelf resources. Sevastopol; 2003. Issue 9. p. 82–94 (in Russian).
60. Korneva LG. Ecological classification of planktonic algae. Algae: taxonomy, ecology and use in monitoring. Ekaterinburg: Ural Division of the Russian Academy of Sciences; 2011. p. 135–9 (in Russian).
61. Kryuchkova TM. Structure of the zooplankton community in water basins of various type. Production and hydrobiological studies of aquatic ecosystems. Leningrad: Nauka; 1987. p. 184–98 (in Russian).
62. Kuznetsov AP. Ecology of the World Ocean bottom communities (Trophic structure of marine bottom fauna). Moscow: Nauka; 1980. 244 p (in Russian).
63. Kulakova II. Features of the distribution of free-living nematodes in the north-western Black Sea: PhD thesis synopsis. Sevastopol; 2002. 20 p (in Russian).
64. Kurashov EA. Meiobenthos as a component of lacustrine ecosystem. St. Petersburg: Alga-Fond; 1994. 224 p (in Russian).
65. Kustenko NG. Effect of environmental stress factors on the reproduction of diatoms. Kiev: Naukova dumka; 1991. 156 p (in Russian).
66. Lebedev VL. Processes in the upper layer of the ocean. In: Duvanin AI, editor. Interaction of the ocean with the environment. Moscow: Publishing House of Moscow State University; 1983. p. 37–53 (in Russian).
67. Leonov AK. Regional oceanography. Part I. Leningrad: Gydrometeoizdat; 1960. 765 p (in Russian).

68. Losovskaya GV. On the changes in the bottom fauna of Sukhoy Liman after its connection with the sea. Biological problems of the oceanography of the southern seas. Kiev: Naukova dumka; 1969. p. 56–9 (in Russian).
69. Losovskaya GV. Feeding habits of some mass species of polychaetes and their distribution in the north-western Black Sea. In: Questions of marine biology: proceedings of the IInd All-Union symposium of young scientists. Sevastopol, 1969. Kiev: Naukova dumka; 1969. p. 74–5 (in Russian).
70. Losovskaya GV. The ecology of Black Sea polychaetes. Kiev: Naukova dumka; 1977. 90 p (in Russian).
71. Losovskaya GV. Changes in species composition, ecological and morphological characteristics of the north-western Black Sea polychaetes over half a century. Ecol Sea 2003;63:41–5 (in Russian).
72. Losovskaya GV. Indicator and tolerant polychaete species: examples from the north-western Black Sea. Ecological safety of coastal and shelf zones and complex exploitation of shelf resource, vol. 1. Sevastopol; 2011. Issue 25. p. 327–34 (in Russian).
73. Losovskaya GV, Sinegub IA. Some peculiarities of the macrozoobenthos of Odessa port. Hydrobiol J. 2007;43(4):43–8 (in Russian).
74. Losovskaya GV, Sinegub IA, Rybalko AA. Comparison of species composition and quantitative development of polychaetes in fouling and benthos based on the example of Odessa port. Marine Ecol J. 2004;3(1):51–8 (in Russian).
75. Lyubin PY. Zoobenthos of the Azov Sea. In: Matishov GG, editor. Modern development of estuarine ecosystems as exemplified by the Azov Sea. Apatity: Publishing House Kola Science Center Russian Academy of Science; 1999. p. 167–94 (in Russian).
76. Makarov KN. Influence of hydrotechnical constructions on drifts' longshore transport and shore dynamics. In: Geology, geography and ecology of the ocean: proceedings of the international scientific conference dedicated to 100 years birthday of D.G. Panov (Rostov-on-Don, 8–11 June 2009). Rostov-on-Don; 2009. p. 215–8 (in Russian).
77. Makarov MV. Seasonal dynamics of Gastropoda in Sevastopol Bay (the Black Sea). Ecological safety of coastal and shelf zones and complex exploitation of shelf resources. Sevastopol; 2004. Issue 10. p. 184–9 (in Russian).
78. Makkaveeva EB. Invertebrates of macrophyte beds of the Black Sea. Kiev: Naukova dumka; 1979. 228 p (in Russian).
79. Matishov GG, Kreneva SV, Muraveyko VM, Shparkovskiy IA, Ilyin GV. Biotesting and forecast. Variability of aquatic ecosystems affected by anthropogenic pollution. Apatity: Publishing House Kola Science Centre Russian Academy of Science; 2003. 468 p (in Russian).
80. Microalgae of the Black Sea: problems of biodiversity, preservation and biotechnology usage. In: Tokarev YN, Finenko ZZ, Shadrin NV, editors. NAS of Ukraine, Institute of biology of the southern seas. Sevastopol: ECOSI-Hidrofisica; 2008. 454 p (in Russian).
81. Milovidova NY. Zoobenthos of harbours of the north-eastern Black Sea: PhD thesis synopsis. Rostov-on-Don; 1967. 18 p (in Russian).
82. Milovidova NY. Zoobenthos significance for sanitary assessment of the sea coastal zone. In: Theory and practice of biological auto-purification of polluted waters. Moscow: Nauka; 1972. p. 175–9 (in Russian).
83. Milovidova NY. Seasonal characteristics of zoobenthos in the harbours of north-eastern Black Sea. Hydrobiological studies of the north-eastern Black Sea. Rostov-on-Don: Publishing House of Rostov University; 1973. p. 73–7 (in Russian).
84. Milovidova NY, Kiryukhina LN. Sanitary and biological aspects of the Black Sea macrozoobenthos. Kiev: Naukova dumka; 1985. 102 p (in Russian).
85. Mironov O.A. Methods and results of oil pollution monitoring in the coastal aquatory of Sevastopol city. Ecological safety of coastal and shelf zones and complex exploitation of shelf resources, vol. 1. Sevastopol; 2011. Issue 25. p. 206–11 (in Russian).

86. Mironov OG. Experimental study of the transformation of some organic matter components in bottom sediments. Long-term dynamics of the Black Sea coastal ecosystems structure. Krasnodar; 1984. p. 47–51 (in Russian).
87. Mironov OG. Biological aspects of marine pollution by oil. Kiev: Naukova dumka; 1988. 148 p (in Russian).
88. Mozgunov GI. Observations on drifts and abrasion activity of the sea on the Odessa coast: an example of the zone Cape Bolshoy Fontan–Sukhoy Liman. In: Meeting on studies of coastal zone and bottom geology of the Black and Azov Seas within the Ukrainian SSR: proceedings (10–20 October 1965). Odessa; 1965. p. 48–51 (in Russian).
89. Mokievsky VO. Spatial structure of the nematode taxocene of the White Sea littoral zone and its seasonal changes. In: IIIrd All-Union conference on marine biology (Sevastopol, 18–20 October 1988). Kiev; 1988. Part 1. p. 228–9 (in Russian).
90. Moroz TG. Macrozoobenthos of Limans and lower river reaches of the north-western Black Sea. Kiev: Naukova Dumka; 1993. 187 p (in Russsian).
91. Morozova-Vodyanitskaya NV. Vegetative fouling in Tuapse port. In: Proceedings of the Novorossiysk biological station. Rostov-on-Don: Publishing House of Rostov University; 1961. p. 11–39 (in Russian).
92. Morozova-Vodyanitskaya NV. On biology and distribution of *Zoostera* in the Black Sea. Hydrobiological studies in the north-western Black Sea. Rostov-on-Don: Publishing House of Rostov University; 1973. p. 20–9 (in Russian).
93. Marine geomorphology. In: Zenkovich VP, Popov BA, editors. Terminological manual. Coastal zone: processes, notions, definitions. Moscow: Misl; 1980. 280 p (in Russian).
94. Moschenko AV. Role of microscale turbidity in the distribution and variability of benthic animals. Vladivostok: Dalnauka; 2006. 321 p (in Russian).
95. Murina VV, Selifonova ZP, Melnik VF. Occurrence of the polychaete *Streblospio* sp. (Polychaeta: Spionidae) in the Novorossiysk port of the Black Sea. Mar Ecol J. 2008;7 (1):46 (in Russian).
96. Nelson-Smith A. Oil pollution and sea ecology. Moscow: Progress; 1977. 297 p (in Russian).
97. Novodran VS. Bottom sediments of mineral resources and their formation in shallow marine waters. In: Meeting on studies of coastal zone and bottom geology of the Black and Azov Seas within the Ukrainian SSR: proceedings of the conference (10–20 October 1965). Odessa; 1965. p. 68–70 (in Russian).
98. Pavlovskaya TV. On the feeding of some Black Sea infusoria species on unicellular algae. In: Questions of marine biology: proceedings of IInd All-Union symposium of young scientists. Sevastopol, 1969. Kiev: Naukova dumka; 1969. p. 96–7 (in Russian).
99. Perfilyev BV. Microzonal structure of silty lake sediments and methods of their studies. Leningrad: Nauka; 1972. 216 p (in Russian).
100. Popova LA. Cilioperiphyton of artificial substrata (hydrotechnical constructions) and its role in the transfer of oil hydrocarbons along the food chain. Ecological safety of coastal and shelf zones and complex exploitation of shelf resources, vol. 1. Sevastopol; 2012. Issue 26. p. 81–7 (in Russian).
101. Propp MV. Metabolism of the coastal marine ecosystems: synopsis of Doctor of Science thesis. Sevastopol; 1982. 26 p (in Russian).
102. Protasov AA. The life in the hydrosphere. Essays of general hydrobiology. Kiev: Academperiodica; 2011. 704 p (in Russian).
103. Reymers NF. The usage of nature: a reference book. Moscow: Misl; 1990. 637 p (in Russian).
104. Role of roughness in the formation of benthos biocenoses of big lakes. Editor Raspopov I. M. Leningrad: Nauka; 1990. 114 p (in Russian).
105. Romanenko VD. Fundamentals of hydroecology. Kiev: Geneza; 2004. 644 p (in Russian).
106. Rubtsova SI, Alemov SV. Influence of bottom cleaning activity on the ecological state of port aquatories. Mar Ecol J. 2011;10(2):81–7 (in Russian).

107. Ryabushko LI. Bottom diatoms of Vostok Bay (the Sea of Japan). In: Biology of the shelf zones of the World Ocean: proceedings of the IInd All-Union conference of marine biology. Vladivostok: FESC AS USSR; 1982. Part 1. p. 61–2 (in Russian).
108. Ryabushko LI. Microphytobenthos of the Black Sea: synopsis of Doctor of Science thesis. Sevastopol; 2009. 44 p (in Russian).
109. Samoilov IV. River Mouths. Moscow: Geografgiz; 1952. 525 p (in Russian).
110. Safyanov GA. The coastal zone of the ocean in XXth century. Moscow: Misl; 1978. 263 p (in Russian).
111. Zaitsev YP, Aleksandrov BG, Minicheva GG, editors. Biology and ecology of the north-western part of the Black Sea. Kiev: Naukova dumka; 2006. 701 p (in Russian).
112. Selifononva ZP. Structure and functioning of the anthropogenically transformed ecosystem of Tuapse Port (Black Sea). In: Proceedings of IVth international conference "Modern problems in hydrobiology" (11–15 October 2010), St. Petersburg, 2010. p. 155 (in Russian).
113. Selifonova ZP. Ecosystems of the aquatories of Novorossiysk and Tuapse ports (Black Sea). St. Petersburg: Nauka; 2012. 227 p (in Russian).
114. Selifonova ZP, Kondakov AA, Kovalenko EP, Biryukova SE. Macrozoobenthos of Novorossiysk Port as an ecological indicator. Bull Moscow Soc Nat. Biol Ser. 2009;114 (3):43–9 (in Russian).
115. Synegub IA. Macrozoobenthos of Odessa port aquatory. Ecological safety of coastal and shelf zones and complex exploitation of shelf resources. Sevastopol; 2007. Issue 15. p. 492–500 (in Russian).
116. Synegub IO. Comparative characteristics of macrozoobenthos in the aquatories of Odessa and Yuzhny ports. In: Proceedings of Ternopil pedagogical university. Series: biology. Special issue: Hydroecology, vol. 3(44). 2010; p. 241–4 (in Ukrainian).
117. Sirenko LA, Gavrilenko MY. Water "blooms" and eutrophication. Kiev: Naukova dumka; 1978. 232 p (in Russian).
118. Sorokin PY. Ecological peculiarities of the north-western Adriatic lagoons under high level of anthropogenic influence: PhD thesis synopsis. Moscow: IO RAS; 2002. 29 p (in Russian).
119. Sorokin YI. The Black Sea. Moscow: Nauka; 1982. 216 p (in Russian).
120. Sorokin YI, Bilio M. Studies of microbial sulphate reduction and of microplankton in the fish culture ponds of the Comacchio region, Italy. Biol Sea 1981;2:50–8 (in Russian).
121. Sorokin YI, Zakuskina OY. Acid-soluble sulphides in the upper layer of the bottom sediments of the north-eastern shelf of the Black Sea: connection with the pollution and environmental consequences. Oceanology 2008;48(2):224–31 (in Russian).
122. Sushchenya LM. Role of amphibiotic crustaceans in the transformation of matter and energy in the coastal zone of the Black Sea. Questions of biological oceanography. Kiev: Naukova dumka; 1967. p. 128–34 (in Russian).
123. Tarasov NI. Biology of the sea and the navy. Moscow: Naval Pub. House of the Peoples Commissariate of the USSR Navy; 1943. 192 p (in Russian).
124. Chebotarev YS. Sulphides content and speed of their formation in bottom sediments of the Black Sea shelf. Long-term dynamics of structure of the coastal ecosystems of the Black Sea. Krasonodar; 1984. p. 57–61 (in Russian).
125. Cherbadzhi II. Influence of ecological factors on the intensity of microalgal photosynthesis in soft sediments. In: Ist All-Union conference on marine biology (Vladivostok, 26–29 Sept 1977). Vladivostok; 1977. p. 148–9 (in Russian).
126. Cherbadzhi II. Production indices of the microphytobenthos of Vostok Gulf (Sea of Japan). In: Biology of shelf zones of the World Ocean: proceedings of the IInd All-Union on marine biology. Vladivostok: FESC AS USSR; 1982. Part 1. p. 178–9 (in Russian).
127. The Black Sea. Collected volume. Varna: Bakalov G. Publishing House; 1978. 635 p (in Bulgarian).

128. Chilikina NS. The biological activity of sea foam. Problems of the World Ocean. In: Proceedings of the conference of young scientists Moscow State University (MSU). Moscow: Publishing House of Moscow State University; 1970. p. 205–10 (in Russian).
129. Shadrin NV. Algal debris in the Black Sea supralittoral: ecological and geochemical role. Proc NAS Ukraine 1998;3:192–5 (in Russian).
130. Shalovenkov NN. The composition of settling detritus in the coastal bottom communities of the Black Sea. In: Biology of the World ocean shelf zones: proceedings of IInd All-Union conference on marine biology. Vladivostok: FESC AS USSR; 1982. Part 1. p. 160–2 (in Russian).
131. Sheremetevsky AM. Role of meiobenthos in the biocenoses of the Southern Sakhalin shelf (Eastern Kamchatka) and Novosibirsk shallow-water. Leningrad: Nauka; 1987. 135 p (in Russian).
132. Shuyskiy YD. Estimation of the coastal state of the Black Sea during forthcoming decades. Ecological problems of the Black Sea. Odessa: OCNTEI; 2001. p. 367–73 (in Russian).
133. Shuyskiy YD, Vykhovanets GV. Drifts distribution between cape Severniy Odessa and cape Adzhiyask. Prichernomorskiy Ecol Bull. Odessa 2004;2–3:64–72 (in Russian).
134. Shuyskiy YD, Zolotov VI. Molluscs productivity in the coastal waters of the Black Sea and its role at calculation of the drifts balance in the coastal zone. In: IInd All-Union conference on shelf biology. Kiev: Naukova dumka; 1978. Part 2. p. 121–2 (in Russian).
135. Shuysky YD, Vykhovanets GV, Bereznitskaya NA. Morphology and dynamics of the shores of Tiligulsky Liman. Prichernomorskiy Ecol Bull. Odessa 2004;2–3:54–63 (in Russian).
136. The ecosystem of Grigorievsky (Maly Adzhalyk) Liman. In: Vinogradov AK, editor. Odessa: Astroprint; 2008. 263 p (in Russian).
137. The ecosystem of the river Neva estuary: biological diversity and ecological problems. In: Alimov AF, Golubkov SM, editors. St. Petersburg; Moscow: KMK Scientific press; 2008. 477 p (in Russian).
138. Belan T. Marine environmental quality assessment using polychaete taxocene characteristics in Vancouver Harbour. Mar Environ Res. 2003;57:89–101.
139. Caspers H. Pollution in coastal waters. An interim report on results of a priority programme of the German Research Society (1966–1974). Boldt. Boppard; 1975. 142 p.
140. Ecology of coastal waters. A systems approach by K. H. Mann. In: Studies in ecology, vol. 8. Berkeley; Los Angeles: University of California Press; 1982. 322 p.
141. Gamenick JA, Jahn A, Vopel K, Giere O. Hypoxia and sulphide as structuring factors in a macrozoobenthic community on the Baltic Sea shore: colonization studies and tolerance experiments. Mar. Ecol. Prog. Ser. 1996;144. p. 73–85.
142. Gerlach SA. Food chain relationships in subtidal silty sand marine sediments and the role of meiofauna in stimulating bacterial productivity. Oecologia (Berl.). 1978;33:55–70.
143. Giere O. Benthic life in sulfidic zones of the sea—ecological and structural adaptations to a toxic environment. Verh Dtsch Zool Ges. 1992;85:77–93.
144. Gray JS. The measurement of marine species diversity, with an application to the benthic fauna of the Norwegian continental shelf. J Exp Mar Biol Ecol. 2000;250:23–49.
145. Jørgensen CB. Mortality, growth, and grazing impact of cohort of bivalve larvae *Mytilus edulis*. In: Droop M, Ferguson Wood E, editors. Advances in Microbiology of the sea, vol. 20(2). New York: Academic Press; 1981. p. 185–92.
146. Marinelli RL, Coull BC. Structural complexity and juvenile fish predation on meiobenthos: an experimental approach. J Exp Mar Biol Ecol. 1987;108:67–81.
147. Sheldon RW, Sutcliffe WH Jr, Paranjape MA. Structure of pelagic food chain and relationship between plankton and fish production. J Fish Res Board Can. 1977;34:2344–53.
148. Sorokin Y, Giovanardi O, Pranovi F, Sorokin P. Need for restricting bivalve culture in the southern basin of the Lagoon of Venice. Hydrobiologia 1999;400:141–8.

149. Tiganus V. Donnees preliminaries sur le zoobenthos du substrate meuble de la zone portuaire de Constanta. Cercetari marina 1982;15:107–14 (in French).
150. Thode H. Comparative studies on the influence of oxygen deficiency and hydrogen sulphide on marine bottom invertebrates. Neth. J. Sea Res. 1973;7:244–52.
151. Vismann B. Sulphide tolerance: physiological mechanisms and ecological implications. Ophelia 1991;34:1–27.
152. Watzin MC. The effects of meiofauna on settling macrofauna: meiofauna may structure macrofaunal communities. Oecologia 1983;59:163–6.

Literatures Chapter 5

1. Aleksandrov BG. Hydrobiological bases of management of the state of coastal ecosystems of the Black Sea. Kiev: Naukova dumka; 2008. 343 p (in Russian).
2. Alimov AF. Studies of biotic balances of freshwater basins ecosystems in the USSR. Hydrobiol J. 1987;23(6):3–9 (in Russian).
3. Alimov AF. Introduction to production hydrobiology. Leningrad: Gydrometeoizdat; 1989. 152 p (in Russian).
4. Alimov AF. Main principles of a theory of the functioning of aquatic ecosystems. Hydrobiol J. 1990; 26(6):3–11 (in Russian).
5. Alimov AF. Seasonal and long-term changes in the biomass of zoobenthos in continental water bodies. Hydrobiol J. 1991;27(2):3–9 (in Russian).
6. Alimov AF. Diversity in animal communities and its preservation. Biological sciences/Uspekhi Biologicheskikh Nauk. 1993;113(6):652–8 (in Russian).
7. Alimov AF. Elements of aquatic ecosystem functioning theory. St. Petersburg: Nauka; 2000. 147 p (in Russian).
8. Alimov AF. Studies on biodiversity in the plankton, benthos and fish communities in ecosystems of freshwater bodies with different productivity. Bull AS of USSR. Ser Biol. 2001;1:87–95 (in Russian).
9. Alimov AF. On theory of aquatic ecosystems functioning. In: Actual problems in the rational use of biological resources of water reservoirs. Rybinsk: Rybinsk publishing house; 2005. p. 18–25 (in Russian).
10. Alimov AF. Changes in the structure of animal communities accompanying eutrophication and pollution of aquatic ecosystems. Proc Acad Sci. General Biol. 2010;433(2):269–72 (in Russian).
11. Alimov AF, Umnov AA. Mathematical modelling of biotic flows of matter and energy in aquatic ecosystems. Application of a mathematical model for studying the diversity of biotic and abiotic conditions. In: Proceedings of the Zoological Institute (ZIN), vol. 272. 1997. p. 308–16 (in Russian).
12. Alimov AF, Bulion VV, Golubkov SM, Krylov PI, Rudenko GP. Fish influence on the structure and functioning of fish farm lakes ecosystems. In: Biotic interrelationships in ecosystems of fish farm lakes. St.Petersburg: Gydrometeoizdat; 1993. p. 299–311 (in Russian).
13. Andronnikova IN. Structural and functional organization of zooplankton in lake ecosystems of various trophyc types. St. Petersburg: Nauka; 1996. 189 p (in Russian).
14. Artemchuk NY. Microflora of USSR's seas. Moscow: Nauka; 1981. 192 p (in Russian).
15. Basic biological studies of Odessa marine port (Aug–Dec 2001): Final report. GloBallast programme Monograph Series of Odessa demonstration centre. Odessa; 2004. Issue 7. 171 p (in Russian).
16. Bizina EV. Correlation of predator pressure and resources in structure and functioning regulation of communities: a review of hypotheses. J General Biol. 1997;58(5):26–45 (in Russian).

17. Biological encyclopedia dictionary. In: Gilyarov SM, editor. Moscow: Soviet encyclopedia; 1986. 831 p (in Russian).
18. Bruevich SV. On the stability of ocean water chemical composition in the Proterozoic period. Oceanology 1968;8(6):939–48 (in Russian).
19. Bruevich SV. Salt composition of the World Ocean waters during the post-proterozoic period. Oceanology 1969;9(5):904–5 (in Russian).
20. Bulyon VV. Structure and function of the microbial loop in plankton of lake ecosystems. Inland Water Bio. 2002;2:5–14 (in Russian).
21. Burkovsky IV. Ecology of free-living infusoria. Moscow: Publishing House of Moscow State University; 1984. 302 p (in Russian).
22. Burkovsky IV. Structural and functional organization and stability of marine bottom communities (an example from the sand littoral of the White Sea). Moscow: Publishing house of Moscow State University; 1992. 208 p (in Russian).
23. Burkovsky IV. Marine biogeocenology. Organization of communities and ecosystems. Moscow: Scientific press KMK; 2006. 285 p (in Russian).
24. Bykova SV, Umanskaya MV. Microscale distribution of protozoa and bacteria in the chemocline of a meromictic reservoir. Water Chem Ecol. 2012;9:43–9 (in Russian).
25. Valyashko MG. Evolution of ocean water chemical composition. In: The history of the World Ocean. Moscow: Nauka; 1971. p. 97–104 (in Russian).
26. Vakhrushev AA, Rautian AS. Historical approach in ecology: main points and perspectives. In: Biological diversity: approaches to the study and preservation. Materials of the conferences BIN Russian Academy of Science and ZIN Russian Academy of Science, 14–15 Feb and 14–15 May 1990, Leningrad; 1992. p. 81–91 (in Russian).
27. Vernadsky VI. Living matter. Moscow: Nauka; 1978. 358 p (in Russian).
28. Vernadsky VI. The chemical structure of Earth's biosphere and its surrounding. Moscow: Nauka; 1987. 339 p (in Russian).
29. Vinogradov AK. Toxicity of highly-mineralized wastes for the marine hydrobionts. Kiev: Naukova dumka; 1986. 160 p (in Russian).
30. Vinogradov AK, Bogatova YI, Sinegub IA. Ecosystems of marine ports aquatories of the Black–Azov Sea basin (Introduction to the ecology of marine ports). Odessa: Astroprint; 2012. 528 p (in Russian).
31. Vinogradov AP. Introduction to ocean geochemistry. Moscow: Nauka; 1967. 213 p (in Russian).
32. Vorobyova LV. Meiobenthos of the Ukrainian shelf of the Black and Azov Seas. Kiev: Naukova dumka; 1999. 300 p (in Russian).
33. Gaevskaya AV. Parasites, diseases and pests of mussels (*Mytilus*, Mytilidae). IV. Viruses. Sevastopol: ECOSI-Hidrofisica; 2007. 96 p (in Russian).
34. Gerlach SA. Marine pollution. Diagnosis and therapy. Leningrad: Gydrometeoizdat; 1985. 263 p (in Russian).
35. Gilyarov AM. Population dynamics of freshwater planktonic crustaceans. Moscow: Nauka; 1987. 191 p (in Russian).
36. Gilyarov AM. Population ecology. Moscow: Publishing House of Moscow State University; 1990. 191 p (in Russian).
37. Gorbenko YA. Ecology and practical importance of marine microorganisms. Kiev: Naukova dumka; 1990. 160 p (in Russian).
38. Gorbunov MY, Umanskaya MV. The chemocline zone of stratified lakes as a microecotone. In: Problems in the studies of biocenoses marginal structures. Materials of 3rd International scientific conference (Saratov, 2–4 October 2012). Saratov: Publishing House of Saratov University; 2012. p. 42–9 (in Russian).
39. Gusev AG. Protection of fisheries from pollution. Moscow: Pishevaya promishlennost (Food industry); 1975. 367 p (in Russian).

40. Gutelmacher BL. Metabolism of plankton as a whole: trophometabolic interaction between zoo- and phytoplankton. In: Proceedings of ZIN AS USSR, vol. 133. Leningrad: Nauka; 1986. 156 p (in Russian).
41. Gutelmacher BL, Alimov AF. Quantitative regularities of filter-feeding of aquatic animals. General fundamentals of aquatic ecosystems study. Leningrad: Nauka; 1979. p. 57–78 (in Russian).
42. Davitaya FF. Atmosphere and biosphere—past, present and future. Leningrad: Gydrometeoizdat; 1976. 36 p (in Russian).
43. Dajoz R. Fundamentals of ecology. Moscow: Progress; 1975. 415 p (in Russian).
44. Dediu II. Ecological encyclopedic dictionary. Kishinev: Chief editorial board of Moldavian Soviet Encyclopedia; 1990. 408 p (in Russian).
45. Emelyanov IG. Diversity and its role in the functional stability and evolution of systems. Kiev: Naukova dumka; 1999. 167 p (in Russian).
46. Zherikhin VV. Use of paleontological data in ecological forecasting. In: Ecological forecasting. Moscow: Nauka; 1979. p. 131–2 (in Russian).
47. Zhirkov IA, with contributions from Azovskiy AI, Maksimova OV. Life at the bottom. Bio-geography and bio-ecology. Moscow: Scientific press KMK; 2010. 453 p (in Russian).
48. Zavarzin GA. Lectures in environmental microbiology. Moscow: Nauka; 2003. 348 p (in Russian).
49. Zaika VE. Specific production of aquatic invertebrates. Kiev: Naukova dumka; 1972. 143 p (in Russian).
50. Zernov SA. General hydrobiology. Moscow; Leningrad: Publishing House of the USSR Academy of Sciences; 1949. 587 p (in Russian).
51. Zilov EA. Hydrobiology and aquatic ecology (structure, functioning and pollution of aquatic ecosystems). Irkutsk: Publishing House of Irkutsk State University; 2009. 147 p (in Russian).
52. Israel YA. Ecology and control of environmental condition. Moscow: Gydrometeoizdat; 1984. 560 p (in Russian).
53. Kiseleva MI. Benthos of soft sediments of the Black Sea. Kiev: Naukova dumka; 1972. 246 p (in Russian).
54. Kommoner B. The closing circle. Leningrad: Gydrometeoizdat; 1974. 276 p (in Russian).
55. Kommoner B. Technology of profit. Moscow: Misl; 1976. 106 p (in Russian).
56. Kondratiev SI, Lyashenko SV. Peculiarities of the distribution of biogenic elements in the water column and in pore waters of Sevastopol Bay. In: Ecological safety of coastal and shelf zones and complex exploitation of shelf resources. Sevastopol; 2003. Issue 9. p. 82–94 (in Russian).
57. Kopylov AI. Nutrition of aquatic infusoria. Inland Water Biol J Inf Bull. 1977;33:19–33 (in Russian).
58. Kopylov AI. Feeding and production of planktonic infusoria of the Rybinsk reservoir: PhD thesis synopsis. Sevastopol; 1983. 23 p (in Russian).
59. Kopylov AI. Role of heterotrophic nanoflagellates in the functioning of marine and freshwater ecosystems: Synopsis of Doctor of Science thesis. Moscow: Institute of Oceanology of Russian Academy of Science (IO RAS); 2003. 40 p (in Russian).
60. Kopylov AI, Romanenko AV. Small-scale changes of picoplankton and heterotrophic flagellates in the coastal waters of Rybinsk reservoir and of the Black Sea. Inland Water Biol. 2004;2:33–8 (in Russian).
61. Kopylov AI, Sazhin AF. Heterotrophic nano- and microplankton in the vicinity of large port cities. Oceanology 1988;28(3):472–7 (in Russian).
62. Kopylov AI, Tumantseva NI. Analysis of the contents of the tintinnids food vacuoles and evaluation of their contribution to the consumption of phytoplankton production off the Peru coast. Oceanology 1987;3:464–9 (in Russian).

63. Kopylov AI, Kosolapov DB, Zabotkina EA. Distribution of viruses and their impact on bacterioplankton in eutrophic and mesotrophic reservoirs. Inland Water Biol. 2008;1:49–57 (in Russian).
64. Kriventsova TD. Bacterial destruction of organic matter at various levels of aquatic environment pollution. Hydrobiol J. 1978;4(6):54–9 (in Russian).
65. Kryuchkova TM. Structure of the zooplankton community in water basins of various type. In: Production and hydrobiological studies of aquatic ecosystems. Leningrad: Nauka; 1987. p. 184–98 (in Russian).
66. Kryuchkova TM. Trophic interrelationships of zoo- and phytoplankton. Moscow: Nauka; 1989. 123 p (in Russian).
67. Kuznetsov EA. Fungi of aquatic ecosystems: Synopsis of Doctor of Science thesis. Moscow; 2003. 65 p (in Russian).
68. Kuznetsova IA. Energy flow in barnacle population. In: Ist All-Union conference on marine biology (Vladivostok, 26–29 September 1977). Vladivostok; 1977. p. 88–9 (in Russian).
69. Kushing DG. Modelling the production cycle in the sea. In: Major problems in oceanology. Moscow: Nauka; 1968. p. 122–37 (in Russian).
70. MacIntyre F. The top millimetre of the ocean. In: Science about the sea. Moscow: Progress; 1981. p. 195–218 (in Russian).
71. Matishov GG, Kreneva SV, Muraveyko VM, Shparkovskiy IA, Ilyin GV. Biotesting and forecast. Variability of aquatic ecosystems affected by anthropogenic pollution. Apatity: Publishing House Kola Science Centre Russian Academy of Science; 2003. 468 p (in Russian).
72. Odum Y. Fundamentals of ecology, vol. 1. Moscow: Mir; 1975. 743 p (in Russian).
73. Odum Y. Ecology, vol. 2. Moscow: Mir; 1986. 376 p (in Russian).
74. Oshurkov VV. Succession of epibenthic communities on newly created substrata. Rus J Mar Biol Biol Sea 1994;4:247–51 (in Russian).
75. Petipa TS. Trophic dynamics of copepods in marine planktonic communities. Kiev: Naukova dumka; 1981. 245 p (in Russian).
76. Pianka E. Evolutionary ecology. Moscow: Mir; 1981. 399 p (in Russian).
77. Piontkovskiy SA, Seregin SA. Copepod behaviour. Sevastopol: ECOSI-Hydrofisica; 2006. 148 p (in Russian).
78. Pokarzhevsky AD. Information in ecosystems, how to measure it? In: Bioindication and biomonitoring. Moscow; 1991. p. 80–5 (in Russian).
79. Pokarzhevsky AD, Krivolutsky DA. Resources and information flows in ecosystems of forest steppe and meadow steppe: importance of phosphorus. Proc Acad Sci. 1992;326 (6):1102–5 (in Russian).
80. Posokhov EV. Chemical evolution of the hydrosphere. Leningrad: Gydrometeoizdat; 1981. 285 p (in Russian).
81. Protasov AA. The life in the hydrosphere. Essays of general hydrobiology. Kiev: Academperiodica; 2011. 704 p (in Russian).
82. Reymers NF. The Usage of Nature: a reference book. Moscow: Misl; 1990. 637 p (in Russian).
83. Reymers NF. Ecology (theories, laws, rules, principles and hypotheses). Moscow: Publishing House Rossiya Molodaya; 1994. 376 p (in Russian).
84. Ricklefs R. Fundamentals of general ecology. Moscow: Mir; 1979. 419 p (in Russian).
85. Romanenko VD. Fundamentals of hydroecology. Kiev: Geneza; 2004. 644 p (in Russian).
86. Sadchikov AP. The feeding of some Cladocera of a mesotrophic lake by phyto- and bacterioplankton. 1. Epilimnal *Bosmina coregoni* and *Diaphanosoma brachyurum*. Biol Sci. 1983;(8):64–70 (in Russian).
87. Sadchikov AP. Significance and role of zooplankton in the transformation of organic matter. 1. Trophic interrelations in planktonic communities (review). Biol Sci. 1993;3–4:5–23 (in Russian).
88. Satklif DF. Absorption of mineral salts by plants. Moscow: Mir; 1964. 350 p (in Russian).

89. Svirezhev YM, Logofet DO. Stability of biological communities. Moscow: Nauka; 1968. 352 p (in Russian).
90. Seki H. Organic matter in aquatic ecosystems. Leningrad: Goskomhydromet; 1986. 199 p (in Russian).
91. Selifonova ZP. Ecosystems of the aquatories of Novorossiysk and Tuapse ports (Black Sea). St. Petersburg: Nauka; 2012. 227 p (in Russian).
92. Selyutin VV. Turnover of matter and energy flow in ecological systems: from the model of a system towards the system of the models. Rev Appl Ind Math. 1994;1(6):957–73 (in Russian).
93. Selyutin VV. System aspects of mathematical modelling and information of marine ecosystem studies. Shelf zoobenthos studies. Informatic support of ecosystem studies. Apatity: Publishing House Kola Science Centre Russian Academy of Science; 2004. p. 289–99 (in Russian).
94. Sorokin YI. The role of bacterioplankton in the biological productivity of the tropical waters of the Pacific Ocean. In: Functioning of pelagic communities in the tropical regions of the ocean. Moscow: Nauka; 1971. p. 92–122 (in Russian).
95. Sorokin YI. Primary production of the seas and oceans. In: Biological productivity of water basins. Moscow: VINITI; 1973. p. 3–30 (in Russian).
96. Sorokin YI. Heterotrophic microplankton as a component of marine ecosystems. J Gen Biol. 1975;36(5):716–30 (in Russian).
97. Sorokin YI. Production of the microflora. In: Biology of the ocean, vol. 2. Moscow; 1977. p. 209–33 (in Russian).
98. Sorokin YI. The Black Sea. Moscow: Nauka; 1982. 216 p (in Russian).
99. Sorokin YI. Bacterioplankton. In: Man-induced pollution and processes of natural purification of the near Caucasus zone of the Black Sea. Moscow: Nedra; 1996. p. 266–76 (in Russian).
100. Sorokin YI. Microzooplankton. In: Man-made pollution and processes of natural self-cleaning of the near Caucasus zone of the Black Sea. Moscow: Nedra; 1996. p. 277–90 (in Russian).
101. Sorokin YI, Bilio M. Studies of microbial sulphate reduction and of microplankton in the fish culture ponds of the Comacchio region, Italy. Biol Sea 1981;2:50–8 (in Russian).
102. Sorokin YI, Burkatskii ON. The content of labile sulphides in the bottom sediments of the central part of the Sea of Azov: impact on benthic communities. Oceanology 2007;47 (5):729–35 (in Russian).
103. Sorokin YI, Zakuskina OY. Acid-soluble sulphides in the upper layer of the bottom sediments of the north-eastern shelf of the Black Sea: connection with the pollution and environmental consequences. Oceanology 2008;48(2):224–31 (in Russian).
104. Stepanova OA. Ecology of allochthonous and autochthonous viruses of the Black Sea. Sevastopol: Express-print; 2004. 307 p (in Russian).
105. Strickland JDH. Marine food chains at their primary stages. Evaluation of the modern studies: are they suitable for solving the problems of the future? In: Main problems of oceanology. Moscow: Nauka; 1968. p. 109–21 (in Russian).
106. Teslenko VA. Determination of the state of rhithron of small salmon-rivers influenced by anthropogenic impact on the example of river Rudnya. In: Bottom invertebrates of the Russian Far East rivers and Eastern Siberia: questions of production and bioindication of pollution. Vladivostok; 1987. p. 15–8 (in Russian).
107. Umnov AA. Mathematical modelling of biotic flows of substance and energy in aquatic ecosystems. Moscow: Nauka; 1997. 132 p (in Russian).
108. Umnov AA. Use of mathematic model for the study of ecosystems stability. Reaction of lake ecosystems to changes in biotic and abiotic conditions. In: Proceedings of ZIN, vol. 272; 1997. p. 300–7 (in Russian).
109. Umorin PP. Relationships between bacteria and flagellates in the destruction of organic matter. J Gen Biol. 1976;37(6):831–5 (in Russian).

110. Umorin PP. The role of protozoans in the destruction of dissolved organic matter. Proceedings of the Institute of Internal Waters Biology, USSR Academy of Sciences. 1978;(40):137–58 (in Russian).
111. Umorin PP. Study of the interrelationship between algae and bacteria in perfusion culture. J Gen Biol. 1981;42(6):907–10 (in Russian).
112. Umorin PP. Interrelationships among algae, bacteria and protozoans in the processes of self-cleaning and formation of water quality. In: Proceedings of the Institute of Internal Waters Biology, USSR Academy of Sciences. 1987. № 54/57. p. 132–148 (in Russian).
113. Fedorov VD. Stability of ecological systems and its measurement. Izv. Akad. Nauk SSSR, Biology series. 1974. № 3. p. 402–15 (in Russian).
114. Fedorov VD, Gilmanov TG. Ecology. Moscow: Publishing House of Moscow State University; 1980. 464 p (in Russian).
115. Khaylov KM. Ecological metabolism in the sea. Kiev: Naukova dumka; 1971. 252 p (in Russian).
116. Hutchinson GE. A treatise on limnology, vol. 2. Moscow: Mir; 1966. 1115 p (in Russian).
117. Khlebovich TV. The role of planktonic infusoria in the biological transformation of matter and energy: PhD thesis synopsis. Leningrad; 1986. 19 p (in Russian).
118. Shvarts SS. Evolution ecological regularities. Moscow: Nauka; 1980. 277 p (in Russian).
119. Shilov IA. Ecology. Manual for biological and medical specialized institutions. Moscow: Vysshaya shkola; 2003. 512 p (in Russian).
120. Shuyskiy VF. General patterns of limitation of freshwater macrozoobenthos by ecological factors: Doctor of Science thesis. St. Petersburg; 1997. 639 p (in Russian).
121. Shuisky VF, Evdokimov II, Mikhnin AE, Belov MM. Quantitative estimation of multifactorial impact on macrozoobenthos communities. In: Proceedings of the State Research Institute of Lake and River Fisheries (GosNIORKh). 1995. Issue 314. p. 87–100 (in Russian).
122. The ecosystem of Grigorievsky (Maly Adzhalyk) Liman. In: Vinogradov AK, editor. Odessa: Astroprint; 2008. 263 p (in Russian).
123. The ecosystem of the river Neva estuary: biological diversity and ecological problems. In: Alimov AF, Golubkov SM, editors. St. Petersburg; Moscow: KMK Scientific press; 2008. 477 p (in Russian).
124. Alimov AF. Towards a theory of functioning of aquatic ecosystems. Leiden: Backhuys Publ.; 2003. 130 p.
125. Andersson A, Lee C, Azam F, Hagstrom A. Release of amino acids and inorganic nutrients by heterotrophic marine microflagellates. Mar Ecol Prog Ser. 1985;23:99–106.
126. Azam F, Fenchel T, Field JG, Meyer-Reil RA, Thingstad F. The ecological role of water column microbes in the sea. Mar Ecol Prog Ser. 1983;10:257–63.
127. Baskin Y. Ecologists dare to ask: how much does diversity matter? Science 1994;264(5156):202–3.
128. Brooks JL, Dodson SI. Predation, body size and composition of plankton. Science 1965;150:28–35.
129. Calbet A, Landry MR. Phytoplankton growth, microzooplankton grazing, and carbon cycling in marine systems. Limnol Oceanogr. 2004;49:51–7.
130. Clements FE. Plant Succession and Indicators. New York: Wilson; 1928. 1512 p.
131. Ecology of coastal waters. A Systems Approach by K. H. Mann. In: Studies in ecology, vol. 8. Berkeley; Los Angeles: University of California Press; 1982. 322 p.
132. Fenchel TM. The ecology of marine microbenthos. The food of marine benthic ciliates. Ophelia 1968;5:73–121.
133. Fenchel TM. The ecology of micro- and meiobenthos. Ann Rev Ecol Syst. 1978;9:99–121.
134. Fenchel TM. Suspension feeding in ciliated Protozoa: functional response and particle size selection. Microb Ecol. 1980;6(1):1–11.
135. Fenchel TM. Ecology of heterotrophic microflagellates. II. Bioenergetics and growth. Mar Ecol Prog Ser. 1982;8:225–33.

136. Fenchel T, Finlay B. Oxygen and the spatial structure of microbial communities. Biol Rev. 2008;83(4):553–69.
137. Fowler SW. Density dependence as related to life history strategy. Ecology 1981;62 (3):602–10.
138. Fowler SW, Small LF. Sinking rates of euphausiid faecal pellets. Limnol Oceanogr. 1972;17(2):293–6.
139. Grumbine RE. What is ecosystem management? Conserv Biol. 1994;8(1):27–38.
140. Holling CS. Resilience and stability of ecological systems. Ann Rev Ecol Syst. 1973;4:1–23.
141. Johannes RE. Nutrient regeneration in lakes and oceans. In: Droop M, Ferguson Wood E, editors. Advances in Microbiology of the sea, vol. 1. New York: Academic Press; 1968. p. 203–13.
142. Lampert W. Predictability in lake ecosystems: the role of biotic interactions. Potentials and Limitat. Ecosyst. Anal. Berlin: Heidelberg; 1987. p. 333–46.
143. Lund JWG. Preliminary observations on the use of large experimental tubes in lakes. Verh. Intern. Ver. Theor. und Angew. Limnol.; 1972. Bd. 18. p. 71–7.
144. Molles MC. Recovery of a stream invertebrate community from a flash flood in Tesuque Creek, New Mexico. Southwest Nat. 1985;30(2):279–87.
145. National Academy of Sciences. Assessing potential ocean pollutants: a report of the study panel on assessing potential ocean pollutants to the ocean affairs board. Washington DC; 1975. 438 p.
146. Odate T, Imai K. Seasonal variation in chlorophyll-specific growth and microzooplankton grazing of phytoplankton in Japanese coastal water. J Plankton Res. 2003;25:1497–505.
147. Paine R. Food web complexity and species diversity. Am Nat. 1966;100:65–75.
148. Palomares-Garcia R, Bustillos-Guzman JJ, Lopez-Cortes D. Pigment specific rates of phytoplankton growth and microzooplankton grazing in a subtropical lagoon. J Plankton Res. 2006;28:1217–32.
149. Pomeroy LR. The ocean's food web, a changing paradigm. Bioscience 1974;9:499–504.
150. Pomeroy LR, Mathews HM, Min HS. Excretion of phosphate and soluble organic phosphorus compounds by zooplankton. Limnol Oceanogr. 1963;8:50–5.
151. Porter JW. Autotrophy, heterotrophy, and resource partitioning in Carribean reef-building corals. Am Nat. 1976;110:731–42.
152. Porter KG. Selective grazing and differential digestion of algae by zooplankton. Nature 1973;244(5412):179–80.
153. Porter KG, McDonough R. The energic cost of response to blue-green algal filaments by cladocerans. Limnol Oceanogr. 1984;29(2):365–9.
154. Porter KG, Sherr EB, Sherr BF, Pace M, Sanders RW. Protozoa in planktonic food webs. J Protozool. 1985;32(3):409–15.
155. Reynolds CS. Temporal scales of variability in pelagic environments and the response of phytoplankton. Freshwater Biol. 1990;23:25–53.
156. Schrodinger E. What is Life. The Physical Aspects of the Living Cell. Cambridge: Univ. Press; 1944. 91 p.
157. Sherr EB, Sherr BH. Heterotrophic dinoflagellates: a significant component of microzooplankton biomass and major grazers of diatoms in the sea. Mar Ecol. 2007;352:187–97.
158. Sieburth JM, Smetacek V, Lenz J. Pelagic ecosystem structure: heterotrophic compartments of the plankton and their relationship to plankton size fractions. Limnol Oceanogr. 1978;23 (16):1256–63.
159. Stone L, Weisburd RSJ. Positive feedback in aquatic ecosystems. Trends Ecol Evol. 1992;7 (8):263–7.
160. Sutcliff WH, Baylor ER, Menzel DW. Sea surface chemistry and Langmuir circulation. Deep Sea Res. 1963;10:233–43.
161. Van Voris P, O'Neill RV, Emanuel WR, Shugart HH. Functional complexity and ecosystem stability. Ecology 1980;61:1352–60.

162. Webster KE, Peters RH. Some size-dependent inhibitions of larger cladoceran filterers in filamentous suspensions. Limnol Oceanogr. 1978;23(6):1238–45.
163. Wetzel R. Limnology. Philadelphia; New York; Chicago: Saunders College Publ.; 1983. 860 p.
164. Wetzel R. Limnology. Lake and river ecosystems. San Diego: Academic Press; 2001. 1006 p.

Literatures Conclusion

1. Zenkovich VP. The principles of seashore evolution theory. Moscow: Publishing House of the USSR Academy of Sciences; 1962. 710 p (in Russian).
2. Maksimchuk VL, Sidorchuk VN. Studies of coastal protection of the Black Sea near Yalta city. In: Meeting on studies of coastal zone and bottom geology of the Black and Azov seas within Ukrainian SSR: proceedings (10–20 October 1965). Odessa; 1965. p. 34 (in Russian).
3. Sirenko LA, Gavrilenko MY. Water “blooms” and eutrophication. Kiev: Naukova dumka; 1978. 232 p (in Russian).
4. Shubakov IV, Shubakov DI, Buskov IP, et al. Oceanography. Odessa: Abrikos; 2003. 332 p (in Russian).

MIX
Papier aus verantwortungsvollen Quellen
Paper from responsible sources
FSC® C105338

If you have any concerns about our products,
you can contact us on
ProductSafety@springernature.com

In case Publisher is established outside the EU,
the EU authorized representative is:
Springer Nature Customer Service Center GmbH
Europaplatz 3, 69115 Heidelberg, Germany

Printed by Libri Plureos GmbH
in Hamburg, Germany